IoT Sensors

This book introduces the basics of the Internet of Things (IoT) and explores the foundational role of sensors in IoT applications. The IoT is a network of devices and objects: sensors, actuators, hardware, software, human beings, domestic appliances, health monitoring equipment, and other things connected to the internet, which is designed to operate in a coordinated fashion to receive, process, and interpret signals and take appropriate action.

It provides a seamless real-time interface between the physical and digital worlds by integrating sensors with networking, computation, and actuation facilities. This book sketches a perspective of the IoT with sensors as the focus of attention. Diverse applications of the IoT that are destined to make an impact on our everyday lives in the near future are discussed.

It presents a comprehensive overview of the most recent sensor technologies used in the IoT to keep the reader abreast of the current advances at the frontiers of knowledge.

The book will cater to student and professional audiences, and will be useful for postgraduate and Ph.D. students studying physics, engineering, and computer science as well as researchers, engineers, and industrial workers engaged in this fast-progressing field.

Key Features:

- Explains the basic concepts and important terms of 'Internet of Things' in simple language
- Provides an up-to-date coverage of the key sensors used in IoT applications
- Explores IoT applications in smart cities, smart agriculture, smart factories, and many more

Vinod Kumar Khanna is an independent researcher from Chandigarh, India. He received a his Ph.D in physics from Kurukshetra University, Kurukshetra, India, in 1988. He is a Retired Chief Scientist and Head of MEMS & Microsensors Group, at the CSIR – Central Electronics Engineering Research Institute, Pilani, India, and a Professor with the Academy of Scientific and Innovative Research, Ghaziabad, India. He is a former Emeritus Scientist, CSIR, and Emeritus Professor, AcSIR. He has worked at CSIR-CEERI for more than 37 years on the design, fabrication, and characterization of power semiconductor devices and micro-/nanoelectronic sensors. He has authored 20 books and 6 chapters in edited books. He has authored/coauthored 194 research papers in refereed journals and conference proceedings. He also has five patents to his name.

IoT Sensors

An Exploration of Sensors for Internet of Things

Vinod Kumar Khanna

CRC Press
Taylor & Francis Group
Boca Raton London New York

CRC Press is an imprint of the
Taylor & Francis Group, an **informa** business

Designed cover image: © Shutterstock

First edition published 2025
by CRC Press
2385 NW Executive Center Drive, Suite 320, Boca Raton FL 33431

and by CRC Press
4 Park Square, Milton Park, Abingdon, Oxon, OX14 4RN

CRC Press is an imprint of Taylor & Francis Group, LLC

© 2025 Vinod Kumar Khanna

Library of Congress Cataloging-in-Publication Data
Names: Khanna, Vinod Kumar, 1952- author.
Title: IoT sensors : an exploration of sensors for Internet of Things / Vinod Kumar Khanna.
Description: First edition. | Boca Raton, FL : CRC Press, 2024. | Series:
Series in sensors | Includes bibliographical references and index. |
Summary. "This book will introduce the basics of internet of things
(IoT), and bring out the foundational role of sensors in
internet-of-things applications. Internet of things is a network of
devices and objects: sensors, actuators, hardware, software, human
beings, domestic appliances, health monitoring equipment, and other
things connected to the internet, which is designed to operate in a
coordinated fashion to receive, process and interpret signals and take
appropriate action. "-- Provided by publisher.
Identifiers: LCCN 2024022105 | ISBN 9781032447803 (hbk) | ISBN
9781032449029 (pbk) | ISBN 9781003374442 (ebk)
Subjects: LCSH: Internet of things. | Intelligent sensors.
Classification: LCC TK5105.8857 .K43 2024 | DDC 004.67/8--dc23/eng/20241009
LC record available at https://lccn.loc.gov/2024022105

ISBN: 978-1-032-44780-3 (hbk)
ISBN: 978-1-032-44902-9 (pbk)
ISBN: 978-1-003-37444-2 (ebk)

DOI: 10.1201/9781003374442

Typeset in Times
by SPi Technologies India Pvt Ltd (Straive)

Dedication

To my parents Late Shri Amarnath Khanna and
Shrimati Pushpa Khanna
Whose blessings always inspire me in all the ways
To my grandson Hansh, daughter Aloka, and wife Amita
For the happy moments in life that brighten my days
And elevate the spirits high
Filling life with the rainbow colors of the sky.

Contents

About the Author

INTRODUCTION

Vinod Kumar Khanna is an independent researcher from Chandigarh, India. He is a retired Chief Scientist from the Council of Scientific and Industrial Research (CSIR) – Central Electronics Engineering Research Institute (CEERI), Pilani, India, and a retired Professor from the Academy of Scientific and Innovative Research (AcSIR), Ghaziabad, India. He is a former Emeritus Scientist, CSIR, and Professor Emeritus, AcSIR, India. His broad areas of research include the design, fabrication, and characterization of power semiconductor devices and micro- and nanosensors.

ACADEMIC QUALIFICATIONS

He received an M.Sc. Degree in Physics with a specialization in Electronics from the University of Lucknow in 1975 and a Ph.D. degree in Physics from Kurukshetra University in 1988 for the thesis titled, 'Development, Characterization and Modeling of the Porous Alumina Humidity Sensor'.

RESEARCH/TEACHING EXPERIENCE AND ACCOMPLISHMENTS

His research experience spans over a period of 40 years, from 1977 to 2017. Starting his career as a Research Assistant in the Department of Physics, University of Lucknow, from 1977 to 1980, he joined CSIR-CEERI, Pilani (Rajasthan), in April 1980. There, he worked on several CSIR-funded as well as sponsored research and development projects. His major fields of research included power semiconductor devices and microelectronics/MEMS and nanotechnology-based sensors and dosimeters.

In the power semiconductor devices area, he worked on the high-voltage and high-current rectifier (600A, 4,300V) for railway traction, high-voltage TV deflection transistor (5A, 1,600V), power Darlington transistor for AC motor drives (100A, 500V), fast-switching thyristor (1,300A, 1,700V), power DMOSFET, and IGBT. He contributed toward the development of sealed tube Ga/Al diffusion for deep junctions, surface electric field control techniques using edge beveling and contouring of large-area devices, and floating field limiting ring design. He carried out an extensive characterization of minority-carrier lifetime in power semiconductor devices as a function of process steps. He also contributed toward the development of the P-I-N diode neutron dosimeter and PMOSFET-based gamma-ray dosimeter.

In the area of sensor technology, he worked on the nanoporous aluminum oxide humidity sensor, ion-sensitive field-effect transistor-based microsensors for biomedical, food, and environmental applications; microheater-embedded gas sensor for automotive electronics, MEMS acoustic sensor for satellite launch vehicles, and capacitive MEMS ultrasonic transducer for medical applications.

As an AcSIR faculty member, he was the course coordinator of MEMS/IC Technology for the advanced semiconductor electronics program (2011–2013) and taught 'MEMS Technology' to students pursuing M.Tech. degree. As an adjunct faculty, BESU, Kolkata, he taught 'MEMS Technology & Design' to M.Tech. (Mechatronics) students. He was invited by IIT, Jodhpur, to deliver lectures on 'Semiconductor Fundamentals and Technology' to B.Tech. students during February 2011. He guided B.Tech./M.Tech. theses of students from BITS, Pilani; VIT, Vellore; and Kurukshetra University. He also guided a Ph.D. thesis on 'MEMS acoustic sensor', MNIT, Jaipur.

SEMICONDUCTOR FACILITY CREATION AND MAINTENANCE

He was responsible for setting up and looking after diffusion/oxidation facilities, edge beveling and contouring, reactive sputtering, and carrier lifetime measurement facilities. As the Head of the MEMS and Microsensors Group, he looked after the maintenance of a six-inch MEMS fabrication facility for R&D projects as well as the augmentation of processing equipment under this facility at CSIR-CEERI.

SCIENTIFIC POSITIONS HELD

During his tenure of service at CSIR-CEERI from April 1980 till superannuation in November 2014, he was promoted to various positions including one merit promotion. He retired as a Chief Scientist and Professor (AcSIR) and as the Head of MEMS and Microsensors Group. Subsequently, he worked for three years as an Emeritus Scientist, CSIR, and Emeritus Professor, AcSIR, from November 2014 to November 2017. After the completion of the emeritus scientist scheme, he now lives in Chandigarh. He is a passionate author and enjoys reading and writing.

MEMBERSHIP OF PROFESSIONAL SOCIETIES

He is a Fellow and Life Member of the Institution of Electronics and Telecommunication Engineers (IETE), India. He is a life member of the Indian Physics Association (IPA), Semiconductor Society, India (SSI), and Indo-French Technical Association (IFTA).

FOREIGN TRAVEL

He has traveled widely. He had participated in and presented research papers at the IEEE Industry Application Society (IEEE-IAS) Annual Meeting in Denver, Colorado, USA, in September–October 1986. His short-term research assignments include deputations to Technische Universität Darmstadt, Germany, in 1999; at Kurt-Schwabe-Institut fur Mess-und Sensortechnike e.V., Meinsberg, Germany, in 2008; and at Fondazione Bruno Kessler, Trento, Italy, in 2011, under collaborative programs. He was a member of the Indian Delegation to the Institute of Chemical Physics, Novosibirsk, Russia, in 2009.

SCHOLARSHIPS AND AWARDS

He was awarded a National Scholarship by the Ministry of Education and Social Welfare, Government of India, on the basis of a Higher Secondary result, 1970; CEERI Foundation Day Merit Team Award for projects on fast-switching thyristor (1986); for power Darlington transistor for transportation (1988), for P-I-N diode neutron dosimeter (1992); and for high-voltage TV deflection transistor (1994); Dr. N. G. Patel Prize for best poster presentation in the 12th National Seminar on Physics and Technology of Sensors, 2007, BARC, Mumbai; CSIR-DAAD Fellowship in 2008 under Indo-German Bilateral Exchange Programme of Senior scientists, 2008. He is featured in the Stanford–Elsevier prestigious list of the world's top 2% scientists (2022, Elsevier Data Repository, V4, doi:10.17632/btchxktzyw.4).

RESEARCH PUBLICATIONS AND BOOKS

He has published 194 research papers in leading peer-reviewed national/international journals and conference proceedings. He has authored 20 books and has also contributed 6 chapters to edited books. He has five granted patents to his credit, including two US patents.

About the Book

Leveraging the connectivity provided by the Internet of Things (IoT) with the information acquired by sensors has expanded the capabilities of sensors exponentially. The powerful synergy of sensors and IoT technologies has offered novel opportunities to develop innovative IoT applications for creating smart homes, offices, factory automation, traffic control on busy roads, maximizing the energy efficiency of the power distribution grid, facilitating the provision of human healthcare, ensuring best agricultural practices, securing safe transportation and sale of goods and services, and, above all, for developing smart cities. Sensors are the eyes, ears, noses, tongues, and skin perception devices in these systems which supply the input signals to the IoT, and, in this way, serve as the main components determining the overall performance of an IoT project.

This research and reference book sketches a perspective of the IoT with sensors as the focus of attention. Diverse applications of the IoT that are destined to make an impact on our everyday lives in the near future are discussed, ranging from the fundamentals to state-of-the-art developments to facilitate easy understanding. The book is organized into 11 chapters whose contents are summarized below:

Chapter 1 introduces the reader to the IoT ecosystem built from a network of interrelated devices in which sensors play the leading role.

Chapter 2 provides a framework for the IoT applications, giving an overview of standard operating procedures and facilities available to set up IoT systems.

Chapter 3 discusses the IoT capabilities that can be incorporated into new dwellings and offices or during the renovation of existing ones.

Chapter 4 considers the use of IoT sensors for factory automation, primarily in tracking components, checking gaps and alignments, and detecting vibratory and angular motions.

Chapter 5 fosters factory automation still forward by presenting the deployment of sensors for controlling vital parameters such as temperature, pressure, vacuum, force, fluid flow, and level.

Chapter 6 presents the widespread utilization of IoT sensors in traffic control, principally magnetic sensors, microwave RADAR and LiDAR, and ultrasonic and infrared vehicle detection devices.

Chapter 7 shows the use of IoT sensors in enhancing the operational efficiency of the power grid, such as voltage and current sensors, transformer winding and oil temperature sensors, dynamic line rating sensors for wind speed and direction, and solar irradiance sensors. Smart electronic energy meters, phasors, synchrophasors, and phasor measurement units are discussed.

Chapter 8 sheds light on the use of IoT sensors for promoting and restoring human health and for coping with disability. Notable examples include the PPG sensors, 1-lead ECG devices, arm-cuff and watch-type wrist cuff oscillometric sensors for blood pressure measurements, respiration sensors, pulse oximeters, painless glucose monitoring, and multiplexed perspiration analysis.

Chapter 9 describes the application of IoT sensors for real-time data collection on soil moisture content, and plant macronutrients. Sensors for the detection of plant diseases and pest threats are outlined. Also, sensors for farm animal health and welfare are presented.

Chapter 10 highlights the impact of GPS on logistics and fleet management. The GPS tracking over long distances is supported by BLE beacon technology in indoor positioning, particularly in big shopping malls and retail stores.

Chapter 11 brings together the ideas developed in preceding chapters regarding smart homes, offices, factories, traffic control, power grid, medical care, agriculture, goods and cargo transportation, and shopping practices with the concepts of air pollution, noise control, garbage disposal, and water quality monitoring to evolve the smart city paradigm, the ultimate goal of IoT sensors and analytics.

Thus, the different chapters of the book provide a broad perspective of the scenario emerging by blending sensing and networking technologies, emphasizing their impact on human life and society at large. In comprehensive developmental planning, unique challenges specific to particular regions and implementation hurdles must be paid due attention.

Preface

Sensors are the key components of control systems. They acquire information about the environment producing the input signals for these systems. Working on this information, these systems can react to the changes in the environment to effectively implement the necessary corrective action in the most appropriate manner. However, sensors alone are insufficient to cope with practical situations. The signal must be processed carefully in order to achieve the desired results. Soon, it was felt that the signal conditioning circuit must be integrated with the sensors, leading to the concept of 'smart sensors'. Several smart sensor chips were designed and fabricated. The Internet Revolution ushered in the era of fast data communication. In the wake of this revolution, there was not much beating around the bush to appreciate that the sensor capabilities must be further augmented by building sensor networks to rapidly disseminate signals to central stations for data analysis and activating control systems. Therefore, sensors with connectivity facilities were introduced. It was realized that sensors can be associated with various things in everyday use to construct a network known as the Internet of Things, laying down the foundation of large-area sensor networks which provide essential services at a higher efficiency. Thus, the new slogan became 'sensors plus signal processing plus connectivity'. The IoT paves the way for the futuristic dreams of the superset of the IoT involving people, processes, data, and things called the Internet of Everything (IoE) and the artificial intelligence-enhanced IoT, known as artificial intelligence of things (AIoT).

In this book, we restrict ourselves to the IoT because IoE and AIoT are very vast subjects. Accordingly, our aim is to look at sensors from a 'connectivity perspective'. Hitherto, sensors are classified as active or passive devices or in terms of the types of signals that they detect, namely, physical, chemical, and biological. Therefore, the exploration of sensors from an IoT viewpoint will keep the applications of the internet to sensors at the center, and try to identify the sensors that can be used to meet specific identified applications. Some application areas of sensors immediately come to our minds. Our homes and offices are the places where routine services can be greatly ameliorated. Therefore, we first look at sensors for homes and offices. Our factories are the places where sensors make a serious impact on the reproducibility and yield of industrial manufacturing processes; in fact, they are already the abodes of a variety of sensors. In metropolitan cities, traffic congestion and jams are becoming serious problems, demanding the widespread deployment of sensors for traffic monitoring. The power distribution grid can provide real-time power supply conforming to load requirements by incorporating sensors in the network, besides quick fault diagnosis and repairs. Human health is an issue of primary concern globally. The sensors can facilitate patient and doctor communication to meet the demands of urgent patient care. Sensors can bring unprecedented changes in our agricultural methods, whose effect on society will be huge because good quality food is a primary need for all mankind. No less important is the role of sensors in logistics, fleet management, and retail store services. Looking at the overall perspective of improvements brought by sensors in our lifestyles, and applying this expertise in solving problems confronting city dwellers in their day-to-day activities, leads to the conceptualization of a city well equipped with a diversity of sensors to make human lives easier and comfortable. A smart city is envisioned.

The book begins with an introduction to the basic terminology of the Internet of Things, followed by a description of the IoT architecture, and the regulatory protocols and platforms facilitating the building of applications. Then, it takes the readers step-by-step on a journey through the vast panorama of sensors that have been, and are likely to be used in evolving novel and innovative IoT solutions. The sensors for homes and offices are presented. The applications of sensors in factories are explained. The ensuing chapters discuss sensors for city traffic control, give glimpses of sensor applications in power distribution networks, and treat sensors for human health and well-being. Sensors for agriculture are briefly touched upon. The use of sensors for logistics and retail stores is also addressed. Finally, an all-inclusive sketch of the state-of-the-art IoT developments is drawn by surveying the sensors deployed to make our cities smarter and more comfortable.

It is hoped that this book will bring together a wide assortment of sensors catering to different applications where the connectivity established by communication networks will make our lives easier and more convenient. The book will be of immense value to all the students, engineers, and scientists engaged in the collaboration between sensors and the internet.

Vinod Kumar Khanna
Chandigarh, India

Acknowledgments

I am thankful to Almighty God for giving me the strength and wisdom to undertake and complete this task.

I express my sincere appreciation to the authors of the research papers, review articles, books, and patents cited in the bibliographies at the end of the chapters of the book. Their intelligence and hard work have brought us to the present state of advanced knowledge and technological expertise.

I owe a debt of gratitude to the editor and staff at CRC Press for their kind cooperation and support throughout this project.

A 'special thank you' to my wife and family who have always stood with me in all my activities. Their continuous encouragement and proactive backing greatly boosted my confidence.

Vinod Kumar Khanna
Chandigarh, India

Acronyms, Abbreviations, and Chemical Symbols

AC	Alternating current
ACD	Activated charcoal detector
ACK	Acknowledgment
ADC	Analog-to-digital converter
AES	Advanced encryption standard
AF	Atrial fibrillation
AFL	Atrial flutter
Ag	Silver (Argentum)
AgCl	Silver chloride
AIoT	Artificial intelligence of things
Al	Aluminum
AlGaN	Aluminum gallium nitride
AlN	Aluminum nitride
Al_2O_3	Aluminum oxide or alumina
ALOHA	Advocates of Linux open-source Hawaii association
Alumel	A 95% nickel, 2% aluminum, 2% manganese, and 1% silicon alloy
Am-241 or ^{241}Am	Americium-241
Am-241/Be	Americium-241/beryllium
AMQP	Advanced message queuing protocol
AMR	Anisotropic magnetoresistance
API	Application programming interface
AQI	Air quality index
Ar	Argon
ATD	Alpha-track detector
ATR	Attenuated total reflectance (spectroscopy)
ATR-FTIR	Attenuated total reflectance – Fourier-transform infrared (spectroscopy)
AU	Attenuation unit
Au	Gold (Aurum)
Au-Rh-Ir	Gold–ruthenium–iridium
AWS	Amazon Web Services, Inc.
B	Boron
BA	Benzyl acetate
BBA	Brainbean Apps
BD	Bulk density
BDD	Boron-doped diamond
Be	Beryllium
BER	Bit error rate
BLE	Bluetooth Low Energy
BOX	Buried oxide
BP	Blood pressure
Bpm	(Heart) Beats per minute
bps	Bits per second
BPSK	Binary phase shift keying
Bqh/m^3	Becquerel hours per cubic meter

BSR	Basal stem rot
bTB	Bovine tuberculosis
BTS	Binary tree search
C	Coulomb
°C	Degree Centigrade
CA	Certificate authority
Ca	Calcium
C/A	Coarse acquisition (code)
cbar	Centibar
CBC	Citrus bacterial canker
CCD	Charge-coupled device
CCG	Cold-cathode (ionization) gauge
CCTV	Closed-circuit television
cd	Candela
CDMA	Code-division multiple-access
CdS	Cadmium sulfide
CdSe	Cadmium selenide
CFI	Chlorophyll fluorescence imaging
$C_6H_{10}O_6$	Glucono-δ-lactone
$C_6H_{12}O_6$	Glucose
CJK	Chinese, Japanese, and Korean
ClO^-	Hypochlorite, or chloroxide anion
cm	Centimeter
CMOS	Complementary metal-oxide semiconductor
CMUT	Capacitive MEMS ultrasonic transducer
CNN	Convolutional neural network
CNT	Carbon nanotube
CO	Carbon monoxide
Co	Cobalt
CO_2	Carbon dioxide
CoAP	Constrained application protocol
CoFe	Cobalt–iron
CoFeB	Cobalt–iron–boron
CONNACK	Connection acknowledge
Constantan	55% copper and 45% nickel alloy
cP	CentiPoise
Cr	Chromium
CRM	Continuous radon monitor
^{137}Cs	Caesium-137or cesium-137or radio caesium
CsI	Cesium iodide
Cu	Copper
CuO	Copper (II) oxide or cupric oxide
CV	Cyclic voltammetry
2D	Two-dimensional
3D	Three-dimensional
DAC	Digital-to-analog converter
dB	Decibel
dBm	Decibel meter
DBP	Dibutyl phthalate
DBPSK	Differential binary phase-shift keying

DC	Direct current
DDS	Data distribution service
Deg-hr^{-1}	Degrees per hour
Deg-s^{-1}	Degrees per second
DFT	Discrete Fourier transform
D-glucosamine-HCl	D-glucosamine hydrochloride
D. I.	Deionized (water)
DLR	Dynamic line rating
DNS	Domain name service
DOS	Dioctyl sebacate
DRIE	Deep reactive ion etching
DSSS	Direct sequence spread spectrum
DT	Disruptive Technology
DTLS	Datagram transport layer security
e-	Electron
ECG or EKG	Electrocardiogram
EDA	Electrodermal activity
EEPROM or E^{2}PROM	Electrically erasable programmable read-only memory
EIC	Electret ion chamber
EID	Electronic integrating device
EMF	Electromotive force
EMG	Electromyogram
EPFA	Equivalent power factor angle
ESD	Electron stimulated desorption
F	Farad
FAU	Formazin attenuation unit
FDR	Frequency domain reflectometry
Fe	Iron
Fe(III)$_4$[Fe(II)(CN)$_6$]$_3$	Iron (III) hexacyanidoferrate (II) (Prussian blue)
FET	Field-effect transistor
FHSS	Frequency-hopping spread-spectrum
FMCW	Frequency modulated continuous wave
fs	Femtosecond
fT	FemtoTesla
FTU	Formazin turbidity unit
g	Gram
1G, 2G, 3G, 4G, 5G	1st, 2nd, 3rd, 4th, 5th generations
GaAs	Gallium arsenide
GaP	Gallium phosphide
GATT	General attribute
GC-MS	Gas chromatography-mass spectrometry
GCP	Google Cloud platform
Ge	Germanium
GFSK	Gaussian frequency shift keying
GHz	Gigahertz
GO$_x$	Glucose oxidase
GPS	Global positioning system
GSM	Global system for mobile communications
gSMS	Gamma soil moisture sensor

GUI	Graphical user interface
H	Hydrogen
h	Hour
Hb	Hemoglobin
HbO$_2$	Oxy-hemoglobin
HCG	Hot-cathode (ionization) gauge
HCl	Hydrochloric acid
HD	High definition
HDR	High dynamic range
He	Helium
HF	Hydrofluoric acid
HFCVD	Hot-filament chemical vapor deposition
Hg	Mercury (Hydrargyrum)
HNO$_3$	Nitric acid
H$_2$O$_2$	Hydrogen peroxide
hPa	Hectopascal = 100 Pascal
HPO$_4^{2-}$	Hydrogen phosphate or monohydrogen phosphate anion
H$_2$PO$_4^-$	Dihydrogen phosphate anion
HRMTD	High-resolution micro-level traffic data
HRV	Heart rate variability
H$_2$S	Hydrogen sulfide
HSI	Hyperspectral imaging
H$_2$SO$_4$	Sulfuric acid
HTTP	Hypertext transfer protocol
HTTPS	Hypertext transfer protocol secure
HVAC	Heating, ventilation, and air conditioning
Hz	Hertz
IBM	International Business Machines Corporation
IC	Integrated circuit
ID	Identification
IDE	Integrated development environment
IEC	International Electrotechnical Commission
IEEE	Institute of Electrical and Electronic Engineers
IETF	Internet Engineering Task Force
IF	Intermediate frequency
InAs	Indium arsenide
Industry 1.0	First Industrial Revolution
Industry 2.0	Second Industrial Revolution
Industry 3.0	Third Industrial Revolution
Industry 4.0	Fourth Industrial Revolution
InGAs	Indium gallium arsenide
InGaAsP	Indium gallium arsenide phosphide
InGaN	Indium gallium nitride
InSb	Indium antimonide
iOS	iPhone operating system
IoT	Internet of Things
IP	Internet protocol
IPS	Indoor positioning system
IR	Infrared
IRI	Innovative Routines International, Inc.

IrMn	Iridium manganese
ISE	Ion-selective electrode
ISF	Interstitial fluid
ISFET	Ion-sensitive field-effect transistor
I-signal	In-phase signal
ISM	Industrial, scientific and medical
ISO	International Organization for Standardization
ISO/IEC	International Organization for Standardization/International Electrotechnical Commission
ISV	Independent software vendor
IT	Information technology
316i Ti	Titanium-stabilized 316 molybdenum-bearing Cr-Ni austenitic stainless steel
K	Kelvin, potassium
K$^+$	Potassium cation
^{40}K	Potassium-40
kB	Kilobytes
kbar	Kilobar
kbits^{-1}, kbps	Kilobits per second
KCl	Potassium chloride
kg	Kilogram
kG	Kilogauss, a unit of magnetic induction (flux density)
KH$_2$PO$_4$	Monopotassium phosphate
kHz	Kilohertz
KNN	K nearest neighbor (classification)
KOH	Potassium hydroxide
kPa	Kilopascal
kV	Kilovolt
kW-h	Kilowatt-hour
L	Liter
L1	1,575.42 MHz
L2	1,227.6 MHz
LA	Left-arm
LAN	Local area network
LC	Inductance–capacitance
LCD	Liquid crystal display
LDR	Light-dependent resistor
LE	Low energy
1L-ECG	Single-lead electrocardiogram
12L-ECG	12-lead electrocardiogram
LED	Light-emitting diode
LIDAR	Light detection and ranging
LiNbO$_3$	Lithium niobate
LLC	Limited liability company
LO	Local oscillator
LoD	Limit of detection
LoRa	Long range
LoRaWAN	Long-range wide area network
LPCVD	Low-pressure chemical vapor deposition
LPN	Low power node

LTCC	Low-temperature cofired ceramic
LTE-M or LTE-MTC	Long-term evolution-Machine type communication
LVDT	Linear-variable differential transformer
m	Meter
mA	Milliampere
MAA	Maximum amplitude algorithm
MAC	Media access control
Magmeter	Magnetic flow meter
mbar	Millibar
MBps	Megabytes per second
Mbps, Mbits^{-1}	Megabits per second
MCU	Microcontroller unit
MEMS	Micro-electromechanical systems
MeV	Mega electron volt
meV	Milli electron volt
Mg	Magnesium
mgL^{-1}	Milligram per liter
MgO	Magnesium oxide
MHz	Megahertz
min	Minute
mM	Millimolar
mm	Millimeter
m: m	Many-to-many
mmol L^{-1}	Millimole per liter
Mn	Manganese
Mo	Molybdenum
MODEM	Modulator-demodulator
MOM	Message-oriented middleware
MOS	Metal-oxide semiconductor
MOSFET	Metal-oxide semiconductor field-effect transistor
MPa	MegaPascal
MQTT	Message queuing telemetry transport
ms^{-1}	Meter per second
MTD	Maximum temperature difference
MTJ	Magnetic tunnel junction
MU-MIMO	Multiple user-multiple in/multiple out
MVAR	Megavolt ampere reactive
mVg^{-1}	Millivolt per gram
MW	Megawatt
mW	Milliwatt
MWCNT	Multi-walled carbon nanotube
N	Nitrogen
Na^{+}	Sodium cation
NaI-Tl	Thallium-activated sodium iodide
NaOH	Sodium hydroxide or caustic soda
NAP	Network access point
Na$_2$WO$_4$	Sodium tungstate
nBA	Poly (n-butyl acrylate)
NB-IoT	Narrowband Internet of Things
NCO	Numerically controlled oscillator

NDIR	Non-dispersive infrared
N-doped PPy	Nitrogen-doped polypyrrole
Ne	Neon
NFC	Near-field communication
NH$_3$	Ammonia
Ni	Nickel
NiCr	Nichrome (80% nickel and 20% chromium alloy)
Nicrosil	Nickel (84.1%)–chromium (14.4%)–silicon (1.4%)–magnesium (0.1%) alloy
NiFe	Nickel–iron
NiO	Nickel (II) oxide
Nisil	Nickel (95.5%)–silicon (4.4%)–magnesium (0.1%) alloy
nm	Nanometer
NO$_2$	Nitrogen dioxide
NO$_3$-	Nitrate anion
Np-237	Neptunium-237
NPOE	2-Nitrophenyl octyl ether
NTC	Negative temperature coefficient (thermistor)
NTP	Network time protocol
NTRU	Nephelometric turbidity ratio unit
NTU	Nephelometric turbidity unit
O	Oxygen
Oe	Oersted
OFDMA	Orthogonal frequency-division multiple access
OLED	Organic light-emitting diode
OMG®	Object Management Group®
OP-AMP	Operational amplifier
OQPSK	Offset quadrature phase shift keying
ORCL	Oracle
O-RIE	Oxygen reactive ion etching
OSI	Open-system interconnection (Model)
OTA	Over-the-air (updates)
P	Phosphorous
PA	Pedometric activity
Pa	Pascal
PaaS	Platform as a Service
PB	Prussian blue
PC	Personal computer
pC	Pico Coulomb
PCB	Printed circuit board
PDF	Portable document format
PECVD	Plasma-enhanced chemical vapor deposition
PEDOT: PSS	{Poly(3,4-ethylenedioxythiophene) polystyrene sulfonate}
PET	Polyethylene terephthalate
pF	Picofarad
PGA	Programmable gain amplifier
pH	Potential of hydrogen
PHY	Physical
PI	Polyimide
PIC	Peripheral interface controller
PIR	Passive infrared

PLC	Power line carrier
PLO	Phase-locked oscillator
PM	Particulate matter
PMU	Phasor measurement unit
PMUT	Piezoelectric MEMS ultrasonic transducer
PolySi	Polysilicon
ppb	Parts per billion
PPG	Photoplethysmography
ppm	Parts per million
PPS	Pulses per second
PRN	Pseudorandom noise
PSA	Pressure-sensitive adhesive
Pt	Platinum
PTC	Parametric Technology Corporation, positive temperature coefficient (thermistor)
PtMn	Platinum manganese
PTP	Precision time protocol
PT100 RTD	Resistance temperature detector with a resistance of 100 Ω at 0°C and 138.5 Ω at 100°C
PU	Polyurethane
PVA	Polyvinyl alcohol
PVB	Polyvinyl butyral
PVC	Polyvinyl chloride
PVDF	Polyvinylidene fluoride
PWM	Pulse width modulation
PZT	Lead zirconate titanate
Q-signal	Quadrature signal
RA	Right arm
Ra	Radon
RBF	Radial basis function
RC	Resistance–capacitance
RCA	Radio Corporation of America
REST API	Representational state transfer application programming interface
RF	Radiofrequency
RFID	Radio-frequency identification
RGB	Red (R), green (G), and blue (B)
Rh	Rhodium
^{222}Rn	Radon-222
RPM	Revolutions per minute
RS	Recommended standard
RSSI	Received signal strength indicator
RTD	Resistance temperature detector
RTSP	Real-time streaming protocol
RTTR	Real-time thermal rating
Ru	Ruthenium
RVDT	Rotary-variable differential transformer
S	Sulfur, Siemen
s	Second
SaaS	Software-as-a-service
SAE	Simultaneous authentication of equals

SAF	Synthetic antiferromagnet
ScAlN	Scandium aluminum nitride
SD	Secured digital (card)
SDK	Software development kit
SDS	SecureDataShot (protocol), Sodium dodecyl sulfate
SI	*Système international d'unités* (International System of Units)
Si	Silicon
Si_3N_4	Silicon nitride
SiO_2	Silicon dioxide
SMO_x	Semiconductor metal oxide
SMTP	Simple mail transfer protocol
SnO_2	Tin (IV) oxide or stannic oxide
SO_2	Silfur dioxide
SO_3^{2-}	Sulfite anion
SO_4^{2-}	Sulfate anion
SOI	Silicon-on-insulator
SOM	Soil organic matter
SONAR	Sound detection and ranging
SPAD	Single-photon avalanche diode
SPDT	Single-pole/double-throw (switch)
SPE	Solid polymer electrolyte
SpO_2	Peripheral oxygen saturation
SRD	Short-range device
SSL	Secure sockets layer
SVM	Support vector machine
SYN	Synchronize
T	Tesla, the SI unit of magnetic field intensity or flux density
Ta	Tantalum
Ta_2O_5	Tantalum pentoxide
3T-APS	Three-transistor active pixel sensor
TCP	Transfer control protocol
TCR	Temperature coefficient of resistance
TDC	Time-to-digital converter
TDDA-NPOE	Tetradodecylammonium o-nitrophenyl octyl ether
TDMA	Time-division multiple-access
TDR	Time domain reflectometry
TDS	Total dissolved solids
^{232}Th	Thorium-232
Ti	Titanium
TiO_2	Titanium dioxide, or titanium (IV) oxide or titania
TLS	Transport layer security
TMR	Tunnel magnetoresistance
$TOF_{X \to Y}$	Time of flight from X to Y
$TOF_{Y \to X}$	Time of flight from Y to X
TTL	Transistor-transistor-logic
TTN	The things network
^{238}U	Uranium-238
UAV	Unmanned aerial vehicle
UDP	User datagram protocol
UHF	Ultrahigh frequency

URL	Uniform resource locator
USB-PD	Universal serial bus-power delivery
UTC	Universal time coordinated
UUID	Universal unique identifier
UV(A)	Ultraviolet (315–399 nm)
UV(B)	Ultraviolet (280–314 nm)
UV(C)	Ultraviolet (100–279 nm)
V	Volt
Va	Vanadium
VCO	Voltage-controlled oscillator
VICS	Vehicle information and communication system
VIP	Video image processor
Vis-IR	Visible-infrared (spectroscopy)
Vis-NIR spectroscopy	Visible-near infrared (spectroscopy)
VOC	Volatile organic compound
VPC	Virtualized packet core
W	Watt, Tungsten
WAMS	Wide area monitoring system
WAP	Wireless access point
WHO	World Health Organization
Wi-Fi	A group of wireless networking protocols
WLAN	Wireless local area network
Wm^{-2}	Watt per square meter
WO$_3$	Tungsten (VI) oxide or tungsten trioxide
WO$_3$. 2H$_2$O	Tungsten oxide dihydrate
WPA3	Wi-Fi Protected Access 3
WPAN	Wireless personal area network
Xcc	*Xanthomonas citri* subsp. *citri*
ZigBee	Zonal intercommunication global standard where battery life was long
ZnS:Ag	Zinc sulfide:Silver
ZrO$_2$	Zirconium oxide
Z-Wave	A wireless communication protocol
μA	Microampere
μcurie	Microcurie
μg/m^3	Microgram per cubic meter
μM	Micromolar
μm	Micrometer
μS	MicroSiemen

Mathematical Notation

A	Cross-sectional area
$A, B, and C$	Coefficients
B	Magnetic field
BW	Bandwidth
C	Capacitance
c	Velocity of light
D	Distance of the target
d	Distance
df/dt	Frequency shift per unit time
E	Electric field
E_{cell}	Cell potential in volts
E_{cell}^{0}	Standard reduction potential of the couple (X/X⁻) in volts
F	Force, Faraday's constant (96,485 Coulombs mol⁻¹)
f	Frequency
f_b	Beat frequency
f_D	Doppler frequency
f_t	Transmitted signal frequency
G	Gauge factor
I	Current
I_0	Peak current
I_C	Collector current of a bipolar junction transistor
I_P	Primary coil current
I_S	Reverse saturation current of a transistor, Secondary coil current
$I(t)$	Instantaneous current at time instant t
j	Imaginary number $= \sqrt{-1}$
K	A calibration factor, cell constant
k	A constant
k_B	Boltzmann constant
L	Inductance
l	Distance, length
L_V	Loop inductance when the vehicle is moving over the loop
L_{WV}	Loop inductance when there is no vehicle over the loop
N	Number of wavelengths, number of turns in the coil, number of channels
n	Environment-dependent path loss exponent
N_P	Number of turns in the primary coil
N_S	Number of turns in the secondary coil
P_1	Pressure acting on the upper surface of the diaphragm
P_2	Pressure acting on the lower surface of the diaphragm
Path loss (d_0)	Path loss in dB at a distance d_0 where d_0 is a reference distance
Path loss (d)	Path loss in dB at an arbitrary distance $d>d_0$
$P_{Differential}$	Differential pressure acting on the diaphragm

Q	Charge
R	Distance from the object to radar, the universal gas constant ($8.315\,\text{J K}^{-1}\,\text{mol}^{-1}$)
$R(0)$	Resistance of platinum resistor at temperature $T = 0°\text{C}$
R_0	Resistance of the thermistor at the reference temperature T_0, in Kelvin scale
R_1, R_2, R_3, R_4	Resistances
$R_\text{Antiparallel}$	Resistance of magnetic tunnel junction in the anti-parallel state
R_L1, R_L2	Resistances of lead wires 1 and 2
R_Parallel	Resistance of magnetic tunnel junction in a parallel state
$R_\text{RTD}\,(T)$	Resistance of a platinum resistor with temperature T in $°\text{C}$
R_Series	A fixed series resistor
R-squared or R^2	Correlation coefficient
RSSI (d_0)	RSSI at a distance d_0
RSSI (d)	RSSI at a distance d
$R(T)$	Resistance of a thermistor at a temperature T, in Kelvin scale
$R_\text{Thermistor}$	Resistance of the thermistor
R_x	Resistance at temperature T_x
S	Seebeck coefficient, Sensitivity of the sensor
$s(t)$	An analog signal s as a function of time t
t_1, t_2	Time instants
t_EW	Time taken by ultrasonic waves sent from the transmitter at the east to reach the receiver at the west
t_R	Time of receiver clock
$T_\text{Reference}$	Temperature of the reference junction
T_s	Period of sawtooth wave
T_Sensing	Temperature of sensing junction
t_T	Time of transmitter clock
t_WE	Time taken by ultrasonic waves sent from the transmitter at the west to reach the receiver at the east
V	Voltage, volumetric flow rate
$\mathbf{v}$	Velocity
v	Resultant wind velocity
V_0	Peak voltage
V_BE	Base–emitter voltage of a transistor
V_LA	Potential of the left arm
$V_\text{OUT+}, V_\text{OUT-}$	Voltages at the output terminals
$V_\text{OUT}|_\text{Full Bridge}$	Output voltage of the full bridge
$V_\text{OUT}|_\text{Half Bridge}$	Output voltage of the half-bridge
V_Output	Output voltage
$V_\text{OUT}|_\text{Quarter Bridge}$	Output voltage of the quarter bridge
v_r	Radial velocity
V_RA	Potential of the right arm
$V_\text{Reference}$	Reference voltage
V_Reset	Reset voltage
V_Signal	Signal voltage
V_Supply	A constant voltage applied across the thermistor
$V(t)$	Instantaneous voltage at time instant t

V_{Th}	Threshold voltage
$V_{Thermistor}$	Voltage at the point between thermistor and resistor R_{Series}
v_x **and** v_y	Wind velocities along the X- and Y-directions
$[X^-]$, $[X]$	Concentrations of X^- and X in moles L^{-1}
x_0	Peak voltage V_0 or current I_0
$x(t)$	Instantaneous voltage $V(t)$ or current $I(t)$

Greek Symbols

β	A material constant
Δf	Frequency difference
ΔL	Change in length, change in inductance caused by the vehicle
ΔR	Change in resistance
ΔT	Difference between two temperatures
Δt	Time difference
ΔV	Voltage difference
ΔV_{BE}	Difference between base–emitter voltages
ε	Mechanical strain
θ	Angle
λ	Wavelength
ρ	Resistivity, pseudorange
σ	Standard deviation, conductivity
ϕ	Phase angle
χ	A zero-mean Gaussian distributed random variable, expressed in dB
Ω	Ohm, unit of resistance
ω	Angular frequency

1 IoT Architecture

1.1 INTRODUCTION

In this chapter, we familiarize the reader with the basic ideas and goals of the Internet of Things (IoT), define the main terms necessary to follow the subject matter, and especially the architecture of IoT, viz., the complex tangle comprising sensors, devices, actuators, the set of regulating rules known as protocols to be followed for guaranteeing proper operation, and the cloud services consisting of the infrastructure, platforms, and software hosted by various parties utilizing which IoT applications can be easily constructed by interested users.

1.2 INTERNET OF THINGS (IoT)

IoT is a broad term that has been defined by different authors in various ways (Sethi and Sarangi 2017). A holistic idea about IoT can be gathered from some simple definitions. IoT is a concept about objects that are connected to each other through the internet and communicate mutually, and with the people to deliver useful services. We know that the internet is a worldwide network of computers working under standard communication protocols: TCP (transfer control protocol)/IP (internet protocol), a suite of rules and procedures governing communication among computers. A network refers to a group of computers connected together to share resources such as files or for communication of information.

To understand how the IoT concept is realized practically, let us incorporate sensors, along with software and hardware in the network. So, IoT is a system of internet-connected physical objects or things that is embedded with sensors, software, and hardware, enabling communication and exchange of information between the participating objects.

To know more about the 'things' in IoT, let us clarify that the object or thing can be a cardiac patient with a heart monitoring device, an animal in a research laboratory being experimented upon to collect data about the influence of a drug, a domestic appliance with built-in sensors or an automobile moving on the road and fitted with sensors, etc. Thus, the IoT is a collective network of interrelated computing devices, digital or mechanical, objects, people, and animals uniquely recognized with identifiers and capable of processing and interchanging information with or without the intervention of humans.

Since IoT originated from the internet, we can say that IoT is an extension of the internet or other networks to include different things furnished with sensors, and hence capable of gathering data from their environments and sharing it with others. In this view, IoT is an umbrella or catch-all term for the proliferating electronic devices on the internet used to send or receive data or instructions.

To emphasize that things are materialistic or physical objects while software and data are digital entities, it can be said that IoT is an interaction between the physical and digital worlds through sensors and actuators.

Overall, IoT is a comprehensive network of sensor-equipped objects which can organize on their own, share data and resources among themselves, and auto-act and react to changes in the environment. It is a paradigm in which objects communicate with each other on a global scale through sensors, actuators processors, and transceivers to achieve meaningful defined goals (Kumar and Mallick 2018). In extension to IoT the Industrial internet of things (IIoT) connects machines and devices in industries; and the Internet of Everything (IoE) relates all the entities: people, process, data and things (Lee and Lee 2015).

DOI: 10.1201/9781003374442-1

1.3   DATA NETWORK TERMINOLOGY

We define a few basic terms as prerequisites to understanding the open-system interconnection (OSI) model: In IoT and data communication, the following terms are widely used (Figure 1.1).

> IoT gateway: It is a device that connects dissimilar networks by translating communications from one protocol or format to the required one. It collects large volumes of data from connected sensors, devices, and machines in the IoT network through Bluetooth, ZigBee, and other IoT protocols. It transmits this information to internet/cloud platforms for analysis and interpretation. It also transmits processed information back to devices and machinery to control their actions.
>
> Node: A node is a physical electronic device or equipment connected to a network, e.g., a computer, printer, modem (modulator-demodulator), switch, or router.

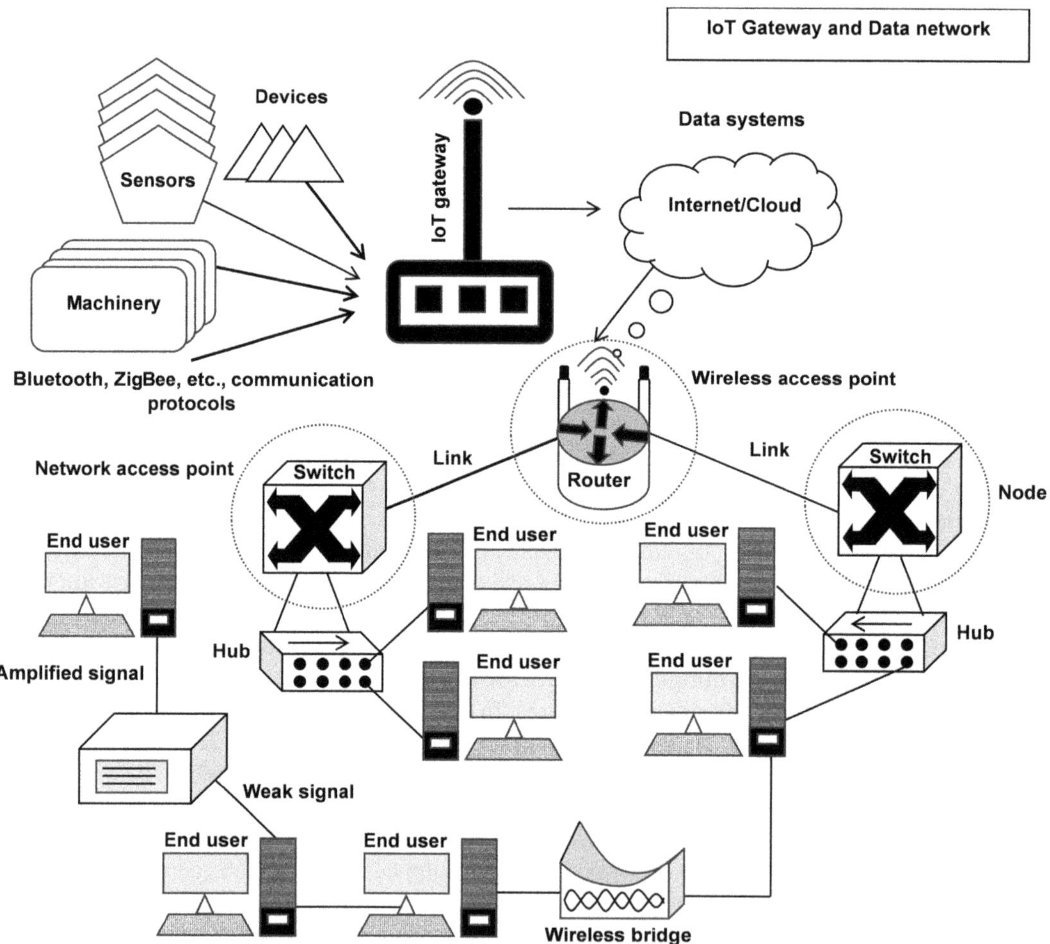

FIGURE 1.1   A typical IoT gateway, data network, and their components: IoT gateway, internet/cloud, sensors, devices, machinery, router, switches, hubs, bridge, and end-user computers. Nodes, links, network, and wireless access points are marked. The gateway is connected to the internet via Bluetooth, ZigBee, and other communication protocols.

Link: A link is a connecting medium between two nodes, either a hardware link through a wire or cable or a wireless link, e.g., Wi-Fi.

Bridge: It is a networking device to convert multiple communication networks into a single network.

Network access point (NAP): It is a device connecting wireless-capable devices to a wired network.

Wireless access point (WAP): It is a hardware appliance which provides Wi-Fi capability to a wired network by bridging traffic from a wireless station into a wired network.

Hub: It is a networking device used to connect multiple devices in the network.

Repeater: It is a booster device to amplify or regenerate an incoming signal traveling over a long distance.

Router: It is a traffic management device in packet-switched networks that forwards packets to the intended destination address and allows several devices to use the same connection.

Switch: It is a piece of hardware for connecting devices on a network by using packet switching to receive data and to send or forward it to the destination.

Switching: It is an action whose job is moving data from one port to another to guide it to the correct destination.

Port or port number: It is a unique number assigned to a connection endpoint for identification.

Session: It is an interaction between an application and a visitor.

1.4 OPEN-SYSTEM INTERCONNECTION MODEL

It is a conceptual model of the flow of information from one computer to another through a physical medium (Figure 1.2). The task of information flow is subdivided into seven smaller tasks which are assigned to seven layers. A layer in the model represents a category or group of activities, behavior, and functionality. The seven abstraction layers and their responsibilities are (Imperva 2022):

Layer 7: Application layer: It is the layer of human interaction with a computer. It is used by the end-user software such as web browsers and email clients to provide protocols for network services.

Layer 6: Presentation layer: It defines encoding, encryption, and compression of data to ensure that it is correctly understood by the receiver.

Layer 5: Session layer: It creates channels for communication between computers. The time between opening and closure of channels is known as a session.

Layer 4: Transport layer: It breaks down the data into chunks called segments at the transmitting end, and recombines the data segments to get back data at the receiver. It also controls data flow and errors.

Layer 3: Network layer: It fragments data segments into smaller units called packets at the transmitting end and reassembles them at the receiver; it also routes packets to the best path to their destination address.

Layer 2: Data link layer: It facilitates data transfer between two devices of the same network by breaking the packets into smaller pieces known as frames. It looks after flow control and error control in intra-network communication in the same way that the transport layer performs this duty in inter-network communication.

Layer 1: Physical layer: It provides the hardware, cable, switches, or wireless medium through which the signal propagates between nodes of the network in the form of a bit stream, a string of 0s and 1s.

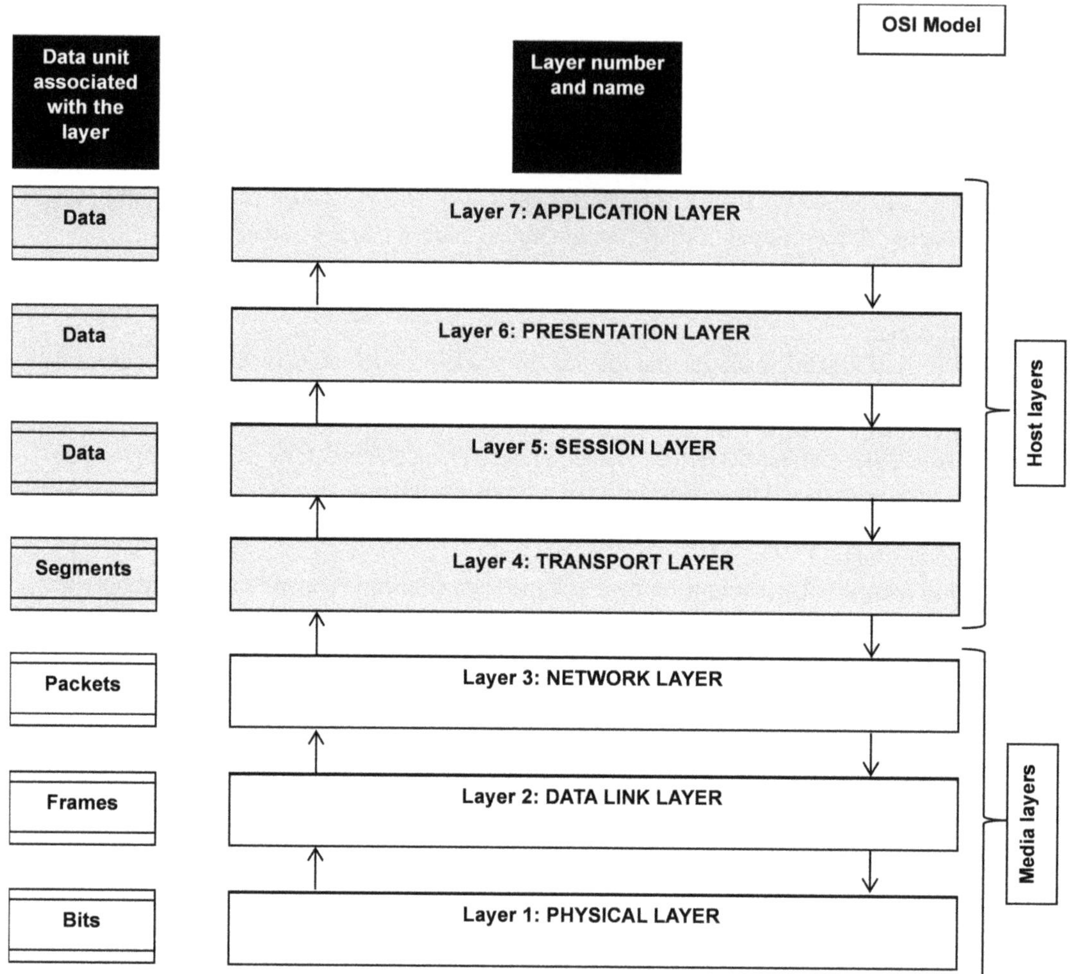

FIGURE 1.2 Open-system interconnection model consisting of seven layers: (1) physical layer, (2) data link layer, (3) network layer, (4) transport layer, (5) session layer, (6) presentation layer, and (7) application layer. The number and name of each layer are given. The data units associated with respective layers are also mentioned, e.g., bits for layer 1, frames for layer 2, packets for layer 3, segments for layer 4, and data for layers 5–7. The layers 1–3 are the media layers, and layers 4–7 are the host layers.

1.5 THE TCP/IP MODEL USED BY THE MODERN INTERNET

The present-day internet is not based on the OSI model. It is based on the four-layer TCP /IP model which combines several OSI layers into a single layer (Figure 1.3):

   (i) Layers 7, 6, and 5 of the OSI model are merged into one application layer in the TCP/IP model.

  (ii) Similarly, layers 2 and 1 of the OSI model are blended together into one network access layer in the TCP/IP model.

 (iii) The OSI model is a generic model for describing all types of network communication. It is independent of protocols. The TCP/IP model is a functional model for solving specific communication issues. It is founded on standard protocols.

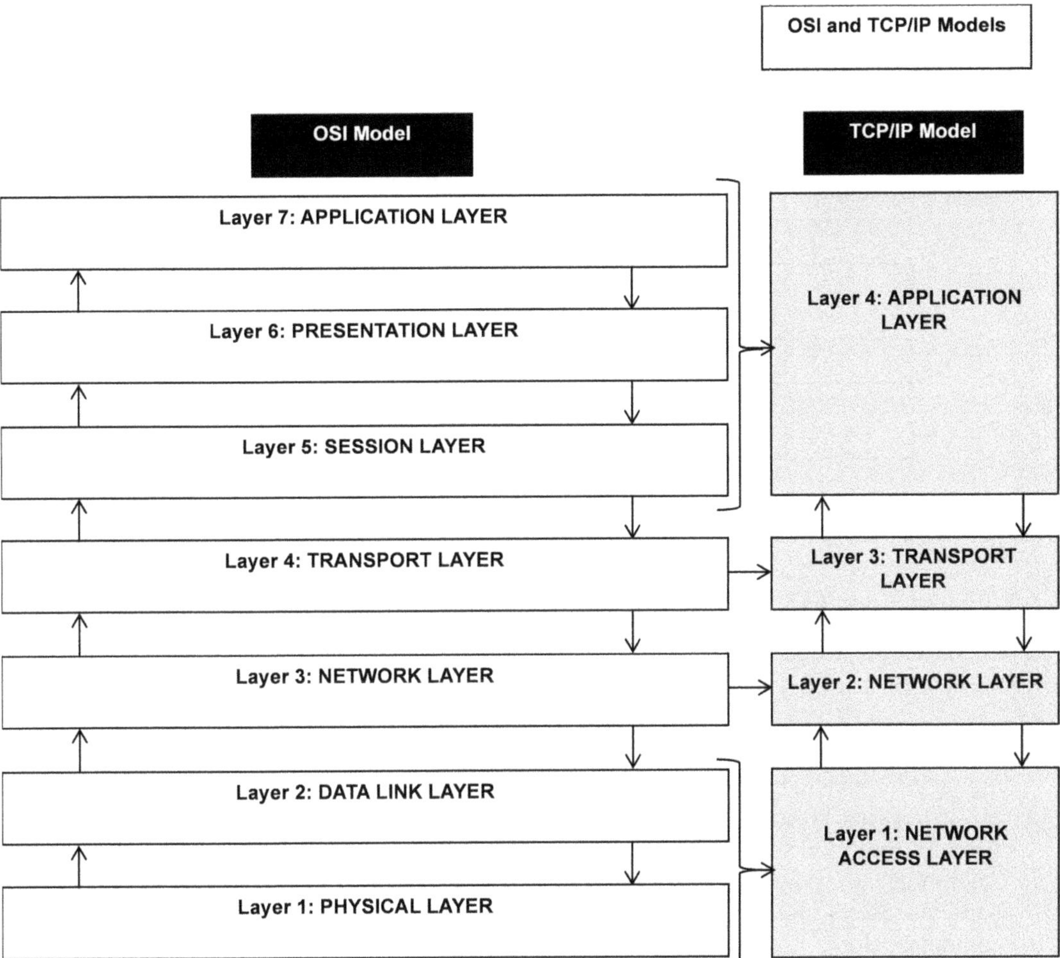

FIGURE 1.3 OSI model vs. TCP/IP model. On the left-hand side of the diagram, the OSI model is drawn showing its seven layers: (1) physical layer, (2) data link layer, (3) network layer, (4) transport layer, (5) session layer, (6) presentation layer, and (7) application layer. On the right-hand side of the diagram, the TCP/IP model is sketched for comparison. It is important to note the blending together of layers 1 and 2 of the OSI model into a single layer in the TCP/IP model; this layer is layer 1: the network access layer of the TCP/IP model. Similarly, layers 5–7 of the OSI model are combined into a single layer in the TCP/IP model known as layer 4: the application layer. Layer 3 of the OSI model and layer 2 of the TCP/IP model are the same, the network layer. Likewise, layer 4 of the OSI model is equivalent to layer 3 of the TCP/IP model. These are transport layers.

(iv) In the OSI model, layers 3, 2, and 1 are essential for any data communication. All seven layers are not utilized in elementary communication. Contrarily, a majority of applications use all four layers of the TCP/ICP model.

1.6 THREE-, FOUR, FIVE-, AND SEVEN-LAYER IoT ARCHITECTURES

IoT architecture is a system comprising diverse elements including sensors, actuators, software, and protocols, i.e., the mix-up of components that work together to build IoT, together with the knowledge about the several ways in which the IoT devices are structured to fulfill the needs of customers. There is no consensus of opinion about a solo universally agreed IoT architecture. The IoT architecture is similar to the OSI model in network and data communications (Madakam et al. 2015).

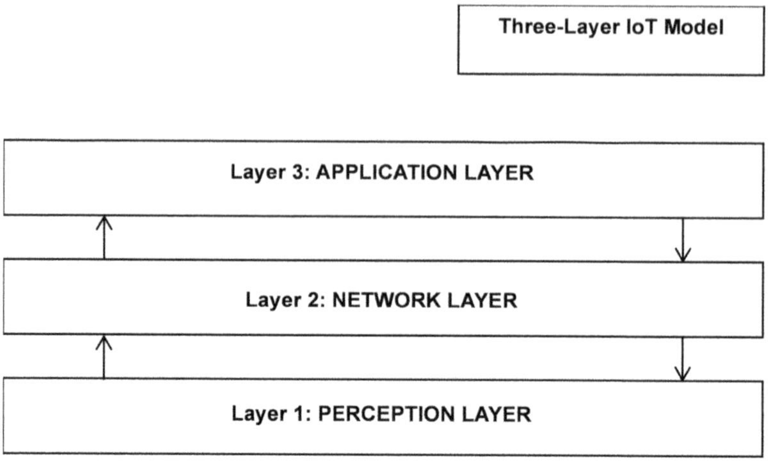

Three-Layer IoT Model

Layer 3: APPLICATION LAYER
↑ ↓
Layer 2: NETWORK LAYER
↑ ↓
Layer 1: PERCEPTION LAYER

FIGURE 1.4 Three-layer IoT architecture model consisting of three layers: (1) perception layer, (2) network layer, and (3) application layer.

1.6.1 THREE-LAYER IoT ARCHITECTURE

The three-layer architecture (Figure 1.4) is easy to implement but has several shortcomings, principally the lack of security (Lombardi et al. 2021, Itsekson 2023).

Layer 1: Perception layer: It contains the sensors to transform environmental parameters into digital information; hence it is also called the sensor layer.

Layer 2: Network layer: It is the layer in which data collected by sensors is transmitted by wires or radio links, e.g., ethernet (a reliable, secure connection using wire), Wi-Fi (a small-range low-power wireless connection commonly at 2.4, 5, or 6 GHz and 60 GHz for particular cases), and Bluetooth (battery-powered 2.45 GHz wireless link for short distance communication).

Layer 3: Application layer: It is the layer at which end users interact with the devices in the network. It includes cloud networking (an IT infrastructure in which the network capabilities and resources of an organization are partially or wholly hosted in a platform, either public or private, and are available on requirement) and web servers (computer software and underlying hardware connected to the internet for supporting interchange of data with other devices on the internet).

1.6.2 FOUR-LAYER IoT ARCHITECTURE

This architecture (Figure 1.5) includes an extra layer between the network and application layers known as the support layer to enhance the weak security of the three-layer architecture (Burhan et al. 2018). The layers are:

Layer 4: Application layer: This layer performs authentication of the user and protects privacy.

Layer 3: Support layer: It uses passwords or shared keys to verify whether the information obtained from sensors has come from an authentic user. After verification, the information is sent to the network layer. The support layer provides secure cloud computing and anti-virus facilities.

Layer 2: Network layer: This layer is engaged in the authentication of user identity and data encryption.

Layer 1: Perception layer: It works on data encryption and key agreement, and provides protection of sensor data.

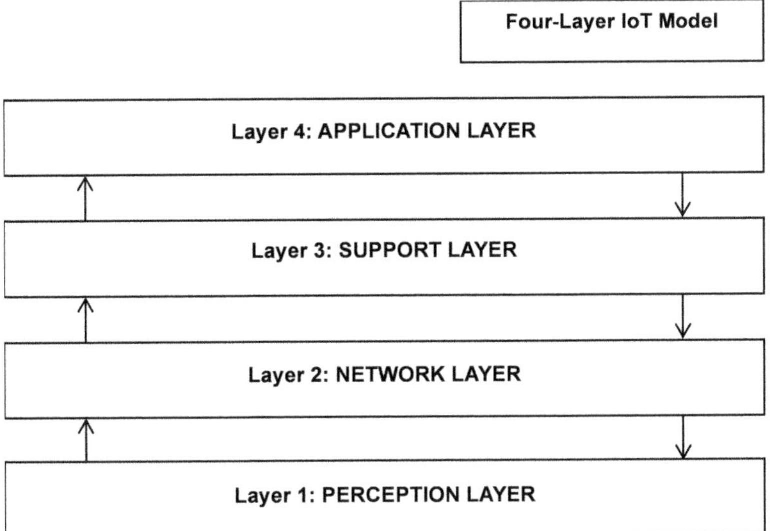

FIGURE 1.5 Four-layer IoT architecture model consisting of four layers: (1) perception layer, (2) network layer, (3) support layer, and (4) application layer.

1.6.3 FIVE-LAYER IoT ARCHITECTURE

It is not an improved version of the four-layer architecture. Rather it is an iterated version of the three-layer architecture incorporating two additional layers, namely, the processing and business layers. The constituent layers are (Figure 1.6):

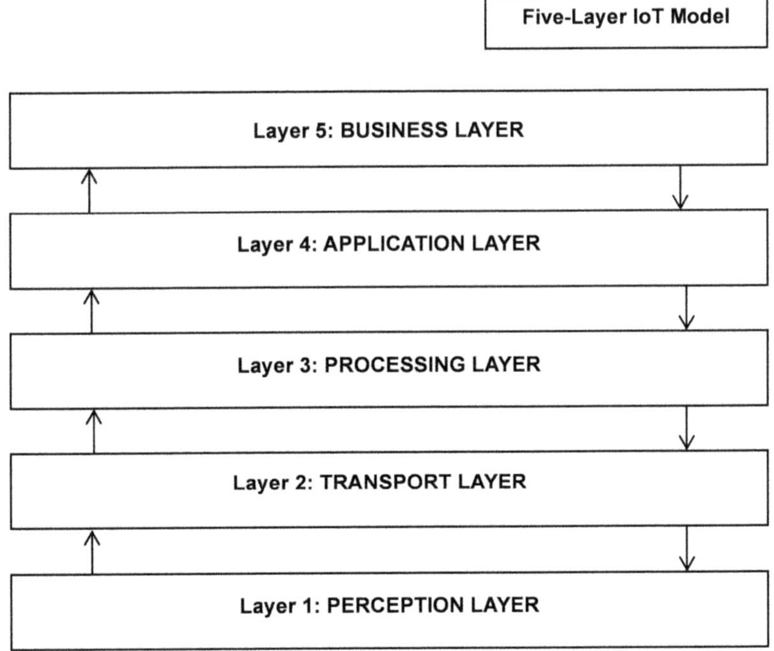

FIGURE 1.6 Five-layer IoT architecture model consisting of five layers: (1) perception layer, (2) transport layer, (3) processing layer, (4) application layer, and (5) business layer.

Layer 5: Business layer: It takes raw data and prepares graphs and charts for visualization and decision-making by stakeholders.

Layer 4: Application layer: Its function is the same as in three-layer architecture.

Layer 3: Processing layer: It works on data accumulation (saving data on the hard disk of the device or other databases for further analysis), and data abstraction (reconciling data to specific formats and looking for helpful insights).

Layer 2: Transport layer: Its function is the same as that of the network layer in the three-layer architecture.

Layer 1: Perception layer: It is similar to the perception layer of the three-layer architecture.

1.6.4 SEVEN-LAYER IoT ARCHITECTURE

It is an ameliorated form of the five-layer architecture in which two more layers are included, namely, edge computing and security layers (Figure 1.7).

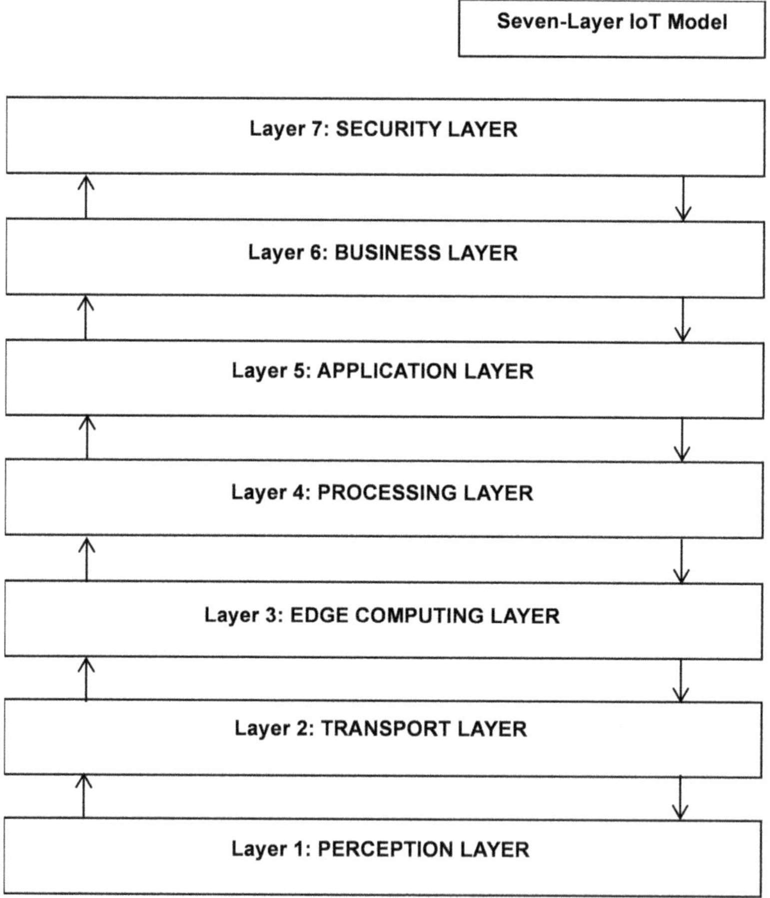

FIGURE 1.7 Seven-layer IoT architecture model consisting of seven layers: (1) perception layer, (2) transport layer, (3) edge computing layer, (4) processing layer, (5) application layer, (6) business layer, and (7) security layer.

Layers 1, 2, 4, 5, and 6: These are the same layers as layers 1, 2, 3, 4, and 5 in the five-layer architecture.

Layer 3: Edge computing layer: It uses the fog computing model which is a decentralized computing infrastructure. In this infrastructure, all the devices need not access the main server for making computations. A large fraction of the information is stored on local devices and most computations are implemented locally, relieving the main server of this load. Consequently, more users can avail the facilities simultaneously. Effectively, the computing resources such as applications have now been placed between the clouds and sources of data. The term 'fog' implies the periphery or edge of the cloud. Hence this model is referred to as fog or edge computing. Fog computing shifts the capabilities of the cloud nearer to locations where data is produced.

Layer 7: Security layer: It is a protection layer for other layers to keep them safe from malicious intruders. All the devices, connections, and clouds are protected by this layer.

1.7 CONCLUDING REMARKS AND PREPARING FOR THE UPCOMING CHAPTER

The OSI model and the TCP/IP protocol of the internet were discussed. Readers were apprised of the salient features of the three-layer, four-layer, five-layer, and seven-layer IoT architectures. In the next chapter, IoT protocols and platforms will be described.

REFERENCES

Burhan M., R.A. Rehman, B. Khan and B.-S. Kim 2018 IoT elements, layered architectures and security issues: A comprehensive survey, *Sensors*, 18, 2796, pp. 1–37.

Imperva; OSI Model©2022. https://www.imperva.com/learn/application-security/osi-model/#:~:text=Security EssentialsProtocols-,What%20Is%20the%20OSI%20Model,companies%20in%20the%20 early%201980s

Itsekson A. 2023 What are the seven layers of IoT architecture? Jelvix©2023. https://jelvix.com/blog/iot-architecture-layers

Lee I. and K. Lee 2015 The internet of things (IoT): Applications, investments, and challenges for enterprises, *Business Horizons*, 58, 4, pp. 431–440.

Lombardi M., F. Pascale and D. Santaniello 2021 Internet of things: A general overview between architectures, protocols and applications, *Information*, 12, 87, pp. 1–20.

Kumar N. M. and P. K. Mallick 2018 The internet of things: Insights into the building blocks, component interactions, and architecture layers, *Procedia Computer Science*, 132, pp. 109–117.

Madakam S., R. Ramaswamy and S. Tripathi 2015 Internet of things (IoT): A literature review. *Journal of Computer and Communications*, 3, 164–173.

Sethi P. and S. R. Sarangi 2017 Internet of things: Architectures, protocols and applications, *Hindawi Journal of Electrical and Computer Engineering*, 2017, pp. 1–25. Article ID 9324035.

2 IoT Protocols and Platforms

2.1 INTRODUCTION

After acquaintance with IoT architecture in the previous chapter, we turn our attention to two very important topics related to IoT. First is the set of rules and standards that must be strictly adhered to when IoT devices communicate among themselves and with other devices over the internet. These are the IoT protocols. Second is the applications or services bridging the IoT sensors and devices with internet. It contains software tools and capabilities that facilitate users to establish a connection with an IoT system.

2.2 IoT PROTOCOLS

These are the set of rules and standards, laid down in an agreement that governs the methods of connecting IoT edge devices, typically sensors with other edge devices and the internet, the ways in which information will be transmitted/received by the devices, and the structural forms of information to ensure that information sent by one device is correctly read by another device and vice versa. Accordingly, they are divided into two categories: network protocols pertaining to the connection of devices, and data protocols focusing on the exchange of information.

2.3 NETWORK PROTOCOLS

2.3.1 HIGH DATA RATE, CLOSE-PROXIMITY COMMUNICATION PROTOCOL: WI-FI

The term 'Wi-Fi' was coined by the Wi-Fi Alliance, a non-profit organization with a worldwide system of companies. Wi-Fi refers to a group of protocols for allowing desktop computers, laptops, tablet computers or tablets, printers, mobiles or smartphones, and other digital devices to communicate with each other and access the internet without using any wires or cables (Figure 2.1). These protocols are based on the IEEE 802.11 network standard, which is a part of the IEEE 802 set of technical standards for local area networks (LANs). The IEEE 802.11 standard lays down the media access control (MAC) and physical layer (PHY) protocols for wireless local area networks (WLANs) (Pahlavan and Krishnamurthy 2021).

Power consumption by Wi-Fi routers is typically 2–20 W. It varies with the model. The sixth generation of protocols trademarked by the Wi-Fi Alliance is Wi-Fi 6 based on the IEEE 802.11ax standard providing high-speed (9,608 Mbps) connections with a coverage range of 45 m and longer battery life (Mozaffariahrar et al. 2022). It can operate between 1 and 7.125 GHz including the 2.4 GHz, 5 GHz, and 6GHz bands.

Wi-Fi uses Orthogonal frequency-division multiple access (OFDMA). Through OFDMA, Wi-Fi 6 divides a wireless channel into several subchannels, each of which carries data for a different device, thereby allowing access points to serve multiple clients simultaneously.

Wi-Fi 6 supports bidirectional MU-MIMO (multiple user-multiple in/multiple out) technology permitting the uploading and downloading of data by many users at the same time.

Wi-Fi 6 has a 'target awake time' feature which enables devices to decide when and how frequently they should wake up for sending or receiving data. This helps in saving power consumption and hence in extending battery life.

Wi-Fi 6-certified devices support the Wi-Fi Protected Access 3 (WPA3) security standard. WPA3 makes weak passwords more resistant to malicious attacks, offers the equivalent of 192-bit cryptographic strength in enterprise mode, and uses an authentication technique called Simultaneous

DOI: 10.1201/9781003374442-2

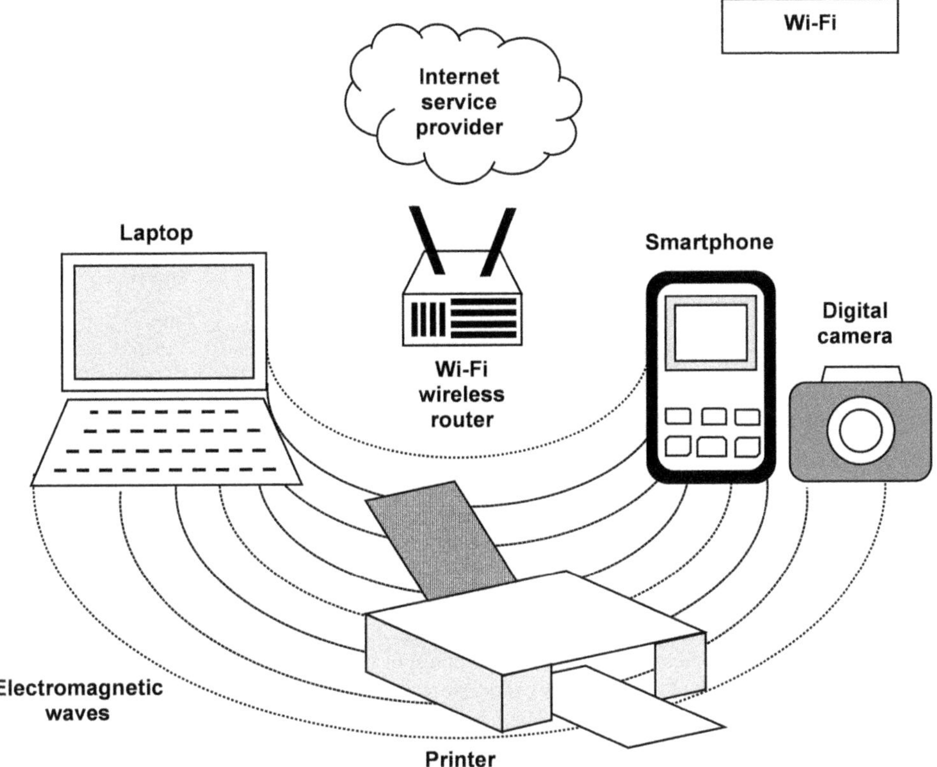

FIGURE 2.1 A Wi-Fi network consisting of a wireless Wi-Fi router connected to an internet service provider. The router broadcasts the received internet signal to Wi-Fi-enabled devices such as a laptop, a printer, a smartphone, and a digital camera.

Authentication of Equals (SAE). The SAE, also called Dragonfly Key Exchange, involves password-based authentication and password-authenticated key agreement to mitigate security issues.

2.3.2 LOW DATA RATE, LOW POWER, CLOSE-PROXIMITY COMMUNICATION PROTOCOLS

2.3.2.1 Bluetooth

It is a wireless technology standard for data transfer over short distances using UHF (ultra-high frequency) electromagnetic waves in the frequency range of 2.402–2.48 GHz (Shorey and Miller 2000, Chew 2019). It is used for file sharing and music streaming. There are several versions of Bluetooth offering different features and capabilities (Zeadally et al. 2019). Bluetooth LE (low energy), i.e., Bluetooth 4 consumes 10–100 mW while traditional Bluetooth uses 1W. This is because traditional Bluetooth operates continuously whereas Bluetooth LE goes into sleep mode during the time span between two connections, and becomes active only when the devices communicate. Thus, by switching on and off, it avoids power wastage. Bluetooth 6 has a range of 240 m. Its maximum rate of data transfer is 100 Mbps (Nair 2023).

The basic structural unit of a Bluetooth system is a piconet or small net, Figure 2.2(a), consisting of one primary or master device and up to seven ad-hoc connected secondary or slave devices within the radio range of the primary or master device <10 m. Master is the device whose clock and frequency hopping sequence are used to synchronize all the remaining devices in the piconet, viz., the slave devices. The master controls the available bandwidth among the slaves and allocates a series of time slots among the slaves for time-division multiplexing.

Often two piconets are connected together for networking several devices over an extended distance. In this interconnection, one device is common to both piconets. This bridging device can be a slave in

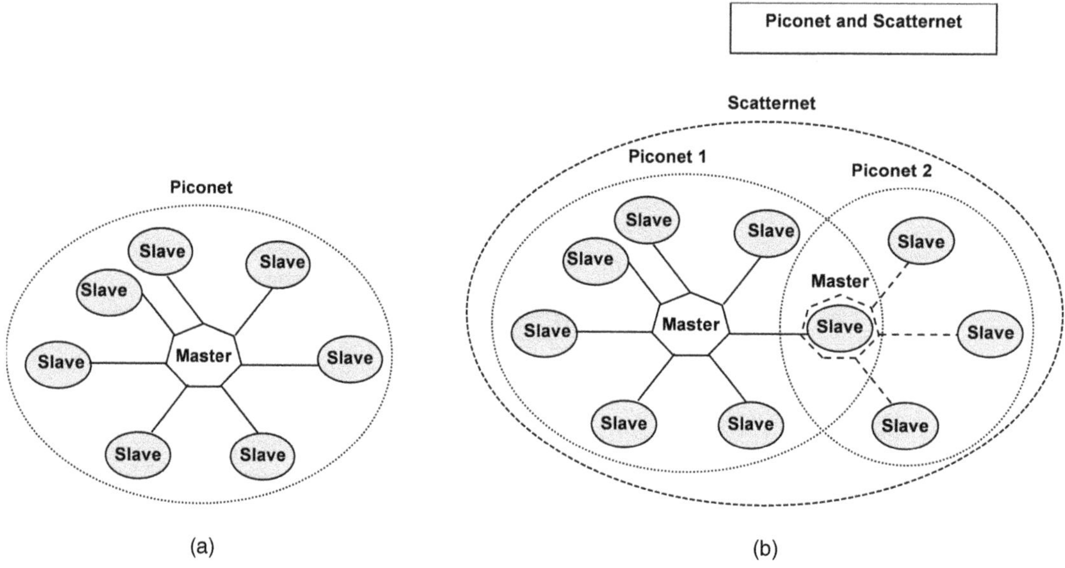

FIGURE 2.2 Connections between Bluetooth-enabled devices: (a) piconet and (b) scatternet. Part (a) shows eight devices connected in the smallest Bluetooth networking unit called a piconet. The eight devices include one master and seven slave devices. It supports a maximum of eight nodes. The piconet is enclosed inside a dotted circle. Part (b) shows an ad-hoc network called the scatternet, made from two piconets, enclosed in two dotted circles, for a larger coverage area and efficient channel width utilization. It supports > 8 nodes. The master and slave devices of both piconets are marked. The device in the overlapping region of the two piconets is a slave in piconet 1 but a master for piconet 2.

one piconet but the master in the other piconet. In this way, multiple piconets are interconnected to form bigger networks. Two or more interconnected piconets constitute a scatternet, Figure 2.2(b).

To cater to the diverse needs of users, Bluetooth technology supports several network topology options (Bluetooth SIG, Inc. 2023).

(i) Point-to-point topology for one-to-one (1:1) device communications, Figure 2.3(a);
(ii) Broadcast topology for one-to-many (1:m) device communications, Figure 2.3(b); and
(iii) Mesh topology for many-to-many (m: m) device communications, Figure 2.3(c).

In the mesh network, the messages are transported by two methods called advertising bearer and GATT (general attribute) bearer. In the advertising bearer method, the messages are transported by periodically sending data packets. In the GATT method, they are transported by transmission and reception in a point-to-point fashion. The node is a basic unit performing a sensing function or an assigned task, and endowed with radio communication capability. It has additional predefined features, and is accordingly named as follows.

(i) Relay node: It transmits the received data.
(ii) Low power node (LPN): It reduces power consumption by remaining in sleeping mode when the hardware is off.
(iii) Friend node: It listens to incoming messages. On awakening, the LPN pings its friend node to check for any new information received.
(iv) Proxy node: It acts as an interface between the mesh network and a non-mesh-supported BLE device, e.g., a smartphone.

The typical mesh network shown in Figure 2.3(c) comprises nodes with different features.

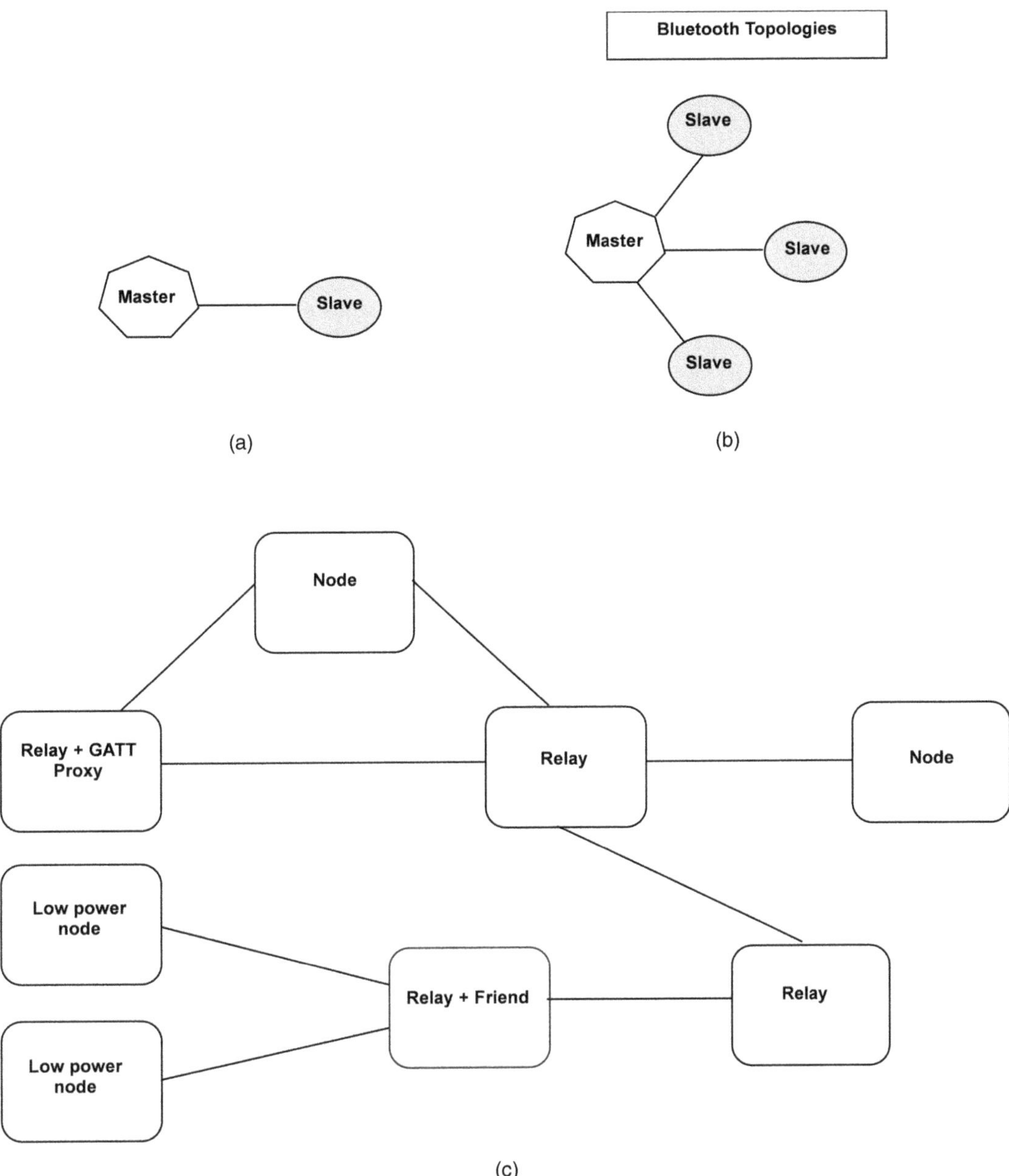

FIGURE 2.3 Three Bluetooth network topologies: (a) point-to-point (P2P), (b) point-to-multipoint (P2MP), and (c) mesh. Part (a) shows a network configuration in which a master device is connected to a single slave device. Part (b) shows a network arrangement featuring a master device joined to three slave devices. Part (c) shows a network connection in which every node has a linkage to every other node in the network. Simple nodes as well as special-feature nodes are included, e.g., relay, relay +friend, relay+ GATT proxy, and low power nodes.

2.3.2.2 ZigBee

ZigBee is a specification created by ZigBee Alliance as a straightforward and less expensive option than Wi-Fi and Bluetooth (Long and Miao 2019, Haque et al. 2022). It is used to form wireless personal area networks (WPANs). It is based on the IEEE 802.15.4 WPAN radio standard. With a typical power of 1 mW or 0 dBm, the range of transmission is restricted to 10 m. But depending on the particular device and environmental conditions, it can be up to 100 m. The data rate is 20–250 kbps. It allows operation in any one channel of the following three frequency bands.

   (i) Channel 0: 868MHz with 20 kbps for Europe,
  (ii) Channel 1–10: 915MHz with 40 kbps for the USA and Australia, using binary phase shift keying (BPSK) modulation, and
 (iii) Channel 11–26: 2.4 GHz with 250 kbps across the world using offset quadrature phase shift keying (OQPSK). It has sixteen 5-MHz operation channels.

The network topologies supported by ZigBee are (E-SPIN 2023) as follows.

   (i) Star Topology: It is the simplest configuration consisting of a coordinator and several end devices connected to it (Figure 2.4(a)),
  (ii) Tree Topology: It consists of a coordinator, several routers, and end devices (Figure 2.4(b)). The routers acting as intermediaries extend the network coverage, consequently allowing effective communication with devices at remote locations from the coordinator, and
 (iii) Mesh or Peer-to-Peer Topology: It is similar to the tree topology with the difference that each device can communicate with any other device (Figure 2.4(c)). If a transmission path fails, the affected node finds an alternative path. This capability of self-healing increases the reliability of the network.

ZigBee offers a range of security features. However, weaknesses in ZigBee implementations against real-life attacks have been examined, and vulnerabilities have been studied (Ren et al. 2023).

2.3.2.3 Z-Wave

It is a wireless communications protocol developed by Zensys, Copenhagen, Denmark (Linh and Kim 2018). The Z-Wave operating frequency band is 868–869 MHz in Europe. In North America, the frequency bands are 908–916 MHz with mesh topology and 912–920 MHz with star topology in long-range mode. In the Z-Wave mesh network topology, each device acts as a node. The devices connect with each other forming a web of interconnected devices. There can be up to 232 nodes in a Z-Wave network.

The Z-Wave frequencies are much lower than the 2.4 GHz, 5 GHz, and 6 GHz bands of Wi-Fi. The data rates provided by Z-Wave are 9.6 kbps, 40 kbps, or 100 kbps. Because of its lower frequency band, the Z-Wave carries less data but it has a longer range, around 200m at line-of-sight outdoors, and 50 m in line-of-sight indoors, and is less prone to obstruction by physical barriers. It works at 1 mW or 0 dBm output power. The Wi-Fi working on a higher frequency band carries more data but provides a shorter range, and shows a greater susceptibility to physical hindrances. By working at sub-GHz frequencies against the 2.4 GHz for Wi-Fi, the Z-Wave has the advantages of low power and long communication range. Moreover, it is less upset by RF interference in distinction to Wi-Fi and Bluetooth working in the crowded 2.4 GHz band carrying a lot of traffic.

Strong AES 128-bit encryption bestows protection to Z-Wave against brute force attacks (Badenhop et al. 2017, Lilli et al. 2021). The AES stands for advanced encryption standard. The AES-128 performs encryption and decryption of messages in blocks of 128 bits using a 128-bit symmetric key. Once a device is paired with a central hub, the unpairing and grabbing control of the device by a mischievous assailant is exceedingly arduous. Wi-Fi is more likely to fall prey to security breaches than Z-Wave.

ZigBee and Z-Wave are used in similar applications. ZigBee is more versatile and can be configured for more applications. However, its complexity leads to a longer development time. The Z-Wave allows faster implementation.

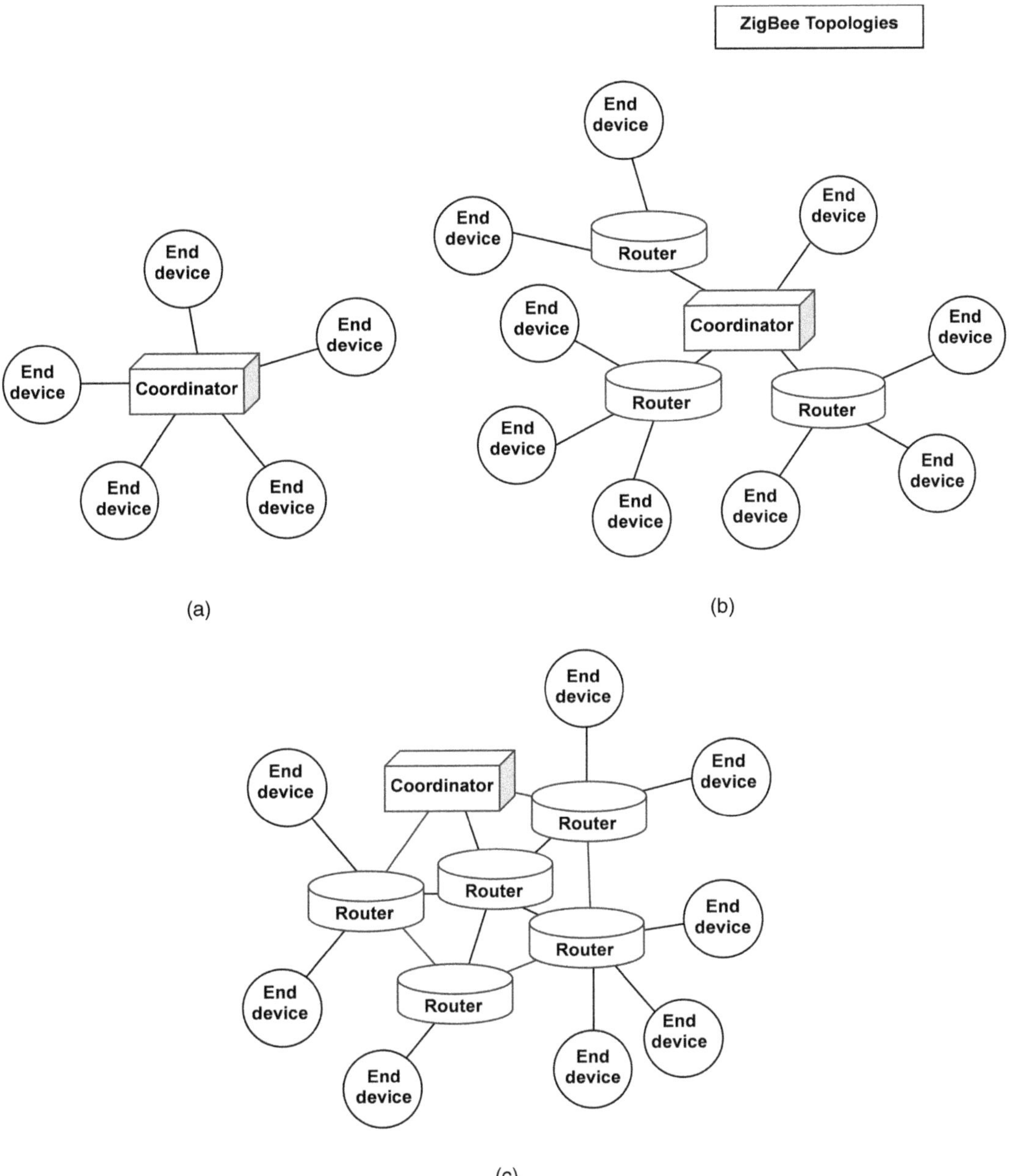

FIGURE 2.4 ZigBee network topologies: (a) star, (b) tree, and (c) mesh. Part (a) shows a single coordinator node to which all the end devices are connected. Part (b) shows a single coordinator node, several routers for passing data between devices, and multiple end devices that are to be controlled. Part (c) shows a single coordinator node, several routers, and multiple end devices similar to the network in part (b). But in part (c), an alternative path is found to the destination if a transmission path fails because each device is connected to every other device. The same is not achieved in part (b) because a failure of a parent node disables the associated child nodes. This self-healing capability makes topology (c) more reliable than (b).

2.4 HIGH DATA RATE, LOW POWER, LONG-RANGE COMMUNICATION PROTOCOL: LoRAWAN

The LoRaWAN, with full form 'Long Range Wide Area Network', is a radio frequency protocol under the governance of The LoRa Alliance, San Ramon, California, USA (Haxhibeqiri et al. 2018). It uses open-source technology. An open-source technology is one whose source code is freely available for usage, modification, and redistribution. So, one can develop one's own LoRaWAN solutions. The star network topology is adopted in which all the end nodes are connected to gateways. The gateways themselves are connected to the central server of the network.

The LoRaWAN transmits over frequency bands that do not require a license. So, an individual who owns the LoRAWAN infrastructure need not pay for data use. In Europe, the energy band used is 868 MHz (Saban et al. 2021). In the USA, it is 915 MHz. These sub-GHz frequency bands impart to it a wide coverage along with good indoor penetration capabilities. Its range is much larger than that of Wi-Fi or Bluetooth, typically up to 5 km in dense urban areas, and 15 km in rural environments. Minimal energy is needed for the transmission of its small data packets. Since the transmission is done a few times during the day and the power consumption is in mW when the end devices are sleeping, the battery life is prolonged to the extent of 5–10 years.

The LoRaWAN is a feasible substitute for Wi-Fi or Bluetooth in isolated locations lacking in connectivity. However, the data rate of a LoRaWAN channel is low ~0.3 kbps to 50 kbps. Additionally, the benefit of low overhead costs provided by unlicensed bands is counteracted by the investments incurred in network development and the unavoidable maintenance expenditure.

The LoRaWAN messaging capabilities are extremely limited because only a restricted number of messages can be sent per day to prevent network overcrowding,

2.5 DATA PROTOCOLS

2.5.1 MQTT

MQTT (message queuing telemetry transport) is a lightweight, open-source, and free-to-use, publish-subscribe, messaging protocol for machine-to-machine communication (Masdani and Darlis 2018). In MQTT, there are two types of network entities: brokers and clients. The broker is a server. The client is any device that connects to an MQTT broker over a network. A client is a publisher if it is sending messages. It is a receiver if it is getting messages. The broker receives all messages from the clients and routes them to the appropriate destination clients. It acts as a coordinator of messages between the clients. Until the messages have been delivered to the intended client, they are stored in a queue. The clients connect only with the broker. They never connect with each other.

The connection is initiated by the client. It does so by sending a CONNECT message to the broker. The broker responds with a CONNACK message implying that the connection is set up. In order to receive messages on their topics of interest, the clients send a SUBSCRIBE message to the MQTT broker.

Like HTTP (hypertext transfer protocol) or SMTP (simple mail transfer protocol), MQTT runs over TCP/IP. But unlike HTTP, which is used for internet communication, MQTT works with devices that are connected to each other with a network. This network is generally an IP (internet protocol) network.

For security, MQTT uses the SSL (Secure Sockets Layer) protocol. Before sharing data, it is essential to ensure identification, authentication, and authorization between clients and the broker using certificates and/or allocated passwords.

2.5.2 CoAP

CoAP with the expanded form 'constrained application protocol' is a software protocol for the application layer, web-based usage. Its defining body is the Internet Engineering Task Force (IETF), the leading organization concerned with the formulation of open standards by transparent approaches for improving the functioning of the internet. This specialized protocol is intended for utilization with low-power and low-memory devices with constrained resources. Sensors and actuators in lossy networks fall under this category of devices. Its purpose is to establish their interactive communication via the internet (Jarvinen et al. 2018).

Its necessity arose because the conventional TCP/IP (transmission control protocol/internet protocol) was found to be too heavy for IoT. The CoAP runs on UDP (user datagram protocol). The UDP is a connectionless protocol whereas TCP is a connection-based protocol guaranteeing trustworthy packet transmission. The UDP is less reliable than TCP because it does not confirm receipt of a message. But UDP works faster than TCP because of its smaller packet size and less overhead. It can tolerate packet loss. It can also support real-time data transfer. Thus, CoAP gives most of the data transport features of TCP without demanding equivalent resources. This makes it suitable for real-time IoT applications where fast data transfers in combination with short response times are of greater significance than reliability (Coetzee et al. 2018).

In the CoAP, a client–server model is followed using two types of messages, namely, request and response messages. A CoAP message contains a four-byte binary header followed by header data and payload. In HTTP, there is a polling mechanism in which the client requests the resource at regular intervals to check whether there are any state changes. In CoAP, the server continues replying after the transference of the initial document. In this respect, the CoAP is an improvement over the HTTP model because of its ability of resource observation.

For protecting information, CoAP depends on UDP security features by using datagram transport layer security (DTLS) over UDP. The DLTS is a communication protocol using certificate-based authentication and symmetric encryption to ensure privacy and prevent eavesdropping or data tampering.

2.5.3 AMQP

AMQP standing for 'Advanced Message Queuing Protocol' is a protocol for message-oriented middleware (MOM) (Bahashwan and Manickam 2018). The 'message-oriented middleware' is meant as the software enabling communication via messages between a dispersed team of software systems.

AMQP is an open standard application layer protocol for transferring point-to-point and publish-and-subscribe routing and queueing messages. Routing is the process of selecting a suitable path for traffic in a network. Routing is accomplished in a point-to-point or publish-and-subscribe format. A point-to-point routing format means communication between two routers directly without the engagement of any intermediary host or other network. A publish-and-subscribe routing format involves the exchange of messages between various publishers and subscribers. The publishers are the components that produce and send messages whereas the subscribers are the components which receive and consume messages. A message queue is a lightweight buffer for temporarily storing messages. It also includes the endpoints permitting software components to connect to the queue for transmission/ reception of messages. A crucial role in AMQP facilitation is played by brokers or servers.

The key functions in AMQP are message routing and queuing. In AMQP messaging, the publishers redirect data acquired from exchanges and place them in queues. The brokers receive messages from publishers and route them to subscribers. Brokers have the responsibility of building the connection to ensure proper routing and queuing of data. They check that the message is conveyed directly from the publisher to the subscriber. The subscribers generate queues and acknowledge messages.

The AMQP is a protocol for asynchronous messaging in which the sender has no expectation of getting an immediate response from the receiver. So, the sender does not wait for the response. Irrespective of whether the receiver responds or not, the sender continues the execution of its leftover tasks. As there is no need to wait for a direct live response, the two components participating in the conversation have the liberty to start, stop, and resume messaging at their own discretion.

The AMQP allows interoperability of messaging between systems of different configurations and infrastructures. Interoperability is the ability of various computer systems to readily connect and exchange information among themselves. The AMQP allows developers to put all the protocol-approved compliant components and brokers into action.

2.5.4 DDS

The DDS representing 'Data Distribution Service' is a middleware protocol. It is developed by Object Management Group® (OMG®), an international computer technology standards consortium in Needham, MA, USA. The middleware is a software layer with a scalable architecture. It is located between the operating system and applications. Its purpose is to bring together the components of the system for the facilitation of reliable, low-latency, and interoperable communication/sharing of data in real time.

The middleware follows a publish–subscribe pattern of operation. The publish-subscribe model coordinates its activities through three principal components, namely, a publisher, a topic, and a subscriber. In this setup, the publisher conveys the message to a topic. In turn, the topic delivers the message to all the subscribers. During the transaction, the publisher and the subscriber are completely isolated from each other. The publisher does not know where the message is being used. Also, the subscriber does not know about the publisher.

The publish-subscribe model differs from the model using message brokers as mediators where a message queue batches the separate messages until receipt of a request from a user for these messages, following which the messages are retrieved by the user.

The publish-subscribe model is a boon for distributed systems because it practically gets rid of intricate network programming This advantage enables software developers to create decoupled applications by concentrating on the targeted goals of their applications instead of worrying about the mechanics of conveying information between applications and systems.

The DDS has been applied in the construction of high-performance distributed systems in the defense, automotive, and finance sectors. Various obstacles faced and open research issues have been identified (Köksal and Tekinerdogan 2017).

2.5.5 HTTP

HTTP (hypertext transfer protocol) is concerned with hypertext and hyperlinks (Gourley et al. 2002). Since 'hyper-'means 'above or beyond', hypertext is a text beyond normal text. So, it is a text which has more capabilities than ordinary text. This capability allows the facility of clicking hypertext with a mouse, whereupon the hypertext immediately takes us to the text in one of the numerous electronic sources connected in a network. This capability is built into the hypertext by incorporating hyperlinks in the text. Hence, the hypertext is text containing hyperlinks. A hyperlink is meant to be an icon, graphic, or text that provides a linkage to another file or object. It is a digital reference to data to which the user is guided and is, therefore, able to access by clipping or tapping on it with a mouse. It points to a complete document or portion of the document. Hyperlinks are widely used in the World Wide Web to establish linkages among a colossus collection of files and pages on the internet. They are mostly embedded links. The URL (uniform resource locator), the address of a resource on the web, is embedded inside the link. It is not straightforwardly evident on the web page. A hyperlink can be inserted for any resource on the web with the intent that the browser moves to the content of that resource upon activation.

Regarding its communication mode, the hypertext transfer protocol is a request–response protocol. In this client–server model (Figure 2.5), the client dispatches a request to the server querying for a resource. Upon receiving the query, the server collects the requested resource and responds back to the client. In case of failure to find the requested resource, the server sends an error message to the client.

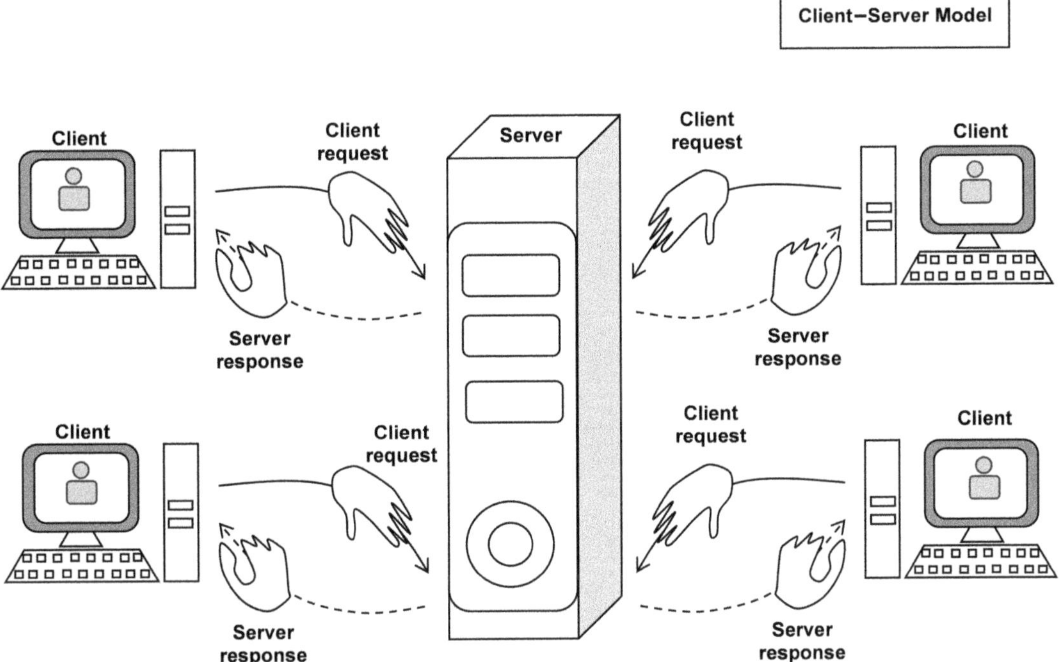

FIGURE 2.5   The client–server model is a model of interaction between a server and a client. The centralized provider of a service is called the server, and the users of the service are known as clients. In this model, the tasks or workloads are partitioned between the client and the server. It works on a request–response pattern. For its execution, the client sends a request to the server to access a service, and the server replies with a response.

Websites are accessed through an application called the web browser. The browser is the client. The user browsing the web requests a webpage from a particular website. The server is an application running on a computer hosting the website. The browser retrieves the requested file from the server and displays its contents on the computer screen of the user who is browsing the internet.

2.5.6   HTTPS

HTTPS is the shortened form of 'hypertext transfer protocol secure'. The necessity of HTTPS arose from the drawback of an HTTP connection to be impaired by a malevolent person. HTTPS is the same protocol as HTTP with the exception that all the messages exchanged between the web browser and the web server are made impervious to attacks. Indeed, HTTPS is a secure version of HTTP, specially enhanced for safeguarding against attacks. The invulnerability to attacks is achieved using the TLS (transport layer security) protocol for encryption (hiding the data), authentication (ensuring the genuineness of message-exchanging parties involved), and integrity (verifying that the data is not forged) (Ristic 2022). The primary feature of data encryption and decryption in the transport layer security protocol is the involvement of the asymmetric public key infrastructure (PKI) using two keys. One key is a private key, which is controlled and maintained by the owner of the website. The other key is a public key, which is available to users. The TLS handshake is preceded by a TCP handshake. The TCP handshake and the ensuing TLS handshake steps are (Figure 2.6) as follows.

Step (i): Three-Way Transmission Control Protocol (TCP) Handshake: The TCP connection between the client's device and the web server is established (Figure 2.6(a)). In this process, the client sends a data packet, the SYN (synchronize sequence number) to the server,

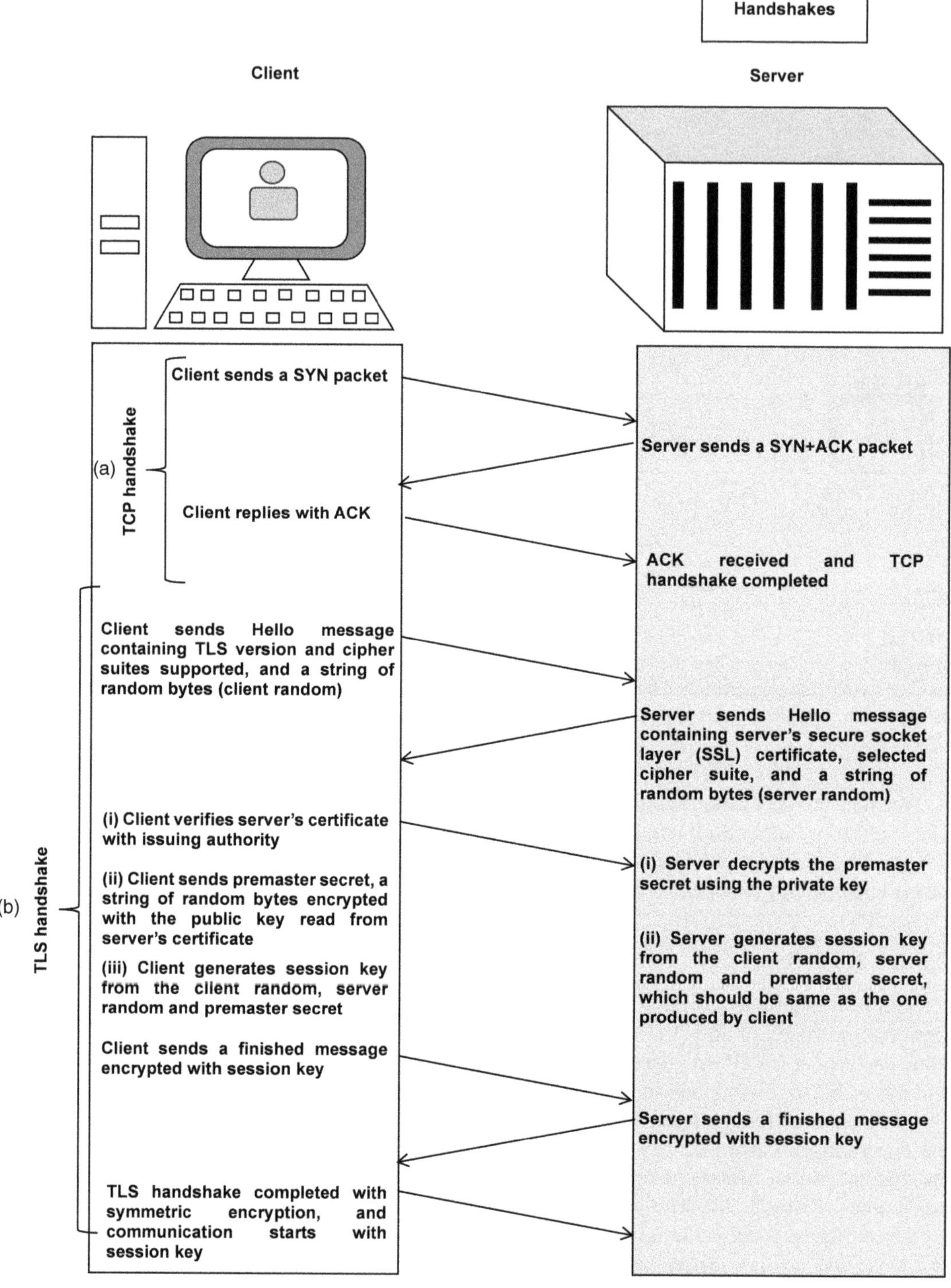

FIGURE 2.6 Handshakes: (a) TCP and (b) TLS. Part (a) shows a three-way handshake between a client and a server consisting of SYN by the client, SYN-ACK by the server, and ACK by the client, in which the two sides synchronize and acknowledge each other. Part (b) shows the process of starting communication between a client and server by establishing an encrypted link for secure transmission of data through the steps of negotiation of cryptography, verification of server from its SSL certificate, exchange of acknowledging messages, setting up of cryptographic algorithms and the agreement of session keys.

while the server sends a SYN (synchronize sequence number)/ACK (acknowledgment sequence number)/ packet to the client. The client replies with an ACK packet.

Step (ii): Initiation of Transport Layer Security (TLS) Handshake: The client informs the server of its desire to make a TLS connection with it, mentioning the version of TLS protocol and the encryption techniques to be used (Figure 2.6(b)). The client does this by presenting a list of supported cipher suites in the form of ciphers and hash functions. The cipher suites comprise a set of algorithms specifying the shared encryption keys, or session keys to be used.

Step (iii): Confirmation of Protocol by the Server: If the server supports the requested version of TLS protocol as revealed by picking a cipher and hash function supported by it from the client-supplied list, it sends a confirmation to the client along with a digital certificate containing the name of the server, the certificate authority (CA) and the server's public encryption key.

Step (iv): Authentication of Server's Identity by the Client: If the client is unable to verify the certificate, the connection is aborted. If the client is able to verify the server's identity, it moves forward to the next step.

Step (v): Generation of Session Keys: The client encrypts a random number from the public key of the server. This random number is sent to the server. It is used by both parties to produce a unique session key, which is applied for the ensuing encryption/decryption of messages.

Step (vi): Conclusion of TLS Handshake and Session Start: The handshake ends with the previous step, and the session begins.

2.5.7 WEBSOCKET

It is a bidirectional, full-duplex computer communications protocol for client–server communication in which the client is the web browser (Skvorc et al. 2014). Whenever the client and the server make a handshake, the connection called a WebSocket is kept alive unless it is terminated by any one of them. Like HTTP, the WebSocket is located at layer 7 in the OSI model. Further, it follows TCP at layer 4. However, it differs from HTTP in several respects:

(i) The WebSocket is an event-driven protocol using events (changes in the state of an object, e.g., placement of an item on a cart in a shopping mall) to trigger communication between decoupled services. In WebSocket, the services asynchronously publish and subscribe to events via an event broker, thereby allowing them to interact with each other in real time or update data in real time. HTTP is a request–response protocol in which a client, i.e., a browser sends a request to the web server. The web server runs an application to process the request and returns a response (output) to the browser.

(ii) The WebSocket handles real-time and continuously updated dynamic data. The HTTP processes static and stagnant data.

(iii) The WebSocket is bi-directional. Simultaneous to-and-fro data transmission can occur enabling two-way communication between the client and the server. The HTTP is unidirectional. At any instant of time, either a request or a response is taking place.

(iv) The WebSocket provides full-duplex communication whereas the HTTP gives half-duplex communication.

(v) The WebSocket connection can continue for an indefinite period. The HTTP connection ends as soon as a response is made.

(vi) The WebSocket is a stateful protocol. By 'stateful protocol' is meant a protocol in which the server saves the status and session data. On the contrary, the HTTP is a stateless protocol treating each request and response as an independent pair. Hence, the server does not retain information about the session or the status of clients in multiple requests.

(vii) The WebSocket has high speed because the connection lasts as long as one party does not end it. The HTTP is slow because a new connection has to be set up for every request requiring sufficient time.

(viii) The WebSocket is associated with lower operational overheads because one request can lead to several responses. The HTTP needs higher operational overheads as a fresh request is generated every time.

2.6 IoT PLATFORMS

After briefly discussing the IoT protocols in the preceding sections of this chapter, let us now move to the IoT platforms. An IoT platform is an on-premises, installed, and configured software suite or a cloud facility (platform as a service, PaaS) containing built-in tools that allow manufacturers and other users to connect to the IoT ecosystem, and hence to manage sensors, collect data from remote, perform data analysis and application enablement. It acts as a bridge for seamless integration of sensors with data networks. It is the controlling center of every IoT deployment interlinking the sensors, connectivity, hardware, software, and applications.

2.6.1 GOOGLE CLOUD PLATFORM

It is a group of cloud computing services from Google LLC, Mountain View, CA, USA. The cloud is a cluster of servers that can be accessed over the internet for information management, processing, and storage. It helps users to rent out computer resources housed in Google's data centers from any place in the world. The users can work from a remote computer by storing and accessing data and programs online without spending on expensive hardware and software. Besides cost savings, the efficiency of companies is considerably enhanced.

The Google Cloud Platform (GCP) is a part of Google Cloud which includes Google Workspace and application programming interfaces (APIs). The GCP runs on the same multi-layer secure infrastructure of a global network as other Google services such as Google Search and Gmail. It provides web-based services of data management, storage, computing, data analysis, AI and machine learning, natural language processing, and image recognition to its customers like the Amazon AWS and Microsoft Azure platforms.

The GCP offers the benefit of scalability to companies. They can start with a small investment and scale up as they grow. Further, the GCP is protected with proper security features. These are data encryption and two-factor authentication. The intrusion detection system monitors network traffic and alerts about suspicious activities.

2.6.2 IBM WATSON

It is a data analytics processor, developed as a question-answering computing system by IBM, The International Business Machines Corporation, Armonk, New York, USA, using natural language processing, reasoning, and machine learning technologies, and named after its founder and first CEO Thomas J. Watson. Workload-optimized on massively parallel processors, it is deployed on a cluster of servers. It works by finding statistically related phrases. This is done by parsing questions into keywords and fragmented sentences. By leveraging speech analysis to understand the syntax, context, and meaning of the language, it is able to give practical answers assisting medical, economic, and retail managers to make correct business decisions by infusing artificial intelligence into its workplan.

2.6.3 AMAZON AWS IoT CORE

The AWS (Amazon web services, Inc., Seattle, WA, USA) IoT Core is a platform providing secure bi-directional communication for sensors, actuators, and other devices linked to the internet by connecting to AWS cloud using protocols such as HTTPS, MQTT, and LoRaWAN. It helps users process and analyze data for building intelligent IoT applications by furnishing the necessary tools and services. Its main features are as follows.

(i) The AWS Cloud: It is a technology using remote servers equipped with services such as storage, computing, databases, data analytics, and artificial intelligence.

(ii) The AWS IoT Device SDK (software development kit): This kit assists the devices in connection, authentication, and exchanging messages with AWS IoT Core. The MQTT, HTTP, or WebSocket protocols are used.

(iii) The Device Advisor: It is a fully managed cloud-based test capability. Its responsibility is to help developers in the validation of their IoT devices to make reliable and secure connections with AWS IoT Core.

(iv) The Device Gateway: It is the entrance point for IoT devices. It manages all the active device connections.

(v) The Message Broker: It is a publisher/subscriber messaging agent working on the MQTT standard.

(vi) The Registry: It establishes an identity for devices and tracks their attributes and capabilities.

(vii) The Device Shadow: It virtually represents the physical device. It can make the state of the device available irrespective of whether it is connected or not. It does so via MQTT topics.

(viii) The Rules Engine: It helps in building IoT applications that act on data generated by connected devices at an international scale without managing any infrastructure.

2.6.4 Microsoft Azure IoT Hub

It is a service hosted and fully managed by the Microsoft Corporation, Redmond, WA, USA in the cloud for the provision of secure and reliable communication between IoT applications and the devices attached to them.

A major communication capability offered in this service is telemetry. In telemetry, the data is automatically measured and transmitted from remote or inaccessible sources between devices and the cloud. Other services include file uploading from devices, and the control of devices through request–response messages from the cloud.

Integration of Azure IoT Hub with other Azure services helps in availing the following facilities.

(i) Azure Stream Analytics: It is a cloud-based, real-time stream processing engine for analyzing high volumes of IoT data for triggering actions or building highly scalable analytical solutions. It deals with data in motion whereas conventional analytics studies static data.

(ii) Azure Time Series Insights: It is a cloud-based service for ingestion, modeling, querying, and visualization of fast-moving time-series data produced by IoT devices.

(iii) Azure Machine Learning: It is a cloud solution for applying machine learning and deep learning models without specialized expertise in data science.

(iv) Azure Logic Apps: It is a cloud-based platform used to automate tasks and workflows

(v) Azure Event Grid: It is a cloud service for fast and reliable event routing. An event is the data/information describing the occurrence of an action.

2.6.5 Oracle Internet of Things Cloud Service

It is a platform of the Oracle Corporation (ORCL), Austin, Texas, USA, in which the IoT devices connect to the cloud, either directly or indirectly through a gateway to establish a reliable bidirectional communication with the cloud. Each device is assigned a unique identity. While making such connections, authentication and authorization are strictly followed. The cross-protocol functionality employed enables directly addressing any particular cloud-connected device disregarding the network protocols and restrictions imposed by firewalls. The incoming stream of data is correlated, aggregated, and filtered. Data visualization and analysis help in providing the required visibility and understanding for the following.

(i) Smart Manufacturing: This is achieved by continuously tracking the production lines of factories for assessment of performance.
(ii) Predictive Maintenance: This is done through surveillance of machinery health, location, utilization, and envisaging maintenance needs and minimizing downtime.
(iii) Controlled Logistics: This is implemented by following the products via sensor signals across transportation and warehouse systems and managing fleet and shipment.
(iv) Workplace Safety Monitoring: Sensor data from employee wearables are examined to track worker movement to avoid unsafe acts such as entering hazardous or unauthorized areas to ensure regulatory compliance and enforcement of safety policies.

2.6.6 THINGWORX INDUSTRIAL IoT SOLUTIONS PLATFORM

It is a technical system of the PTC (Parametric Technology Corporation Boston, Massachusetts, USA), offering a rich set of capabilities to provide solutions for various industrial operations such as design, manufacturing, and service. It has the complete set of tools and processes to connect contrasting devices and applications for accessing multiple data sources. It can build complete industrial IoT solutions and quickly endow augmented reality experience by blending the existing world environment with computer-generated virtual information. It analyzes complex data gathered from sensors fixed in the industry and makes recommendations for improvements. It engages with physical objects and devices in a contextualized manner to initiate suitable actions. Advanced artificial intelligence techniques are applied for the diagnosis and resolution of problems, and for preventing interruptions in work. However, all the suggestions are given in an easy-to-understand format for non-expert workers.

2.6.7 SALESFORCE IoT CLOUD

It is the platform of Salesforce.com, a software-as-a-service (SaaS) provider in San Francisco, USA. It is primarily developed for the storage and processing of data from devices in the Internet of Things. It is driven by a scalable, real-time event processing engine known as the Thunder which collects, filters, and promptly answers to events. Data acquired from sensors and IoT applications is applied to initiate necessary actions. To consider a few examples, a malfunctioning or failure of any IoT-connected device is detected by sensors to stimulate action for servicing, adjustments are made in wind turbines in accordance with prevailing weather conditions, and businessmen get a clear view of customers' responses without the services of an expert in data analysis.

The Thunder is based on open-source tools such as a messaging system Apache Kafka; the Big Data event processing platforms, Apache Storm and Apache Spark; and a distributed database management system, Apache Cassandra, all run on a proprietary platform of the Salesforce. An IoT platform such as AWS manages the data and transfers it to the Salesforce IoT cloud. A convenient web services interface for interaction with Salesforce is REST API (representational state transfer API).

The algorithm for implementation of the Salesforce IoT cloud platform comprises the following steps.

(i) Plan: It involves the identification of event triggers.
(ii) Connect: It is concerned with the finalization of data sources.
(iii) Transform: It entails the conversion of customer scenarios to technical language.
(iv) Build: It relates to the incorporation of orchestration rule, and
(v) Deploy: It means activation of the orchestration rule.

Orchestration refers to the automatic configuration, organization, and coordination of computer systems, applications, and services. Orchestration seeks to streamline and optimize the execution of recurrent, and repeatable processes. The assistance given enables data teams to easily manage complex tasks and workflows.

2.6.8 PARTICLE

It is an open-source, scalable, and reliable IoT platform of Particle Industries, Inc., San Francisco, USA, for connecting devices to web and mobile applications. It is an integrated Platform-as-a-Service (PaaS) in which IoT hardware, software, and connectivity work together in harmony to build intelligent, interactive machines that optimize industrial operations and accelerate business growth.

Particle's Hardware Tool: The main tool is the particle photon, a Wi-Fi, and a small memory-equipped microcontroller. It acts as a centralized control and analysis center for IoT devices.

Particle's Operating System: A lightweight operating system includes an integrated development environment and programming framework for creating applications using IoT devices.

Particle's Data Management System: It contains the asset tracking system for pursuing decisive asset data and IoT data pipeline service for extracting data from IoT devices and analyzing them.

2.6.9 IRI VORACITY

The IRI (Innovative Routines International, Inc., Melbourne, Florida) Voracity is an independent software vendor (ISV) data management platform for discovering, integrating, migrating, governing, and analyzing data. The processing engine used in IRI Voracity is the CoSort, a data manipulation software capable of sorting large multi-gigabyte files allowing companies to quickly classify, join, and protect data given in several CJK (Chinese, Japanese, and Korean) formats across Unix, Linux and Windows systems on its multi-CPU servers. The graphical user interface (GUI) for IRI Voracity is the IRI workbench. It is a plug-in to the Eclipse Integrated Development Environment (IDE) designed for the user-friendly development of applications using various programming languages, e.g., Java, C/C++, and Python. With its help, Voracity users can manipulate, migrate, and make multiple uses of structured/unstructured sources of data to solve practical problems of Big Data, namely, its volume, diversity, accuracy, velocity, and worth.

2.6.10 BLYNK

It is a software collection of Blynk Inc., New York, USA, for performing different operations on connected electronic devices. These operations are prototyping, deploying, and managing from a distance. Using the Blynk platform, the users can connect their hardware with the cloud to build iOS (iPhone Operating System), Android, and web applications. They can control their devices remotely. Real-time as well as historical data analysis can be done.

The Blynk IoT platform consists of the following components.

(i) Blynk.Console is a web application for configuring and managing the devices, the data, and their overall monitoring.

(ii) Blynk.App is an iOS and Android mobile application assigned with the responsibilities of remote monitoring of devices, configuring mobile user interfaces, and automating the work.

(iii) Blynk.Edgent is a facility for managing connectivity with Wi-Fi, cellular, and Ethernet.

(iv) Blynk Library is a hardware library from which hardware models can be connected to the Blynk.Cloud.

(v) Blynk.Cloud is a server infrastructure.

(vi) Blynk micro-services is a software module such as Blynk.Inject for claiming ownership of device, and supplying Wi-Fi credentials, Blynk.R for user registration, and Blynk.Air for firmware over-the-air (OTA) updates.

2.6.11   Cisco IoT

It is a convenient IoT platform from Cisco Systems, Inc., San Jose, CA, USA, for managing and storing data in the cloud. It works by obtaining data from sources, processing it on disseminated nodes, and sharing it among distributed applications to provide industry-focused solutions for enhancing business efficiency and productivity. The Cisco Virtualized Packet Core (VPC) based on StarOS software is a technology catering to the essential services for small cell networks, Wi-Fi, 2G, 3G, and 4G.

2.6.12   Bosch IoT Platform

It is an open-source solution from the Robert Bosch GmbH, Gerlingen, Germany, providing middle-ware services for developing applications. It acts as a hidden translation-layer software between the operating system and the applications. It helps the applications to connect intelligently for efficient data management and communication supported with strong protection and privacy. Being pliable and adaptable to other platforms, smooth IoT unification is achievable between applications of different domains. Besides providing connectivity and interoperability for building prototypes of applications, the Bosch platform offers data analysis in real time.

2.6.13   Brainbean Apps

The Brainbean Apps (BBA) is a modular IoT platform from the Brainbean Apps LLC, Tallinn, Estonia. It offers an all-inclusive data control panel to analyze inputs from customers for making decisions leveraging machine learning. Trusted data exchange is accomplished between devices through Wi-Fi and Bluetooth Low Energy (Bluetooth LE or BLE) assuring data security and protection. It supports Amazon Alexa, a virtual technology assistant, and Samsung SmartThings, an automation platform to connect devices from different companies. Another vital feature is the provision of open-source APIs and SDKs, two indispensable software development tools. APIs are an agreed set of rules and protocols to be followed for communication and data/functionality sharing between different computer programs. SDKs (software development kits) contain implements, libraries, and other resources to create mobile applications, including debugging, monitoring, and optimization components.

2.7   CONCLUDING REMARKS AND PREPARING FOR THE UPCOMING CHAPTER

After reading this chapter, the reader will know the structure of an IoT system, the rules of the system, and the set of facilities that enable users to build IoT applications. In the next chapter, we start our journey of sensors that are essential to realize the different goals and fulfill the dreams of IoT promises.

REFERENCES

Badenhop C. W., S. R. Graham, B. W. Ramsey, B. E. Mullins, and L. O. Mailloux 2017 The Z-Wave routing protocol and its security implications, *Computers & Security*, 68:112–129.
Bahashwan A. A. O. and S. Manickam 2018 A brief review of messaging protocol standards for internet of things (IoT), *Journal of Cyber Security and Mobility*, 8(1):1–14.

Bluetooth SIG, Inc. 2023 Learn about Bluetooth topology options, https://www.bluetooth.com/learn-about-bluetooth/topology-options/

Chew D. 2019 Protocols of the wireless internet of things, in: *The Wireless Internet of Things: A Guide to the Lower Layers*, Wiley-IEEE Standards Association, IEEE Press, John Wiley & Sons, Inc., NJ, USA, pp. 21–45.

Coetzee L., D. Oosthuizen, and B. Mkhize 2018 An analysis of CoAP as transport in an internet of things environment, In: P. Cunningham and M. Cunningham (Eds.), *IST-Africa 2018 Conference Proceedings, IIMC International Information Management Corporation*, pp. 1–7.

E-SPIN 2023 Exploring Zigbee network topology: Unveiling the three types for robust communication, https://www.e-spincorp.com/zigbee-network-topology/#:~:text=To%20sum%20up%2C%20Zigbee%20network,lacks%20redundancy%20and%20alternative%20paths

Gourley D., B. Totty, M. Sayer, A. Aggarwal and S. Reddy 2002 *HTTP: The Definitive Guide*, O'Reilly Media, Sebastopol, CA, 656 pages.

Haque K. F., A. Abdelgawad, and K. Yelamarthi 2022 Comprehensive performance analysis of Zigbee communication: An experimental approach with XBee S2C module, *Sensors (Basel)*, 22(9):3245, pp. 1–23.

Haxhibeqiri J., E. De Poorter, I. Moerman and J. Hoebeke 2018 A survey of LoRaWAN for IoT: From technology to application, *Sensors (Basel)*, 18(11):3995, pp. 1–38.

Jarvinen I., L. Pesola, I. Raitahila, Z. Cao and M. Kojo 2018 Performance evaluation of constrained application protocol over TCP, *IEEE 88th Vehicular Technology Conference (VTC-Fall)*, 27–30 August, Chicago, IL, pp. 1–7.

Köksal Ö. and B. Tekinerdogan 2017 Obstacles in data distribution service middleware: A systematic review, *Future Generation Computer Systems*, 68:191–210.

Lilli M., C. Braghin and E. Riccobene 2021 Formal proof of a vulnerability in Z-Wave IoT protocol, In *Proceedings of the 18th International Conference on Security and Cryptography, Volume 1: SECRYPT*, SciTePress, pp. 198–209.

Linh An P. M. and T. Kim 2018 A study of the Z-wave protocol: Implementing your own smart home gateway, *2018 3rd International Conference on Computer and Communication Systems (ICCCS)*, Nagoya, Japan, 27–30 April, pp. 411–415.

Long S. and F. Miao 2019 Research on ZigBee wireless communication technology and its application, *2019 IEEE 4th Advanced Information Technology, Electronic and Automation Control Conference (IAEAC)*, 20–22 December, Chengdu, China, pp. 1830–1834.

Masdani M. V. and D. Darlis 2018 A comprehensive study on MQTT as a low power protocol for internet of things application, *3rd Annual Applied Science and Engineering Conference (AASEC 2018), IOP Publishing IOP Conf. Series: Materials Science and Engineering*, 434, 012274, pp. 1–7.

Mozaffariahrar E., F. Theoleyre and M. Menth 2022 A survey of Wi-Fi 6: Technologies, advances, and challenges, *Future Internet*, 14, 293, pp. 1–52.

Nair J. 2023 Bluetooth versions and their differences, https://www.linkedin.com/pulse/bluetooth-versions-differences-jinu-nair#:~:text=Bluetooth%206.,transmission%2C%20and%20improved%20security%20features

Pahlavan K. and P. Krishnamurthy 2021 Evolution and impact of Wi-Fi technology and applications: A historical perspective. *International Journal of Wireless Information Networks*, 28:3–19.

Ren M., X. Ren, H. Feng, J. Ming, and Y. Lei 2023 Security analysis of Zigbee protocol implementation via device-agnostic fuzzing, *Digital Threats: Research and Practice*, 4(1):pp. 9:1 to 9:24. Article 9.

Ristic I. 2022 *Bulletproof TLS and PKI: Understanding and deploying SSL/TLS and PKI to secure servers and web applications*, Feisty Duck, London, UK, 512 pages.

Saban M., O. Aghzout, L. D. Medus, and A. Rosado 2021 Experimental analysis of IoT networks based on LoRa/LoRaWAN under indoor and outdoor environments: Performance and limitations, *IFAC-PapersOnLine*, 54(4):159–164.

Shorey R. and B. A. Miller 2000 The Bluetooth technology: merits and limitations, *2000 IEEE International Conference on Personal Wireless Communications. Conference Proceedings (Cat. No.00TH8488)*, 17–20 December, Hyderabad, India, pp. 80–84.

Skvorc D., M. Horvat and S. Srbljic 2014 Performance evaluation of Websocket protocol for implementation of full-duplex web streams, *37th International Convention on Information and Communication Technology, Electronics and Microelectronics (MIPRO)*, 26–30 May, Opatija, Croatia, pp. 1003–1008.

Zeadally S., F. Siddiqui and Z. Baig 2019 25 years of Bluetooth technology, *Future Internet*, 11(9), 194, pp. 1–24.

3 IoT Sensors for Smart Homes and Offices

3.1 INTRODUCTION

Home is not just a physical structure of a residence or a dwelling place but a place where a person or family lives, feeling protected and secure, enjoying life, filled with love and emotions. It is a place where everyone desires to be at the end of the day, a place where one feels the utmost comfort and peace of mind, the safe refuge or shelter where people spend a large chunk of their lives.

An office is a set of rooms or buildings where employees of an organization work at their desks to perform their duties, administrative, clerical, scientific meetings or others, in a coordinated manner to implement the tasks to be executed by the organization. If a person is in service and not at home, then office is the place where he/she is most likely to be present. So, office comes next to home as a place of occupancy in terms of the hours of life of a person.

Looking at the times we spend at home or office, it is not difficult to comprehend that these are two areas where sensors and the internet can work together to improve our quality of life. So, we start our exploration of sensors beginning from homes and offices.

3.2 SMART HOME

Smart home technology, known as home automation or domotics from the Latin word *domos* for home, is a technology for building a smart home. The smart home is a residence in which internet-connected sensors and actuators enable remote monitoring and control of appliances, often using voice assistants, e.g., Google Nest (Google), Alexa (Amazon), Bixby (Samsung), and Siri (Apple). Usually, a smart home app on the owner's mobile phone is applied for managing home activities. The intention is to provide security together with comfortable and convenient living to dwellers enabled by efficient use of energy.

A smart office is a workplace using the Internet of Things to connect sensors and systems to assist employees in seamless communication and collaboration, thereby creating an efficient working environment for enhancing employee comfort and productivity.

3.3 INDOOR/OUTDOOR VIDEO AND AUDIO SURVEILLANCE SYSTEMS, AND SECURITY DEVICES

3.3.1 CCTV AUDIO SECURITY CAMERAS

It is a CCTV camera system with 360° view and night vision capability, a built-in microphone for recording sound in conjunction with images, a speaker for listening to the recorded sound, and capability for human/pet motion detection. It is installed in homes, banks, retail stores, passages, offices, and other places to prevent thefts and protect residents/employees from intruders. Prior information about an employer's policy to its employees is mandatory because its installation infringes on their privacy. Cameras with wireless connectivity avoid the inconvenience of cluttered wires. Those compatible with voice assistants can be used with smart home devices. Water-proof cameras with HD resolution and motorized pan and tilt options are available. The recording is stored either in in-built memory or on a payment basis in internet-accessible remote cloud servers.

DOI: 10.1201/9781003374442-3

Devices: Motion sensor (passive infrared), CCD or CMOS image sensor with microphone and speaker

Communication Protocols: Internet Protocol (IP) for data transmission over Ethernet links, Real Time Streaming Protocol (RTSP) for viewing live video stream; RSTP uses Transmission Control Protocol (TCP) connection

3.3.2 Helping the Homeowner to Watch and Talk with Any Visitor: Video Doorbell

It is a doorbell with a built-in video camera providing real-time video and two-way audio, and a motion sensor (Figure 3.1). When someone comes near the front door or presses the doorbell, a notification is sent to the house owner on his/her smartphone. On receiving the alert, the owner can see the live video of the visitor at the door and decide on opening the door or keeping it shut. All the doorbells respond to the movements of objects. Some models can differentiate between people, pets, package deliveries, and vehicles passing by.

Devices: Motion sensor (passive infrared), CCD or CMOS image sensor with microphone and speaker

Network connection: HTTP, HTTPS, DNS, TCP, NTP, Wi-Fi, Z-Wave, LoRa.

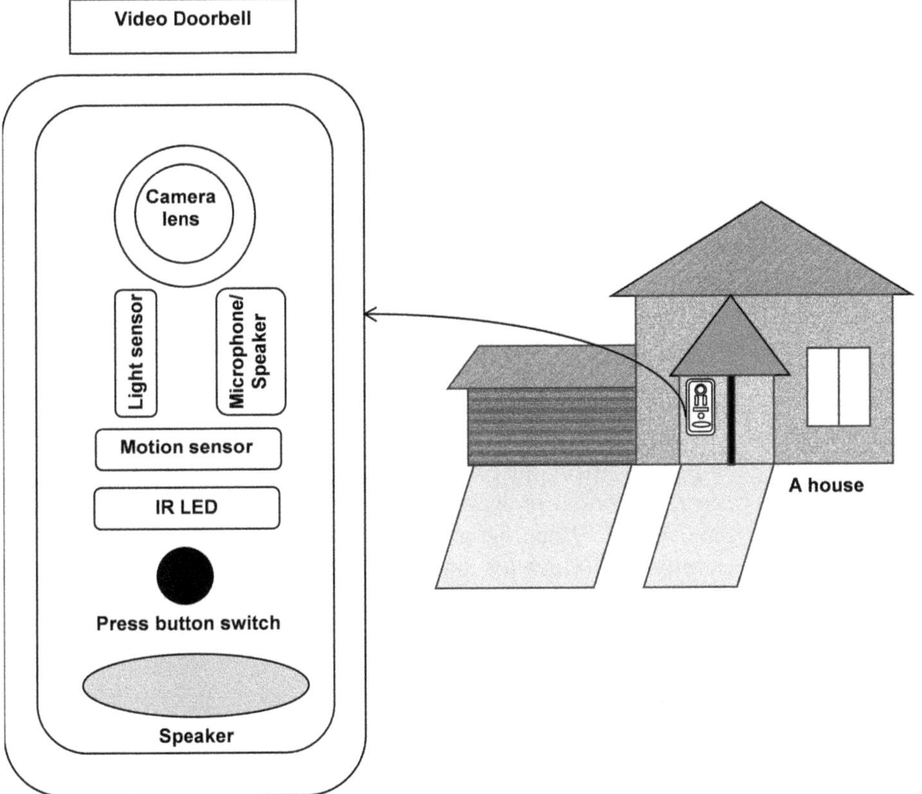

FIGURE 3.1 The video doorbell shows the camera lens, the light and sound sensors, the IR LED, the press button switch, and the speaker, mounted on the front wall of a house.

3.3.3 DETECTING HUMAN OR DOMESTIC PET MOVEMENT IN HOME: MOTION SENSOR

Human beings and animals release infrared radiation. This radiation consists of electromagnetic waves having wavelengths greater than visible light but less than microwaves. It is the radiation emitted by an object by virtue of its temperature. Hence, IR radiation is called heat or thermal radiation. Although invisible to the human eye, its presence can be felt as heat.

Human beings and animals can be considered as heat sources. When these heat sources move through a space in which a passive infrared sensor can sense heat changes, the sudden changes in heat caused by human/animal movement in the sensor zone known as their heat signatures are detected by the sensor, and notified to the home security to sound an alarm. Pet-immune motion detectors use weight filtering algorithms to ignore pets by taking the weight of the pet as a primary feature. More complex algorithms take into consideration many other parameters into account, e.g., speed and patterns of movement to arrive at a decision.

The pyroelectric PIR sensor has two slots constituting two halves of the sensor. In the idle condition, both slots receive equal quantities of background IR radiation from the surrounding room walls or from outdoor objects. When a warm body passes by, one-half of the PIR sensor is first affected. This event is registered as a positive differential change between the two halves. When the warm body leaves the area covered by the above slot, a negative differential signal is recorded. This will be elaborately described in Section 3.8.3.

Devices: Pyroelectric PIR sensor
Protocol: Wi-Fi

3.3.4 NOTIFYING THE HOMEOWNER ABOUT THE OPENING OR CLOSING OF DOORS AND WINDOWS: DOOR/WINDOW CONTACT AND GLASS BREAK SENSORS

A system is installed in the home to alert the owner immediately when some intruder tries to open any door/window of the premises or attempts to break any glass. Three types of sensors are widely used for sending alerts: contact door/window sensors, glass break sensors, and motion sensors. The contact door/window sensors (Figure 3.2) consist of two parts:

(i) A magnet fixed on the moving door or window, and
(ii) A reed switch mounted on the frame of the door or window in alignment with the magnet.

When the door or window is closed, the magnet is in contact with the reed switch. On opening the door or window, the magnet loses contact with the reed switch. The breaking of the magnetic field flips the reed switch sending an electrical pulse to trigger an alarm. These contact sensors are either surface-mounted or recessed, hardwired or wireless. Glass break sensors are small, unobtrusive devices adhered to the glass surface. When the glass is shattered, the surface tension of the glass changes, and vibrations are produced which are detected and sent to the alarm system. Motion sensors focused toward doors or windows trigger an audible alarm and send an alert to the owner's smartphone whenever there is an unusual human activity close to it. The law enforcement authorities can be immediately informed about theft or burglary.

Devices: Door or window opening alarm sensors
Protocols: Z-Wave, ZigBee

3.3.5 GIVING THE HOMEOWNER REMOTE ACCESS TO THE DOOR LOCK: SMART LOCK

It is an electromechanical lock activated for locking/unlocking by a Bluetooth or Wi-Fi connection from a registered mobile phone, a numeric keypad, a biometric sensor, or an access card. A lock

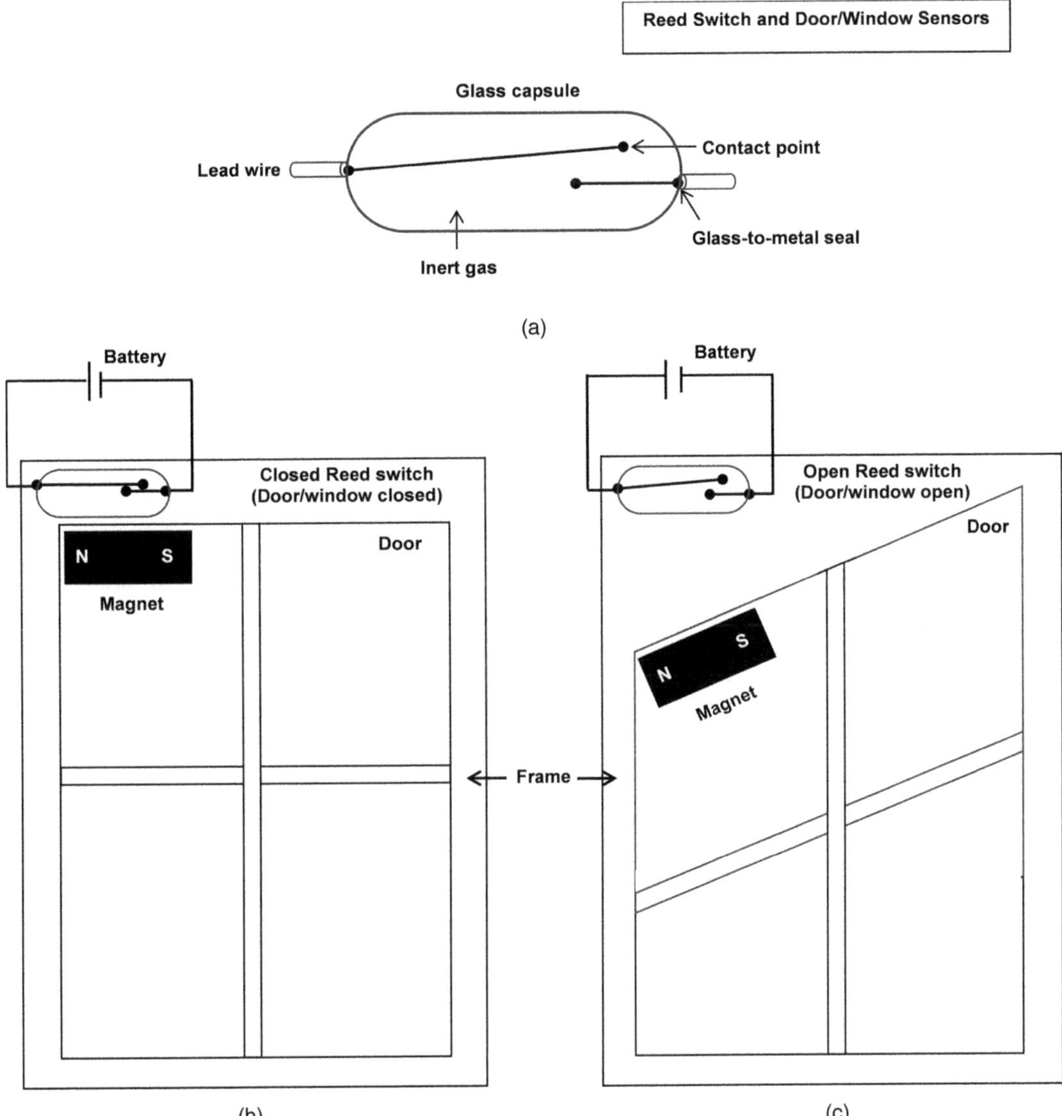

FIGURE 3.2 Door/window opening alarm sensors; (a) Reed switch, (b) closed door condition, and (c) open door condition. (a) shows a glass capsule filled with an inert gas in which two magnetic reeds are suspended for making and breaking contact. Lead wires are attached to these reeds by glass-to-metal sealing. In parts (b) and (c), reed switches are fitted on the frames of the doors and the magnets are mounted on the doors. These fittings are done in such a way that the reed switch and the magnet are in proximity and aligned with each other in the closed position of the door. The lead wires of the switches are connected to batteries. (b) shows the alignment of the magnet with the reed switch when the door is closed. Then the switch is under the influence of the magnet, the two contact points of the reed switch are touching each other and the switch is closed. (c) shows that the magnet is misaligned with the reed switch and also at a great distance from the switch when the door is opened. As the magnet fixed on the door has moved away from the reed switch, the switch is outside the magnetic field of the magnet, and the two contacts in the switch move apart. As they are no longer touching each other, the circuit is broken and the reed switch is opened.

remotely accessed by a digital key through a smartphone app in place of the physical key of a traditional lock is called a 'keyless lock'. Biometric locks use fingerprint, facial or iris scanners, or voice recognition software for authentication of identity before releasing the locking mechanism.

> Devices: Mechanical lock, user interface (electronic keypad for feeding the digital key); biometric sensors that convert biometric traits into electrical signals, and the sensed traits are digitized according to a software-defined template; cameras or time-of-flight sensors for facial recognition, e.g., time taken by light to travel to a feature point such as nose tip and back for high-resolution depth 3D visualization; Hall-effect sensor for rotational sensing to check whether the deadbolt is locked, unlocked or somewhere in midway, and for tamper detection.
>
> Protocol: Bluetooth, Wi-Fi, ZigBee, or Z-Wave.

3.3.6 Opening and Closing the Garage Gate from any Remote Place: Smart Garage Control

A smart garage door opener system consists of a door motor and a belt rail installed in the garage, along with cameras and a wireless internet connection to remotely see whether the door is open or shut, receive alerts on the smartphone if there is a problem with the door, and control the door opener through an app. To actuate the door opener, a coded message is sent from the phone to the circuit to trigger the relays for powering the motor.

Garage door safety sensors are a pair of infrared devices, one infrared transmitter and the other infrared receiver (Figure 3.3(a)). One infrared device is installed on one side of the door and the other exactly on the opposite side, both less than 15cm off the ground. The two devices face each other to create an invisible IR line joining them by the IR signal propagating along the line. The pair acts as a photo-eye. If the invisible line is broken due to the presence of an obstacle, e.g., a car parked under the door, or a person standing underneath (Figure 3.3(b)), then the door stops closing and moves back avoiding damage to the car or injury to the person.

> Devices: Door motor, belt rail, cameras, infrared emitter and receiver, relays
> Protocol: MQTT

3.4 AMBIENT LIGHT SENSORS

It is a device to assess the amount of light falling on an instrument such as a smartphone or a laptop and to adjust the screen brightness suitably, thereby preventing it from becoming too dim in outdoor use or too bright in a dark room. It increases the length of time for which the battery can be used without recharging.

It is called by various names such as light sensor, optical sensor, photosensor, or photodetector (Bielecki et al. 2022). It is of three types: photoresistor, photodiode, and phototransistor.

3.4.1 Photoresistor

It is a light-dependent resistor (LDR) made of CdS, CdSe, etc. (Figure 3.4). Its resistance changes when illuminated with light. An LDR is also made of undoped or doped semiconductors, e.g., Ge, Si, and GaAs. The intensity of light or illuminance is measured in units of lux. 1 lux = 1 lumen per square meter. Lumen is the amount of light emitted per second from a uniform source of 1 candela in a solid angle of 1 steradian. 1 candela is the luminous intensity of a monochromatic radiation source in a given direction such that the source emits radiation of frequency 540×10^{12} Hz with an intensity of 1/683 watt per steradian in the specified direction.

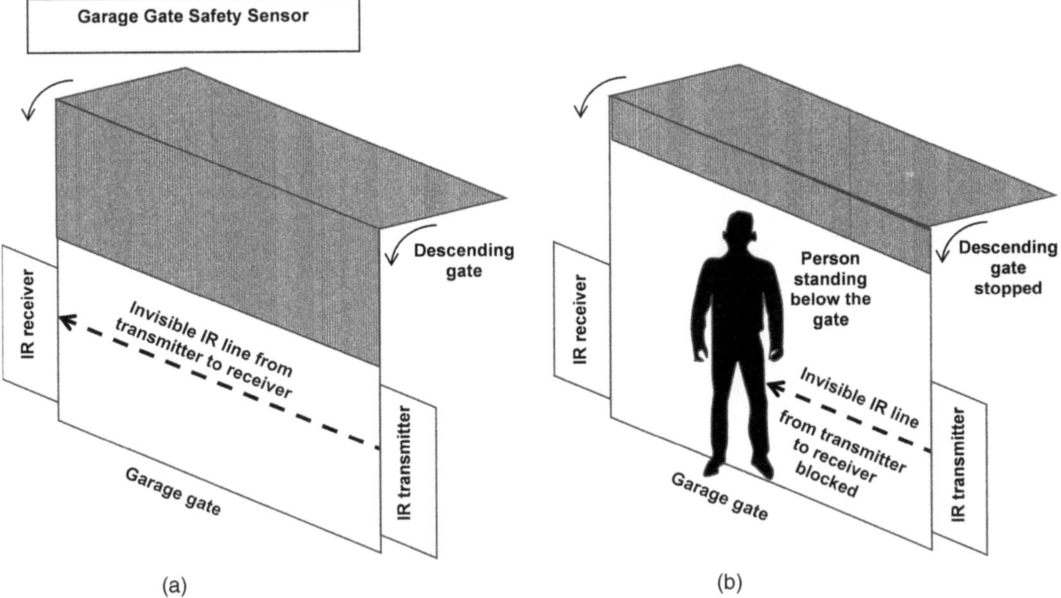

FIGURE 3.3 Garage gate accident prevention sensor: (a) when nobody is standing under the gate so that it is safe to close the gate, and (b) when a person is standing below the gate so that the gate will fall on the person's head if it is closed, and so must be stopped from moving down. (a): A garage gate is shown. At a small distance from the floor of the garage, PIR devices are fixed face-to-face opposing each other, one on the left side of the garage door and the other on its right side. One device is the IR transmitter and the other is IR receiver. The IR radiation emitted by the transmitter is detected by the receiver. Thus, an invisible IR line is imagined, joining the IR transmitter with the receiver. (b): When a person is standing under the gate or a vehicle is parked below the gate, the IR line connecting the transmitter with the receiver is intercepted. The IR from the transmitter is no longer able to reach the receiver. When the receiver does not get the IR signal, its output signal falls and the relay activating the gate motor is turned off immediately preventing the gate from coming down. This prevents the person standing below the gate from injury due to crushing under the gate. Any vehicle standing there is similarly protected from damage. At the same time, an alarm rings warning the person to move away or to remove the vehicle from that place.

The sensitivity of an LDR depends on the wavelength of light. LDRs also show time latency. It takes ~ 10 ms for resistance to decrease after light falls on the photoresistor. Recovery is slow taking ~ 1s for resistance to return to its initial value. This makes photoresistors unsuitable for applications involving rapid fluctuations in light intensity.

3.4.2 PHOTODIODE

3.4.2.1 P-N Junction Diode

A P-N junction diode (Figure 3.5(a)) is made with Ge, Si, InGAs, InGaAsP, or other semiconductors. A transparent window is built into the package to allow light to fall on the junction. The window sometimes has a lens to focus light on the junction. In a normal diode, the junction is protected from light.

It can be used in three modes: photovoltaic mode (without applied bias), photoconductive mode (under reverse bias), and avalanche mode (under high reverse bias). As a photodetector, it is generally used in the photoconductive mode by creating a depletion region across the junction.

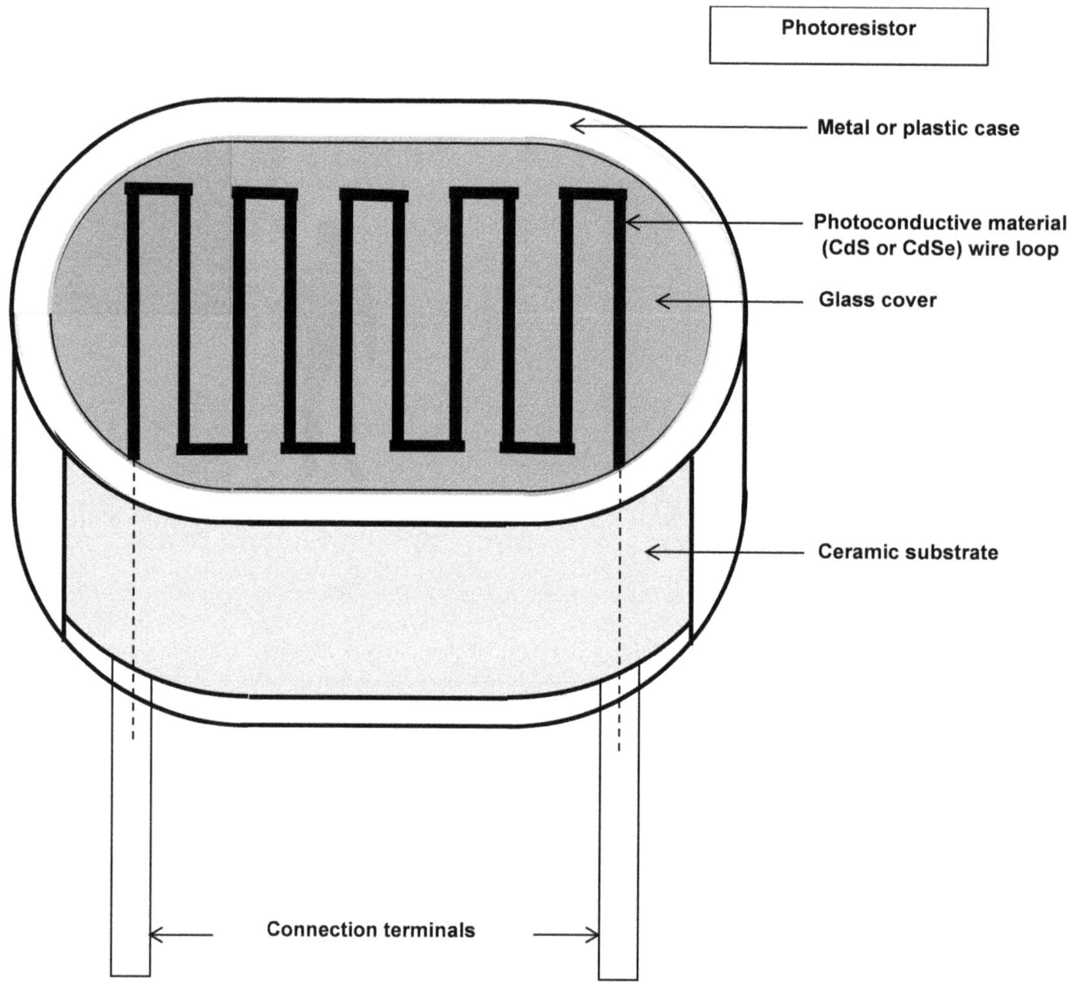

FIGURE 3.4 Photoresistor made with CdS or CdSe spiral on a ceramic substrate. A loop of wire is made on a ceramic substrate using a photoconductive material, e.g., CdS or CdSe. It is packaged in a metal or plastic case with a transparent glass cover and two connection terminals.

3.4.2.2   P-I-N Photodiode

A favorite structure for the photodiode sensor is the PIN diode (Figure 3.5(b)) containing an intrinsic region sandwiched between highly doped P and N regions. Due to the high resistance of the intrinsic region, the depletion region stretches to a larger thickness in this region at a given reverse bias voltage. This makes a larger volume of semiconductor available for photon-to-electron conversion because electrons produced within the depletion region or within a distance up to the diffusion length of carriers from the depletion region are not lost by recombination, and therefore contribute to reverse current. Furthermore, the junction capacitance decreases with an increase in depletion region thickness, thereby enhancing the switching speed of the diode and improving its frequency response.

3.4.3   PHOTOTRANSISTOR

It is a bipolar junction transistor made of Ge, Si, GaAs, etc. (Figure 3.6). There is no base terminal in the two-lead version. The base is at a floating potential. (Three-lead versions are also available for normal use). In place of the base current, the base input is light energy. Another difference from a

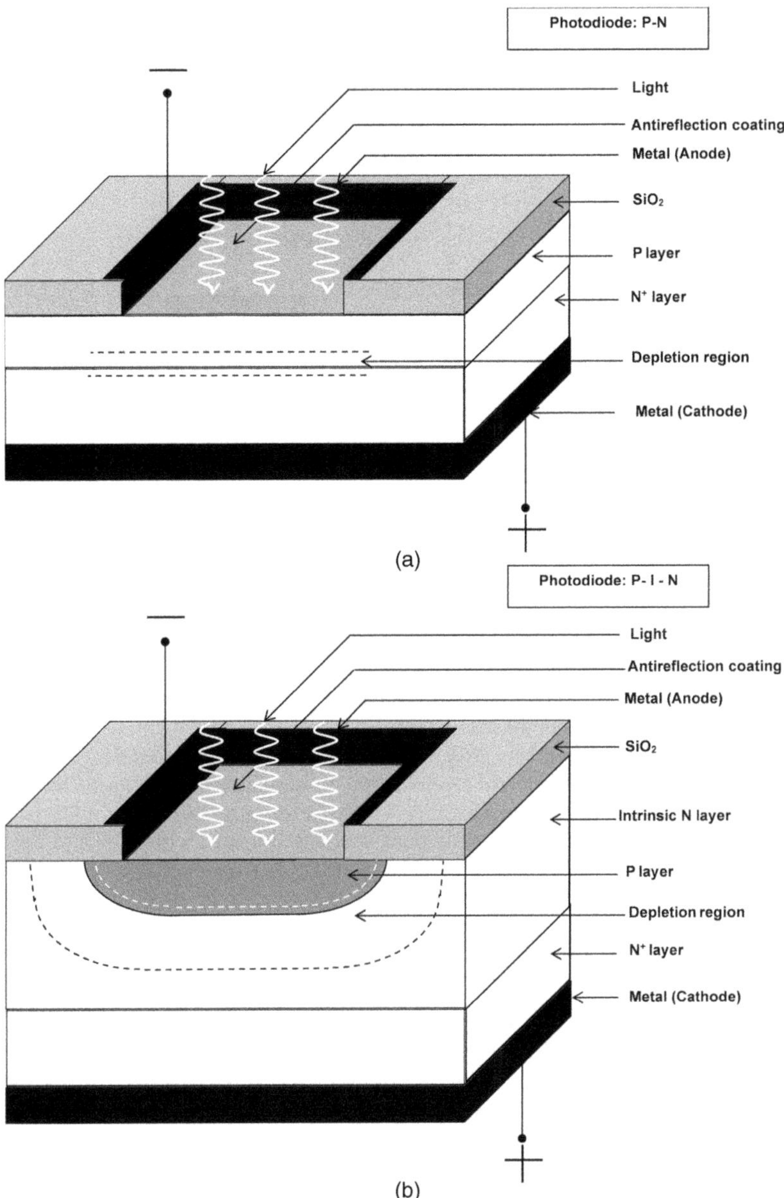

FIGURE 3.5 Photodiodes: (a) P-N junction and (b) having a P-I-N structure. (a): A P-N$^+$ junction diode is covered on its top surface with an SiO_2 film having a rectangular opening to receive light. The P-layer is coated on its boundaries with a metal film deposited around the walls of the opening and overlapping the SiO_2 film for making electrical contact. Its central portion is left uncovered so that light can pass through it. This P-layer serves as the anode of the diode. The lower surface of the N-layer is metallized to act as the cathode. A depletion region is seen at the junction between P- and N-layers. (b): A diode with the structure: P-layer/ intrinsic N-layer/N$^+$ layer, is shown. The upper surface of the diode is covered with an SiO_2 film having a rectangular opening to receive light. As in (a), the P-layer of the diode is coated on its borders with a metal film deposited around the walls of the opening, and overlapping the SiO_2 film for making electrical contact. As its central portion is left uncovered, light can pass through it and reach the underlying layers. The P-layer serves as the anode of the diode. A depletion region is seen at the junction between P- and intrinsic N-layers. A metal film is deposited on the lower surface of the N-layer of the diode. The N-layer acts as the cathode. A depletion region can be seen at the junction between P-layer and intrinsic N-layer. The width of depletion region is much larger inside the intrinsic N layer than the P-layer because of the comparatively low doping of intrinsic N-layer.

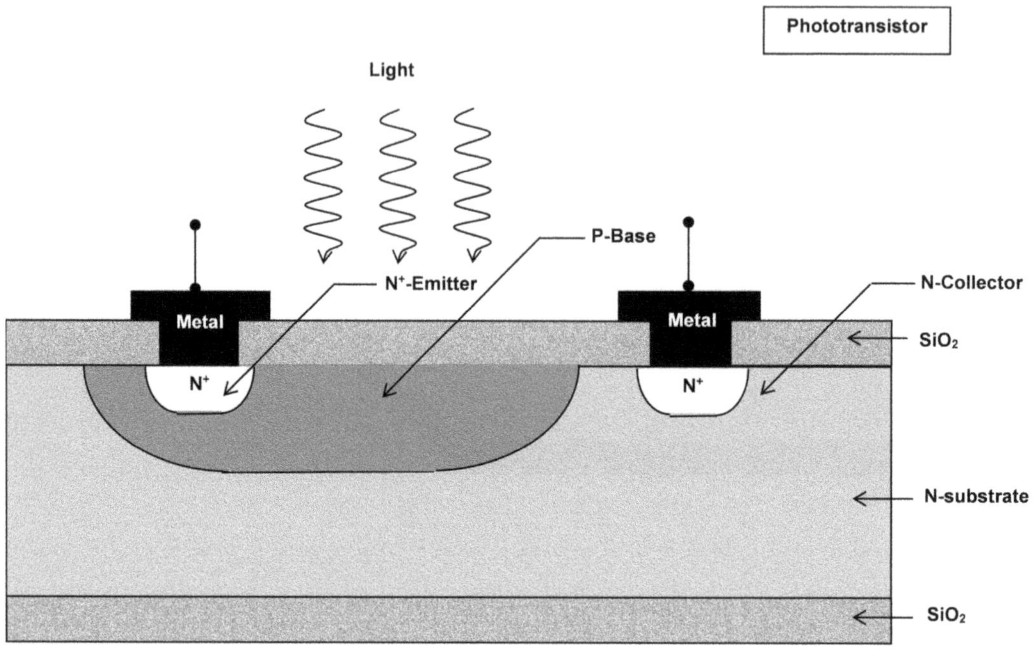

FIGURE 3.6 Phototransistor. A planar transistor is shown with an N$^+$ emitter, P-base, and N-substrate collector. The top and bottom surfaces of the transistor are covered with a SiO_2 film with necessary openings in the top surface for depositing metal contacts. The metal is directly deposited on the N$^+$ emitter for making an emitter electrode. But for making the collector electrode, an N$^+$ region is diffused inside the collector through the SiO_2 film, and the metal film is deposited on this N$^+$ region to act as the collector electrode. There is no electrode for base contact. The upper surface of the transistor is illuminated with light.

normal transistor is that during packaging, a small hole is made over the collector-base junction. A lens is fixed over this hole to focus light on this junction. Further, the photogeneration takes place in the collector-base region. Hence, a large collector-base region must be provided for incident light. Therefore, a phototransistor is designed by taking a large area of this region.

In the absence of light, there is only a small collector-to-emitter leakage current flow. This is the dark current. It is produced by the thermal generation of carriers. When light falls on the collector-base junction, a base current is produced. By amplification, this base current leads to a larger collector current which is proportional to the intensity of light. So, a phototransistor performs light detection combined with current amplification. Amplification of current increases its sensitivity to light. A phototransistor produces more current than a photodiode when exposed to light of equal intensity. However, it is slower than the photodiode providing an inferior high-frequency response. A photodiode is more efficient for high-frequency operations.

3.5 IMAGE SENSORS

An image sensor is a device that converts the information in an image into electronic signals that can be used to transmit and rebuild the image. Image sensors work on the photoelectric effect for the conversion of light into electricity.

Two types of image sensors in widespread use today are CCD (charge-coupled device) and CMOS (complementary metal-oxide-semiconductor [MOS]) sensors (Zhang et al. 2008, RadhaKrishna et al. 2021). CCDs offer the advantages of high sensitivity and low noise due to more efficient utilization of area. However, they suffer from the disadvantages of high-power consumption, slow readout, and the inability to directly access individual pixels because the data is read serially. CMOS image sensors provide high-speed imaging with low power consumption. Image data can be accessed randomly. Selective readout is also available. On the downside, they are less sensitive and noisier than CCD image sensors.

3.5.1 CCD (Charge-Coupled Device) Image Sensor

3.5.1.1 The CCD Pixel

The fundamental structural unit of the CCD is a light-sensitive picture element or pixel (Lesser 2014). This sensor is a two-dimensional array of MOS capacitors (Figure 3.7(a)) acting as receptacles for photogenerated electrons to convert the incident photons into electrons.

Each pixel (Figure 3.7(b)) of a CCD has three gate electrodes. Of these electrodes, one electrode is used to create a potential well under the silicon dioxide layer. The other two electrodes are used for transferring the charge stored in the potential well to other electrodes of the structure. The size of a pixel is 10 μm × 10 μm (typically). A CCD with a specification of 9-megapixel contains 9×10^6 pixels. These pixels are arranged in a pattern comprising 3×10^3 horizontal rows and 3×10^3 vertical columns in a chip of size $9 \times 10^6 \times 10$ μm $\times 10$ μm $= 9 \times 10^8$ μm$^2 = 9 \times 10^8$ μm$^2 = 9 \times 10^8$ μm$^2/1 \times 10^8 = 9$ cm^2.

The MOS capacitor array of Figure 3.7(a) is redrawn in Figure 3.7(c), in order to present a clear illustration of the constituent MOS capacitors having metal-SiO$_2$-P-Si structure, and their gate terminals.

3.5.1.2 Potential Well Formation and Charge Storage

To investigate the process of potential well formation, we concentrate our attention on the gate electrode in the middle of Figure 3.7(b). This central electrode is connected to the positive terminal of a voltage source to apply a potential +V. When light falls on the pixel, and the energy of the photon is greater than the energy gap of the semiconductor, electrons are dislodged from their shells in the atoms of the semiconductor by the impact of photons forming electron–hole pairs. Under the influence of a positive gate voltage, electrons are attracted toward the gate electrode while holes are repelled away from this electrode. Hence, a region that is rich in electrons is created under the gate electrode. The electrons are trapped in this region by the positive gate voltage. So, this region acts as a potential well from which the electrons cannot escape unless the positive gate voltage is removed. The gate terminals on the two sides of the central gate electrode are kept at zero potential. As the semiconductor substrate is grounded, no current flows from these electrodes toward the substrate nor do electrons confined in potential wells below the central gate electrode move toward the side regions. So, these regions act as potential barriers surrounding the potential well. The potential wells can be called capacitive bins storing photogenerated electrons.

3.5.1.3 Charge Transportation

How can the electrons in capacitive bins in a semiconductor substrate be transported from one place to another? The packets of photogenerated electrons are transferred in a controlled manner between capacitive bins by application of the correct sequence of voltage pulses to the gate electrodes of the MOS capacitors. In Figure 3.8, it is shown how the electrons are transferred in the horizontal direction from the potential well under a gate electrode to the potential well under an adjoining electrode by simultaneously biasing the adjoining electrode at a positive electrode and reducing to zero the gate potential of the electrode under which the potential well was initially localized. By the same procedure, the electrons are transferred from the potential well under one gate to that under another gate electrode in the vertical direction (Figure 3.9). Here the gate electrodes are oriented in the perpendicular direction to that in the previous case.

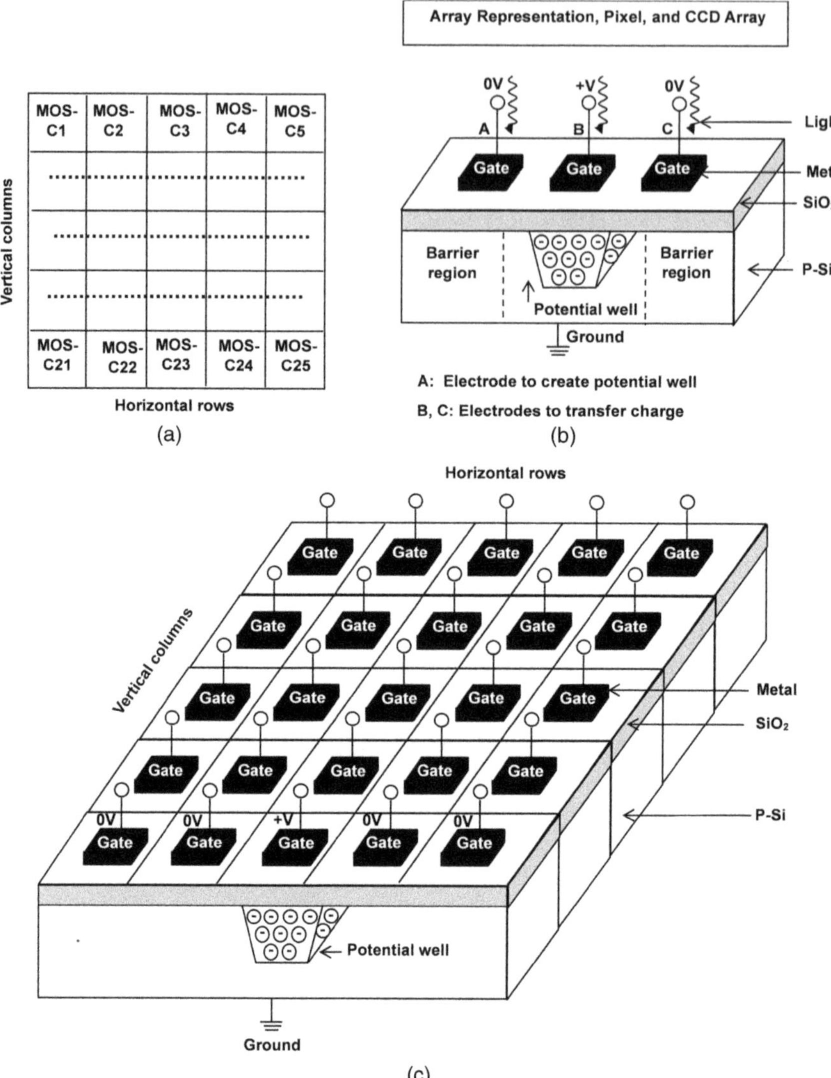

FIGURE 3.7 Charge-coupled device: (a) arrangement of pixels in a 5 × 5 pixel array, (b) the construction of a single pixel with the potential well, and (c) sketch of the 5 × 5pixel array of Figure (a) showing the MOS capacitors with their gate electrodes. (a): An array is constructed by dividing a square into 5 horizontal rows and 5 vertical columns to form 5 × 5 = 25 squares. Each square represents a MOS capacitor. The capacitors are numbered as MOS C-1, MOS C-2, MOS C3, …, MOS C25. (b): A single pixel consisting of three layers: a metal electrode, a SiO₂ dielectric layer, and a P-Si semiconductor layer is illuminated with light, producing electron–hole pairs in P-Si due to photon bombardment. The pixel has three gate electrodes A, B, and C. Initially, electrode A is used to create a potential well in the underlying P-Si by applying a positive potential +V to A, while electrodes B and C are held at 0V. The P-Si substrate is grounded. Electrons are attracted toward the positive gate potential creating an electron-filled region. Next, the potential of electrode A is reduced to zero while the potential of electrode B is raised to +V. The electrode C is kept at 0V. When these gate voltages are applied, the potential well formed under electrode A is shifted to the P-Si region under electrode B because the electrons are pulled toward the positively charged gate electrode B. This is the condition in which the pixel is shown in the diagram with a potential well formed under electrode B, and barrier regions under electrodes A and C. (c): The 3D view of the array with each square of Figure (a) replaced by the equivalent MOS capacitor. The array consists of 5 horizontal rows containing 5 MOS capacitors and 5 vertical columns with 5 MOS capacitors in each column. The potential +V is applied to the third gate electrode in the first horizontal row while the gate electrodes of the 1st, 2nd, 4th, and 5th rows are kept at 0V. The P-Si substrate is kept at ground potential. The potential well is formed under the electrode carrying a positive potential.

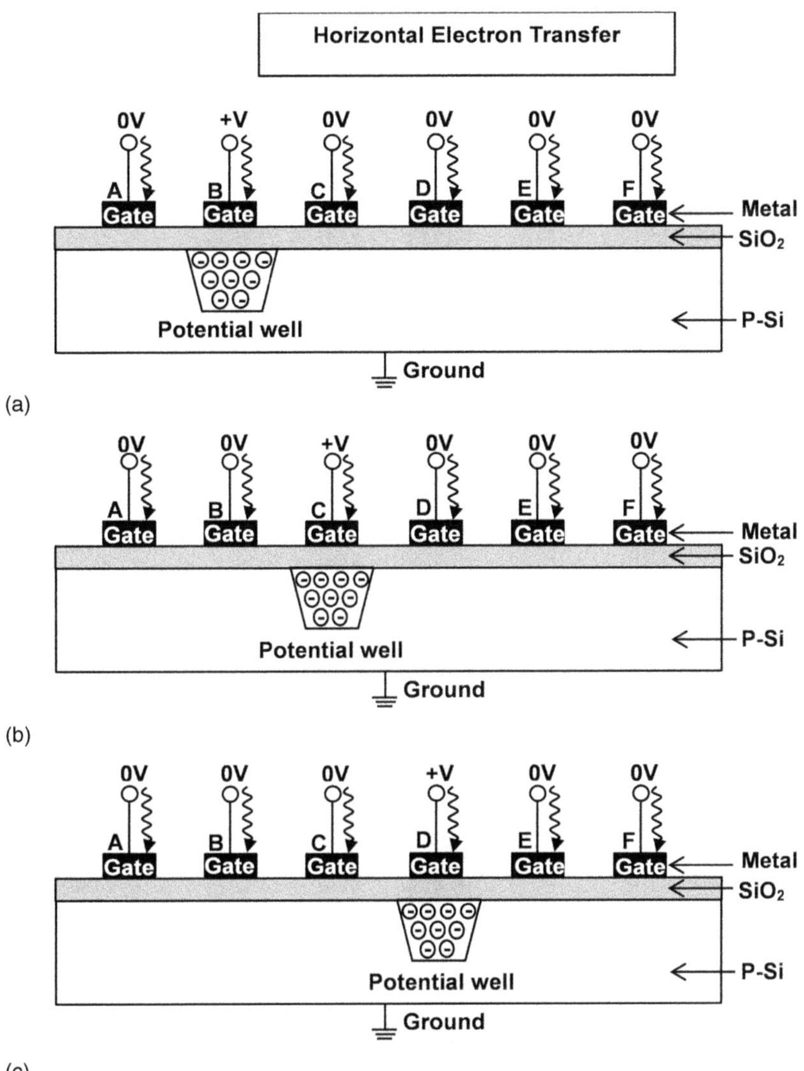

FIGURE 3.8 Transportation of electric charge in the horizontal direction with potential wells under (a) electrode B, (b) electrode C and (c) electrode D. A MOS capacitor is drawn with six gate electrodes labeled as A, B, C, D, E, and F. The P-Si substrate is grounded. (a): Electrode B is at +V potential while electrodes A, C, D, E, F are at 0V. The electrons pile up under electrode B forming a potential well under this electrode. (b): Potential of electrodes A, B, D, E, F is at 0V while that of electrode C is kept at +V. The electrons collected under electrode B are transported below electrode C, forming a potential well under this electrode. (c): The potential of electrodes A, B, C, E, and F is at 0V while that of electrode D is kept at +V. At this sequence of potentials, the electrons gathered under electrode C move below electrode D, forming a potential well under this electrode.

3.5.1.4 Full Frame CCD

To use a CCD as an image sensor, it is positioned at the focal plane of the lens of a camera. In a full-frame CCD, the CCD surface is exposed by opening an electromechanical shutter. A focused image of the pictured scene is produced on the CCD surface. The shutter is opened for a limited time to expose the CCD. During the exposure time, each pixel of the CCD acquires a charge whose magnitude is proportional to the intensity of light falling on that pixel. These charges are constrained in the potential wells of the respective pixels. So, a charge distribution replica of the scene is stored in the CCD. The next step is to transfer these charges to a preamplifier in the correct order and transform

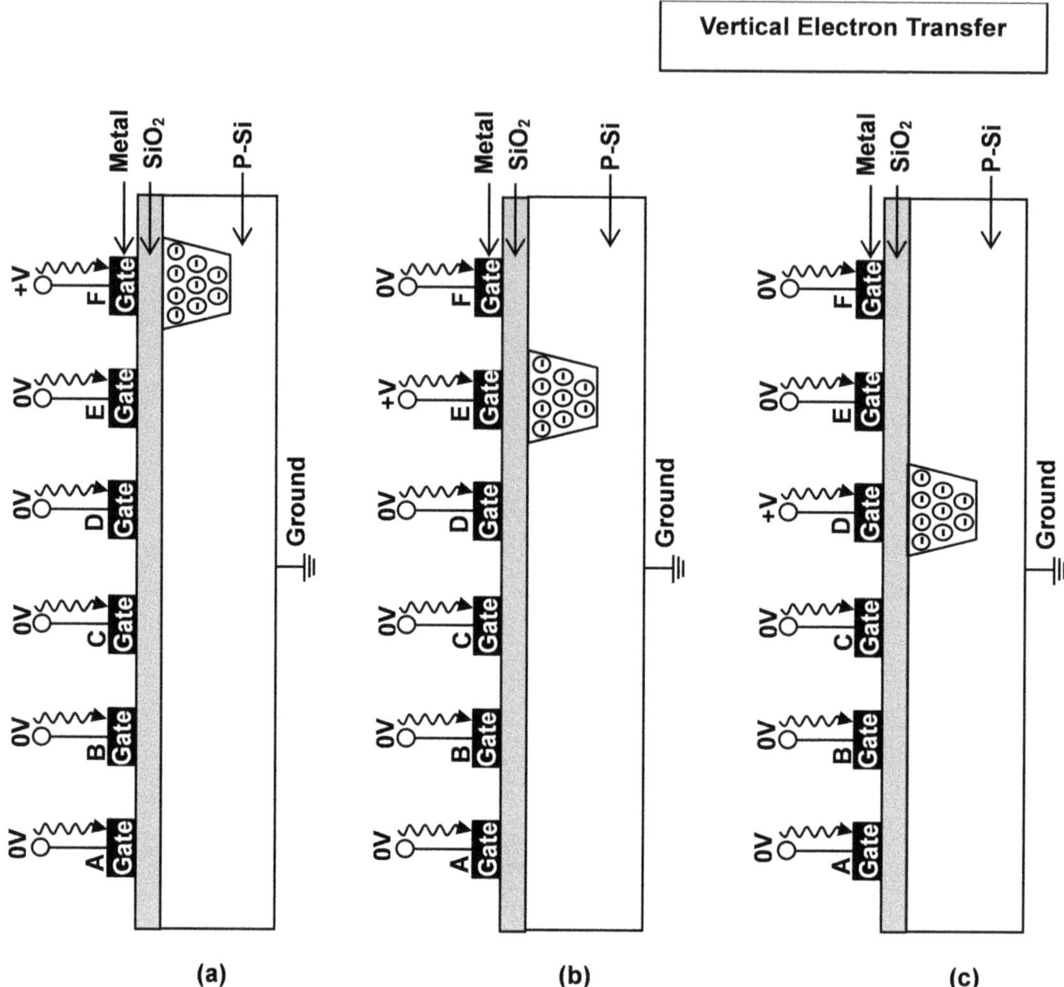

FIGURE 3.9 Transportation of electric charge in the vertical direction with potential wells under (a) electrode F, (b) electrode E and (c) electrode D. A MOS capacitor is drawn with six gate electrodes labeled as A, B, C, D, E, and F. The P-Si substrate is grounded. (a): Electrode F is at +V potential while electrodes A, B, C, D, and E are at 0V. The electrons accumulate under electrode F, forming a potential well under this electrode. (b): The potential of electrodes A, B, C, D, and F is at 0V while that of electrode E is kept at +V. The electrons congregated under electrode F are conveyed below electrode E forming a potential well under this electrode. (c): The potential of electrodes A, B, C, E, and F is at 0V while that of electrode D is kept at +V. At this sequence of potentials, the electrons assembled under electrode E move below electrode D, forming a potential well under this electrode.

them into voltage signals so that they can be combined together to reproduce the image. These operations are executed using readout circuits.

The sequence of readout operations can be understood from Figure 3.10. In Figure 3.10(a), a 3 × 3-pixel segment of the CCD is shown in the unexposed state. All the pixels and the readout register are empty. Figure 3.10(b) shows the same segment after optical exposure where it is seen that each pixel now has electrons in its potential well, the number of electrons varying with the intensity of illumination of the concerned pixel. The readout register is still vacant. In Figure 3.10(c), the bottom row of the potential wells has moved downward into the readout register as all the horizontal rows have moved downward by one-pixel step. Figure 3.10(d) shows that the rightmost potential well has

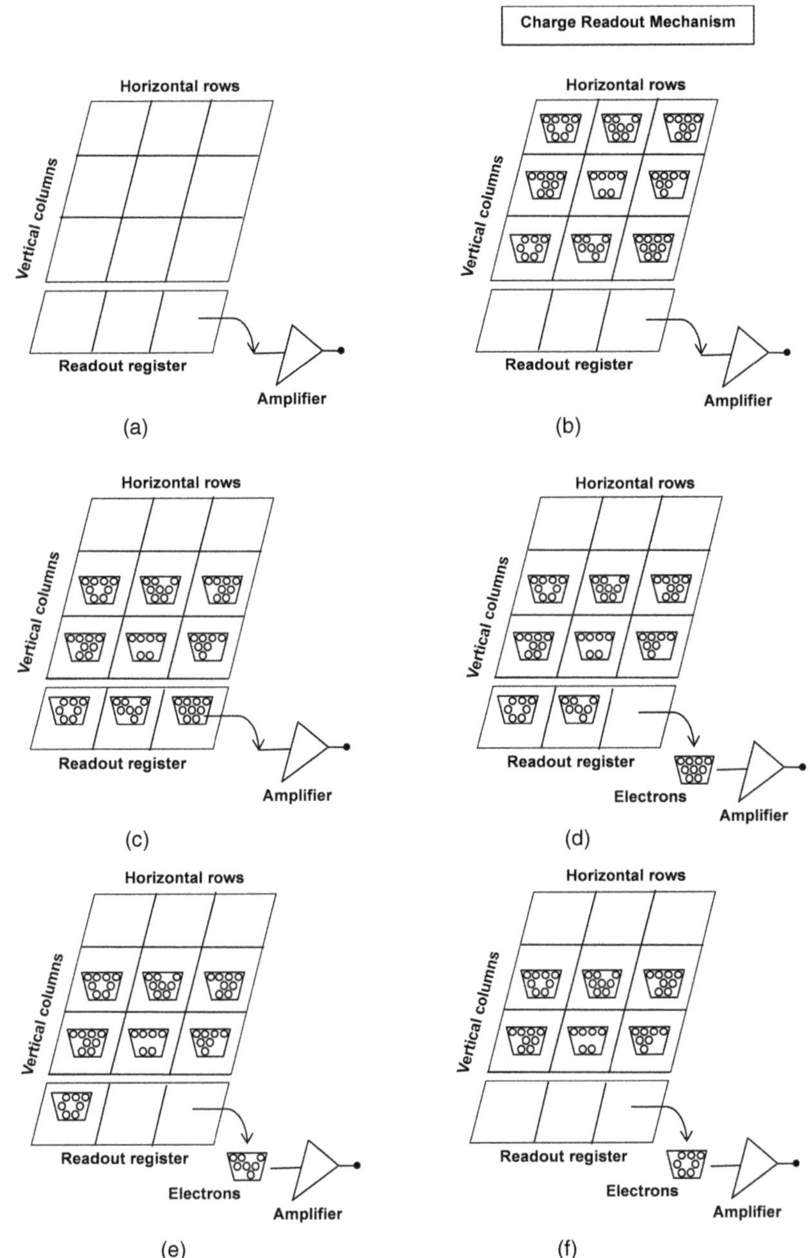

FIGURE 3.10 Mechanism of readout of charges produced by incident light from a picture on a CCD array comprising three horizontal rows and three vertical columns, and a 3-element register feeding an amplifier, from exposure of scene to transfer of charges and creation of voltages: (a) initial unexposed state, (b) after exposure of the three rows and columns, (c) electrons transferred downward by one row, (d) one element of the register emptied with electrons flowing into the amplifier, (e) two elements of the register emptied with electrons drained into the amplifier, and (f) all the three elements of the register vacated with electrons moving into the amplifier. (a): All rows, columns, and register elements are empty. (b): All rows and columns are filled with electrons but register elements are empty. (c): The first row is empty while remaining rows are filled with electrons. All register elements are also filled with electrons. (d): The first row is empty while remaining rows are filled with electrons. One register element is vacant. (e): The first row is empty while the remaining rows are filled with electrons. Two register elements are vacant. (f): The first row is empty while remaining rows are filled with electrons. All three register elements are vacant.

loaded its electrons into the amplifier. Figure 3.10(e) and (f) show that the electrons in the middle and leftmost potential wells are successively dumped into the amplifier. The readout register is now empty to accept the next row of potential wells. After all the three pixels of the bottom row have emptied their electrons into the amplifier, the next row of pixels moves downward and the same process of loading the electrons from each potential well into the amplifier is repeated. In the same manner, the third row delivers its electrons in potential wells to the amplifier. Thus, all three rows have given their electrons to the amplifier through one-by-one vertically downward movements of the rows followed by pixel-by-pixel horizontal movement of potential wells toward the amplifier. The series of voltage signals from these electronic charges constitutes an array of numbers representing the brightness of pixels. These numbers are fed to software to reconstruct the image of the scene.

3.5.1.5 Full Frame vs. Frame and Interline Transfer CCDs

The CCD architecture described in Section 3.5.1.4 was the full-frame type. There are three types of CCD architecture: full-frame, frame, and interline transfer. In a full-frame CCD, the complete CCD array surface is photosensitive. Exposure and readout are separate operations. The CCD illuminated during exposure is protected from light when readout is in progress. When exposed, the CCD stores charge. During a readout operation, the charge is transferred and measured. The exposure and read-out operations do not occur simultaneously. Therefore, the image frame rate is determined by the speed of the shutter, the rate of data transfer and its reading.

Frame-transfer CCDs allow overlapping of exposure and readout processes. Simultaneously performing the two operations saves time, enabling faster image frame rates to be achieved. In these CCDs, no mechanical shutter is used. One-half of the pixel array is left unmasked for photoelectron generation. The other half is covered with an opaque mask and is used for storage of photoelectrons produced in the first half. Obviously, a much larger chip size is necessary in a frame-transfer CCD to make the area of photogeneration equivalent to that of a full-frame CCD.

Interline transfer CCDs are faster than both the above types. Here, the parallel register array is organized with the unmasked pixels (for electron generation) and masked pixels (for electron storage and transfer), alternating over the full array. As the charge transfer paths lie next to the columns of unmasked pixels, the accumulated charges need to be shifted only one column into the transfer path. Thereafter, the stored information is read by sequential parallel moves into the serial register. During this period, the unmasked pixels are exposed to the succeeding image. By electronically controlling the intervals of image exposure, the integration periods can be drastically reduced.

Using a 4.5 μm interline transfer pixel, the KAI-43140 image sensor packs 43-megapixel CCD pixels in a 35 mm optical format, providing high resolution and image uniformity (DeLuca 2018).

3.5.2 CMOS IMAGE SENSOR

This sensor is an array comprising a large number of three-transistor active pixel sensors (3T-APSs), arranged in rows and columns (Bigas et al. 2006). A pixel sensor representing the circuit of a single pixel measures the intensity of light at a single point of the object. Unlike CCD image sensors, where the pixel only performs the photon-to-electron conversion and the electron-to-voltage conversion is done separately, in CMOS image sensors, the pixel performs both the operations of photon-to-electron and electron-to-voltage conversions.

The sensing component of the 3T-APS circuit is a reverse-biased photodiode. The circuit shown in Figure 3.11(a) contains three transistors: a reset transistor M_{Reset}, a transistor M_{SF} acting as a source follower amplifier and used as a voltage buffer, and a row select transistor M_{Select}. The photo-diode and transistors are fabricated by complementary metal-oxide semiconductor (CMOS) technology; hence it is a pixel circuit of a CMOS image sensor.

Transistor M_{Reset} is switched. If V_{Th} is the threshold voltage of M_{Reset}, a voltage V_{Reset} is applied to the photodiode through the transistor M_{Reset}. This voltage

$$V_{\text{Reset}} = V_{\text{DD}} - V_{\text{Th}} \tag{3.1}$$

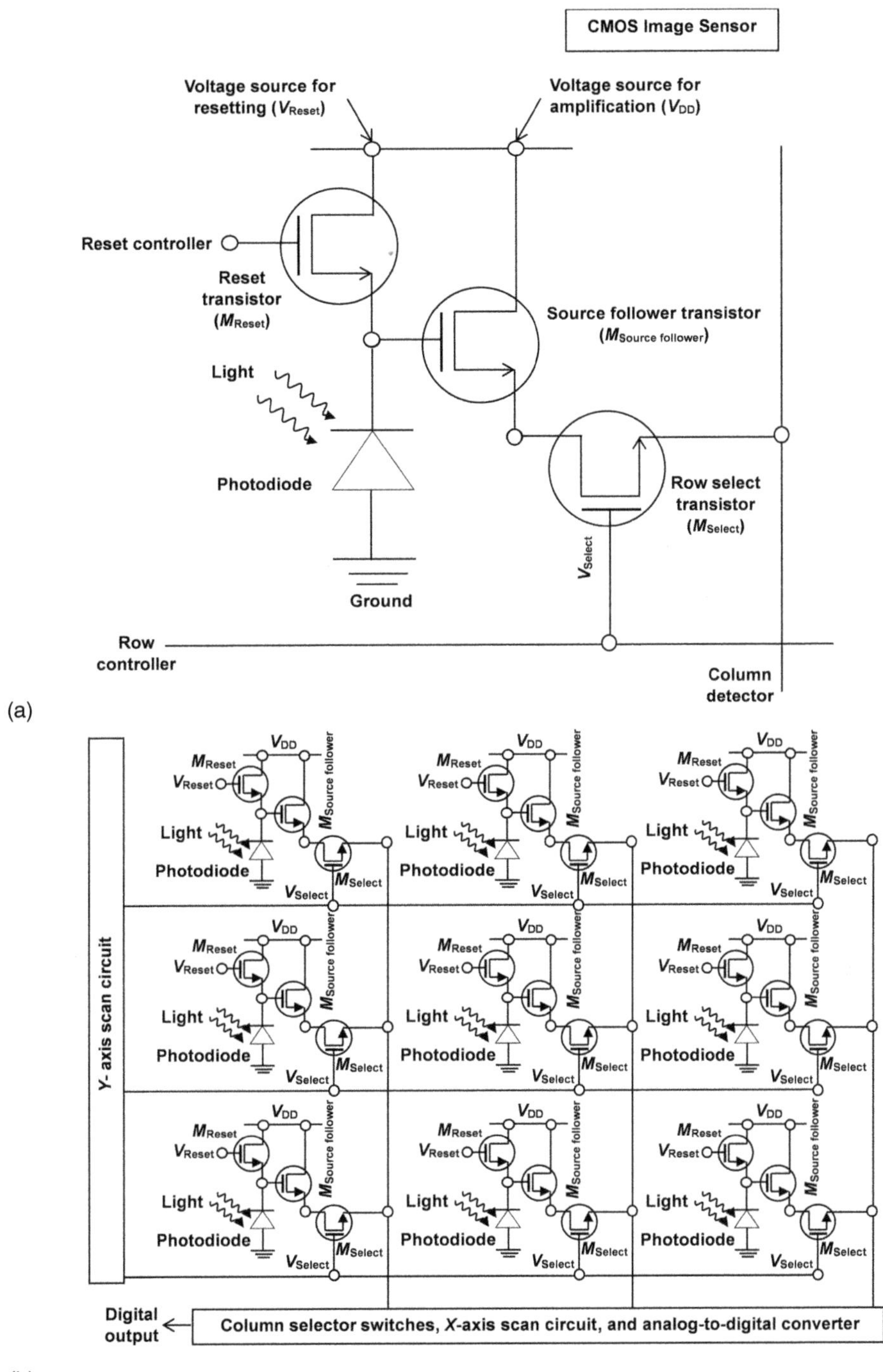

FIGURE 3.11 CMOS image sensor: (a) single pixel and (b) 3 × 3 pixel array. (a): Circuit of a single pixel containing three transistors: a reset transistor (M_{Reset}), a source follower transistor ($M_{Source follower}$) and a row select transistor (M_{Select}); supported by a voltage source for resetting (V_{Reset}), a voltage source for amplification (V_{DD}), a reset controller, a row selector and a column detector. Light from the imaged object is falling on the photodiode whose one terminal is grounded. (b): An array of pixels of part (a) is constructed. It consists of three rows and three columns. Each pixel is represented by its circuit given in part (a). On the left side of the array, the Y-axis scan circuit is connected. On its bottom side, column selector switches, X-axis scan circuit and analog-to-digital converter are connected. The digital output point is indicated.

reverse biases the photodiode. A depletion region is created across the P-N junction. If the incident light crossing the depletion region has an energy exceeding the energy gap of the semiconductor, electrons are transferred from the conduction band to the valence band of the semiconductor by the impact of incoming photons. Holes are left behind in the conduction band of the semiconductor. Hence, electron–hole pairs are generated. The electric field of the junction separates the electrons and holes produced. These charge carriers are collected at the opposite electrodes of the photodiode, resulting in the flow of a photocurrent. The photocurrent thus obtained is proportional to the intensity of light received from the object. The photocurrent can be measured directly. Otherwise, we can measure the voltage drop across the junction. The charges accumulated alter the voltage of the diode depending on the intensity of light. Let us explain the procedure for voltage drop measurement.

We emphasize that the photocurrent discharges the diode whereby its depletion region width is reduced, lowering the sensitivity of the detector. Further, its magnitude is in the range of femto- to pico-amperes. As this current is too small for measurement, the photodiode is put in reverse bias by applying an inverse voltage and exposed to light for a pre-decided period of time. Then the photodiode is left floating. During this time, the charge is stored in the capacitance of the photodiode. This charge storage yields an accumulated current signal of increased intensity. The associated averaging of current helps in acquiring a more faithful signal representing closely the true intensity of light in the object. The voltage dropped across the photodiode reckoned from the application of reverse bias to the end of the time interval during which the junction was left floating, is proportional to the integral of all the charges generated during this interval. Therefore, this time interval is called the integration time.

After the integration time has ended, the M_{Select} transistor is switched on and a constant current is maintained with the help of transistors M_{Select} and M_{SF}. The transistor M_{SF} acts as a voltage follower. It drives the pixel output independent of the photodiode. Thus, the voltage is read at the output of the pixel. It is the voltage V_{Signal}.

As soon as the voltage across the photodiode is transferred outside the pixel, one cycle is completed. The transistor M_{Select} is turned off and the transistor M_{Reset} is again turned on. The photodiode is reset to commence a new cycle, and the foregoing cycle of events is repeated.

A CMOS image sensor is a rectangular matrix built with several pixels (Figure 3.11(b)). Pixels are read row by row. Considering any particular row, the M_{Select} transistors of that row are switched on. A fixed current is imposed on each column line. At the end of each column, the voltage of the corresponding pixel of the chosen row is read out.

An ultra-high resolution CMOS image sensor has 19,568 (horizontal) × 12,588 (vertical) = 246.32 × 10^6 pixels = 246.32 megapixels. The pixel size is 1.5 μm × 1.5 μm, and the sensor size is 9.35 mm × 18.88 mm. The frame rate is 5fs (Canon: Ultra-high 250MP Resolution, 2022).

3.6   SOUND SENSORS

The sound sensor is an acoustic sensor or microphone, a device used for the conversion of sound energy into electrical energy; hence it is more appropriately called a sound transducer, rather than a sound sensor. It detects various types of sound signals such as voice, music, claps, and knocks. As the audible sound ranges in frequency from 20 Hz to 20 kHz, the sound sensor should have a frequency response within this range.

3.6.1   CARBON GRANULES MICROPHONE

In this microphone (Figure 3.12), the space between two metal plates is filled with carbon granules. Sound waves cause vibrations in the front thin plate called the diaphragm. The front plate presses against the carbon granules, thereby compressing them (Figure 3.12(a)). When the carbon granules

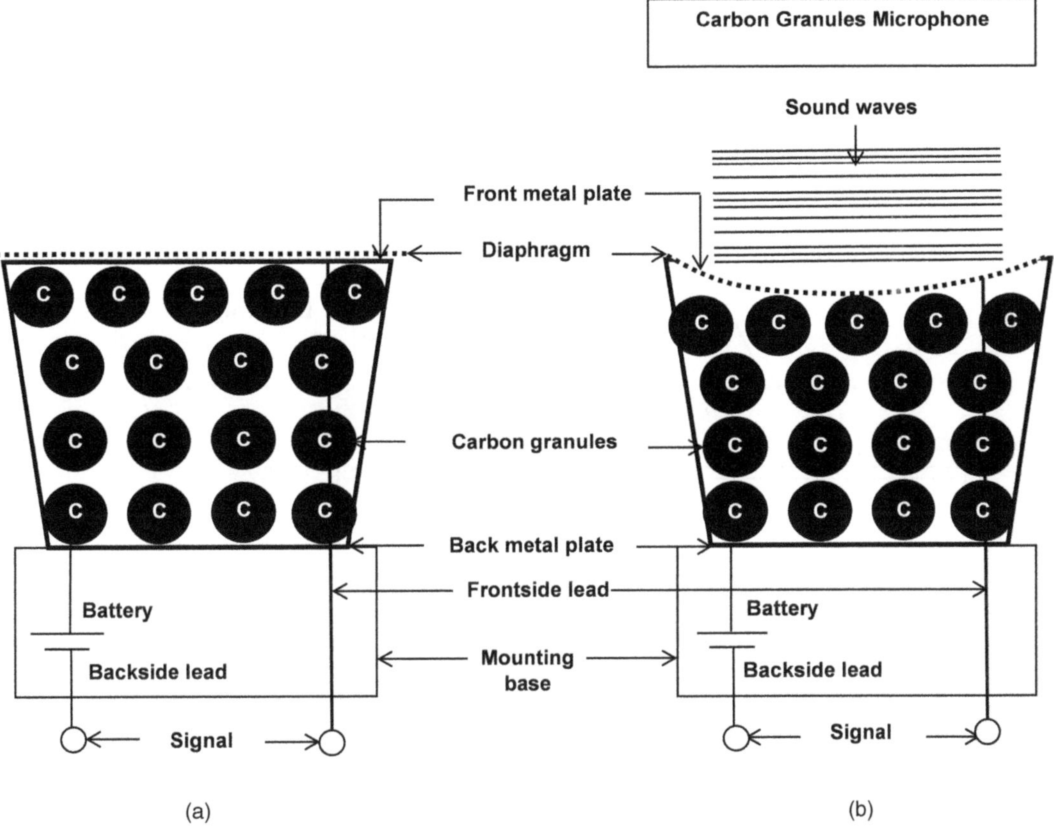

FIGURE 3.12 Carbon granules microphone: (a) in the absence of sound, with carbon granules lying apart, and (b) when impacted by sound waves, with carbon granules coming close together. Carbon granules, shown as black circles with the letter 'C' inscribed, are packed between a front metal plate (diaphragm) and a back metal plate. The device is kept on a mounting base. The frontside lead is tied to the front metal plate. A battery is connected to the backside lead from the back metal plate. Sound waves impinge on the diaphragm. The resulting electrical signal is obtained across the marked terminals.

come closer together, their resistance to current flow decreases (Figure 3.12(b)). When the compressing force is released, the carbon granules move apart, resulting in an increase of resistance. Thus, the sound pressure waves produce a varying current signal by changing the effective resistance due to carbon granules.

3.6.2 Moving Coil Microphone

Also called a dynamic microphone, it consists of a movable coil placed in the magnetic field of a permanent magnet (Figure 3.13). The coil is attached to a diaphragm. Sound waves falling on the diaphragm cause it to vibrate. The attached coil also vibrates. It moves back and forth cutting the lines of force of the magnet. An electric current is induced in the coil by electromagnetic induction. This microphone does not require any batteries or power supply for its operation. It is rugged and reliable.

3.6.3 Ribbon Microphone

A corrugated strip of aluminum, the aluminum ribbon, with contacts at its ends is suspended between the poles of a permanent magnet (Figure 3.14). Sound waves striking the aluminum ribbon make it

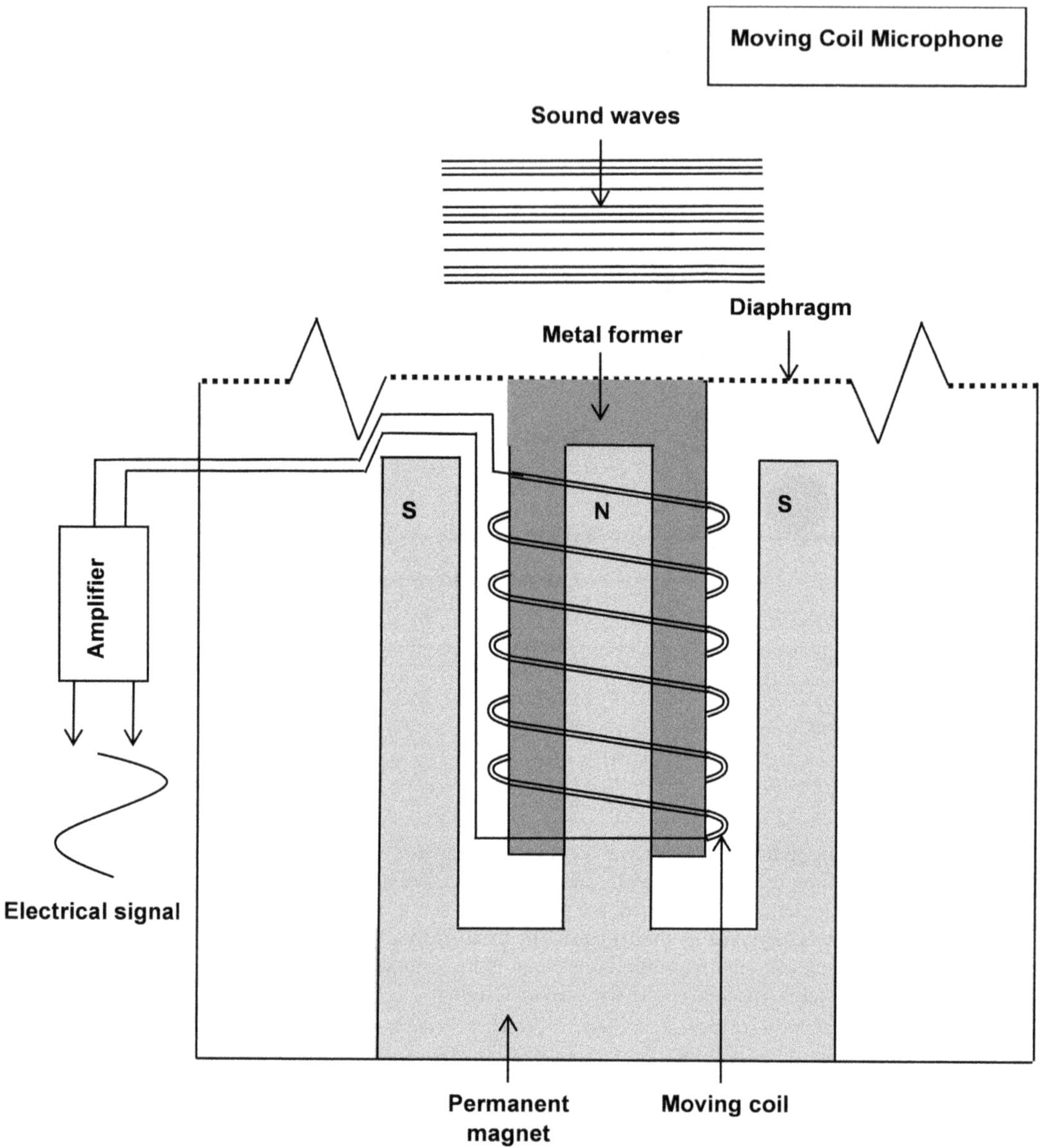

FIGURE 3.13 Moving coil microphone. A coil is wound around a metal former. Having the ability to move, it is called a moving coil. The former with the coil forms the central polepiece of a permanent magnet marked as the north pole. The north pole is surrounded on its two sides by south poles. A diaphragm is fixed over the metal former. When sound waves strike the diaphragm, it is set into vibration. Along with the diaphragm, the metal former vibrates, and so also the coil wound around it. As the vibrating coil moves between the polepieces of the magnet, a current is induced in the coil. The induced current flows to an amplifier, from whose output terminals the electrical signal produced by sound is withdrawn.

vibrate. A voltage is induced in the ribbon in a direction perpendicular to the direction of ribbon velocity and that of the magnetic field. This microphone is called a velocity microphone because the induced voltage is proportional to the velocity of air particles and hence the ribbon. In contrast, the voltage produced in a moving-coil microphone is proportional to the displacement of air particles and therefore the diaphragm.

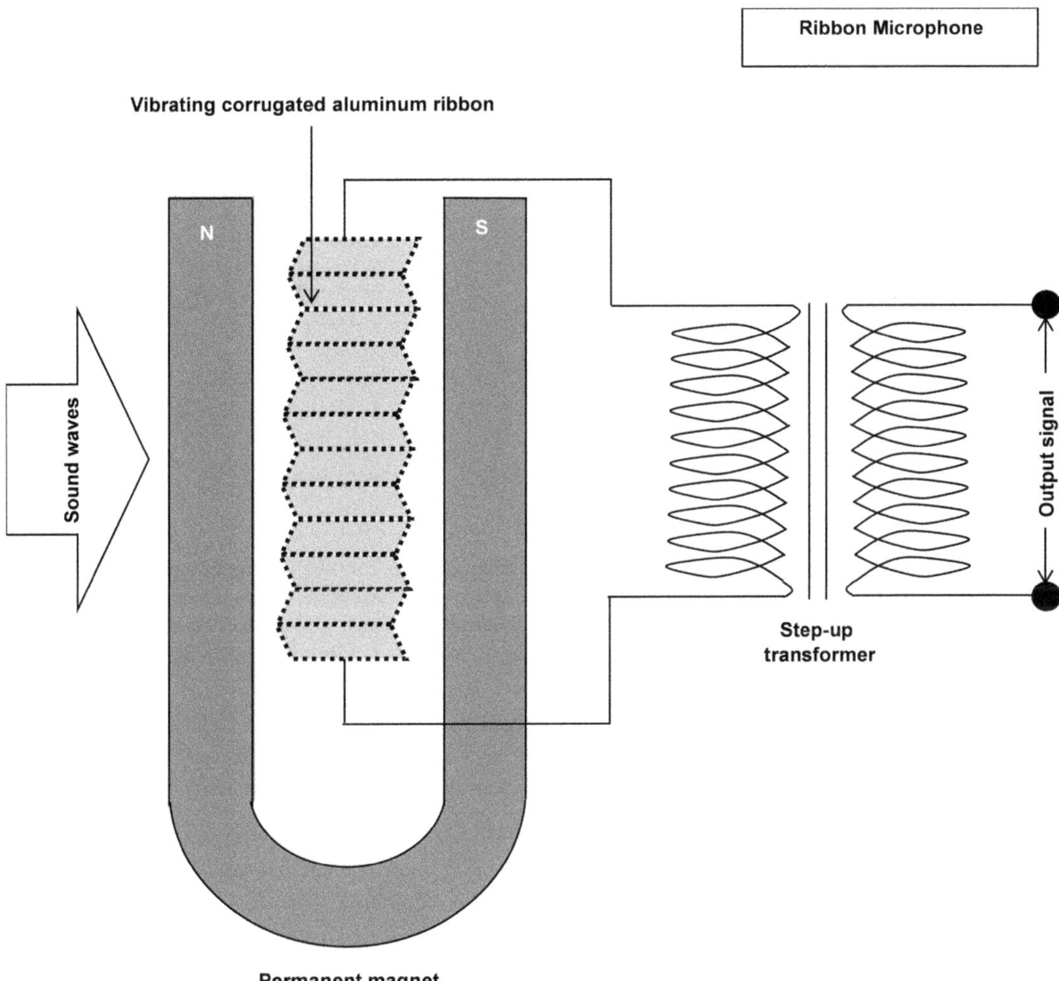

FIGURE 3.14 Ribbon microphone. Sound waves set a corrugated aluminum ribbon into vibrations. As the ribbon is hanging between the N and S pole pieces of a horseshoe magnet, its oscillations induce an electric current within its body which is fed to a transformer to step up the voltage and supply the desired output signal.

3.6.4 CONDENSER MICROPHONE

It is a parallel plate capacitor (Figure 3.15). One plate of the capacitor is movable and acts as the diaphragm, vibrating in response to sound waves. The other plate is a fixed plate called the backplate. In a DC-biased microphone, the capacitor is charged by a fixed quantity of charge Q. Sound-induced vibrations of the diaphragm cause variations in the distance between it and the backplate. The capacitance C of the capacitor changes, and so the voltage V across the capacitor varies according to the equation

$$V = Q/C \tag{3.2}$$

3.6.5 ELECTRET MICROPHONE

It is a variant of the condenser microphone in which either the diaphragm plate or the backplate is made of an electret material with a sputtered gold film (Figure 3.16). An electret is a dielectric

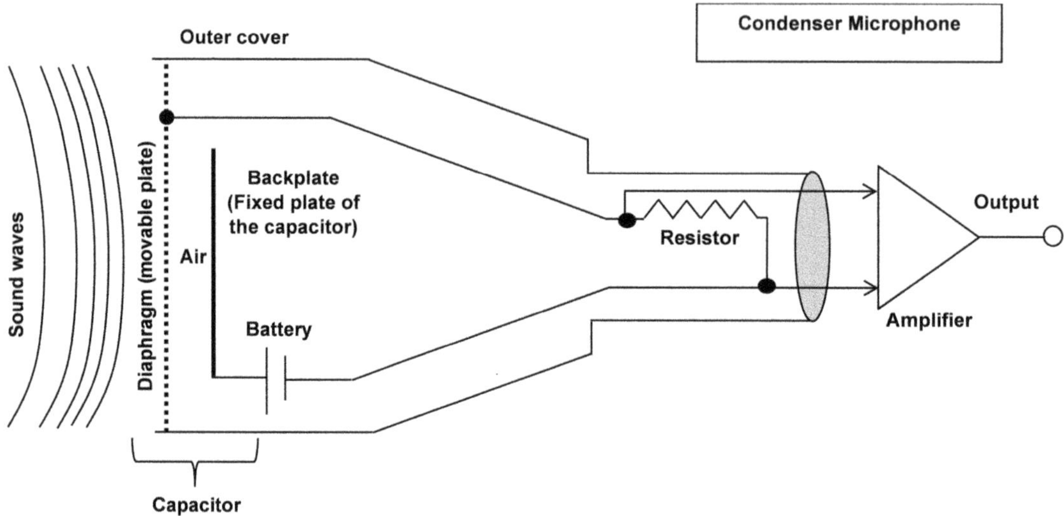

FIGURE 3.15 Condenser microphone. A capacitor is made with a conductive diaphragm as the movable plate, air as the dielectric medium, and a stationary backplate. The backplate is connected to one pole of a battery. The other pole of the battery is joined to the diaphragm through a resistor. The voltage across the resistor is amplified to give the output signal for the sound waves that are vibrating the diaphragm.

material having a quasi-permanent electric charge or dipole moment and hence exhibits an electric field in the same way that a permanent magnet displays a magnetic field. The advantage gained is that no external power source is needed for this microphone although it is necessary for the associated circuitry.

3.6.6 MEMS MICROPHONE

It is a condenser microphone fabricated by MEMS technology using a silicon wafer (Loeppert and Lee 2004, 2006, Jung et al. 2012, Hur et al. 2012). It consists of a movable diaphragm suspended behind a perforated backplate (Figure 3.17). In the absence of perforations, a squeezed film is trapped between the backplate and the diaphragm which dampens the vibrations of the diaphragm and hence reduces the sensitivity of the microphone. Further, a vent hole is made in the diaphragm to connect the cavity behind the diaphragm with the atmosphere.

A fixed electrical charge is applied between the diaphragm and the backplate. An incoming sound wave enters through the holes in the backplate and builds up longitudinal compressions and rarefactions in the air film enclosed between the backplate and the diaphragm. The diaphragm begins to vibrate and the amplitude of vibrations of the diaphragm varies in proportion to the intensity of sound waves. These vibrations alter the distance between the diaphragm and the backplate, which in turn changes the capacitance of the microphone. Since a fixed charge has been placed on the capacitor plates, the voltage changes across the capacitor plates follow the pattern of variations in the impinging sound waves (Torkkeli et al. 2000).

MEMS microphones are used in mobile phones, smart watches, tablets, and medical devices (Zawawi et al. 2020).

3.7 SPEAKERS

3.7.1 ELECTRODYNAMIC SPEAKER

A lightweight cone or diaphragm is connected to a support chassis by a flexible suspension known as a spider (Figure 3.18). The cone is attached to a coil carrying the voice signal, which is free to

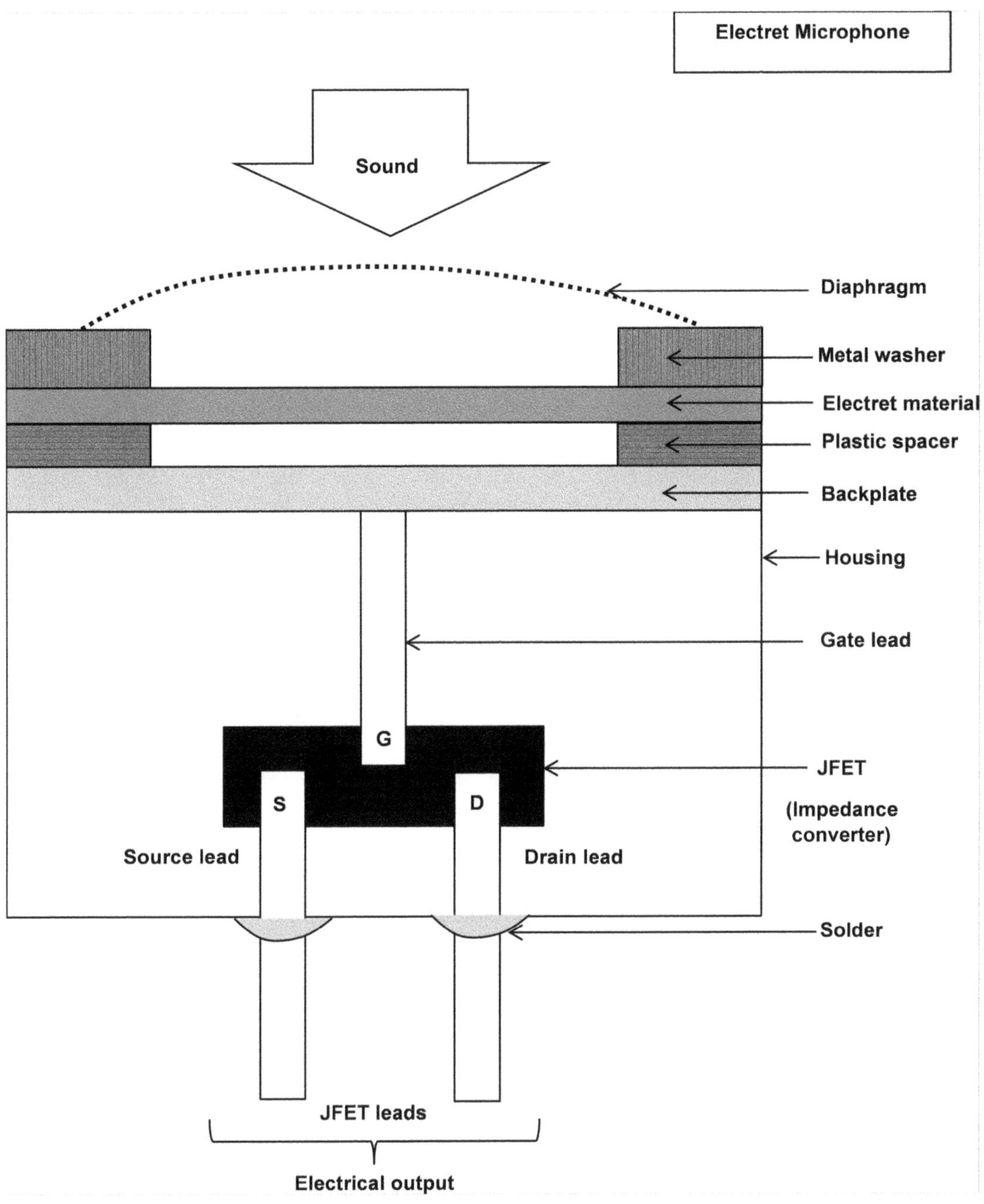

FIGURE 3.16 Electret microphone. Sound waves impact a diaphragm. The diaphragm rests on metal washers at its two ends. The washers lie on an electret film, and the electret film itself is placed over plastic spacers on a backplate. To the backplate, the gate lead of a JFET is attached. Holes are made in the housing. Through these holes, the source and drain terminals of the JFET come outside the housing. The output electrical signal is delivered across these terminals.

move axially on a metallic former in a cylindrical gap between the north and south poles of a permanent magnet. When the voice signal is supplied to the coil, a magnetic field is produced around the coil according to the current variations in the circulating voice signal. Thus, it becomes a variable electromagnet. Now two magnetic fields are simultaneously present, one due to the current in the coil and the other due to the permanent magnet. The two fields interact with each other inducing

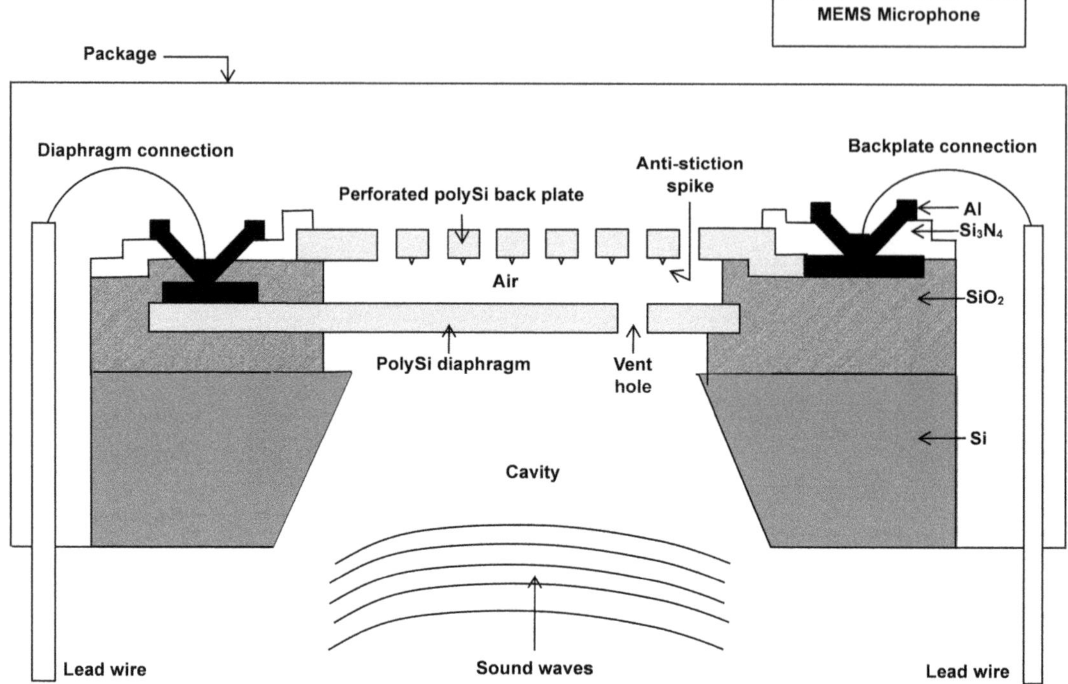

FIGURE 3.17 MEMS microphone. A cavity is etched in silicon. A polysilicon diaphragm with a vent hole is suspended over the cavity from the SiO_2 film. Sound waves are hitting the diaphragm causing it to vibrate. At a small distance above the diaphragm, a perforated silicon backplate is placed over the SiO_2 film. It has anti-stiction spikes on its surface facing the diaphragm to prevent it from sticking to the diaphragm. Aluminum contacts are made through an Si_3N_4 layer deposited on the SiO_2 film to provide diaphragm and backplate connections to the external lead wires. The microphone is housed inside a package with an opening below the cavity.

vibrations in the coil. The cone attached to the coil moves along with it and undergoes vibrations. These vibrations are transmitted through the surrounding air medium as sound waves, realizing the function of an electroacoustic transducer.

3.7.2 Electrostatic Speaker

A flat diaphragm made of a polyester film coated with a conductive material such as graphite is suspended between two electrically conducting grids in the form of perforated metal sheets (Figure 3.19). A DC bias supply maintains the diaphragm at a high potential of several kilovolts with respect to the grids. The two grids known as the front and rear grids are driven in anti-phase by the voice signal, thereby producing a uniform electrostatic field proportional to the current variations in the audio signal. The force exerted on the electrically charged diaphragm by two grids is a linear force acting along a straight line because of the cancellation of the voltage-dependent non-linearity. So, harmonic distortion is avoided. The motion of the diaphragm due to the force exerted on it by the grids causes movement of the air on its two sides, which is responsible for creating sound. Electrostatic speakers provide highly directional sounds in narrow beams with minimal distortion. But they are humidity-sensitive and lack bass response.

3.7.3 Magnetostatic Speaker

It is a dipole loudspeaker identical to the electrostatic loudspeaker. It consists of permanent magnets for producing a static magnetic field, and a thin diaphragm to which the current-carrying wires or

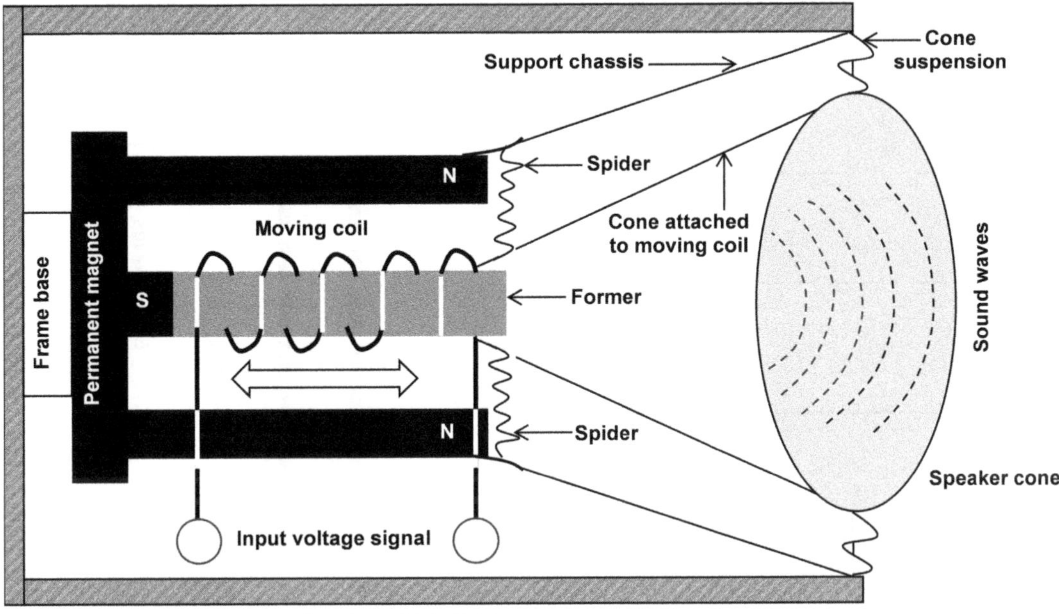

FIGURE 3.18 An electrodynamic loudspeaker consisting of a moving coil wound around a metallic former on the south pole piece of a permanent magnet with the north pole pieces on its two opposite sides. The coil is fixed to a cone which hangs by a cone suspension from the support chassis. The cone is tied to flexible springs called spiders for free movement. The entire assembly of the magnet and the coil is enclosed inside a loudspeaker cover for protection. The coil is energized by the input voltage signal. Its consequent vibrations are coupled to the cone whose motion produces the sound waves.

strips are bonded. However, unlike the electrostatic loudspeaker working at high voltages, the magnetostatic loudspeaker operates at high currents. Magnetostatic speakers offer good-quality sound with satisfying bass response. Their disadvantages are large size and slow response.

3.7.4 SPEAKER FREQUENCY RANGES

3.7.4.1 Sub-Woofer

A subwoofer or sub-bass speaker is a loudspeaker that is designed to produce low-frequency sounds. These frequencies are in the range of 20–80 Hz. Hence, it takes care of the lowest 2–3 octaves of the 10 audible octaves.

3.7.4.2 Woofer

A woofer or bass speaker is a loudspeaker designed for the reproduction of sounds from 20 Hz to a few hundred Hz. Challenges in woofer design are concerned with cone movement such as slowing down the cone motion when the input signal falls to zero and controlling its excursions for loud sounds.

3.7.4.3 Mid Frequency: Squawker

A squawker is a loudspeaker for intermediate frequencies between those of a woofer and a tweeter.

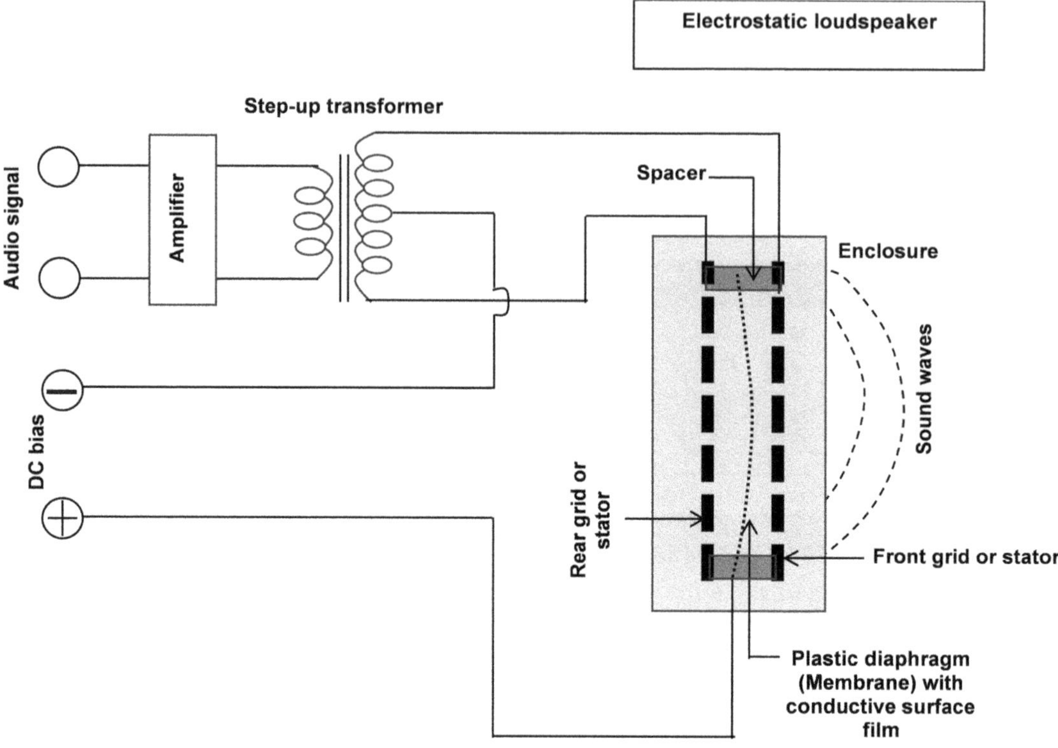

FIGURE 3.19 An electrostatic loudspeaker consisting of a plastic diaphragm with a conductive coating fixed between two grids or stators, one on each side with the help of spacers. The grids are metal sheets with perforations. The assembly is placed inside an enclosure. A DC bias is applied to polarize the diaphragm. The audio signal is amplified and its voltage is stepped up with a transformer. The two terminals of the transformer's secondary winding are connected to the two grids. Forces produced by interaction between the grid signal and the potential on the diaphragm cause vibrations of the diaphragm which generate sound waves.

3.7.4.4   High Frequency: Tweeter

A tweeter is a loudspeaker that is specially designed to produce high-frequency sounds up to frequencies around 100 kHz. A polymer resin-impregnated polyester film or fabric is used to make the diaphragms of soft-dome tweeters, while the diaphragms of hard-dome tweeters are made from Al, Al-Mg alloys, or Ti.

3.8   IR SENSORS

3.8.1   IR Photodiode

Regular diodes produce electric current signals in response to visible light. Unlike these visible light sensors, IR diodes perform the same work when showered with IR photons. They are able to do so because they are specially designed for this purpose. This is done by interposing a filter in the path of incoming light (Figure 3.20). The photodiode is placed below the filter. This filter allows only the electromagnetic waves of infrared wavelengths to pass through but blocks all the remaining wavelengths of visible light. So only the IR photons reach the photodiode and produce electron–hole pairs. Thus, the filter makes sure that the ambient light in the room or sunlight does not contribute to current except for its IR portion. So, the function of the sensor is insignificantly interfered by these portions of the visible spectrum. The fact that IR radiation is invisible to humans is also beneficial because the sensor can operate without disturbing human activities.

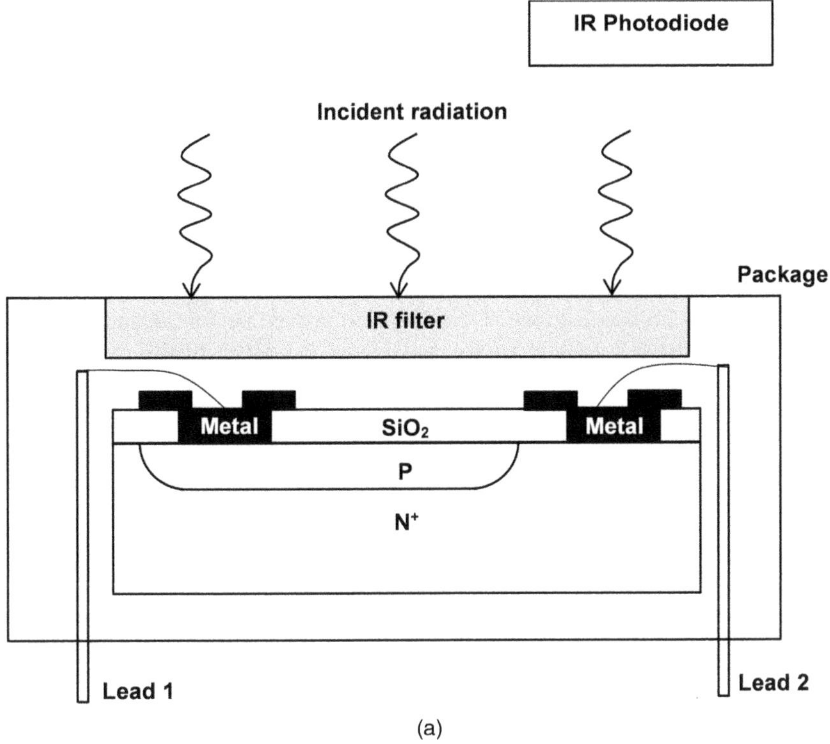

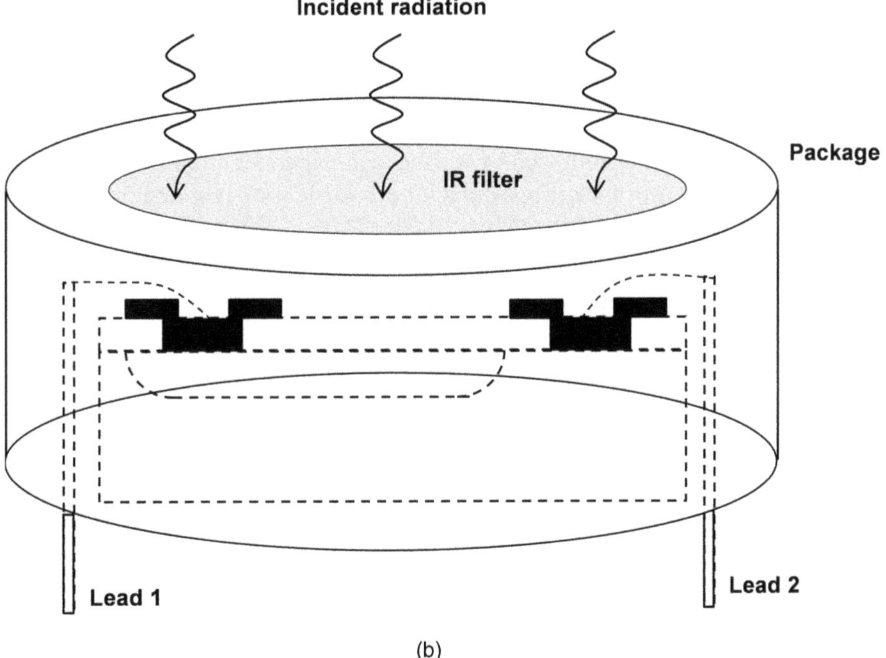

FIGURE 3.20 Infrared photodiode: (a) unpackaged and (b) packaged. Part (a): A P-N⁺ diode is made with metal contacts through SiO_2 film, and wire-bonded to leads 1 and 2 for external connection. It is enclosed in a package with a central hole covered with an IR filter. Only the IR component of the incident radiation passes through the filter. Part (b): External appearance of part (a) with the diode shown by dotted lines.

3.8.2 IR Thermopile Sensor Array

An IR thermopile sensor array is a serial connection of a few hundred IR thermopile sensor elements, e.g., an array of 256 = 16 × 16 sensor elements (Tanaka et al. 2014). A packaged IR thermopile sensor array is shown in Figure 3.21(a). The infrared radiation is focused by a silicon lens on the IR thermopile sensor array.

An IR thermopile sensor element consists of thermocouples connected in series. Figure 3.21(b) shows the cross-sectional diagram of a single IR thermopile sensor element. In this diagram, the thermocouples are made by forming junctions between N^+-doped and P^+-doped polysilicon with Cr/Au metal. Semiconductors have higher Seebeck coefficients than metals. N^+ and P^+ doped polySi have Seebeck coefficients with opposite signs leading to improved performance. The hot junctions are coated with an IR-absorbing layer and are located exactly below an IR filter window.

3.8.3 Pyroelectric Sensor and Human/Pet Motion Detection Circuit

This sensor uses a pyroelectric material, e.g., gallium nitride, polyvinylidene fluoride (PVDF), lithium niobate, lithium tantalate, and tourmaline. An output voltage is produced by heating or cooling the material. Pyroelectric materials are naturally electrically polarized exhibiting spontaneous polarization without an externally applied electric field. Any change in ambient temperature alters the positions of atoms within the crystal. The relative changes in atomic positions vary the polarization state of the material. Hence, a temporary voltage appears across the material depending on the thermal effects.

The heat radiated by a human being or animal is adequate to produce a pyroelectric potential enabling the use of these sensors in human or animal motion sensing. The incoming IR radiation is focused by a plano-convex lens known as the Fresnel lens on the pyroelectric crystal of the sensor (Figure 3.22). In a practical implementation (Figure 3.23), two oppositely connected sensors are used for canceling out spurious signals arising from sunlight, shocks, and vibrations because they influence both sensors identically. However, the body temperature of a person moving near the sensors affects one sensor PIR1 first and then the second sensor PIR2 sequentially. Refering to Figure 3.22, the positive end of PIR1 is connected to the gate of an FET. The FET source terminal is connected through a pulldown resistor to the ground. It is used to maintain the desired state of a logic circuit. It feeds an amplifier followed by a window comparator circuit responding to positive as well as negative transitions of the sensor output signal. It uses two comparators to ascertain whether a signal value lies between two reference voltages.

3.9 MAINTAINING A COMFORTABLE HOME AMBIENCE

3.9.1 Room Humidity and Temperature Monitoring: Humidity and Temperature Sensors

3.9.1.1 Humidity Sensors

Humidity sensors are primarily of three types: capacitive, resistive, and those based on thermal conductivity (Lee and Lee 2005, Kuzubasoglu 2022). The first two sensors measure relative humidity while the third device gives absolute humidity.

3.9.1.1.1 Capacitive Humidity Sensor

The capacitive humidity sensor (Figure 3.24(a)) is a capacitor consisting of a moisture-sensitive polymer sandwiched between a thin water-permeable metal electrode and a bottom electrode. Ambient humidity changes alter the capacitance of the device.

This sensor consists of a hygroscopic dielectric film, either a polymer, e.g., polyimide (dielectric constant $\varepsilon = 2.8$) or cellulose acetate ($\varepsilon = 3$–7), or oxide, e.g., anodic aluminum oxide ($\varepsilon = 8$), sandwiched between a thin metal electrode on the top side to allow moisture to pass through, and a thick bottom electrode.

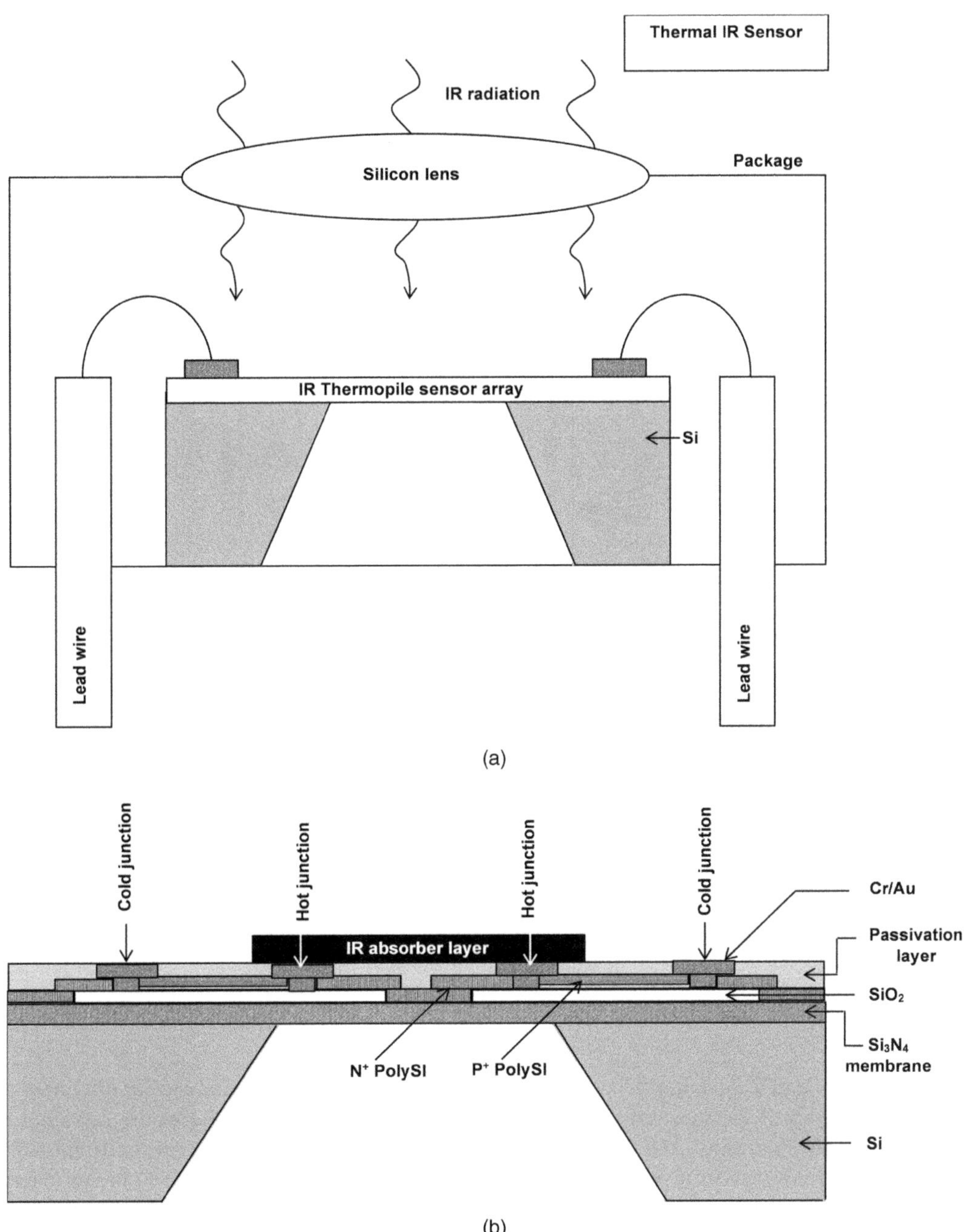

FIGURE 3.21 Schematic representation of (a) a packaged IR thermopile sensor array, and (b) a single IR sensor element. (a): IR radiation is falling on a packaged thermopile sensor array through a silicon lens at the center of the top surface of the package. The thermopile array is mounted on a trapezium-shaped cavity etched in silicon. The array has two metal contacts, one on its left side and the other on the right side. Wires bonded to the metal contacts pass outside the package through its bottom side. These are used for electrical connections to the thermopile array. (b): A silicon nitride membrane is suspended over a trapezium-shaped cavity etched in silicon. An SiO$_2$ film is deposited on the silicon nitride film and patterned to form N$^+$ polysilicon (vertical shading) and P$^+$ polysilicon layers (horizontal shading) joined with Cr/Au metal pads (dotted rectangles). The upper surface of the device is coated with a passivation layer having holes through which the metal pads are contacted. The two N$^+$ polysilicon/metal/P$^+$ polysilicon junctions at the center of the device over the cavity are covered with an IR absorber layer. These are the hot junctions. The two N$^+$ polysilicon/metal/P$^+$ polysilicon junctions on the outer sides over the silicon portion of the silicon platform without cavity are the cold junctions.

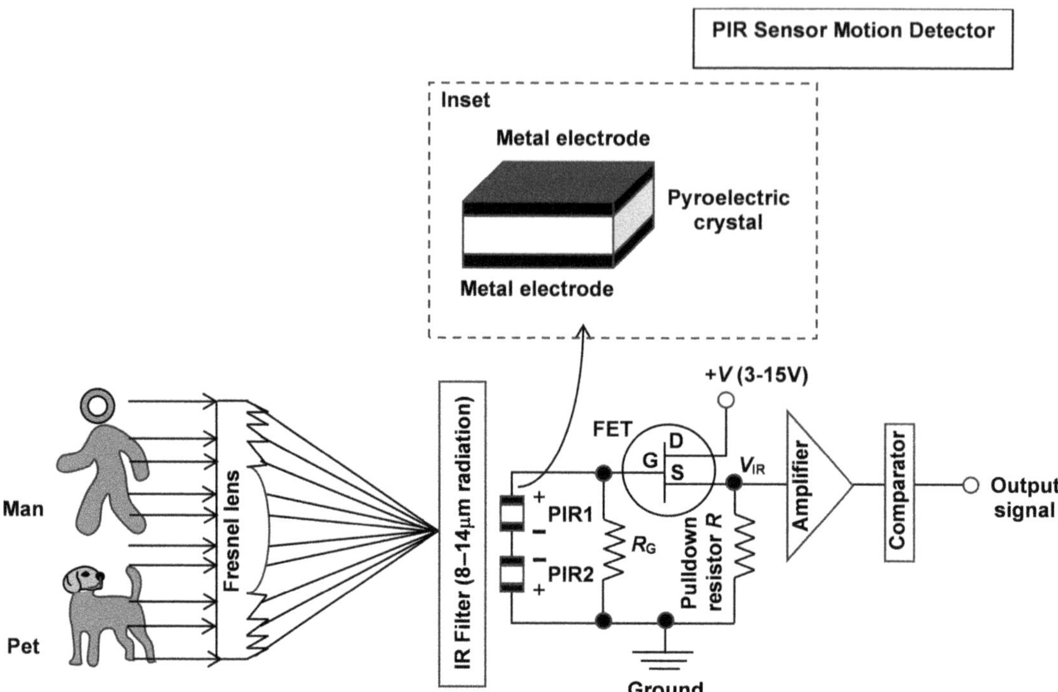

FIGURE 3.22 Human or pet motion detector circuit using two passive pyroelectric infrared sensors PIR1 and PIR2. The electromagnetic radiation from a human being or pet passing near the setup is collected by a Fresnel lens, passed through an IR filter, and focused on the two PIR sensors. The two PIR sensors are connected in opposition with the positive terminal of one tied to the negative terminal of the other. The current from the positive terminal of PIR1 is supplied to the gate of an FET. A resistor R_G is connected between the FET gate and the ground. The drain of the FET is joined to the +V supply while its source is connected via a pulldown resistor to the ground, and also feeds the infrared voltage V_{IR} to an amplifier circuit. The amplified signal is delivered to a window comparator circuit, which provides the detector output. The PIR sensor consisting of a crystal of pyroelectric material sandwiched between two metal electrodes is shown in the inset diagram within a dashed rectangle.

The capacitance of a capacitor depends on the area of its conducting plates, and the thickness and relative permittivity of the intervening dielectric. The geometrical parameters of the capacitor are not influenced by humidity. The humidity-sensitive parameter is the dielectric constant which changes with the adsorption of moisture. The dielectric constant of water ($\varepsilon = 80$) is much larger than that of the dielectrics used. As the humidity rises, the dielectric constant of moisture-filled dielectric and hence the capacitance of the sensor increases. The change in capacitance with humidity indicates the ambient humidity.

Capacitive humidity sensors are also fabricated like resistive sensors in interdigitated electrode geometry. Planar interdigitated Au electrodes are coated with polymide sensing films of different thicknesses and morphologies to make capacitive humidity sensors. The 1.5 μm thick nanograss polyimide film shows low response (18s) and recovery times (31s). It has a sensitivity of 4.75fF $(\%RH)^{-1}$. The flat 4.6 μm thick polyimide film shows a higher sensitivity of 15.2fF$(\%RH)^{-1}$ but its response (54 s) and recovery (80 s) times are much longer (Boudaden et al. 2018).

Capacitive sensors are unsuitable for remote monitoring in which the distance between the sensor and the signaling circuit is large Lead capacitances affect the measurements.

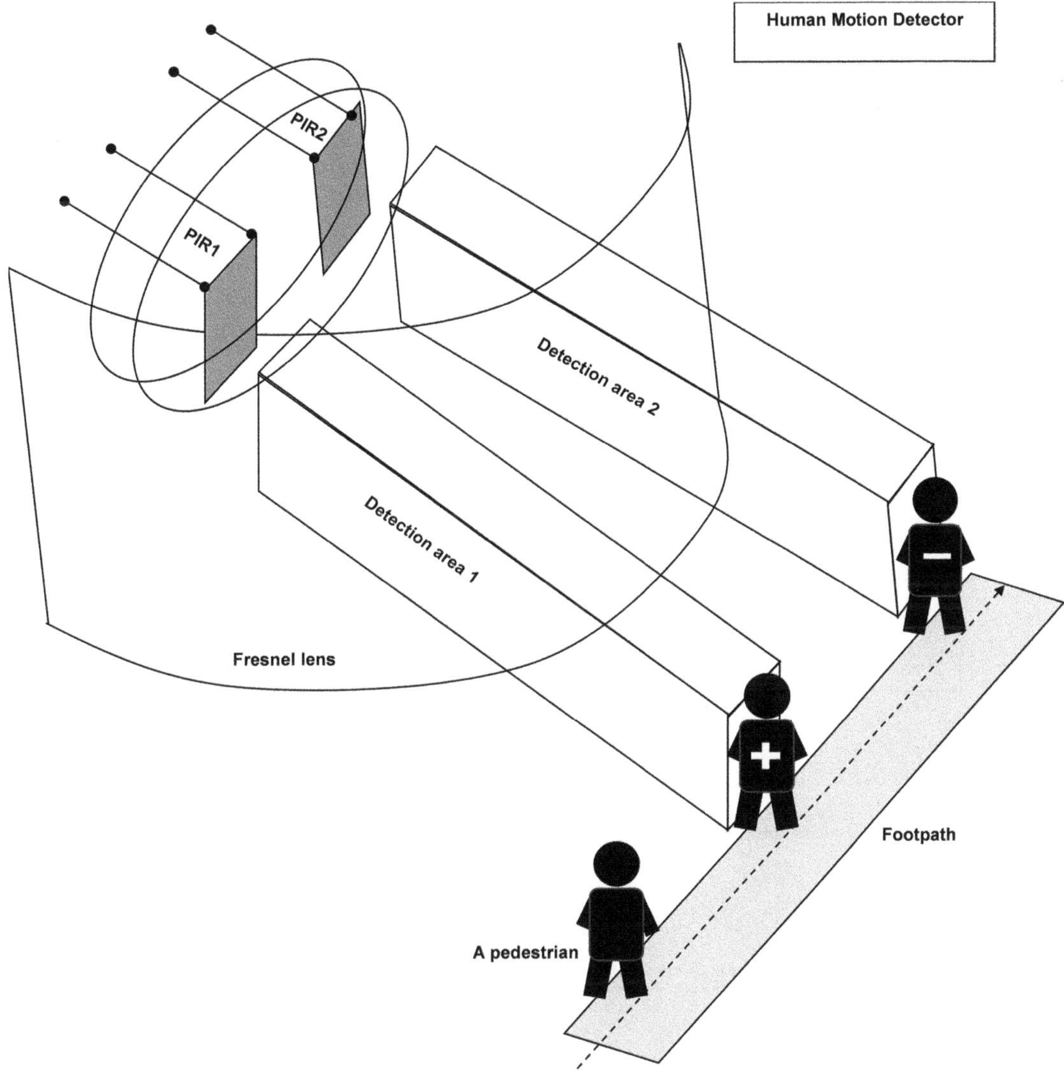

FIGURE 3.23 Practical Illustration of human motion detection with two PIR sensors covered by a Fresnel lens. The detection areas 1 and 2 of the two PIR sensors are defined. A man walking on the street moves across detection area 1. This movement is marked with a plus sign. The man moves further and crosses the detection area 2. This movement is marked with a minus sign to emphasize that the two PIR sensors are connected in opposite directions with the positive polarity of one joined to the negative polarity of the other.

3.9.1.1.2 Resistive Humidity Sensor

In this sensor (Figure 3.24(b)), a moisture-sensitive polymer film, e.g., polystyrene sulfonate or polytetrafluoroethylene, is coated on a ceramic substrate. A pair of interdigitated electrodes is formed over the polymer film using a noble metal such as gold or platinum. The adsorption of moisture by the polymer increases its conductivity by several orders of magnitude.

A humidity sensor made by depositing 10nm thick shellac-derived carbon (SDC) film on carbon interdigitated electrodes shows a dynamic range of 0–90% RH and sensitivity of 0.54 (%RH)$^{-1}$. It responds fast to humidity in 0.14 s and also recovers rapidly in 1.7 s between 0% and 70%RH (Joshi et al. 2020).

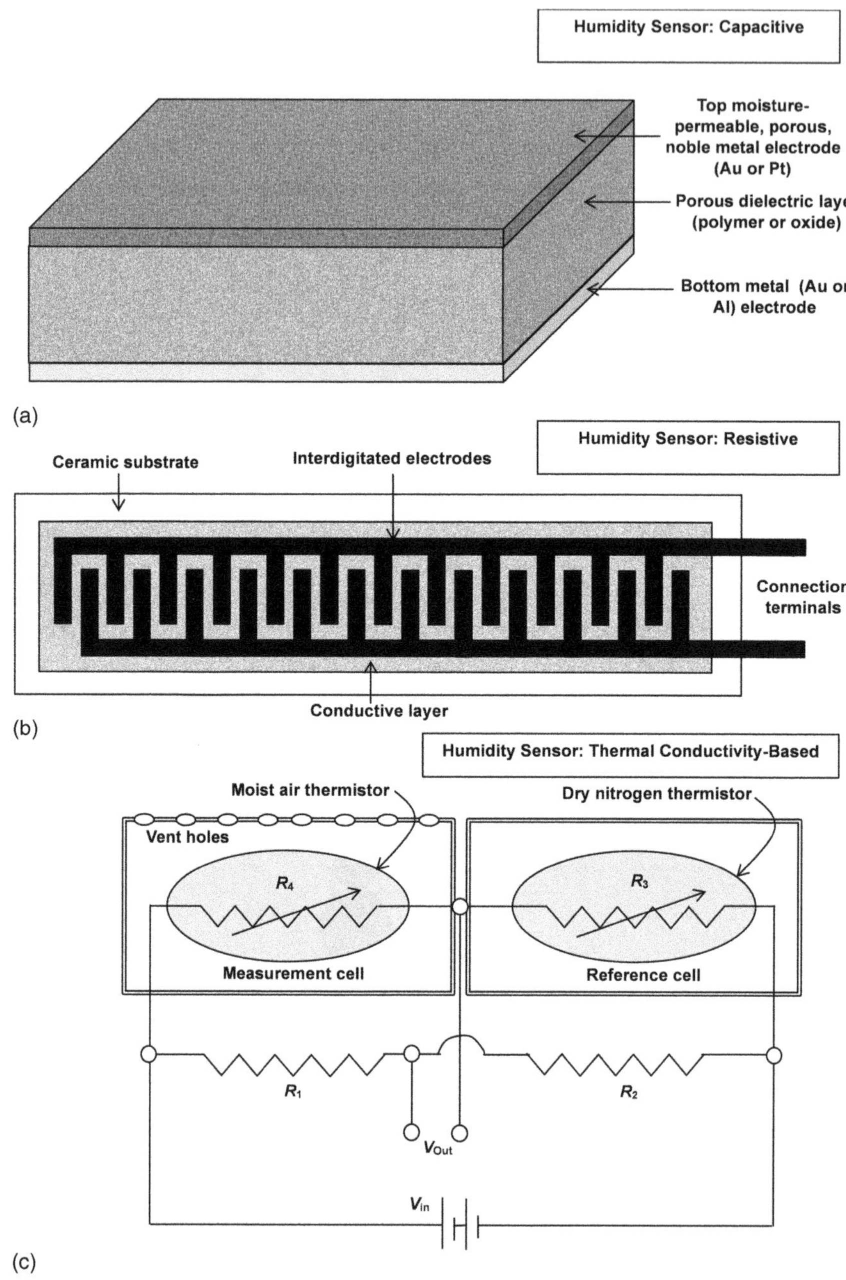

FIGURE 3.24 Humidity sensors: (a) Capacitive, (b) resistive, and (c) thermal-conductivity-based. (a): The sensor consists of a top moisture-permeable noble metal electrode of gold or platinum, a porous dielectric polymeric or oxide layer and a bottom metal electrode of gold or aluminum. (b): A conductive layer is made on a ceramic substrate. Over this conductive layer, a pattern of interdigitated electrodes is defined. Connection terminals are shown. (c): Two cells are made side by side. The first cell has vent holes through which water vapor can enter the cell. The thermistor of resistance R_4 in this cell is named moist air thermistor becuase it responds to moisture-laden air. The second cell filled with dry nitrogen is the reference cell. The thermistor of resistance R_3 in this cell is named dry nitrogen thermistor because it shows the effect of dry nitrogen. Together with resistors R_4 and R_3, two more resistors R_1 and R_2 constitute a Wheatstone bridge circuit. The bridge is powered by the battery V_{in}. The output voltage V_{Out} is measured between two points, one point is the junction of resistors R_4 and R_3. The other point at the junction of resistors R_1 and R_2.

Resistive sensors can be used for remote monitoring from a large distance between the sensor and the signaling circuit.

3.9.1.1.3 Thermal Conductivity Humidity Sensor

The thermal conductivity of humid air differs from that of dry air. Dry air such as in a desert, cools down more rapidly than wet air. Water vapor present in the air retains heat preventing fast cooling. So, if a heater is suspended in a dry environment, it will lose heat at a different rate than one suspended in wet surroundings.

Utilizing this thermal conductivity difference, a humidity sensor is constructed with two thermistors (Figure 3.24(c)), one hermetically sealed in a chamber filled with dry nitrogen gas and the other placed in a chamber with vent holes through which air from a moist environment can gain entry. When current is passed through these thermistors through an electrical bridge circuit, they are self-heated by the current flowing through them. But one thermistor being in dry nitrogen and the other exposed to wet air, their temperatures will differ. The difference in temperatures of the two thermistors and therefore their resistances will depend on the moisture content of the wet chamber. This resistance difference is related to the absolute humidity of the moist chamber. See Sec. 5.2 on temperature sensors.

P-N junction diodes are sometimes used as temperature sensors in place of thermistors. A thermal conductivity-based humidity sensor implemented with 0.6 µm CMOS technology works from a 5V supply by comparing the output voltages of two heated and isolated P-N junction diodes. The sensitivity is 14.3 mV(%RH)$^{-1}$ at 20°C ambient temperature, and increases to 46.9 mV (%RH)$^{-1}$ at 40°C because the quantity of water vapor is larger at a higher temperature for the same relative humidity (Okcan and Akin 2004).

Thermal conductivity sensors are particularly useful for high-temperature or corrosive environments containing chemical vapors.

3.9.1.1.4 Humidity Sensor Specifications

Vital parameters of a humidity sensor are sensing element, e.g., ceramic or polymer, measurement range (0–100% RH), nonlinearity 0.05–1.0%RH, resolution 0.1%, accuracy ± 1% RH, response time 10–30 s for 63% step change of RH. Typical values of these parameters are given by vendors.

3.9.1.2 Temperature Sensor

The temperature sensor is a thermistor with a negative temperature coefficient, i.e., a device whose resistance decreases with rising temperature.

The temperature and humidity sensor, mounted on the wall through adhesive tape or screws, sends its readings through Wi-Fi or ZigBee to the user. It can be used to set to activate an air-conditioning system.

Devices: Humidity sensor, temperature sensor. LCD display
Protocol: Wi-Fi, ZigBee
Further reading: Section 5.2

3.9.2 ALLOWING HOME OWNERS TO PROGRAM THE TEMPERATURE SETTINGS OF THEIR HVAC SYSTEM FROM A REMOTE LOCATION: SMART THERMOSTAT

An internet-connected programmable thermostat is used to adjust the heating, ventilation, and air conditioning (HVAC) system of a house. It aids in heating or cooling of a room by sensing the room temperature and sending a command using a smartphone app or laptop from any distant location to change its settings, or even to automatically schedule the heating/cooling cycles for optimal performance depending on occupancy for ensuring energy savings, and by a machine learning-based system working on the user's heating and cooling preferences.

Devices: Temperature sensor to measure room temperature, conventional bimetallic strip thermostat, PIR occupancy sensor
Protocol: Wi-Fi secured using a strong password and enabling WPA2 (Wi-Fi Protected Access 2) or higher encryption

3.10   FIRE SAFETY AND POISONOUS CO GAS DETECTION

3.10.1   DETECTION OF FIRE: SMOKE/FIRE DETECTOR

A smart smoke/fire detector is an ionization, photoelectric, or combination-type smoke detector connected to a Wi-Fi network through which data is sent to a cloud-hosted software from where the user receives alerts through a smartphone app when away from home in case of a fire breakout, and sends control commands to the device. Photoelectric smoke detectors are more sensitive to fires that begin with smoldering while ionization smoke detectors are more responsive to flaming fires containing dense small particles.

Device: Smoke sensor
Protocol: Wi-Fi

3.10.1.1   Ionization Smoke Detector

It consists of two metal plates, one positively charged and the other negatively charged, placed inside a steel ionization chamber (Figure 3.25(a)). About 1 μcurie of an isotope of americium Am-241 placed underneath the chamber decays by radioactivity into neptunium-237, accompanied by emission of 37,000 alpha particles per second, of energy ~5.4 MeV with a half-life of 432.2 years. In the absence of smoke, the air inside the chamber is clean. The alpha particles strike the air molecules in the chamber to produce positive and negative ions. The positive ions are attracted toward the negative metal plate and negative ions toward the positive metal plate, leading to a flow of a current in the external circuit.

In case of a smoldering or flaming fire (Figure 3.25(b)), smoke particles enter the ionization chamber. They stick to the ions. Since the smoke particles are much larger in size (0.3–10 μm for smoldering stage and 0.01–0.3 μm for flaming stage) than the ions produced from air molecules, the average mass of the ions increases. As a result, the average velocity of the ions decreases causing a fall in current output which triggers the smoke alarm in 30–60 s.

Ionization type detector is more responsive to flaming fires than smoldering fires in comparison to a photoelectric detector. This smoke detector does not pose any environmental threat but americium is harmful if inhaled, ingested, or improperly discarded.

3.10.1.2   Photoelectric Smoke Detector

Light from an LED is focused by a lens on a light catcher such as a mirror, away from the light-sensing photocell (Figure 3.26). When there is no smoke (Figure 3.26(a)), the light propagates directly to the mirror without any sideways deviation. But when smoke enters the chamber (Figure 3.26(b)), it scatters the light in different directions. Some of the scattered light falls on the sensor, increasing the current output and hence activating the alarm.

As already said, a photoelectric detector responds faster to a fire in its early smoldering stage whereas an ionization detector responds more quickly to a flaming or blazing fire. So, a combination of both types of detectors is often used to ensure safety from fire hazards because early warning is essential to save precious lives.

3.10.2   DETECTION OF CARBON MONOXIDE: CO GAS SENSOR

A smart CO sensor (Figure 3.27) is an internet-connected electrochemical cell consisting of three electrodes: a platinum working electrode, a counter electrode made of an inert material such as gold,

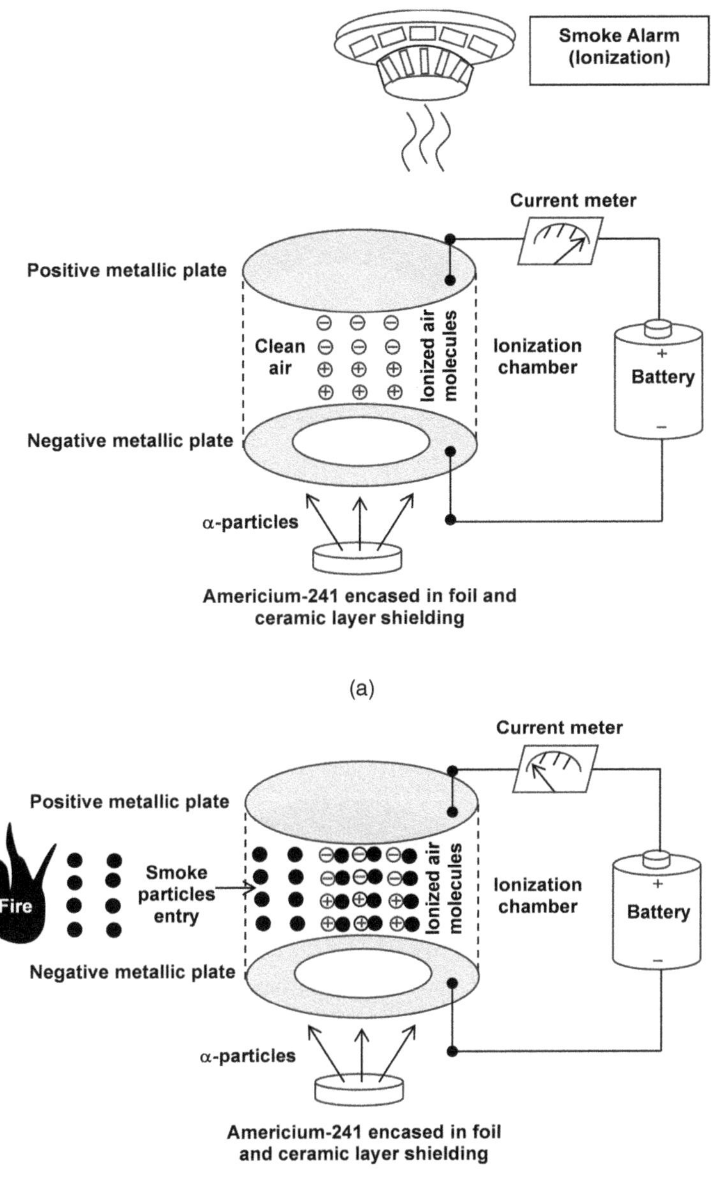

FIGURE 3.25 Ionization smoke detector: (a) without smoke and (b) with smoke inside. Both parts (a) and (b) show an ionization chamber with a positively charged upper metallic plate connected to the positive terminal of a battery, and a negatively charged lower metallic plate joined with the negative terminal of the battery. The lower metallic plate has a hole at its center. Exactly under the hole, a radioactive material americium-241 is placed inside a proper shielding cover. The alpha particles emitted by Am-241 enter the ionization chamber through the hole in the lower metallic plate, and their collisions with the air molecules inside the chamber create positive and negative ions. These ions are attracted to the metallic plates carrying charges of opposite polarity leading to electron flow in the outside circuit which has a current meter to measure the current. (a) shows the situation when there is no fire or smoke. The air inside the chamber is clean and a large electric current is displayed by the current meter. (b) shows the condition when a fire breaks out and the ionization chamber is filled with smoke. The smoke particles adhere with the ions producing larger size charged particles, which are sluggish in movement. Hence, the resulting current in external circuit is lowered which raises an alarm warning the people about fire breakout.

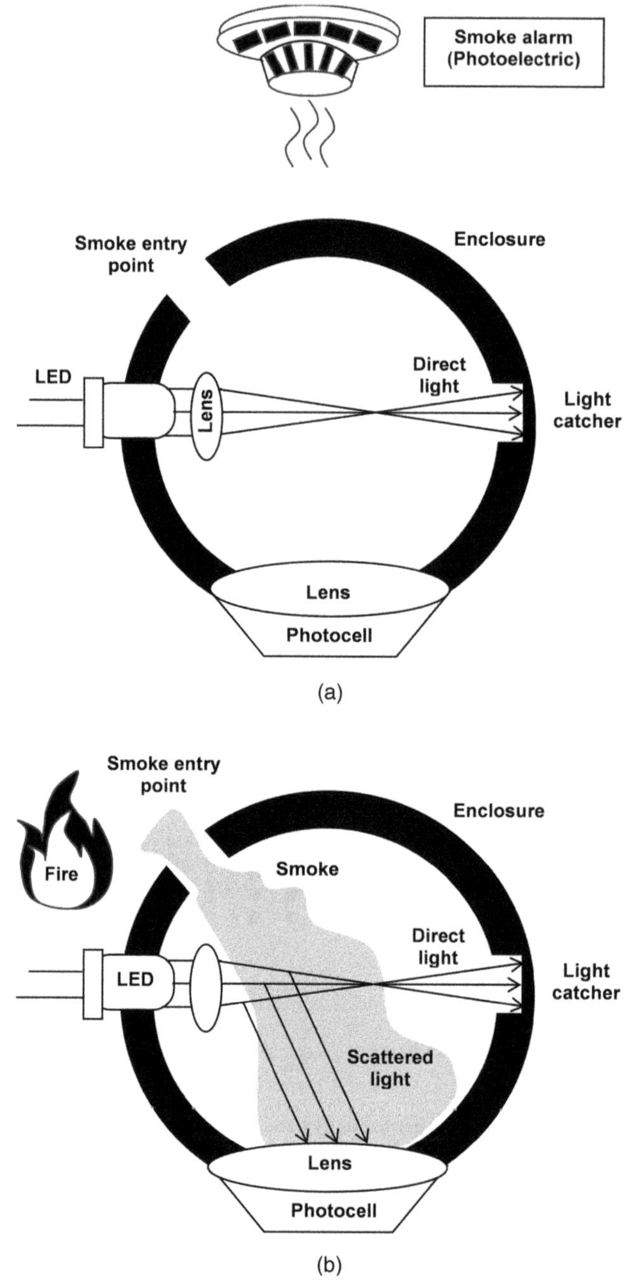

FIGURE 3.26 Photoelectric smoke detector: (a) in the absence of fire smoke and (b) with smoke entering the chamber when a fire accident occurs. Both parts (a) and (b) contain a spherical enclosure. On one side of the enclosure, an LED is mounted. The light from the LED is focused by a lens to fall on a mirror placed exactly on the opposite side of the enclosure, and facing the LED at normal incidence so that it is reflected back in in opposite direction toward the LED without any lateral deflection. A photocell is fixed on the enclosure surface in a perpendicular direction to the direction of light propagation from the LED so that no light falls on it and the current output is low. (a) is concerned with the status of the detector when there is no fire and hence no smoke inside the enclosure. The light from the LED moves in straight lines toward the mirror and is also reflected back straightway toward the LED. The photocell does not receive any light showing a low current value. (b) is related to the fire breakout scenario. Smoke enters the enclosure scattering the light from LED in all directions. The light incident on the photocell produces a photoelectric current. So, the output current of the smoke detector rises and the alarm rings, warning the residents that a fire has broken out.

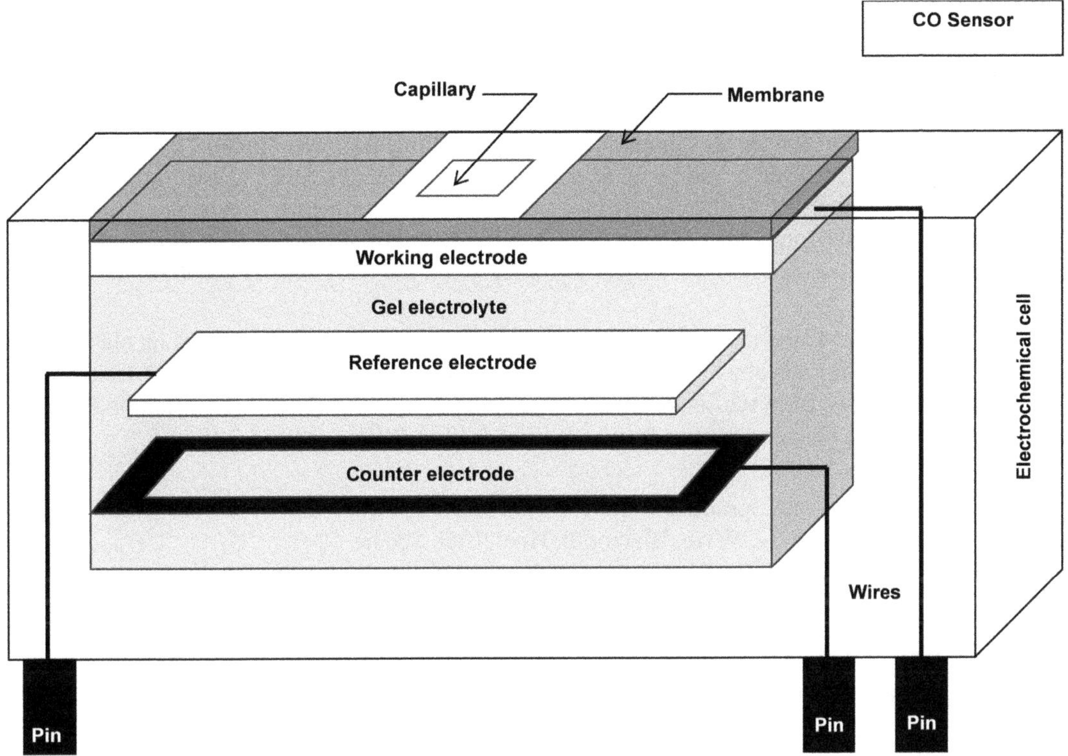

FIGURE 3.27 Carbon monoxide sensor working on the electrochemical principle. The diagram shows an electrochemical cell, the top surface of which is covered with a membrane and has a capillary allowing entry of carbon monoxide. Below the membrane is the working electrode of the three-electrode cell. A gel electrolyte is filled in the cell. Besides the working electrode, the reference and counter electrodes are also immersed in the gel. All the electrodes are connected with wires to pins at the lower surface of the cell for electrical connections.

platinum, graphite, or glassy carbon, and a reference electrode immersed in aqueous 38% sulfuric acid electrolyte or gel electrolyte containing 4.8% SiO_2, 38% sulfuric acid, and 0.005% polyvinyl alcohol (PVA) (Zhang et al. 2020). The gas enters the electrochemical cell from a water-proof gas-permeable membrane underneath the Pt working electrode. Platinum acts as a catalyst for the oxidation of CO to CO_2. So, when CO comes in contact with the platinum working electrode, an oxidation reaction occurs producing carbon dioxide, hydrogen ions, and electrons:

$$CO + H_2O \rightarrow CO_2 + 2H^+ + 2e^-, \text{ at the working electrode} \qquad (3.3)$$

The hydrogen ions and electrons move to the counter electrode where they react with oxygen, producing water.

$$\frac{1}{2}O_2 + 2H^+ + 2e^- \rightarrow H_2O, \text{ at the counter electrode} \qquad (3.4)$$

Adding together Eqs. (3.3) and (3.4), we get

$$CO + \frac{1}{2}O_2 \rightarrow CO_2, \text{ overall reaction} \qquad (3.5)$$

The electrons resulting from the reaction flow from the counter electrode through the external circuit leading to an electric current output proportional to CO concentration.

Other types of CO sensors include the following.

(i) The Biomimetic Sensor: It comprises a gel containing the chromophore cyclodextrins (cyclic oligosaccharides), and metal salts, whose color changes with CO adsorption, mimicking the binding of CO on hemoglobin, which darkens forming carboxyhemoglobin, and

(ii) The Microheater-Embedded Metal Oxide Sensor: It consists of a tin oxide film heated to 400°C. CO is detected through a decrease in the electrical resistance of the SnO_2 film.

The CO sensor sends the measured CO data to a microprocessor which triggers an alarm when the CO concentration rises above a threshold limit, alerting the homeowners of CO leakage and helping them to escape from the silent killer CO gas poisoning. A message is also sent on the owner's mobile about the dangerous situation.

Devices: CO sensor, microprocessor
Protocol: TCP/IP, Wi-Fi, Z-Wave, Bluetooth, HomeLink Secure
Further reading: Section 11.5.4

3.11 DOMESTIC WATER SUPPLY METERING AND WATER LEAKAGE PREVENTION

3.11.1 WATER CONSUMPTION BILLING: ELECTRONIC WATER METER

An electronic water meter uses an electromagnetic or ultrasonic water flow sensor and microcontroller for data processing and internet connectivity to provide real-time information to consumers about water usage and the same to city water supply companies for water supply management and water billing.

Electromagnetic flow sensors (magmeters), restricted in use to conductive liquids only, work on the principle that water is a conducting liquid. So, water passing through a magnetic field applied through current flowing in a set of coils mounted on the water pipe produces a voltage between two sensing electrodes placed in the flowing water according to Faraday's law of electromagnetic induction. The magnitude of this voltage depends on the velocity of water flow enabling the estimation of volumetric water flow based on dimensional parameters of the water pipe.

Ultrasonic flow meters can measure flow rates of conductive/non-conductive liquids, gases, and vapors. They work on the transit-time difference of ultrasonic waves in the direction of flow and against it. This difference arises because ultrasonic waves propagating in the direction of flow move faster than those moving against the flow direction. Therefore, the transit-time difference is proportional to the mean velocity of water flow. The components of this device are two ultrasonic transducers, an ultrasound reflector, and a water pipe. One ultrasonic transducer acts as a transmitter and the other ultrasonic transducer as a receiver. The device operates by alternately transmitting and receiving a burst of ultrasonic pulses from the upstream transducer to the downstream transducer and from the downstream transducer to the upstream transducer, and measuring their transit times for the two cases.

Accuracy of electromagnetic flow meters is ±0.5% while that of ultrasonic flow sensors is ±1–2%

Devices: water flow sensor, microcontroller, GSM modem
Protocols: cellular, Wi-Fi, LoRa
Further reading: Section 11.12

3.11.2 Alerting about Any Water Leakage: Water Leak Sensor

Direct contact water leak detectors or spot leak sensors are mounted at perceived leakage risk locations, e.g., faucets, washing machines, and water heaters. They consist of a set of electrodes or probes extending from a housing of adjustable height toward the floor. Slim-height versions are placed underneath the floor carpeting. In case of leakage, the water comes in contact with the electrodes, and the electrical circuit is completed. An audible alarm is triggered and a message is sent through a Wi-Fi router to the mobile phone. The water supply is cut off using valves placed at critical locations to prevent flooding.

Flow-type water leak sensors function by measuring water flow rate through pipes using upstream and downstream ultrasonic transducers and keeping a vigil on any abnormal or irregular water flow patterns.

Devices: Contact or ultrasonic water leak sensor
Protocols: Wi-Fi, LoRaWAN, NB-IoT (NarrowBand-Internet of Things deployed on 2G and
　4G), LTE Cat M1 [Long term evolution (4G), Category M1].

3.12　HELPING TO TRACK DOWN LOST ITEMS: SMART TAG

A microelectronic smart tag is a small-size integrated circuit device <10 mm^3 in volume, which is used to track an object, e.g., a key, purses, baggage, or person, attached to it by a key chain or strap (Andry et al. 2015). It consists of a processor, a non-volatile memory for storing identity, security, time, and location information; an internal or external source of power, and a transceiver for two-way communication with a reader outside the smart tag. A smart tag communicates through BLE and can be located using a mobile app.

Device: Smart tag IC chip
Protocol: Bluetooth low energy (BLE) with a range of up to 120 m

3.13　DETECTING DESK OCCUPANCY TO MONITOR EMPLOYEE ACTIVITY IN OFFICE: DESK OCCUPANCY SENSOR

A desk occupancy sensor is a tiny PIR device with an adhesive backing for easy peel-and-stick installation on the underside of an office desk or workstation. It has a narrow field of view so that only the movements of persons sitting at the desk are detected. Interferences from the movements of passers-by are minimized. The sensor response is wirelessly transmitted through SecureDataShot (SDS) protocol encrypted by Disruptive Technologies (DT) directly to the user instead of the gateway to update desk use without any risk of man-in-the-middle attacks. The data is applied for optimizing space utilization and activating services, e.g., if no movement is noticed for a preset period of time, the desk is considered empty, and HVAC/lights are switched off to save power.

Device: PIR sensor
Protocol: SecureDataShot (SDS)

3.14　COUNTING THE NUMBER OF VISITORS IN AN OFFICE: VISITOR COUNTER SENSOR

A bidirectional visitor counter sensor is used to count the number of people entering and exiting a room, and hence to determine the number of people in the room at a given instant. It uses a pair of infrared sensors. Each sensor consists of an infrared emitter and a photodiode or phototransistor detector. When the path of the infrared beam of a sensor is interrupted by a person walking past the sensor, a person is counted. Entry of a person is distinguished from exit by the order in which the sensors are triggered. If the first sensor is triggered earlier than the second sensor, the person is entering the room, and the people count is incremented. For a person exiting the room, the second sensor is triggered before the first sensor. Then the count is decremented. The count data is uploaded to an IoT platform such as Blynk for encrypted and secure messaging. Appliances like lights, fans, and coolers can be controlled using visitor counting.

Devices: IR sensors, Microcontroller, LCD display
Protocol: Blynk

3.15   INTELLIGENT HOME LIGHTING

3.15.1   SMART SWITCHES AND BULBS

It involves remotely controlling lights via wireless smartphone apps using Wi-Fi, ZigBee, Bluetooth, or other protocols. A smart switch is a programmable wireless switch enabling remote lighting control and automation from a computer or mobile app. It can be used to make any existing lighting fixture smart.

A color-changing smart bulb is an internet-connected LED bulb with capabilities for light customization, color adjustments, and scheduling from a remote location. The control operations include dimming/brightening light intensities to dynamically adapt light levels according to ambient light levels, color changing, and setting time schedules to turn the lights on and off. These operations can also be implemented by voice commands.

A color-changing bulb, also called an RGB bulb contains three separate LEDs within the same bulb, one red LED, one green LED, and one blue LED. These LEDs are made from different semiconductor materials. The red LED is made of aluminum gallium arsenide (AlGaN). The material used in green LED is gallium phosphide (GaP) while that in blue LED is indium gallium nitride (InGaN). A combination of colors and hues can be generated by changing the luminous intensities of the three LEDs giving rise to amazing visual effects suiting different celebratory and festive occasions.

3.15.2   OCCUPANCY SENSORS

Occupancy sensors keep a watch on whether a person is present in a room or if the room is vacant and turn on/off lights depending on the occupancy of rooms. In case there is no person, the light is automatically switched off saving the power consumption. When a person is away from home, the lights can be randomly turned on to deter any thief from venturing to get inside. The outdoor lights can be turned on if a person comes near the apartment, as done in Section 3.8.3, by detecting human presence by a motion sensor.

Three types of indoor occupancy sensors are used depending on the mounting height, the distance of detection, the space covered, and the sensitivity and power consumption specifications, viz., the PIR, microwave, and ultrasonic (Figure 3.28).

3.15.2.1   PIR Occupancy Sensor

It measures the changes in heat differential in the space under examination (Figure 3.28(a)). It contains a pyroelectric module containing pyroelectric elements sensitive to 7–14 μm radiation connected in opposite polarization and housed in a package covered with a plastic window of Fresnel lens with segmented parabolic mirrors to concentrate IR and filter the incoming radiation for avoiding false positives.

3.15.2.2 Microwave Occupancy Sensor

A monostatic sensor consists of a microwave transceiver (transmitter plus receiver with a common antenna), Figure 3.28(b). Microwaves are emitted in the examined space and the reflected microwaves are observed to ascertain any Doppler shift in frequency from which occupancy is determined. A bistatic sensor having a separate transmitter and receiver is used for probing larger volumes.

In both cases, the occupancy is revealed from a comparison of transmitted and reflected wave patterns. If the reflected wave pattern differs from the incident wave pattern and is constantly changing, the space is considered to be occupied. If the reflected wave pattern remains the same over a preset time, the space is vacant.

(a)

FIGURE 3.28 Occupancy sensors: (a) PIR and (b) Microwave/Ultrasonic. (a) and (b): two persons are sitting in a room. (a): A PIR sensor is installed in the ceiling. The IR radiation emitted by the room occupants is falling on the PIR sensor indicating their presence.

(*Continued*)

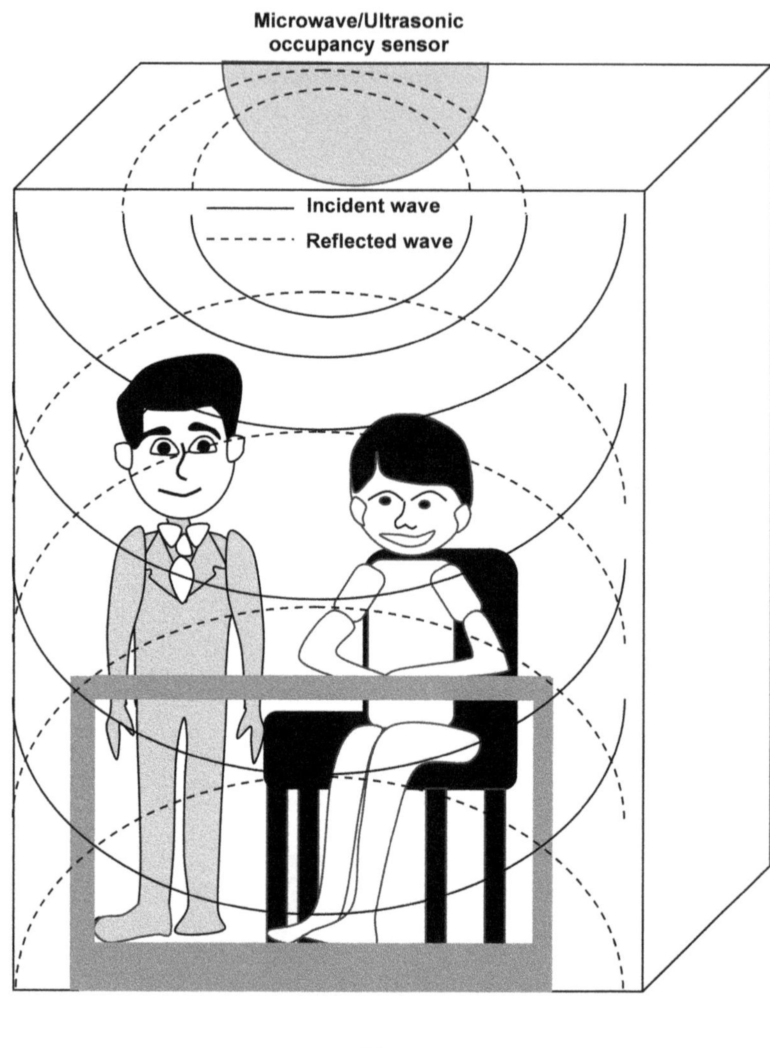

(b)

FIGURE 3.28 (CONTINUED) (b): Microwave/ultrasonic waves from a microwave transceiver or a single-package ultrasonic transmitter/receiver are sent by the respective transmitters and the received echo signals are examined for frequency changes to detect movements.

3.15.2.3 Ultrasonic Occupancy Sensor

It resembles the microwave sensor in principle because it establishes occupancy from the Doppler shift in frequency of ultrasound reflected from moving persons in the room (Figure 3.28(b)). The sensor is an ultrasound transceiver. The ultrasound is continuously emitted by a quartz crystal in the 25–40 kHz range and the returning ultrasound is detected. The transmitted and reflected ultrasonic wave patterns are compared to judge whether the reflected pattern is same as transmitted or shows changes over a given time period. Vacancy and occupancy are judged from similarities or differences noticed between them.

3.15.2.4 Comparison of Occupancy Sensors

The PIR sensor detects changes in temperature. It can detect along the line-of-sight direction, has 90° coverage, is more sensitive at low temperatures, less sensitive at high temperatures, and can be insensitive when moving directly toward the sensor. The microwave/ultrasonic sensors detect frequency changes in microwaves/ultrasound reflected from a person. They have 360° coverage and have consistent performance for all temperatures. Since these sensors continuously emit waves, they consume more power than PIR sensors. They are not limited to line-of-sight operation and can detect motion behind partitions or bookcases.

Smart street lighting will be discussed in Section 11.16.

3.16 CONCLUDING REMARKS AND PREPARING FOR THE UPCOMING CHAPTER

We could touch upon some of the many sensors that are being used to make our home life convenient and office work more efficient. More developments are expected to come in the forthcoming years. Only the sky is the limit. Apart from homes and offices, factories are the places where a large number of people toil hard for long hours to bring various industrial goods to markets. This is the topic of the ensuing chapter.

REFERENCES

Andry P. S., M. Denneau, J. U. Knickerbocker and R. L. Wisnieff 2015 Microelectronics smart tags, *US20170124446A1 International Business Machines Corp*, 2015-10-29.

Bielecki Z., K. Achtenberg, M. Kopytko, J. Mikołajczyk, J. Wojtas and A. Rogalski 2022 Review of photodetectors characterization methods, *Bulletin of the Polish Academy of Sciences Technical Sciences*, 70(2): pp. 1–26. Article number: e140534.

Bigas M., E. Cabruja, J. Forest and J. Salvi 2006 Review of CMOS image sensors, *Microelectronics Journal*, 37: 433–451.

Boudaden J., M. Steinmaßl, H.-E. Endres, A. Drost, I. Eisele, C. Kutter and P. Müller-Buschbaum 2018 Polyimide-based capacitive humidity sensor, *Sensors*, 18(5): 1516, pp. 1–15.

Canon: Ultra-high 250MP Resolution 2022 https://asia.canon/en/campaign/cmos-image-sensors/ultra-high-250mp/ultra-high-250mp-resolution Accessed on 15th December 2022.

DeLuca M. 2018 Onsemi High Resolution and Image Uniformity from the KAI-43140 CCD Image Sensor https://www.onsemi.com/company/news-media/blog/iot/High-Resolution-Image-Uniformity-KAI-43140-CCD-Image-Sensor?utm_source=pr&utm_medium=press_release&utm_campaign=KAI-43140&utm_content=link-kai-43140-blog

Joshi S. R., B. Kim, S.-K. Kim, W. Song, K. Park, G.-H. Kim and H. Shin 2020 Low-cost and fast-response resistive humidity sensor comprising biopolymer-derived carbon thin film and carbon microelectrodes, *Journal of The Electrochemical Society*, 167(14): 147511.

Kuzubasoglu B. A. 2022 Recent Studies on the humidity sensor: A mini review, *ACS Applied Electronic Materials*, 4: 4797–4807.

Lee C.-Y. and G.-B. Lee 2005 Humidity sensors: A review, *Sensor Letters*, 3: 1–14.

Lesser M. 2014 3-Charge coupled device (CCD) image sensors, In: Durini D. (Ed.), *High Performance Silicon Imaging*, Woodhead Publishing, pp. 78–97.

Loeppert P. V. and S. B. Lee 2004 A commercialized MEMS microphone for high-volume consumer electronics, *Journal of the Acoustical Society of America*, 116: 2510.

Loeppert P. V. and S. B. Lee 2006 SiSonic™ – The first commercialized MEMS microphone, Solid-State Sensors, Actuators, and Microsystems Workshop, Hilton Head Island, South Carolina, June 4–8, 2006.

Okcan B. and T. Akin 2004 A thermal conductivity-based humidity sensor in a standard CMOS process, *17th IEEE International Conference on Micro Electro Mechanical Systems*, 25–29 January, Maastricht, Netherlands, Maastricht MEMS 2004 Technical Digest, IEEE, NY, pp. 552–555,

RadhaKrishna M.V.V., M.V. Govindh, and P. K. Veni 2021 A review on image processing sensor, *Journal of Physics: Conference Series*, 1714:012055, 9pp.

Tanaka J., H. Imamoto, T. Seki, and M. Oba 2014 Low power wireless human detector utilizing thermopile infrared array sensor, SENSORS, 2014 IEEE, 2–5 November, Valencia, Spain, IEEE, USA, pp. 462–465.

Torkkeli A., O. Rusanen, J. Saarilahti, H. Seppa, H. Sipola and J. Hietanen 2000 Capacitive microphone with low-stress polysilicon membrane and high-stress polysilicon backplate, *Sensors and Actuators*, 85: 116–123.

Zawawi S. A., A. A. Hamzah, B. Y. Majlis and F. Mohd-Yasin 2020 A Review of MEMS capacitive microphones, *Micromachines*, 11, 484, pp. 1–26.

Zhang L., Y. Jin, L. Lin, J. Li and Y. Du 2008 The comparison of CCD and CMOS image sensors, *2008 International Conference on Optical Instruments and Technology: Advanced Sensor Technologies and Applications*, In: Wang A., Y. B. Liao, A. G. Song, Y. Ishii and X. Fan, Proceedings of SPIE, Vol. 7157, pp. 71570T-1 to 71570T-5 © 2009 SPIE.

Zhang Y., D. Cheng, Z. Wu, F. Li, F. Fang, and Z. Zhan 2020 Fumed SiO_2-H_2SO_4-PVA gel electrolyte co electrochemical gas sensor, *Chemosensors*, 8, 109, pp. 1–14.

Hur S., Y. Jung, Y. H. Lee and J.-H. Kwak 2012 Two-chip MEMS capacitive microphone with CMOS analog amplifier, Sensors, 28–31 October, Taipei, Taiwan, IEEE, pp. 1–4.

Jung Y., Y. H. Lee, J. Kwak, and S. Hur 2012 Development of capacitive type MEMS microphone with simulation and electro-mechanical characterization, XX IMEKO World Congress Metrology for Green Growth, September –14, Busan, Republic of Korea, p. 4.

4 IoT Sensors for Factory Automation-I

Tracking Components, Gaps, Alignments; Sticker Labelling; Proximity and Position Detection

4.1 INTRODUCTION

The factory is a set of buildings where various machines, small and big, are operated by trained skilled workers to manufacture or assemble goods for everyday use or specialized purposes, at a large scale for selling in markets, where they can be accessed by the society as a whole at affordable prices. The mass production of items done at factories significantly lowers costs than if done in small quantities. For the smooth functioning of factories, sensors are used in almost all machines. These sensors are assigned two types of duties, supervisory inspection and process control. In this chapter, we shall look at sensors for supervisory inspection. These include sensors for tracking components, those for detecting whether the machine components and job items are properly aligned to reduce errors due to gaps as well sensors to detect vibrations of machine parts and angular tilts, wherever intolerable.

4.2 SMART FACTORY

A smart factory, sometimes called a digital factory or an intelligent factory is a hyperflexible manufacturing facility built up of an integrated system of physical and computational components to monitor the complete production process of an item from the raw material supply chain to machines, tools, and individual operators on the shop floor. It is an autonomous system combining artificial intelligence and cloud computing with sensors and the Industrial Internet of Things to observe, control, and analyze physical processes and utilize the collected data for self-learning, self-optimization, self-adaptation, and self-correction.

The concept of a smart factory, an offshoot of the fourth industrial revolution or Industry 4.0, is still in its infancy and is a dream of the future. The fourth industrial revolution or Industry 4.0 commenced in the year 2000 and is still in progress. It is striving toward the goal of making production nearly autonomous. It has four key components: cyber-physical systems integrating networking, computing, and physical processes; artificial intelligence and machine learning; two mainstream technologies, cloud computing and Big Data (exabytes and beyond); and the Internet of Things. The third industrial revolution or Industry 3.0, is known as the Digital Revolution (1970–early 2000). It was ushered through the use of semiconductors, personal/mainframe computers, mobile, internet, robotics, and automation. The second industrial revolution or Industry 2.0 (1871–1914) was initiated through the use of electricity and oil power and assembly line or mass production. The first industrial revolution or Industry 1.0 (1760–1840) began through the use of steam and water power and mechanization of production.

DOI: 10.1201/9781003374442-4

4.3 AUTOMATED TRACKING OF MANUFACTURING COMPONENTS AND MATERIALS IN THE WAREHOUSE: RFID, NFC TAGS

4.3.1 RFID TAGGING

4.3.1.1 Need for RFID Deployment

A car manufacturing company uses a large number of components of different models. Tracking of these components is done by an automated technique using an RFID system. RFID is the short form for radio frequency identification. Deployment of RFID sensors in a factory assists in tracking the movements of materials and products, thereby improving asset tracking, and efficiency of manufacturing together with cost reduction.

4.3.1.2 Operating Frequencies and Ranges of RFID

  (i) Low-Frequency: 125–134 kHz, up to 0.5 m
 (ii) High-Frequency: 13.56 MHz, up to 1 m
(iii) Ultrahigh-Frequency: 433 MHz, up to 1 m and 860–956 MHz, 1–10 m
 (iv) Microwave Frequency: 2.45–5.8 GHz, 100 m

4.3.1.3 Components of the RFID System

The RFID system consists of two components (Figure 4.1(a)).

  (i) The RFID Tag or RFID Sensor: It contains an integrated circuit chip in which the tag's alphanumerical ID code (letters + numbers) used as its unique identifier, location, and other information are stored, along with a miniature antenna to receive and transmit radio signals. It acts as a transponder (transmitter + responder), a device which on receiving a signal automatically sends a response signal.

A passive tag has no battery and is powered by the RFID reader. An active tag is battery-powered. The passive tag is smaller and cheaper than the active one. Its range is up to 12 m. The active tag has a much larger range of more than 100 m.

The tag is attached to an object whose movement is to be tracked, e.g., an RFID tag is attached to an automobile part to track its progress through the assembly line.

 (ii) The RFID Reader or Interrogator: It consists of a scanning antenna and a transceiver (transmitter + receiver in one package). It operates at a specified radio frequency and protocol to transmit and receive data from the RFID tag within its read range. It is either fixed or mobile.

4.3.1.4 Working of RFID, and Its Advantages over Barcodes

The RFID reader sends an interrogator electromagnetic pulse to activate the RFID tag. The activated tag sends its identification number back to the reader which is used for tracking it.

The RFID works like a barcode with the difference that the two primary limitations of barcode: line-of-sight scanning, and scanning one item at a time, are removed. A large number of RFID tags can be simultaneously read remotely without line-of-sight access. Hence, RFID is often referred to as a smart barcode.

4.3.1.5 RFID Tag and Reader Collision Avoidance

In an environment containing a high density of RFID tags and readers, e.g., a warehouse, the tags and readers interfere with signals of each other, reducing the efficiency and reliability of the RFID system. This situation called collision is tackled by the application of tag and reader collision avoidance techniques.

Tag collision avoidance techniques are of two types: deterministic and probabilistic (Figure 4.1(b)). The deterministic tag collision avoidance technique uses an algorithm such as BTS, subdividing the tags into two subsets based on their IDs, and then making recursive searches until each tag is recognized. The

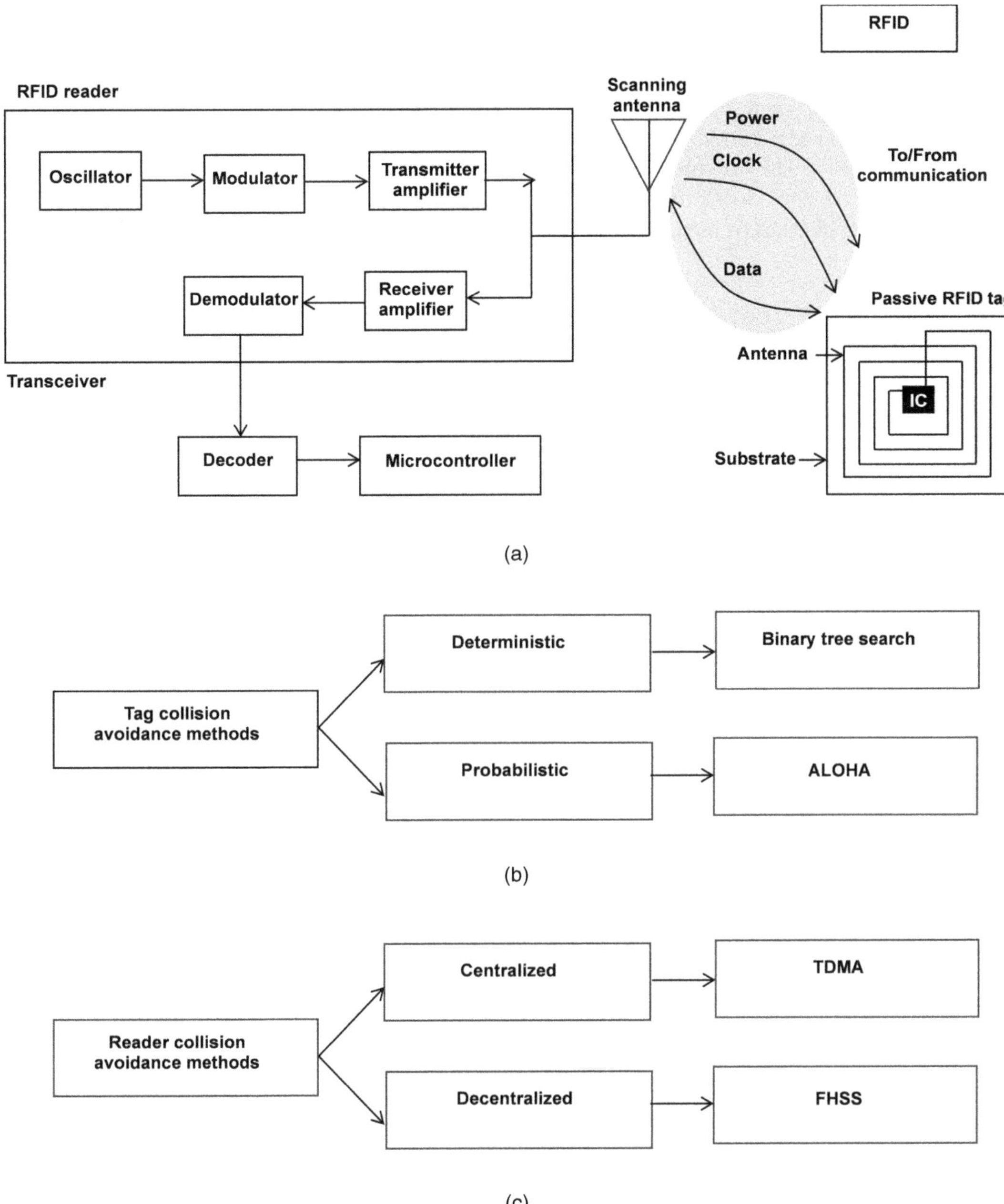

FIGURE 4.1 Radio-frequency identification system: (a) components; (b) and (c) collision avoidance techniques: (b) tag and (c) reader. Part (a) shows the RFID reader consisting of the transceiver and the scanning antenna. The transceiver block comprises a transmitter (made of an oscillator, a modulator, and an amplifier) and a receiver (an amplifier and a demodulator). The demodulated signal is decoded and fed to a microcontroller. The passive RFID tag is an integrated circuit chip bonded to a substrate and surrounded by a spiral antenna. Communication between the tag and the reader takes place through clock, power, and data signals. Part (b) classifies tag collision avoidance techniques into two types of algorithms: deterministic (e.g., binary tree search [BTS] algorithm) and probabilistic (e.g., ALOHA algorithm). Part (c) classifies reader collision avoidance techniques into two types of digital modulation techniques: centralized (e.g., TDMA) and decentralized (e.g., FHSS).

probabilistic tag collision avoidance technique uses an algorithm such as ALOHA in which each tag responds to the reader at once or after a random interval.

Reader collision avoidance techniques are also of two kinds: centralized and decentralized (Figure 4.1(c)). The centralized reader collision avoidance technique uses a method like time-division multiple-access (TDMA) assigning a separate time slot to each reader to prevent overlapping of their signals. The decentralized reader collision avoidance technique applies a method like frequency-hopping spread-spectrum (FHSS) to randomly change between different frequencies, thus reducing the chances of interference.

Devices: RFID tag, RFID reader

Protocol: ISO 18000-63 ((International Organization for Standardization-63), a standard defining the parameters for air interface communication at 860 MHz to 960 MHz

4.3.2 NFC Tagging

4.3.2.1 The NFC Tag

An NFC (near-field communication) tag is used for tracking product location, validating product information for its authentication, detecting any tampering, and for supplying additional data to the manufacturing information system. The NFC tag also helps in asset maintenance, e.g., scanning the NFC tag on a machine provides its specifications, instruction manuals, and repair history.

Generally comprising a 25 mm^2 IC with a 30 mm × 40 mm antenna and an adhesive surface, an NFC tag works by inductive coupling similar to a transformer (Rohde & Schwarz 1MA182_5e n.d.). It functions through electromagnetic induction between two loop antennas at 13.56 MHz frequency and 106–424 kbit-s^{-1} data rate.

4.3.2.2 Operating Modes of the NFC Tag

NFC has three operating modes (Figure 4.2).

(i) Peer-to-Peer Mode: In this mode (Figure 4.2(a)), data exchange takes place between two NFC-enabled devices. Hence, there is a two-way data transmission. Both devices switch between an active role when sending data and a passive role when receiving data. Both the reader and the tag have batteries and are self-powered. So, both devices produce electromagnetic fields to exchange data with each other.

(ii) Reader/Writer Mode (Active Mode): This mode (Figure 4.2(b)), involves reading of the data on the tag/writing of data on the tag, by bringing the NFC reader close to the tag. Only the reader is powered producing an electromagnetic field. The tag receiving this field draws power from it through magnetic induction. Upon reaching the tag, the reader energizes it resulting in the extraction and transference of data in encrypted form from the tag to the reader. As a reader can connect to one tag at a time, the possibility of unintended transactions is minimized. It involves only one-way power and clock transmission.

(iii) Card Emulation Mode (Passive Mode): In this mode (Figure 4.2(c)), the NFC device emulates the function of a card, e.g., a smart contactless credit or debit card for payments, conforming to a legacy standard. In this mode also, there is one-way power and clock transmission.

4.3.2.3 Use of the NFC Tag

When an NFC tag is attached to a component of an equipment being made in a factory, simply tapping with the NFC reader provides all details about its exact location and manufacturing status without manual intervention, thus greatly improving the efficiency of production. The NFC tag is also used to restrict access to sophisticated machinery for trained workers only. It can provide secure pairing with Wi-Fi and Bluetooth, enabling on-site data collection from sensors and configuring them to produce alerts on reaching threshold values, etc.

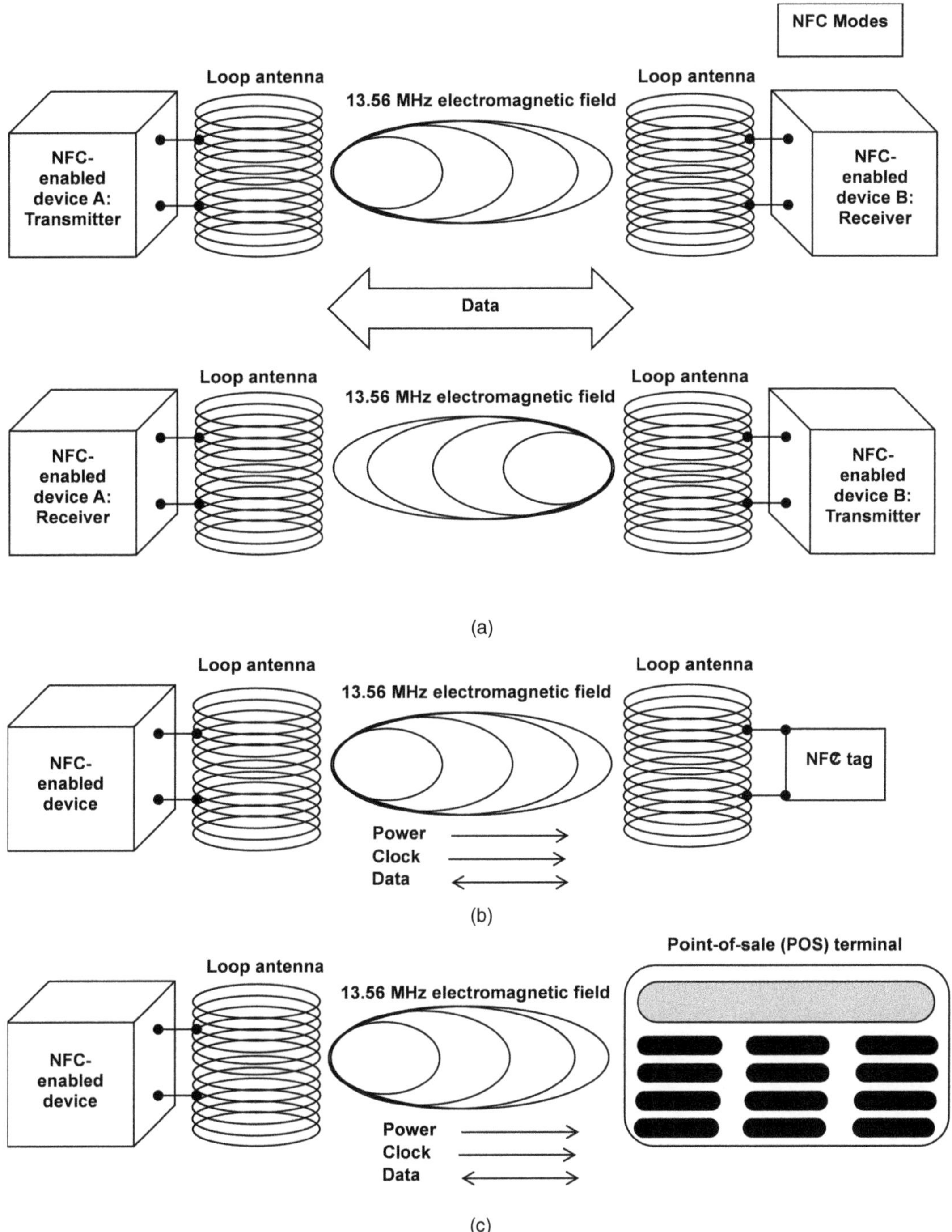

FIGURE 4.2 Operating modes of near-field communication system: (a) peer-to-peer, (b) reader/writer, and (c) card emulation. In part (a), two NFC-enabled devices A and B with their own loop antennas exchange data via a 13.56 MHz electromagnetic field, with either A as a transmitter and B as a receiver or in opposite roles, so that two-way data communication is achieved. In part (b), an NFC-enabled device with a loop antenna sends power and clock signals to an NFC tag with a loop antenna through an electromagnetic field at 13.56 MHz and reads/writes data on the tag. In part (c), an NFC-enabled device with a loop antenna acts as a contactless card, sending power and clock signals to a point-of-sale terminal through an electromagnetic field at 13.56 MHz, and interacting with it as a reader/writer.

4.3.2.4 NFC vs. RFID Tag

While RFID, the parent technology of NFC can work from a long distance, the maximum range of an NFC reader is 10 cm allowing short-range or close proximity communication only. But unlike RFID which is capable of one-way communication, NFC technology can perform two-way or bi-directional communication. Moreover, NFC tags can store more data than RFID tags which hold mainly information for identification. Passive RFID tags have usually 64 bits to 1 kB of non-volatile memory. The storage space in NFC tags is generally 64–888 bytes but some types provide 8 kB, or 32 kB memory. The NFC tags can be read by smartphones in contrast to the expensive readers needed for RFID tags. The NFC technology is used for secure contactless communications while RFID is used for asset tracking and inventory management.

Devices: NFC tag, NFC reader

Protocol: ISO/IEC 14443 (International Organization for Standardization/International Electro-technical Commission 14443), a proximity or contactless smart card communication standard

4.4 TRACKING GAPS AND ALIGNMENTS BETWEEN MOVING COMPONENTS IN PRODUCTION LINES: MAGNETIC FIELD SENSORS

Magnetic sensors are used for the detection of magnetic fields, which can be generated by permanent magnets, electromagnets, or magnetic materials. They can discriminate and localize ferromagnetic and conductive parts of machines, and provide precise inexpensive, contactless linear and angular measurements.

Three types of magnetic field sensors are commonly used:

(i) Hall-effect sensor
(ii) Anisotropic magnetoresistance sensor
(iii) Magnetic tunnel junction sensor

4.4.1 THE HALL-EFFECT SENSOR

When a current-carrying semiconductor is placed in a magnetic field acting in a direction perpendicular to the direction of the flow of current, a force acts on the moving charge carriers constituting the current in the transverse direction, pushing the positively charged carriers toward one side of the semiconductor and negatively charged carriers toward the other side, thereby generating a potential difference between the two sides of the semiconductor. The direction of this potential difference acting sideways, called the Hall voltage is perpendicular to the direction of current flow and the direction of the magnetic field. The magnitude of Hall voltage is proportional to the strength of the magnetic field. Hall effect is the creation of a potential difference in a current-carrying semiconductor in the presence of a magnetic field. It has its origin in the Lorentz force, the force experienced by a charged particle moving with velocity $\mathbf{v}$ through an electric field $\mathbf{E}$ and magnetic field $\mathbf{B}$.

Hall effect forms the basis of the Hall-effect sensor used for the determination of both the strength and direction of a magnetic field (Khan et al. 2021). This sensor is essentially a thin rectangular strip of a P-type semiconductor material, e.g., GaAs, InAs, InSb (Figure 4.3). The output voltage from a Hall-effect sensor is small even when a large magnetic field is applied, making measurements difficult in a noisy environment. The output voltage is increased with the help of a DC amplifier on the output side and a voltage regulator on the input side. The modified sensor can be used over a wide range of power supply voltages. A continuous analog output is obtained. The signal varies linearly with the magnetic flux density. Another way of using a Hall-effect sensor is as a switch with a digital output. A Schmidt trigger threshold detector with built-in hysteresis is added to achieve a non-oscillatory changeover from the OFF to the ON state when the magnetic flux density exceeds a prescribed limit. Thus, the Hall-effect sensor is used as an analog or digital device.

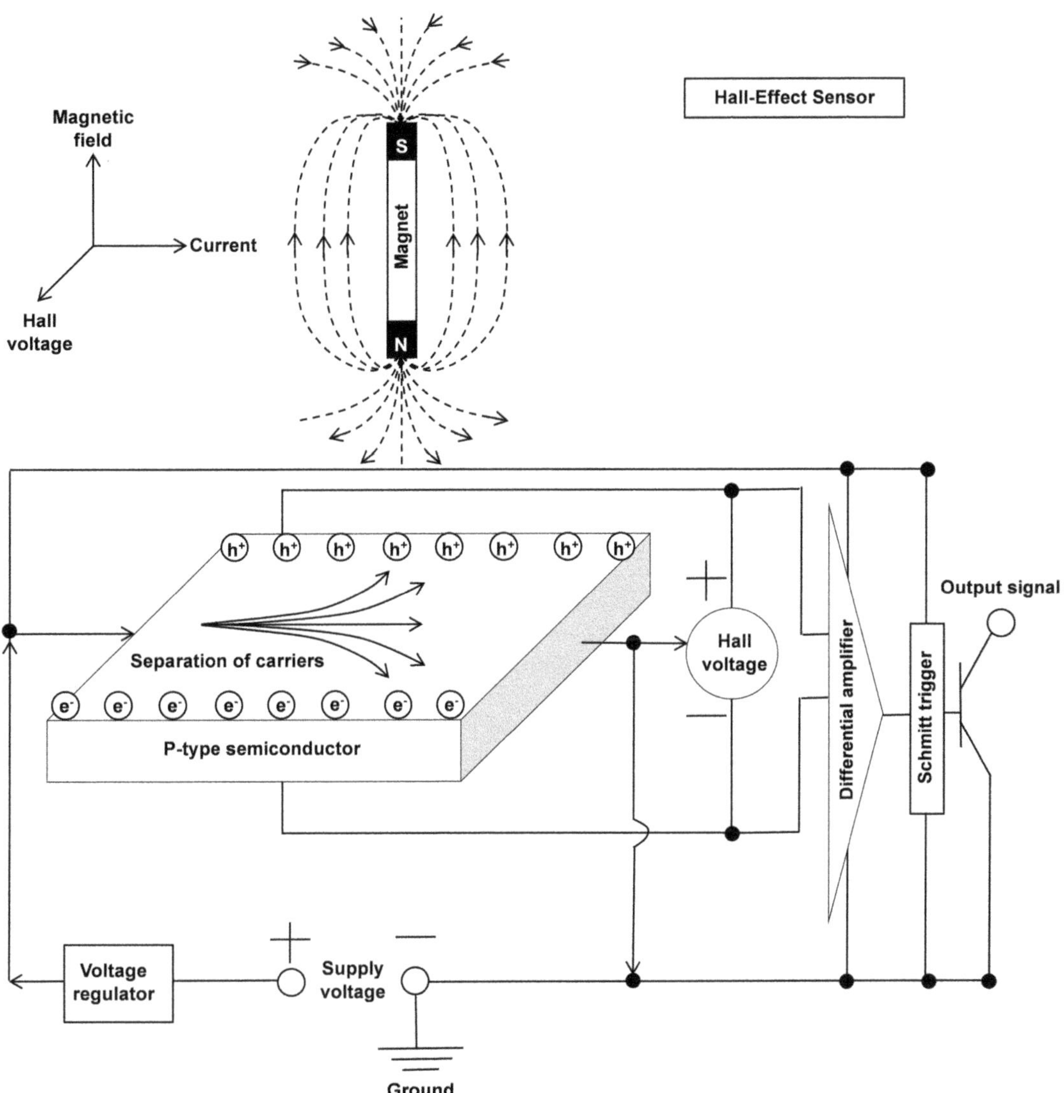

FIGURE 4.3 Hall-effect sensor. The supply voltage is regulated by a voltage regulator, and the regulated voltage is applied across the terminals of a P-type semiconductor bar. A magnetic field is applied by bringing a bar magnet near the P-type semiconductor. The north and south poles of the magnet are marked, and the magnetic lines of force are shown coming out from the north pole of the magnet and entering its south pole. The current flowing in the P-type bar interacts with the magnetic field, separating the electrons e⁻ and holes h⁺ toward opposite sides of the semiconductor. The resulting Hall voltage is fed to a differential amplifier, followed by a Schmitt trigger circuit to actuate a bipolar transistor switch. The directions of the flowing current, the applied magnetic field, and the Hall voltage point toward the three mutually perpendicular axes, as shown in the inset diagram.

4.4.2 Anisotropic Magnetoresistance Sensor

Magnetoresistance is the change in the electrical resistance of a material in an applied electric field in the presence of a simultaneously acting magnetic field (Yang and Zhang 2021). Anisotropic magnetoresistance is a phenomenon typical of ferromagnetic materials in which the magnetoresistance depends on the angle between the direction of electric current and the direction of magnetization. When these directions are parallel to each other, the resistance remains unchanged. This is the maximum value of the resistance. But when these directions are mutually perpendicular, resistance decreases to a minimum value. Ferromagnetic metals and alloys show an AMR effect ($\Delta R/R$) ~2%.

In an AMR sensor (Figure 4.4(a)), four resistors R_1, R_2, R_3, and R_4 made of wires of a ferromagnetic material are connected in a bridge configuration. These resistors are arranged in two groups (R_1, R_4) and (R_2, R_3). In each group of resistors, the current flows in the same direction. Moreover, the direction of current flow in one group is perpendicular to that in the other group.

When the sensor is placed in a magnetic field (Figure 4.4(b)), there is no change in the values of resistors (R_2, R_3) in which the current flow direction is parallel to the magnetic field direction. But the values of resistors (R_1, R_4) decrease to minimum values. This happens because the direction of current flow in these resistors is perpendicular to the direction of the magnetic field. Because of the changes in values of resistors (R_1, R_4), the balance of the bridge is disturbed and an output voltage is observed between points A and B, the midpoints of the two arms of the bridge. This output voltage is fed to an IC comparator, and when it becomes larger than the adjusted threshold setting of the sensor, the ON/OFF digital switching is actuated.

4.4.3 Magnetic Tunnel Junction (MTJ) Sensor

The MTJ is a nanostructured device. The simple construction of an MTJ device is shown in Figure 4.5(a). It consists of a thin insulating tunneling layer called the barrier layer separating two ferromagnetic layers. There are two contact layers, one at the top and the other below the stack of layers. These layers are deposited on a substrate. The ferromagnetic layer above the barrier layer is known as the free layer. It is named so because the magnetization direction of this layer is free to change. The ferromagnetic layer below the barrier layer is the reference layer. The magnetization direction of this layer is not free to change and, therefore, it serves as a reference for the free layer.

The MTJ works on the tunnel magnetoresistance (TMR) effect. The TMR is a magnetoresistive effect. By applying a magnetic field, the magnetization direction of the free layer can be switched between two states, from being parallel to that in the reference layer to becoming antiparallel to it, and vice versa. In the parallel state, the electrons are likely to tunnel from one ferromagnetic layer to the other ferromagnetic layer through the barrier layer. This is a low-resistance state of the junction. Conversely, in the antiparallel state, the tunneling is prohibited so that it is a high-resistance state. Thus, the junction can be switched from a low-resistance state to a high-resistance state, and back from a high-resistance to a low-resistance state by application of a magnetic field. Moreover, the electrical resistance of the junction varies with the relative orientations of the magnetization directions in the two ferromagnetic layers depending on the applied magnetic field. The TMR ratio of the junction is defined by the following equation:

$$\text{TMR} = \frac{\text{Resistance in antiparallel state} - \text{Resistance in parallel state}}{\text{Resistance in parallel state}}$$

$$= \frac{R_{\text{Antiparallel}} - R_{\text{Parallel}}}{R_{\text{Parallel}}} \tag{4.1}$$

In practice, the basic structure is modified into that shown in Figure 4.5(b), e.g., from bottom upward: Si/SiO$_2$ (substrate), Ta/Ru (bottom contact), PtMn or IrMn (antiferromagnetic layer), CoFe/Ru/CoFeB (reference layer), MgO (barrier layer), CoFeB/Ru/NiFe (free layer), and Ta/Ru (top contact) (Ranjbar et al. 2020). The thicknesses of all the layers are a few nm or less. DC sputtering is used to deposit the metallic films while the MgO layer is formed by RF sputtering.

Ta/Au films form the two contact layers, one on each side of the stack. An antiferromagnetic layer such as PtMn or IrMn is inserted below the reference layer to fix the magnetization direction in it. It is called the pinning layer and the reference layer is the pinned layer. Further, the reference layer is often in the form of a synthetic antiferromagnet (SAF) structure comprising two magnetic layers antiferromagnetically coupled through a metal spacer film, e.g., CoFe/Ru/CoFeB with Ru spacer

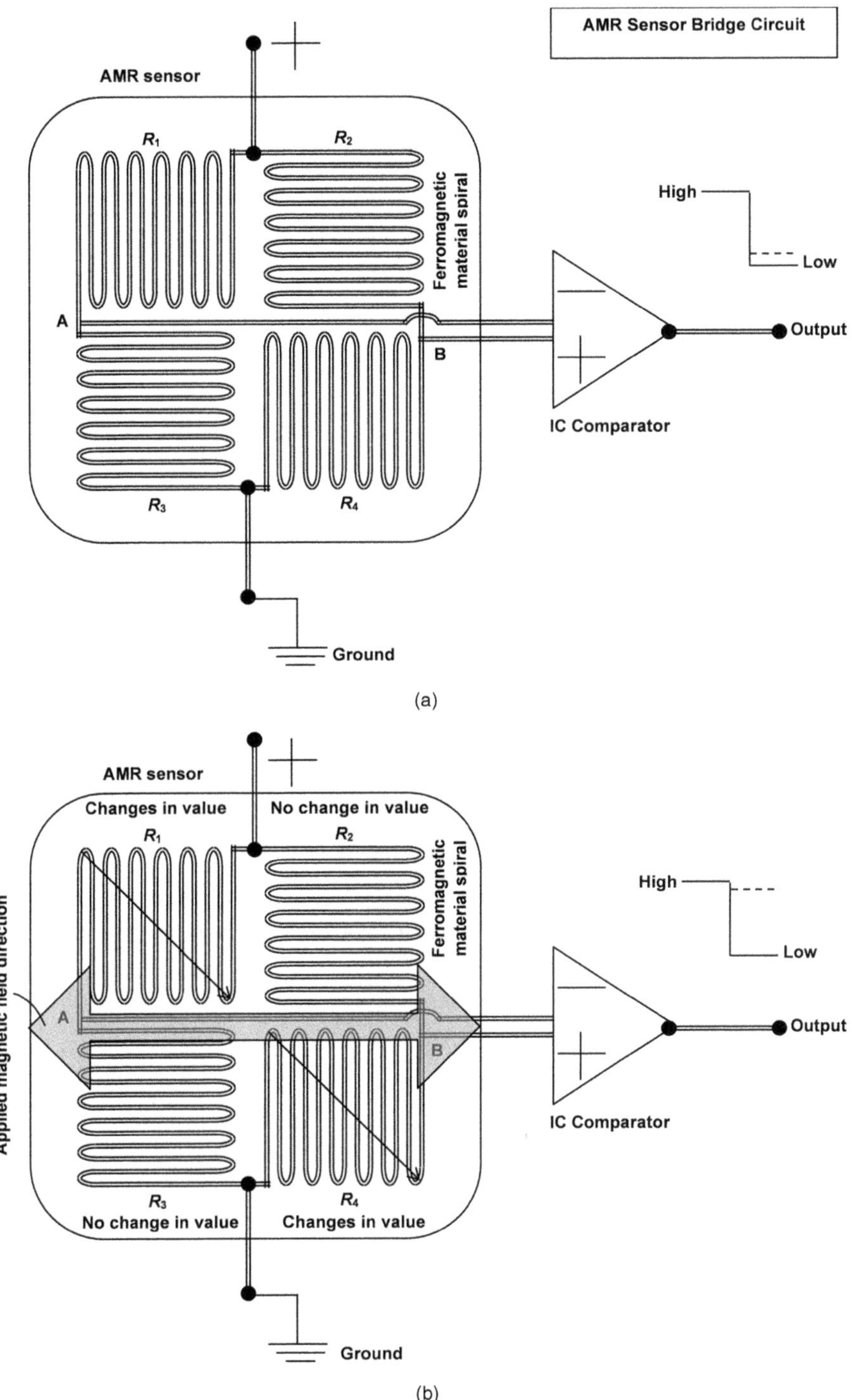

FIGURE 4.4 Two conditions of an anisotropic magnetoresistance sensor bridge circuit: (a) without magnetic field, and (b) under the influence of a magnetic field applied along the direction AB. Four spiral-shaped resistors R_1, R_2, R_3, and R_4 are made from a ferromagnetic material Current between points A and B goes to an IC comparator giving high-low pulse output. A magnetic field is applied along the direction AB. The current in resistors R_1 and R_4 flowing perpendicularly to the magnetic field changes but that in resistors R_2 and R_3 flowing parallel to the field remains unaffected. The bridge balance is disturbed, and the output climbs to a value greater than a preset threshold, causing digital switching.

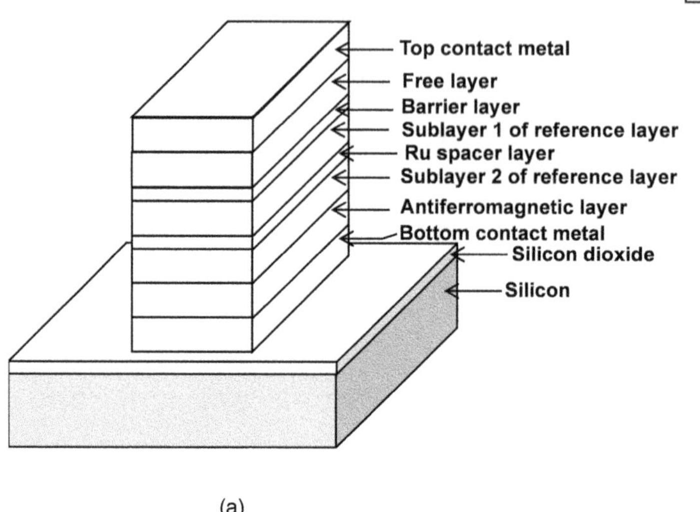

(a)

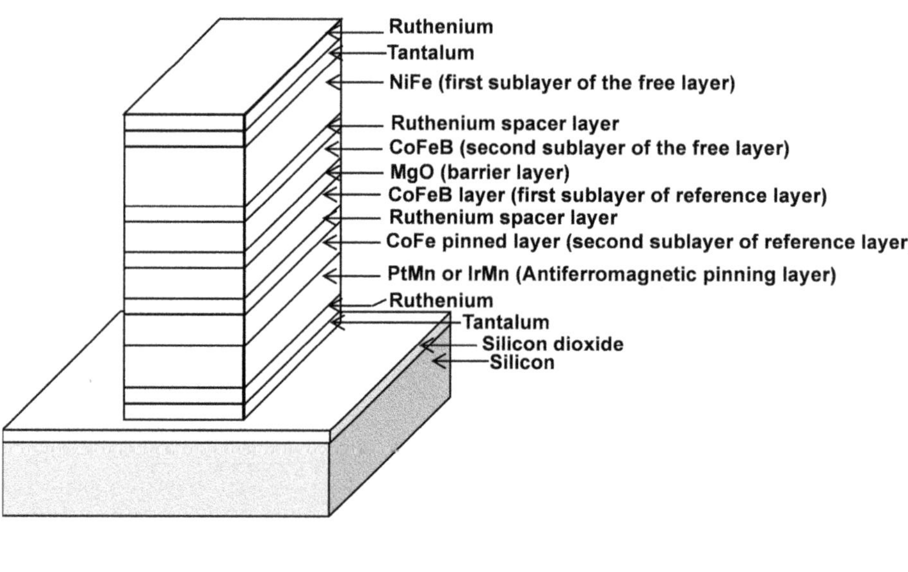

(b)

FIGURE 4.5 Magnetic tunnel junction sensor: (a) basic structure and (b) high-temperature magnetic tunnel junction magnetometer. In both parts (a) and (b), the substrate is SiO_2/Si. The sequence of layers in part (a) from bottom upward is bottom contact metal, antiferromagnetic layer, sublayer 2 of the reference layer, Ru spacer layer, sublayer 1 of the reference layer, barrier layer, free layer, and top contact metal. The layered structure of part (b) from bottom upward is: tantalum, ruthenium, PtMn or IrMn antiferromagnetic pinning layer, CoFe pinned layer, Ru spacer layer, CoFeB layer, MgO barrier layer, CoFeB layer, Ru spacer layer, NiFe layer, tantalum, and ruthenium.

layer. The SAF structure is extremely flexible toward manipulation of magnetic configurations by external excitation such as by an electric field. It provides fast switching with low switching current and hence reduced power consumption. Besides displaying good thermal stability, it is resilient to disturbances from outside magnetic fields (Huang et al. 2022). The free layer is also a composite layer such as a parallelly coupled hard CoFeB/Ru/soft NiFe synthetic structure (Ono et al. 2010). The soft layer aids the hard layer in switching at a lower critical current density (Yen et al. 2008).

The performance of the MTJ device is tested from room temperature to 250°C. For devices using the PtMn pinning layer, the TMR ratio decreases from 133% to 80% while for devices with IrMn pinning layer, it decreases from 146% to 50%. A constant sensitivity of 4–5% $(Oe)^{-1}$ is maintained up to 250°C by the PtMn pinning layer sensors. However, a sharp decline in sensitivity is observed for sensors with an IrMn layer (Ranjbar et al. 2020).

4.5 AUTOMATIC APPLICATION OF STICKER LABELS BY MOVING MAGNET-ATTACHED MACHINE COMPONENTS, AND LABELING MACHINES: HALL-EFFECT AND LABEL SENSORS

4.5.1 HALL-EFFECT SENSORS

In the industrial automation era, proper labeling of products is mandatory. A magnet is attached to a moving machine component. When the component comes near a Hall-effect sensor, its presence is detected and the data is sent to the machine controller. Applying this method to a rotating bottle labeler, the position of the label spindle is accurately controlled and adjusted based on the information received from the Hall-effect sensor. Then the label is automatically applied on the bottle.

4.5.2 LABEL SENSORS

The labeling of a product or container is done using a pre-printed label with pressure-sensitive adhesive (PSA) on its backside and an automatic labeling machine (Figure 4.6). The items to be labelled such as bottles are placed on a conveyor belt moving at a constant speed. They are separated by a fixed distance by a mechanical fixture. The machine has a label drive wheel, a labeling wheel, and a reel. The speed of the labeling wheel is the same as that of the conveyor. When a certain position is reached, the label drive wheel undergoes acceleration to match the speed of the conveyor. To ensure accuracy in the placement of the label, a registration mark on the label is read by a sensor on the labeling machine, and the label drive wheel adjusts its position to avoid errors. The labeling wheel presses the label on the bottle as it passes through it. After label attachment, the bottle travels on the conveyor belt to the ensuing processes.

The label and gap positions are detected by sensors for correct label dispensing. It looks through the backing paper to find liner regions or gaps between labels to cease the dispensing process before the label is wholly removed from the backing layer. There are four types of label sensors, as explained in the subsections below.

4.5.2.1 Optical Sensor

An IR source is placed under the web or roll and an IR detector is placed above it. Due to the difference in their opacities, IR is blocked by the label material but can penetrate through the gap or liner between the labels.

4.5.2.2 Ultrasonic Sensor

They work by measuring the thickness of the web or roll. One ultrasonic transducer (transmitter) is placed below the web and another ultrasonic transducer (receiver) above it. Ultrasound energy passing through the gap is larger than that crossing the label material, giving an indication of their relative thicknesses.

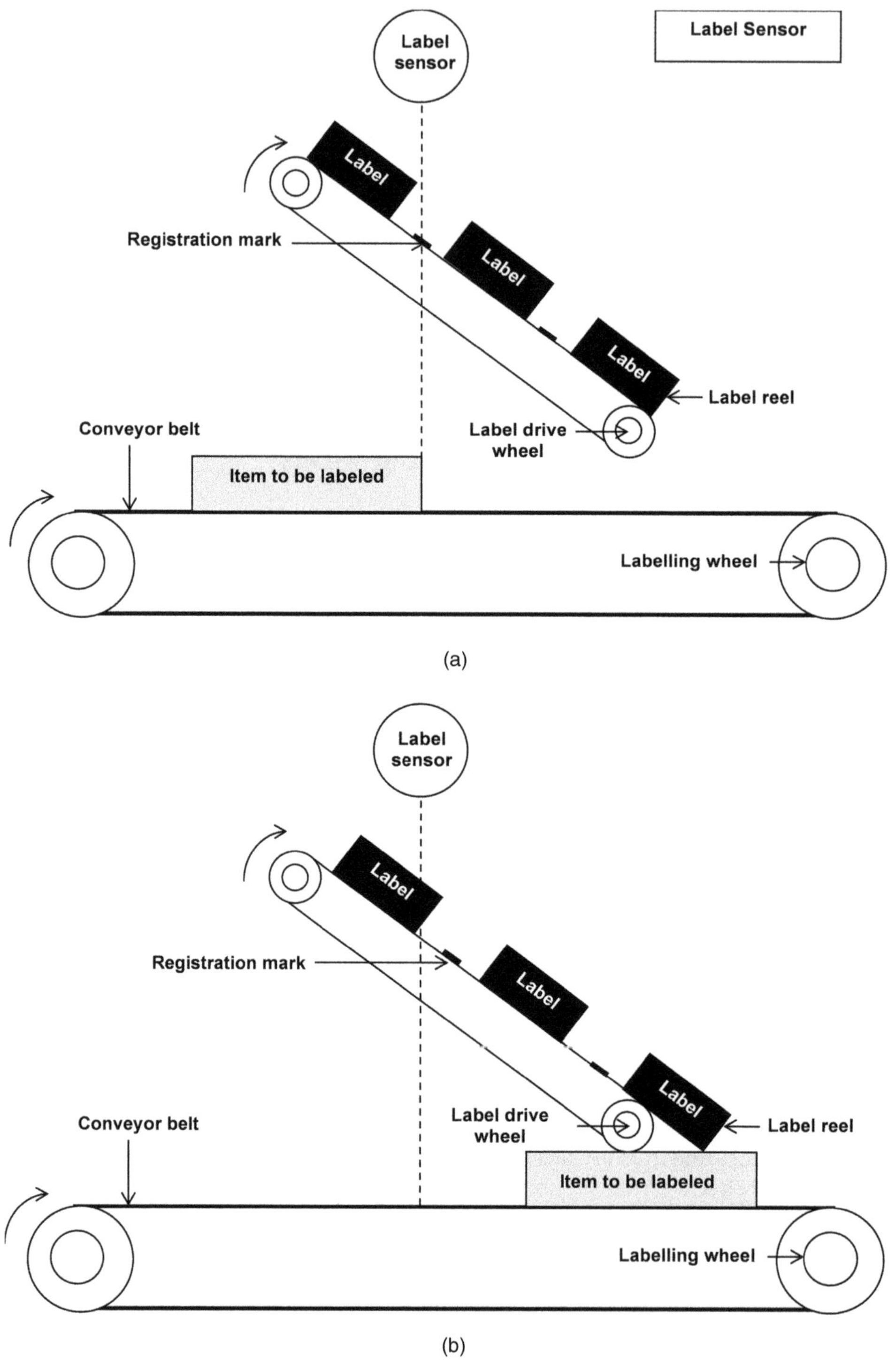

FIGURE 4.6 Label sensor and labeling machine: (a) detection of the item to be labeled by the label sensor and (b) label fixation. Both parts (a) and (b) show a conveyor belt moving on a labeling wheel, a label reel moving on a label drive wheel, and a label sensor watching a registration mark between two adjacent labels. In part (a), the label sensor sees the registration mark aligned with the item to be labeled and gives its output signal. In part (b), the label sensor output adjusts the speed of the conveyor belt to ensure that the item comes under the label at the correct instant, and the label is attached to the item when they come together.

4.5.2.3 Single-Ended Capacitive Sensor

A single sensing element is used to sense the thickness of the web, and it is adjusted such that the liner thickness is less than the trigger point of the sensor but (liner + label) thickness exceeds it. Hence, the sensor is triggered as soon as it finds a thickness > the set trigger-point thickness.

4.5.2.4 Differential Capacitive Sensor

Two sensing elements are used, and their outputs are subtracted. Hence, the device has an output only when one sensing element is above the label and the other above the liner or gap. It triggers only when the outputs of the two sensing elements are different. The advantage derived from the differential approach is that any variations in distance between the sensing elements and baseplate arising from temperature changes or those induced by vibrations are unable to affect the performance of the device.

4.6 PICKING AND PLACEMENT OF COMPONENTS USING MACHINES: PROXIMITY SENSORS

The control system of a pick-and-place machine used for picking a component from one location and placing it at a pre-destined location uses proximity sensors to inform the tool about its distance from the component when picking it and its distance from the destination during placement. The proximity sensor can be looked upon as a non-contact position sensor which detects the presence of an object without making any physical contact and triggers an action based on its distance relative to the object. Proximity sensors are of five types, as given in the subsections below.

4.6.1 CAPACITIVE PROXIMITY SENSOR

A capacitor is a device consisting of two conductive plates separated by a dielectric. Its capacitance, the parameter indicating its ability to store charge depends on the relative permittivity or dielectric constant of the dielectric layer and its thickness, i.e., the distance between the plates besides the area of the conductive plates of the device.

A capacitive proximity sensor (Figure 4.7) works on the mutual capacitance between or among objects close to each other. In a capacitive proximity sensor, one plate is the sensor surface, the other plate is the object to be detected, henceforth called the target, with an imaginary ground, and the dielectric is the air between them. So, it is a capacitor formed between the target and the sensor surface with air as a dielectric. When the target comes close to the sensor surface, the dielectric thickness decreases, hence capacitance increases. When it moves far away, the reverse happens. Hence, the distance of the target from the sensor surface is determined by measuring the capacitance of the sensor. The point at which the sensor can detect an approaching target is the operating point. The release point is the point at which the sensor returns to its initial state when the target moves away.

Capacitive proximity sensors can detect metallic or non-metallic objects in solid or liquid form. Conductive materials form a good capacitor with the sensor surface. Hence, they are more easily detected. Capacitive proximity sensors have a large detection range of ~3–60 mm.

4.6.2 INDUCTIVE PROXIMITY SENSOR

This sensor is based on the phenomena of electromagnetic induction governed by Faraday's and Lenz's laws:

(i) The induction of currents, circulating in loops, in a metallic conductor known as eddy currents in a time-varying magnetic field (Faraday's law), and

(ii) the opposition of the originating magnetic field by the induced eddy currents in accordance with Lenz's law which states that the current due to induced EMF flows in such a direction that its magnetic field acts counter to the root cause which led to its creation.

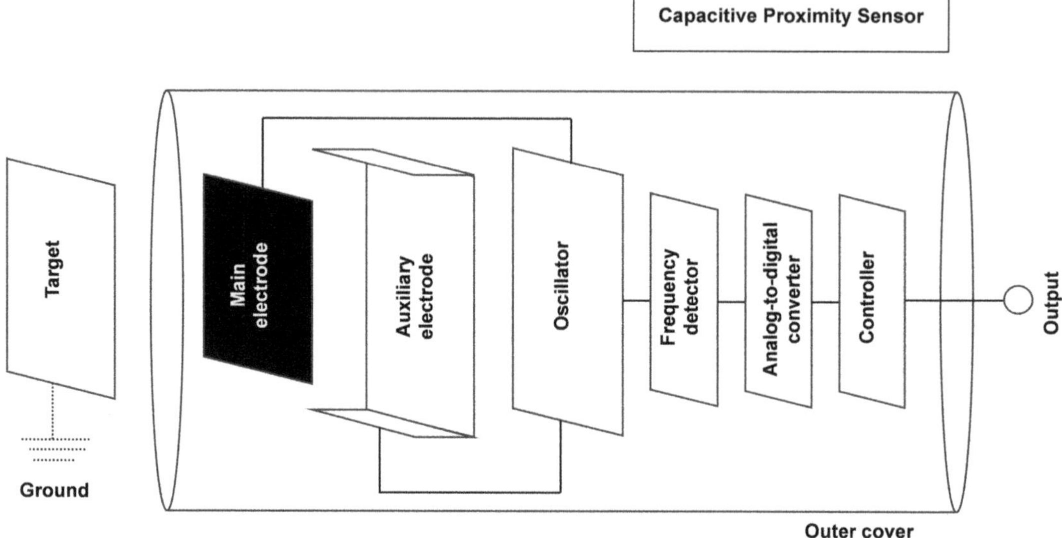

FIGURE 4.7 Capacitive proximity sensor consisting of the target, the main and auxiliary electrodes; the oscillator, the frequency detector, and A-to-D converter circuits. The digitized signal is fed to the controller to deliver the output signal.

The sensor is restricted in use to detection of metallic objects only. It consists of a coil wound around a ferromagnetic core (Figure 4.8). An oscillator circuit supplies an alternating current to the coil to produce a time-varying magnetic field in its surrounding region. When a metallic target comes near the coil, eddy currents are induced in it. These eddy currents flow in such a direction that their magnetic field is opposite to the time-varying magnetic field of the oscillator. The extent of opposition

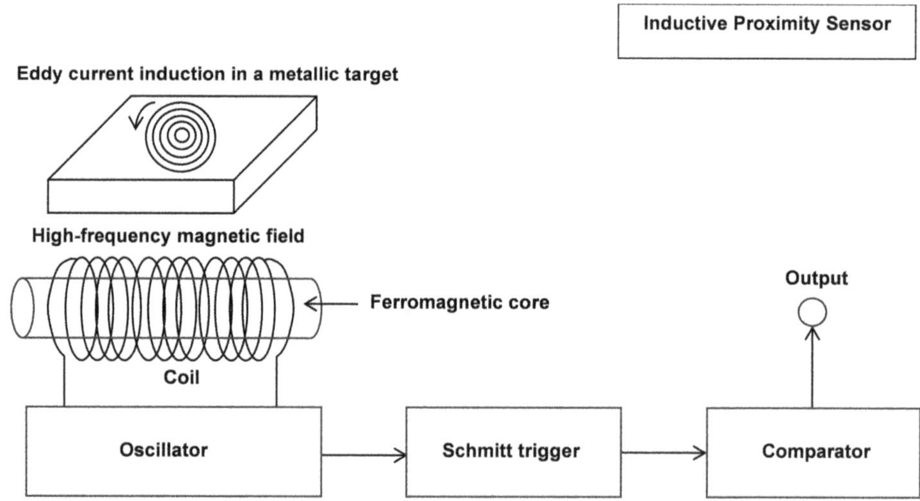

FIGURE 4.8 Inductive proximity sensor. Alternating current from an oscillator circuit is applied to a coil of conducting wire wound around a ferromagnetic core. The high-frequency electromagnetic field produced by the coil induces eddy currents in a metallic target brought near the coil. The magnetic field produced by the eddy currents opposes the magnetic field of the coil responsible for eddy current induction, thereby decreasing the voltage across the oscillator. The decreased voltage acts on the Schmitt trigger. The signal from the Schmidt trigger flows to a comparator circuit from which the output signal is obtained.

and hence the decrease in the originating magnetic field increases as the target moves toward the coil, providing an estimation of the distance between the target and the coil. As this distance decreases, the voltage across the oscillator decreases. A Schmitt trigger (detector circuit) and an amplifier (comparator) monitor the amplitude of the oscillator, and at certain predetermined levels, the output is switched on or off.

4.6.3 Magnetic Proximity Sensor

A common magnetic proximity sensor is an easily operated small-size, maintenance-free device called the reed switch (Figure 4.9(a)). It is an electromechanical switch consisting of two NiFe ferromagnetic blades coated with Au-Rh-Ir films at contact areas for protection from dust, oxidization, and corrosion to ensure its long-life operation. The blades are hermetically sealed inside a matching temperature-coefficient glass envelope filled with an inert gas. When a permanent magnet or electromagnetic coil is brought near the reed switch (Figure 4.9(b)), the individual reeds are magnetized. As the magnetic interaction becomes stronger, the blades bend and touch each other to make contact. When the magnet is withdrawn, the magnetic field produced by the blades is dissipated, and the blades move apart. Thus, the reed proximity sensor is actuated by the presence or absence of a magnet rather than mechanically.

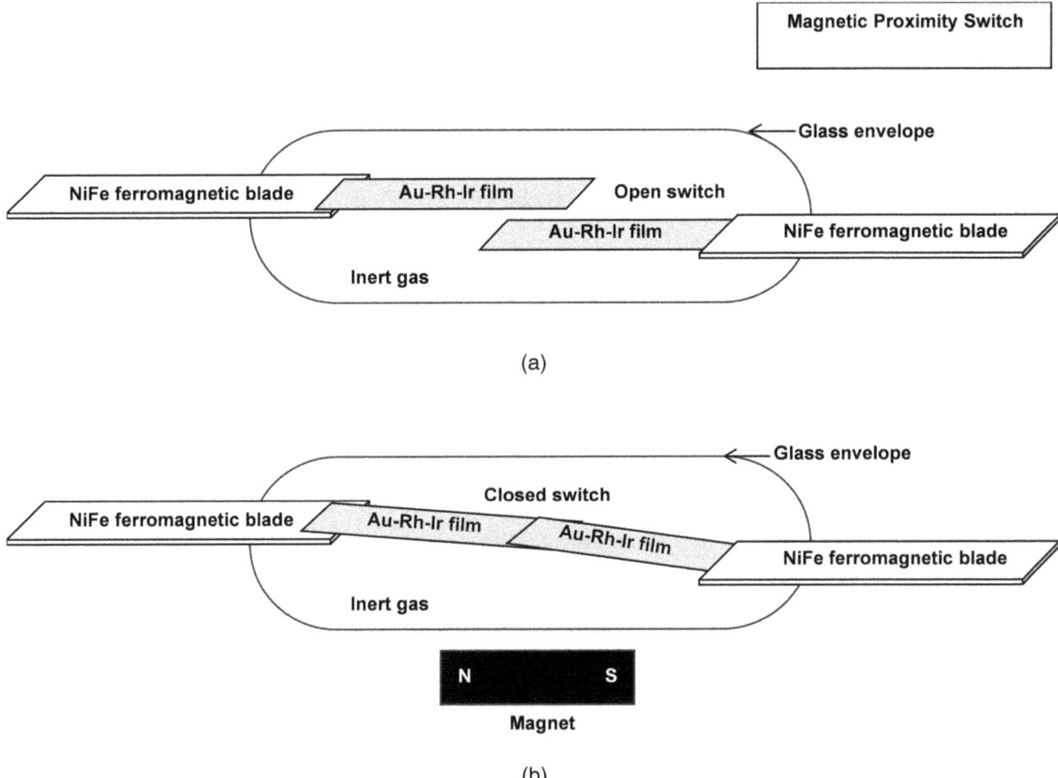

FIGURE 4.9 Magnetic proximity switch in (a) open and (b) closed conditions. Parts (a) and (b) show the magnetic proximity switch consisting of a glass envelope filled with an inert gas in which two electrodes are sealed. These electrodes are made in the form of NiFe ferromagnetic blades and have Au-Rh-Ir film contacts. Part (a): The contacts are far apart when there is no magnet near the switch. As a result, the circuit is open and the switch is OFF. Part (b): A magnet is brought near the switch. Under the influence of the magnetic field, the contacts come together and touch each other to close the circuit and turn the switch ON.

In comparison to the alternative solid-state Hall-effect sensor, the reed switches offer less resistance, provide superior electrical isolation, and can work with a diversity of voltages, loads, and frequencies to cater to various needs.

4.6.4　Optical and Ultrasonic Proximity Sensors

This sensor can work during the day or night time. Unlike the ultrasonic sensor, it is able to detect soft objects also. It consists of an infrared emitter and an infrared receiver (Figure 4.10). Infrared light from the IR emitter strikes the target and is reflected back to the IR receiver as well as transmitted through the target. There are four common modes of operation.

4.6.4.1　Diffuse-Reflective Sensor

Diffuse reflection is the reflection of light from a rough surface such as a wall in several directions. It takes place when the imperfections or irregularities on the reflecting surface are larger than the wavelength of light. Contrarily, in regular or specular reflection, all the reflected rays are in the same direction.

The IR light emitter and receiver are at the same place and frequently in the same package forming a compact assembly (Figure 4.10(a)). The light beam traveling from the emitter to the target is reflected back by the target to the receiver. There is no separate reflector. The total time of travel of transmitted and reflected light is used to find the distance between the emitter and the target.

4.6.4.2　Retro-Reflector Sensor

As in the case of a diffuse-reflective sensor, the IR light emitter and receiver are in the same housing, but a retroreflector is placed to reflect the incident light beam (Figure 4.10(b)). A target entering into the region between the reflector and the receiver obstructs the reflected light beam. This interruption diffuses the light beam, and the receiver detects the change. Thus, the target is detected from the interruption of reflected light.

4.6.4.3　Through-Beam Sensor

It is based on the rectilinear propagation of light in a homogeneous medium. The IR light emitter and receiver are installed facing each other (Figure 4.10(c)). When a target comes between the emitter and receiver obstructing the path of the IR light beam, the output of the receiver changes enabling the detection of the target.

4.6.4.4　Background Suppression Sensor

It works by triangulation. The emitter sends IR light through a lens in a straight line toward the target (Figure 4.10(d)). The target sends back the reflected light to the receiver lens at some angle. This angle depends on the distance between the emitter and the target. The angle is larger when the target is near the emitter than when it is far away.

4.6.4.5　Ultrasonic Proximity Sensor

The ultrasonic waves emitted by the ultrasonic transducer are reflected from the target (Figure 4.11). The time taken for the two-way journey of ultrasound from the transducer to the target and back to the target is used to calculate the transducer-to-target distance. Typical detection range of these sensors is 50–1000 mm. Flat smooth surfaces reflect ultrasound well but soft materials such as foam and those with curved surfaces are not good reflectors. The sensor is able to detect solid, liquid, and granular materials but does not work in vacuum.

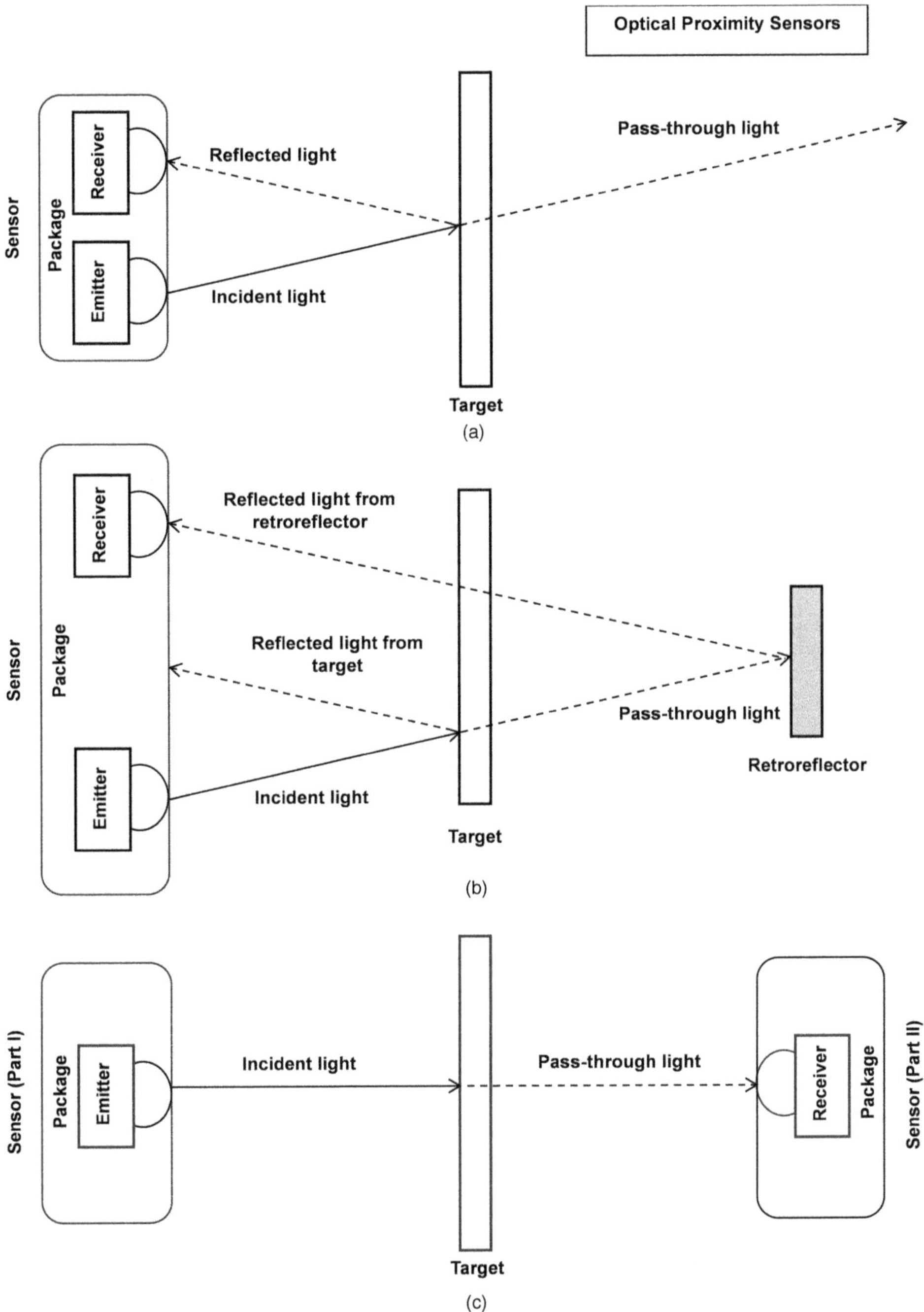

FIGURE 4.10 Optical proximity sensors of various types: (a) diffuse reflective, (b) retroreflector, (c) through-beam, and (d) background suppression. Part (a): The IR emitter and receiver of the sensor are accommodated in one package. The incident light falls from the emitter on the target and the reflected light from the target falls on the receiver. The pass-through light moves to the opposite side of the target. Part (b): The IR emitter and receiver of the sensor are located in one package. The retro-reflector is placed behind the target. The diagram shows the directions of incident light from the emitter, the reflected light from the target to the receiver, and the reflected light obtained from pass-through light crossing the target and falling on the retro-reflector. Part (c): The IR emitter is placed on one side and in front of the target while the IR receiver is placed on the opposite side and behind the target. The IR emitter and IR receiver are kept in separate packages. Pass-through light crossing the target falls on the receiver.

(Continued)

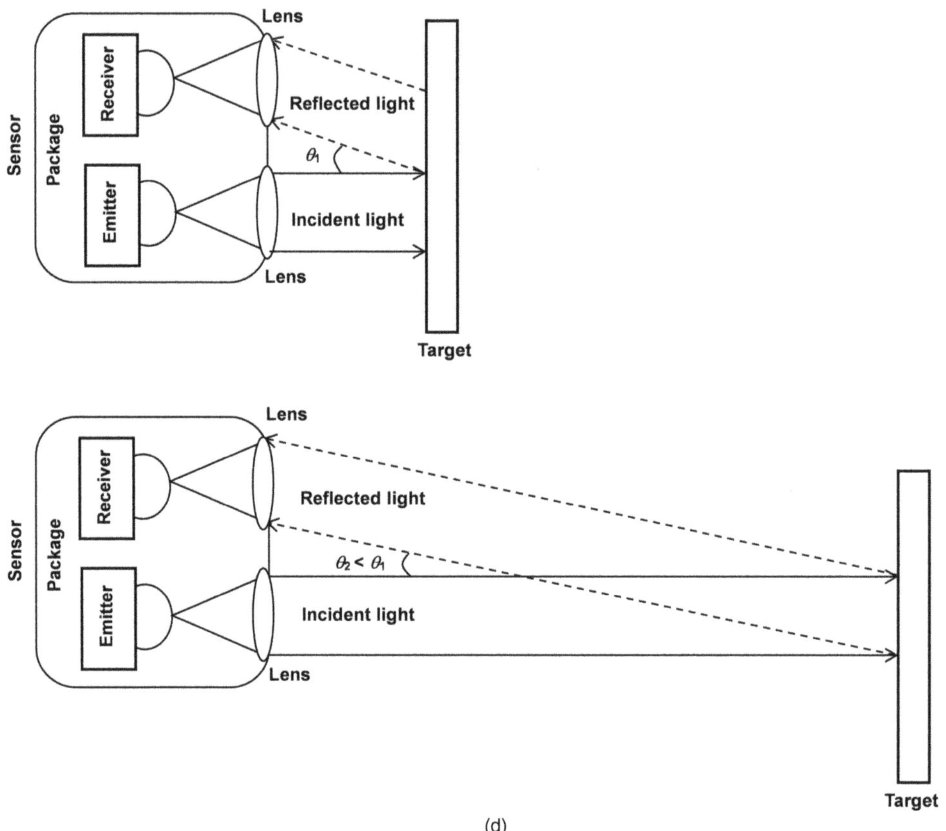

FIGURE 4.10 (CONTINUED)   Part (d): Top drawing: IR emitter and receiver are placed in one package. One lens located in front of the emitter collimates the light and this light is incident on the target. The light reflected from the target is focused by a lens on the receiver. The angle between the incident and reflected light is θ_1. Bottom drawing: The sensor has identical parts to that in the top drawing but the difference is that the target is at a larger distance from the sensor than in the top drawing. The angle between the incident and reflected light is θ_2, which is smaller than the angle θ_1 in the top drawing.

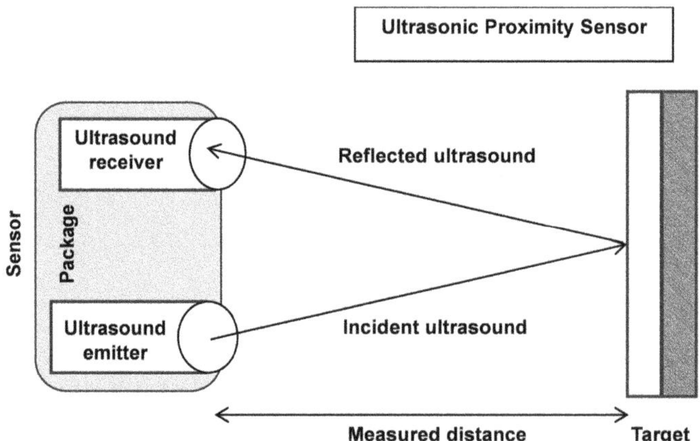

FIGURE 4.11   Ultrasonic proximity sensor. The ultrasound emitter and receiver are enclosed in one package. The incident ultrasound beam from the emitter falls on the target and the reflected ultrasound beam from the target falls on the receiver. The distance between the sensor and the target is calculated by measuring the time taken by ultrasound waves to travel from the ultrasound emitter to the receiver using the known velocity of ultrasound propagation.

4.7 DETERMINING THE ACCURATE POSITION OF MACHINE COMPONENT OR WORKPIECE FOR AUTOMATICALLY RESPONDING OR ACTIVATING ALARMS: POSITION SENSORS

4.7.1 DIFFERENCE BETWEEN THE POSITION SENSOR AND THE PROXIMITY SENSOR, AND ITS USE IN THE INDUSTRIAL PROCESS

As distinguished from a proximity sensor which prompts an action based on the distance of the target, the position sensor, also called the displacement sensor, is a device which works on the determination of the absolute physical location of the target or the distance of the target with respect to a reference point. An industrial process uses a position sensor to make sure that the target object on which a job is to be performed is properly positioned at its place before an automatic process step is executed, such as before spraying paint on an auto body, it must be placed at the required station otherwise the paint will be wasted. Before filling a chemical in a beaker, the beaker must be in the correct place, otherwise the chemical will be spilled around. So, a position sensor detects the target and produces a signal about its position which is utilized to control the process.

Many types of position sensors are used in factories, a few of which are given below.

4.7.2 RESISTIVE OR POTENTIOMETRIC POSITION SENSOR

4.7.2.1 Types, the Principle, and Resistors Used

Types: Linear or rotary (angular).

Principle: A wiper attached to the target moves on a resistive track resulting in linear/rotational displacement (Figure 4.12). The track length traversed, either along a straight line or in angular motion, determines the resistance and hence position of the target. The measurement is done using a voltage divider circuit. A fixed voltage is applied between the two ends of the resistive track. The voltage dropped from the wiper position to one of the tracks indicates the position of the target.

Resistors Used: Carbon film or wire-wound resistor

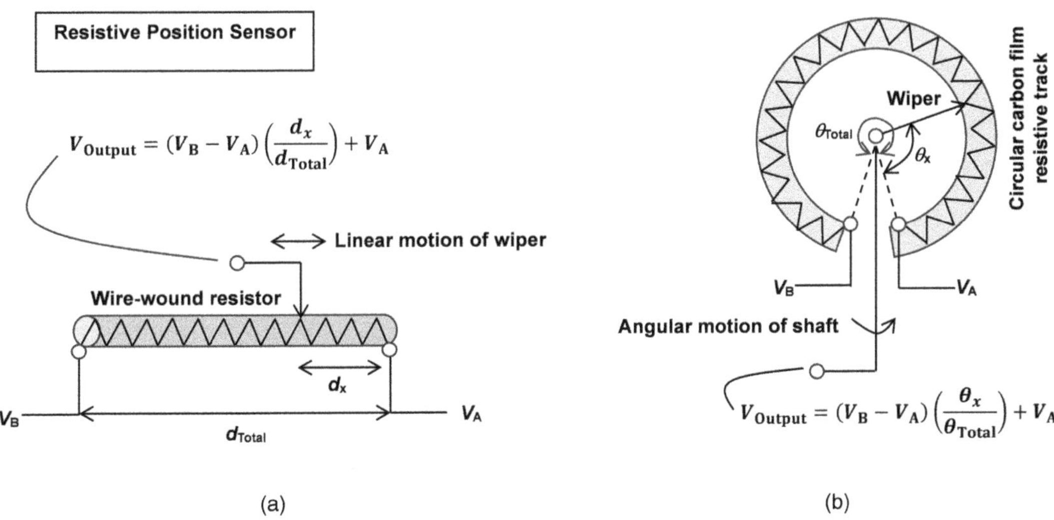

FIGURE 4.12 Potentiometric position sensor: (a) linear and (b) rotary. Part (a): A wire-wound resistor of wire length d_{Total} is connected between points at potentials V_B and V_A. A wiper moves leftward on the resistor over a distance d_x from V_A. The output voltage V_{Output} changes with the linear motion of the wiper. It depends on the [product of $(V_B - V_A)$ with d_x/d_{Total}] added to V_A. Part (b): A wiper moves by angular motion of a shaft on a circular track of resistive carbon film connected between points at potentials V_B and V_A. It traverses an angle θ_x from its initial position at V_A to the new position. The output voltage V_{Output} changes with the rotary motion of the wiper. It depends on the [product of $(V_B - V_A)$ with θ_x/θ_{Total}] added to V_A.

4.7.2.2 Advantages, Disadvantages, and Applications of Resistive Position Sensor

Advantages: Simple technology; easy to use, low-cost devices.

Disadvantages: Wear and tear problems arising from moving parts, limited range of detection depending on wiper movement range, sensitivity to dust and extreme temperature environments, low accuracy, and repeatability.

Applications: Computer game joysticks, robotics

4.7.3 CAPACITIVE POSITION SENSOR

4.7.3.1 Types of Capacitive Position Sensor

It is a non-contact sensor with two versions (Figure 4.13).

(a) Spacing Variation Type: The target is attached to one plate of the capacitor (Figure 4.13(a)). As the target moves, the plate attached to it moves toward or away from the other stationary plate of the capacitor, thereby changing the thickness of the air film acting as the dielectric between the two plates. The position of the target is found from the capacitance change due to target movement.

(b) Overlapping Area Type: Here the moving plate with attached target slides transversely over the fixed plate (Figure 4.13(b)). Hence, the thickness of air dielectric remains constant but the overlapping area of the two plates changes, thus altering the capacitance. The change in capacitance with target motion helps in determining the target position.

4.7.3.2 Advantages, Disadvantages, and Applications of Capacitive Position Sensor

Advantages: Useful for linear and angular motions, and for various materials such as metals, plastics, and liquids.

Disadvantages: Affected by air humidity.

Applications: Accelerometers

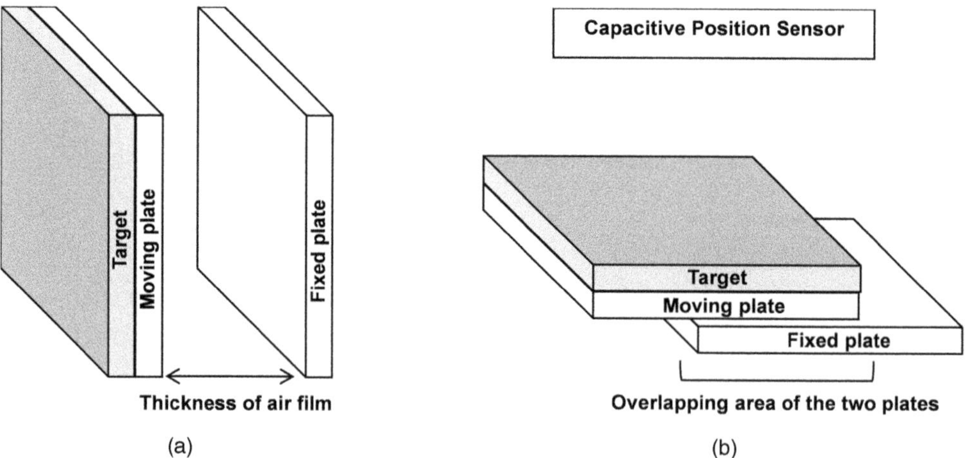

FIGURE 4.13 Capacitive position sensor based on variation of (a) spacing between two plates and (b) area of overlapping of two plates. Part (a): The diagram shows the target, along with the fixed and moving plates separated by an air gap. The thickness of the air film is indicated. Part (b): The diagram shows the target, along with fixed and moving plates. The moving plate is sliding against the fixed plate. By this sliding motion, the overlapping area between the two plates varies, thereby changing the capacitance of the sensor.

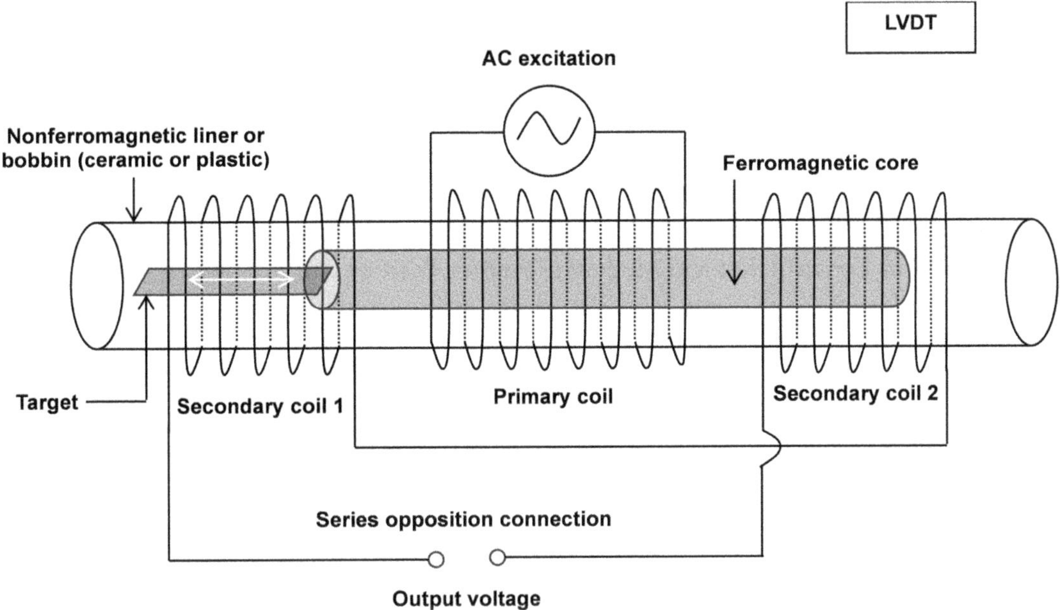

FIGURE 4.14 Linear or rotary variable differential transformer. A non-ferromagnetic liner or bobbin made of a ceramic or plastic material is shown. The target is attached to a ferromagnetic core inside the bobbin. The primary coil is excited by an AC supply. The secondary coil 1 on the left side of the primary coil is connected with the secondary coil 2 on the right side of the primary coil. As the target moves, the output voltage is measured between the two secondary coils connected in series opposition.

4.7.4 LINEAR OR ROTARY VARIABLE DIFFERENTIAL TRANSFORMER (LVDT OR RVDT) POSITION SENSOR

4.7.4.1 Sensor Construction and its Use for Measurement of Displacement

It is an inductive position sensor, which finds the position of the target from the changes in the magnetic field induced in its coils (Figure 4.14). It consists of two transformers sharing a single ferromagnetic core. The two transformers are formed by winding three coils on the core: one primary coil in the center, one secondary coil on the left side of the primary coil, and another secondary coil on its right side. The secondary coils are connected in series and wound in opposite directions so that they are 180° out of phase with each other. They are said to be in 'series opposition'.

The target is attached to the core and a voltage is applied to excite the primary coil. The induced EMFs produced in the two secondary coils act in opposite directions. The difference between the induced EMFs is the electrical output of the sensor.

When the target and hence the core is located at the center position, the induced EMFs in the secondary coils cancel each other, yielding a zero output. This balanced position is the null position. When the target moves away from the null position, the magnitude of the output voltage changes. Its polarity may also change depending on the phase angle. As the target moves across the null point, the phase of the output signal shifts by 180°. These variations give an idea about the magnitude of displacement along with its direction.

4.7.4.2 Advantages, Disadvantages, and Applications of LVDT/RVDT

Advantages: Robust, reliable, heavy-duty; high sensitivity, accuracy, and resolution; can measure displacements from a fraction of a mm to several cm, low energy consumption, good linearity, low hysteresis, high durability with long lifespan because it involves no physical contact, extreme conditions withstanding capability, good adaptability and repeatability.

Disadvantages: Bulky, heavy, and cumbersome; expensive.

Applications: It is an extensively used metrological tool in automotive, aerospace, and pharmaceutical industries; as well as in energy, electronics, and environmental sectors. It is used for inspection of bearing alignment and examination of assembly lines. It can measure fluid levels and test soil strength.

4.7.5 EDDY-CURRENT-BASED POSITION SENSOR

4.7.5.1 Sensor Construction and Method of Finding Position

Similar in operation to the proximity sensor, it consists of a coil through which alternating current is passed to produce a magnetic field (Figure 4.15). This time-varying magnetic field induces eddy currents in a metallic target placed in the vicinity of the coil. The direction of eddy currents is such

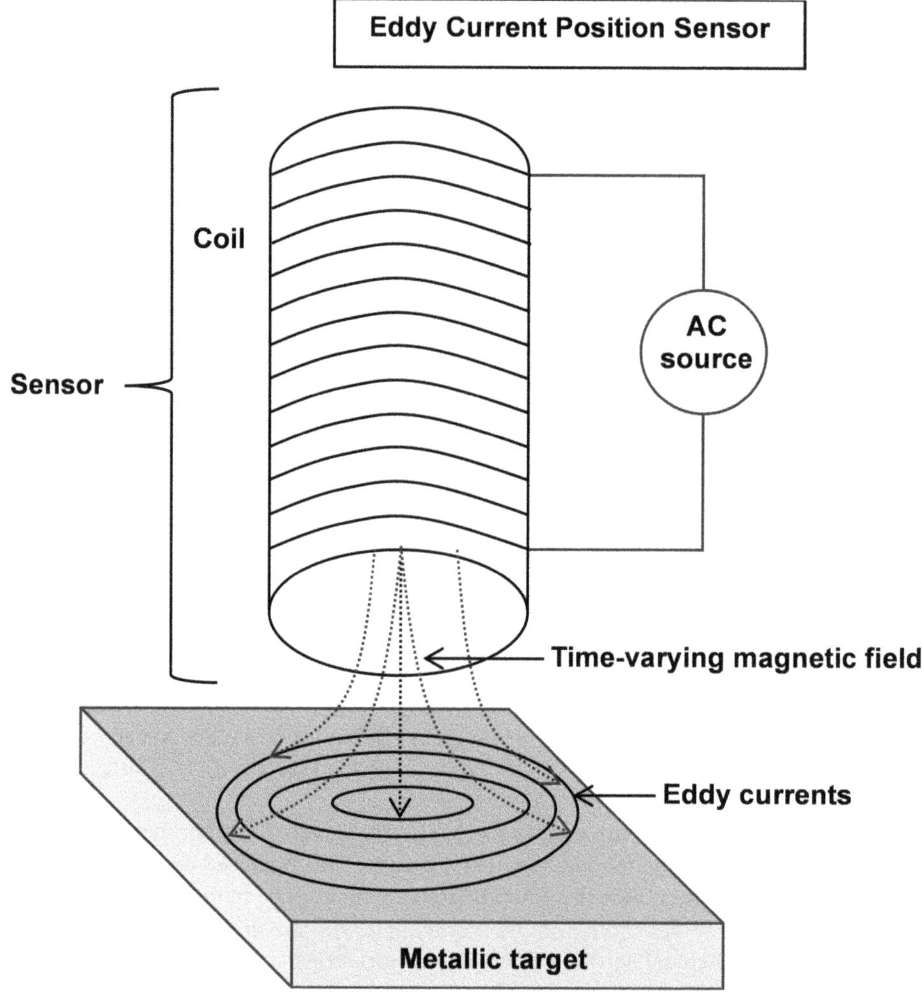

FIGURE 4.15 Eddy-current-based position sensor. An AC source is applied across a coil to produce a time-varying magnetic field around the coil. This magnetic field induces eddy currents in a nearby located metallic target. The impedance of the coil increases as the coil comes closer to the target because the magnetic field of eddy currents flowing in the target opposes the magnetic field of the coil, which was the cause of eddy current induction in the target.

as to oppose the magnetic field responsible for their production, increasing the impedance of the coil. The change in this impedance allows us to know the position of the target because the closer is the target to the coil, the larger its impedance.

4.7.5.2 Advantages, Disadvantages, and Applications of Eddy-Current-Based Position Sensor

Advantages: Inexpensive, can work in dirty environments unaffected by contaminants.

Disadvantages: It can only be used for conductive targets, e.g., Al, Cu, Ti; it has an omnidirectional response; it can measure the relative distance of the target from the coil but not the direction in which the target is located.

Applications: Machine tool mounting, drive shaft monitoring, delicate machine assembly.

4.7.6 MAGNETOSTRICTIVE POSITION SENSOR

4.7.6.1 Magnetostriction

Magnetostriction is a property of ferromagnetic materials. e.g., Fe, Ni, Co, Co-Fe-Va alloy (permendur), Fe-Ni alloy (permalloy), Co-Ni alloy, ferrites, and rare earths, which makes them change their shape or dimensions when subjected to magnetization by an external magnetic field. It occurs due to the rotation of magnetic domains of the material and movement of the domain walls (the boundaries separating domains with different magnetization directions) due to the torque exerted by the field. This domain wall movement causes the growth of domains that are aligned with the applied magnetic field and shrinkage of domains that are arranged in the opposite direction, thereby producing a magnetostrictive strain in the material and its deformation. As a consequence, expansion or contraction of the material takes place. On withdrawal of the magnetic field, the domains remain pinned in their newly acquired orientation.

4.7.6.2 Sensor Operation and Determination of Position

The sensor has a magnetostrictive waveguide in the form of a wire of ferromagnetic material enclosed in a stainless steel or aluminum protective cover (Figure 4.16). A short interrogation electrical pulse is sent by the electronics module into the waveguide to produce a helical magnetic field along the waveguide direction. A movable permanent magnet is attached to the moveable machine part called the target whose position is to be found. It is called the position magnet. It moves outside the waveguide, generating an axial magnetic field whose field lines are coplanar with the waveguide. A momentary interaction takes place between the two magnetic fields, the magnetic field created from the electrical pulse in the waveguide and the magnetic field from the position magnet outside the waveguide. As a result of this interaction of a longitudinal magnetic field and a circular magnetic field created by an electric current, a twisting effect is produced in the waveguide. It is known as the Wiedmann effect. A torsional strain pulse is launched in both directions from the location of the position magnet toward the two ends of the waveguide. This mechanical elastic wave is termed the sonic wave. It propagates at approximately ten times the velocity of sound, which is estimated to be around 3000 ms^{-1}.

At one end of the waveguide, there is a sonic wave detection system or pickup device acting as a read head. It works on the induction of stress by the sonic wave in a piece of magnetostrictive material passing through a coil and magnetized by a permanent magnet. The stress thus created initiates a surge of altered permeability in the magnetostrictive material. Hence, there is a change in its magnetic flux due to which an output voltage is produced across the coil from which the instant of arrival of the sonic wave at the pickup device is noted.

The time interval between the injection of the current pulse in the waveguide and the instant of detection of the sonic wave at the pickup device gives the relative position of the position magnet from the pickup device. The measurement is performed by starting a timer as soon as the current

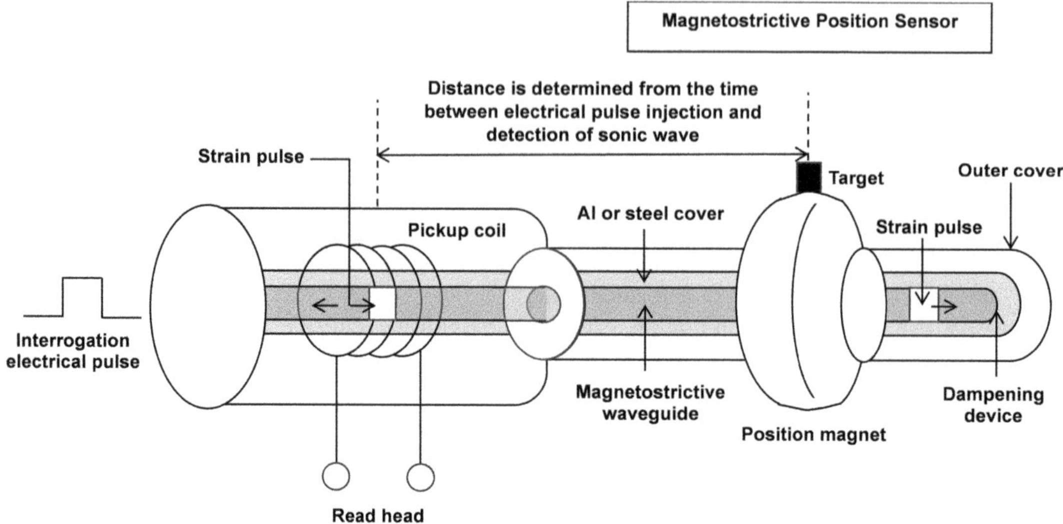

FIGURE 4.16 Magnetostrictive position sensor. A magnetostrictive waveguide with an Al or steel cover and an outer cover carries a position magnet to which a target is attached. A dampening device is placed at the right end of the waveguide. An interrogation electrical pulse is launched from one end of the waveguide toward the target. The launched electrical pulse interacts with the magnetic field of the position magnet. This interaction creates a strain pulse called a sonic wave, which travels toward both sides of the magnet. The sonic wave reaching the right end of the waveguide is damped while the sonic wave moving leftward in the waveguide is detected on reaching a pickup coil. The target position is estimated by measuring the time difference between sonic wave detection and electrical pulse launching, i.e., injection into the waveguide.

pulse is sent in the waveguide, which stops on detection of the sonic wave. Since this time delay from the position magnet to the pickup device is proportional to the distance of the position magnet from the pickup device, it allows the distance between them to be calculated. The electronic circuit presents it in the desired format.

At the opposite end of the waveguide, there is a dampening device. It absorbs the sonic pulse in order that no interfering signal is reflected back toward the pickup device, vitiating the measurements.

4.7.6.3 Advantages, Disadvantages, and Applications of Magnetostrictive Position Sensor

Advantages: It is a compact, and robust non-contact linear position sensor accurate for long lengths (resolution up to 1 µm), operating maintenance-free, unaffected by friction or wear, and also not upset by shocks and vibrations. Other parameters: Fast operating cycles <1 ms, Measurement lengths up to 7620 mm.

Disadvantages: The dead zone on both sides of the sensor cannot be made zero. It is expensive, less accurate for short lengths, and is also temperature-sensitive.

Applications: Automotive suspension and steering system, hydraulic process control, precision aerospace testing, and geoseismic survey.

4.7.7 HALL-EFFECT-BASED POSITION SENSOR

4.7.7.1 Sensor Components and Working

Also called a magnetic position sensor, its main component is a Hall element consisting of a rectangular strip of P-type Gas or InSb through which a continuous current flows from one end to the other along the length direction (Figure 4.17). The Hall voltage is measured between electrodes across the width direction. The target, e.g., the piston moving in a cylinder, is connected to a magnet enclosed in the sensor shaft. The magnetic field of this magnet acts perpendicularly to the direction of current

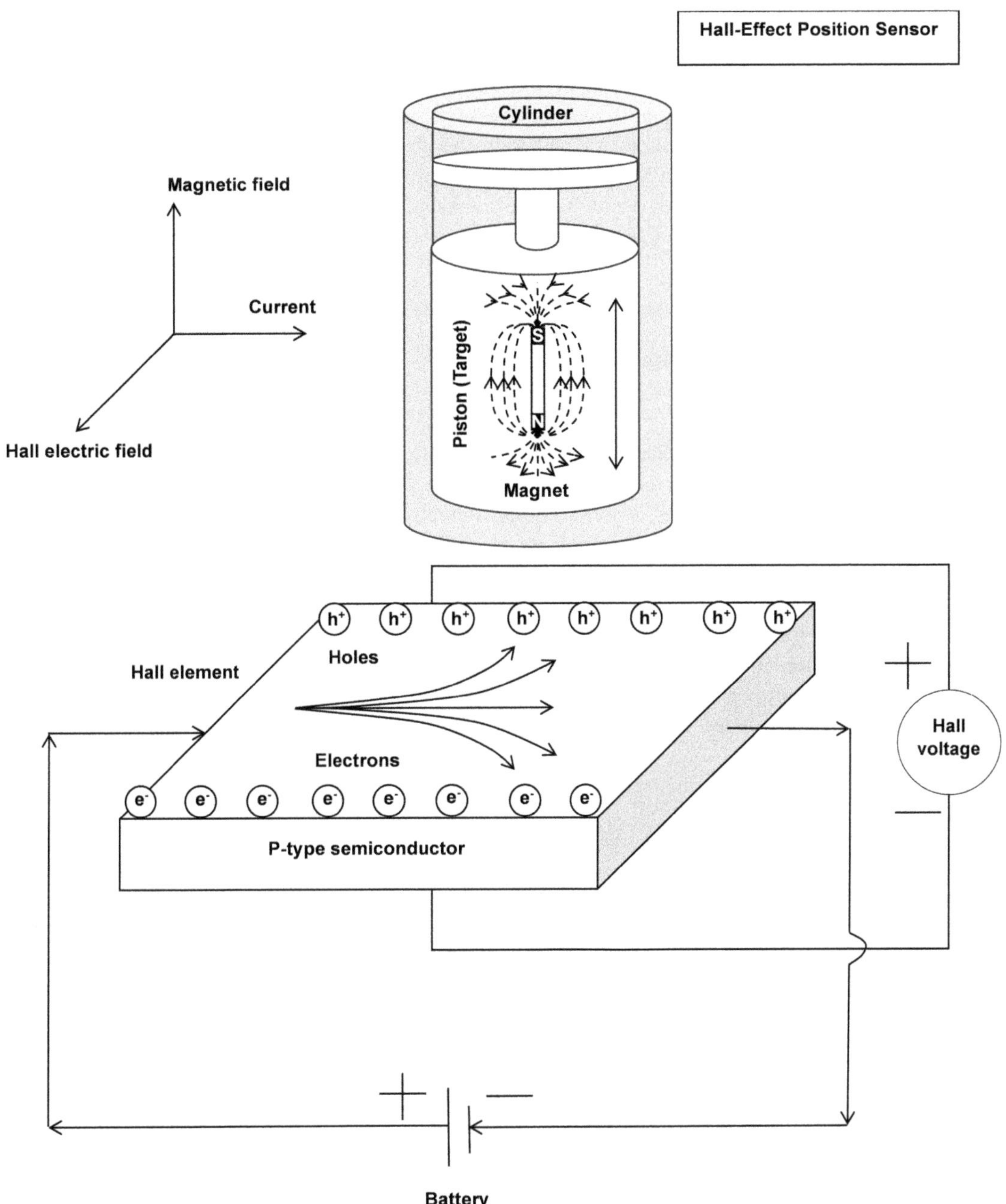

FIGURE 4.17 Hall-effect-based position sensor. A piston-connected magnet moves as the piston moves inside a cylinder. The Hall element consists of a P-type semiconductor with a battery connected between its two ends. Electrons (e−) and holes (h+) drift toward opposite sides of the P-type semiconductor bar. This charge separation induces a Hall voltage between the lateral sides of the P-type semiconductor. Electric current, magnetic field, and Hall electric field are directed along three mutually orthogonal Cartesian axes.

flow in the Hall element. When the piston-connected magnet is near the Hall element, the influence of its magnetic field is strong so that the Hall voltage increases. As the magnet recedes away from the Hall element, its effect on the Hall element weakens, and therefore the Hall voltage falls. Thus, the Hall voltage is an indicator of the distance or position of the piston.

4.7.7.2 Advantages, Disadvantages, and Applications of Hall-Effect-Based Position Sensor

Advantages: Heavy-duty devices, insensitive to liquids or dust particles.

Disadvantages: Susceptibility to magnetic interference, especially from neighboring current-carrying wires, accuracy prone to hysteresis effects, and disturbed by impacts.

Applications: Brushless DC motors for detecting the permanent magnet position, monitoring valve position, finding the RPM of a rotating shaft, timing of ignition in internal combustion engines by sensing the angular position of the crankshaft for adjusting the firing angle of spark plugs, sensing the positions of seat and safety belt for controlling the air-bags in automotive.

4.7.8 OPTICAL POSITION SENSOR

4.7.8.1 Types of Optical Position Sensor

This sensor is of three types.

(a) Light-Transmission-Based Sensor: Light is transmitted from an emitter on one side of the sensor and is detected by a receiver on its opposite side (Figure 4.18). An example is the transmissive optical encoder. It is an emitter/detector module in which an infrared LED illuminates a rotating code wheel with alternating opaque and transparent segments. The light passing through the transparent regions of the wheel is received by a photodiode on the opposite side of the wheel. Thus, the rotary motion of the code wheel is translated into a digital waveform by the LED/code wheel/photosensor sandwich.

(b) Light-Reflection-Based Sensor: Light emitted from a source is reflected by the target and the changes in properties of the reflected light from the target, e.g., intensity, wavelength, phase or polarization are used to find the position of the target. An example is the reflective

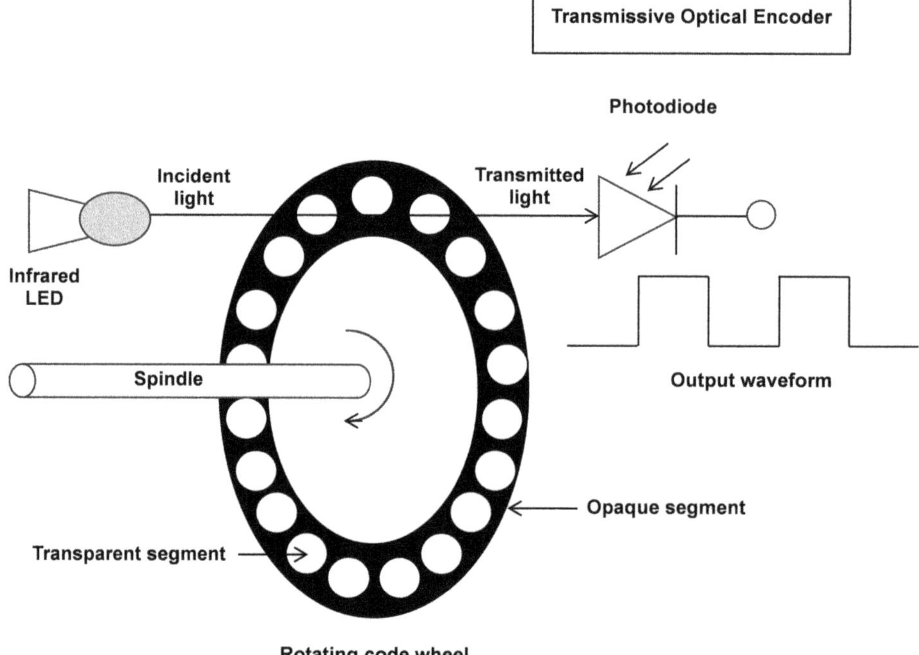

FIGURE 4.18 Transmissive optical encoder. A code wheel with transparent and opaque segments is rotating on a spindle. On one side of the wheel is an infrared LED and exactly aligned on its opposite side is a photodiode. A rectangular pulse waveform is recorded as the transparent and opaque segments of the wheel alternatively come in alignment with the line joining the infrared LED and the photodiode.

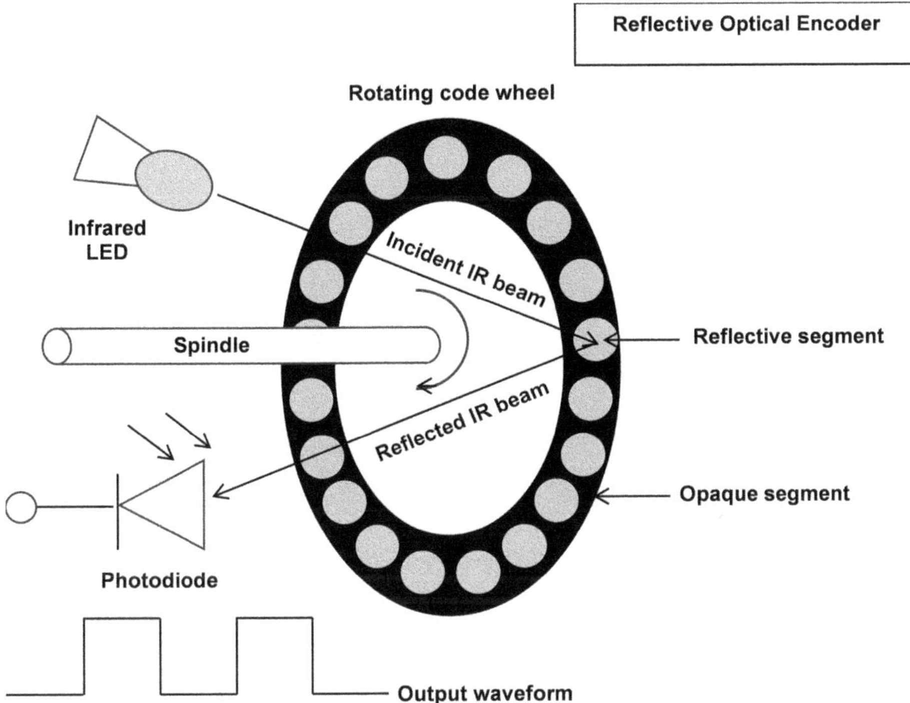

FIGURE 4.19 Reflective optical encoder. A code wheel with reflective and opaque segments is rotating on a spindle. Both infrared LED and photodiode are placed on the same side of the wheel. A rectangular pulse waveform is generated as the reflected light from the wheel falls on the photodiode when light strikes a reflective segment of the wheel and no reflected light is received by the photodiode when light falls on an opaque segment of the rotating wheel.

optical encoder (Figure 4.19). Here the LED and the photodiode are placed on the same side of the code wheel. The light reflected by the code wheel is detected by the photodiode.

(c) Light-Interference-Based Sensor: In an interferential encoder (Figure 4.20), a divergent beam from a coherent laser source illuminates a diffraction grating pattern created by depositing chrome on a glass scale or laser writing on a metallic tape scale. Diffraction of light by the periodic grating generates a high-contrast interference pattern on a detector array. By the Talbot effect, also termed self-imaging or lens-less imaging, the image of the grating is repeated at regular distances called Talbot length, away from the plane of the grating. These repeated images are known as Talbot planes of interference patterns or Talbot images. The discrete Talbot planes are labeled as 1st, 2nd, 3rd, ..., planes. Position measurement is based on the principle that a change in a relative position of the scale and detector produces a translation of the diffraction pattern across the detector array causing a sinusoidal change in each detector cell.

An interferential sensor can be made in compact size because it needs minimal optical components but it requires stricter environmental conditions for proper functioning. Without interpolation, an order of magnitude higher resolution is obtained with an interferential sensor than with transmissive or reflective sensors. The fidelity of sine and cosine signals enables high interpolation making possible nanoscale resolution. So, the interferential sensor leads to precision over competing devices. Reflective sensors can also be in small size and low weight. The transmissive sensor generally lies in an enclosure. So, it can be made more rugged depending on the enclosure's endurance and durability.

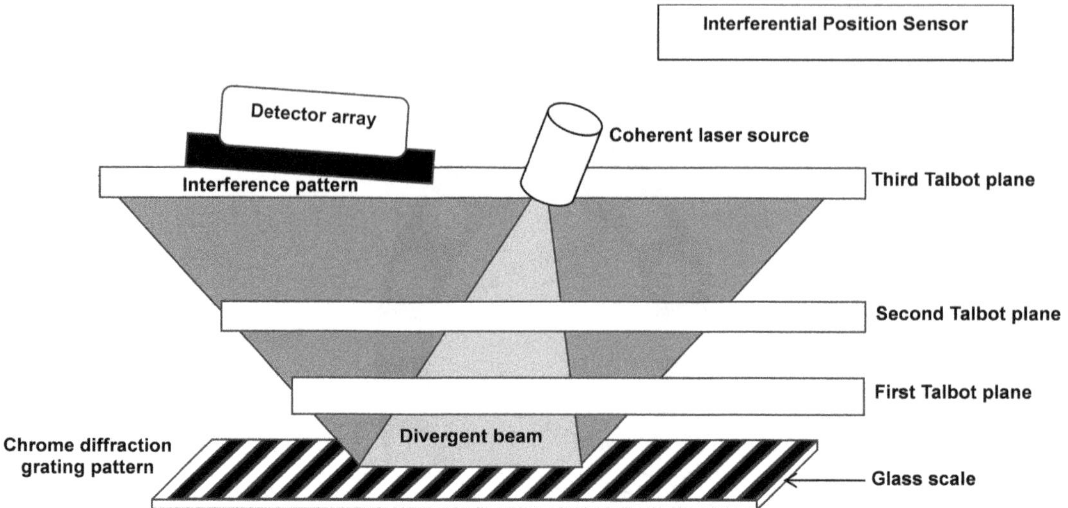

FIGURE 4.20   Inferential position sensor. A grating diffraction pattern is made with chrome on a glass scale. It is illuminated by a divergent beam from a coherent laser source. The resulting interference pattern is recorded by an array of detectors. Repeated images of interference patterns called 1st, 2nd, and 3rd Talbot planes are shown.

4.7.8.2   Advantages, Disadvantages, and Applications of Optical Position Sensor

Advantages: High accuracy and resolution, no hysteresis, and not disturbed by magnetic fields.
Disadvantages: Fragility, sensitivity to dust, impact, vibration, and extreme temperatures.
Applications: Induction motors and furnaces, electric brakes, speedometers, bottle or jar filling by telling the machine about the positions of the containers, servo labeling system for controlling the timing and speed of bottle rotation, printing jobs for activating the printhead to produce a mark at a specified position, and in large cranes to provide positioning information to the crane for picking or releasing its load.

4.7.9   ULTRASONIC POSITION SENSOR

4.7.9.1   Operating Modes

Ultrasonic sensors operate in three modes (Figure 4.21).

(a) Diffuse Mode Ultrasonic Position Sensor: It is the most commonly used mode of operation. A piezoelectric ultrasonic transducer is used as a transmitter and receiver of ultrasonic waves (Figure 4.21(a)). A high-frequency (40–70 kHz) ultrasonic pulse is emitted from the transducer. The pulse is reflected by the target toward the sensor. The time difference between the emission of the ultrasonic pulse and the receipt of the echo pulse is measured to find the distance between the sensor and the target. This measurement technique is called transit time or time-of-flight method.

Distance of the target from the sensor

$$= \frac{\text{Velocity of ultrasound in air}\left(344\text{ms}^{-1}\right) \times \text{Total time taken by ultrasound in traveling from the ultrasonic transducer to the target and back}}{2} \quad (4.2)$$

The output signal is either in analog, digital, or binary switching form.

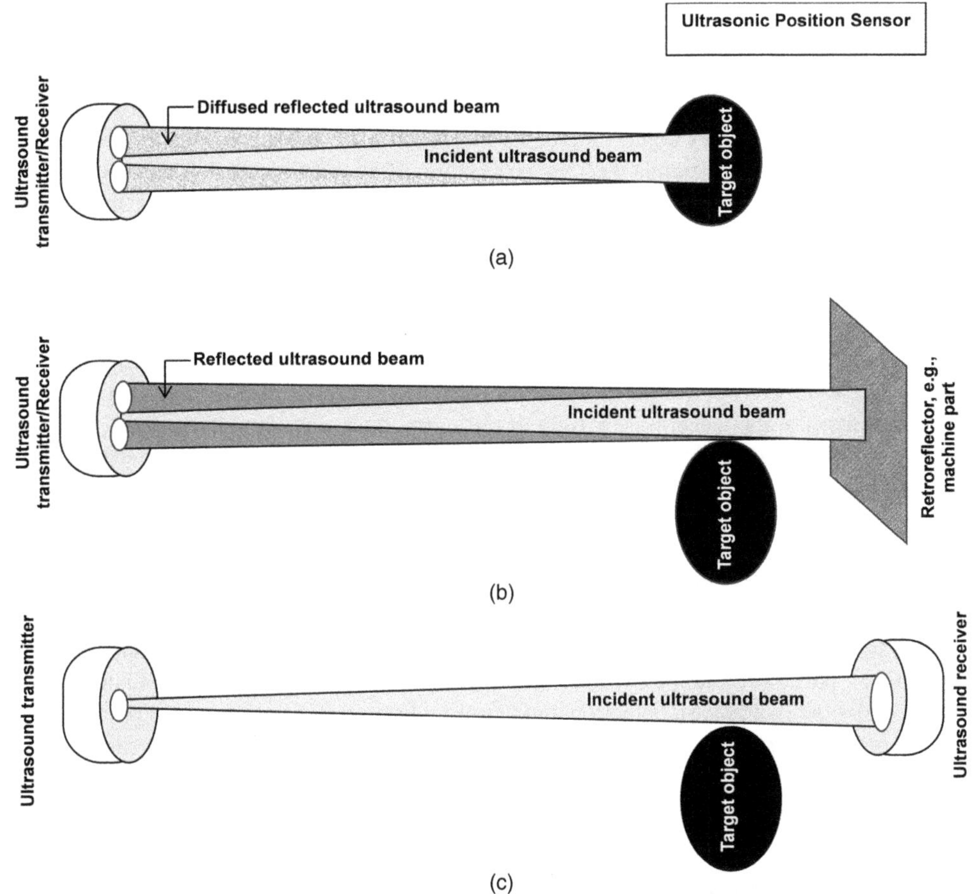

FIGURE 4.21   Ultrasonic position sensor operating in (a) diffuse, (b) retroreflector, and (c) through-beam modes. Part (a): An ultrasound beam from the ultrasound transmitter/receiver falls on the target. This is the incident ultrasound beam. Diffuse reflected ultrasound beam from the target strikes the receiver. Part (b): An ultrasound beam from the ultrasound transmitter/receiver falls on the retroreflector, which can be part of a machine. This is the incident ultrasound beam. The reflected ultrasound beam from the retroreflector is detected by the receiver when the target is away from the incident ultrasound beam and so does not block the path of the ultrasound beam. Part (c): An ultrasound beam from an ultrasound transmitter is incident on the ultrasound receiver if it is not blocked by any target between them. Otherwise, it is stopped, and the non-receipt of the beam by the receiver indicates that it has been intercepted by the target. Hence, the presence of the target is inferred.

(b) Retroreflector Mode Ultrasonic Position Sensor: Like the diffuse mode sensor, the same ultrasonic transducer acts as the emitter and receiver of ultrasonic waves (Figure 4.21(b)). An ultrasound reflector (metal, plastic, wall, floor conveyor belt, etc.) permanently installed facing the ultrasound emitter and aligned to it, is called the reference reflector. Ultrasonic pulses are continuously emitted by the ultrasonic transducer. These pulses are reflected back toward the emitting ultrasonic transducer as long as there is no object between them to block the reflected pulse. As soon as a target object comes between them, it interrupts the reflected pulse and prevents it from reaching the ultrasonic transducer. Immediately, the output signal of the ultrasonic transducer changes causing the sensor to switch its state. The sensor-to-target distance is determined from the propagation time of the ultrasonic pulse. The detection range is 20 cm to 16.5 m. All types of targets are detected irrespective of their material, surface features, and color.

(c) Through-beam Mode Ultrasonic Position Sensor: Unlike the previous two modes, two ultrasonic transducers are used (Figure 4.21(c)). One acts as the ultrasound emitter and the other as the ultrasound receiver. They are placed facing each other. Normally, a continuous signal from the emitting ultrasonic transducer is detected by the receiving ultrasonic transducer. But when a target object intercepts the path of the ultrasonic beam, the receiving ultrasonic transducer responds with the production of an output signal. The sensor is operated either as ultrasound-received-ON or ultrasound stopped-ON. Its detection range is twice as large as that of the diffuse or retroreflector mode sensors.

4.7.9.2 Advantages, Disadvantages, and Applications of Ultrasonic Position Sensor

Advantages: It is a completely non-contacting device without any moving parts providing wear-free, long-life operation. Its function is neither affected by the color, transparency, or opaqueness of the target nor by light or darkness, dust, dirt, rain or moisture, and snow in the environment. It does not pose any danger to users. It is easy to interface with a microcontroller.

Disadvantages: It does not work in vacuum. It is less accurate for soft material, curved or rough surface targets. Irregular surface patterns create dead zones. Ultrasound cannot penetrate grainy materials or cast iron. The sensor is sensitive to temperature variations.

Applications: Measuring liquid level in storage tanks, detection of breaks in wires or threads, object sorting and classification, high-speed counting, monitoring waste levels in rubbish containers, pallet detection for forklifts, accident prevention (vehicle anti-collision system, crane collision avoidance), detection and counting bottles at different points on a machine, object presence sensing to automatically halt the machine movement on interruption of the sensing field, creating a virtual barrier around a hazardous area by creating machine access and perimeter guards (light curtains).

4.8 DETECTING VIBRATIONS AND ANGULAR TILTS IN MOVING PARTS, MACHINES AND WORKSHOP STRUCTURES: MEMS ACCELEROMETERS

4.8.1 PRINCIPLE AND TYPES OF ACCELEROMETERS

The accelerometer is an electromechanical device. It is used for measuring the linear acceleration of a body, i.e., the rate of change of its linear velocity. Low-cost accelerometers made as tiny chips by MEMS technology are favored over conventional bulky devices.

The accelerometer consists of a mass called the seismic or proof mass suspended by a spring. The direction in which the seismic mass is allowed to move is known as the axis of sensitivity of the accelerometer. When the accelerometer device is subjected to an acceleration, the seismic mass in it is displaced to one side depending on the direction of acceleration. The displacement of the seismic mass from its original position is proportional to the applied acceleration. The shift in the position of the seismic mass is determined by the change in resistance of a resistor made of piezoresistive material, the change in the capacitance of a capacitor, or the voltage produced in a piezoelectric crystal. Accordingly, there are three common types of accelerometers: piezoresistive, capacitive, and piezoelectric. In general terms, the piezoresistive accelerometer is most suited for shock testing, a capacitive accelerometer for low-frequency applications such as slow human movements, and a piezoelectric accelerometer for vibratory motion.

4.8.2 PIEZORESISTIVE ACCELEROMETER

In this accelerometer (Figure 4.22), the seismic mass is suspended at its four corners by flexible beams or flexures (Sankar and Das 2009). Diffused piezoresistors are formed at the joints of the

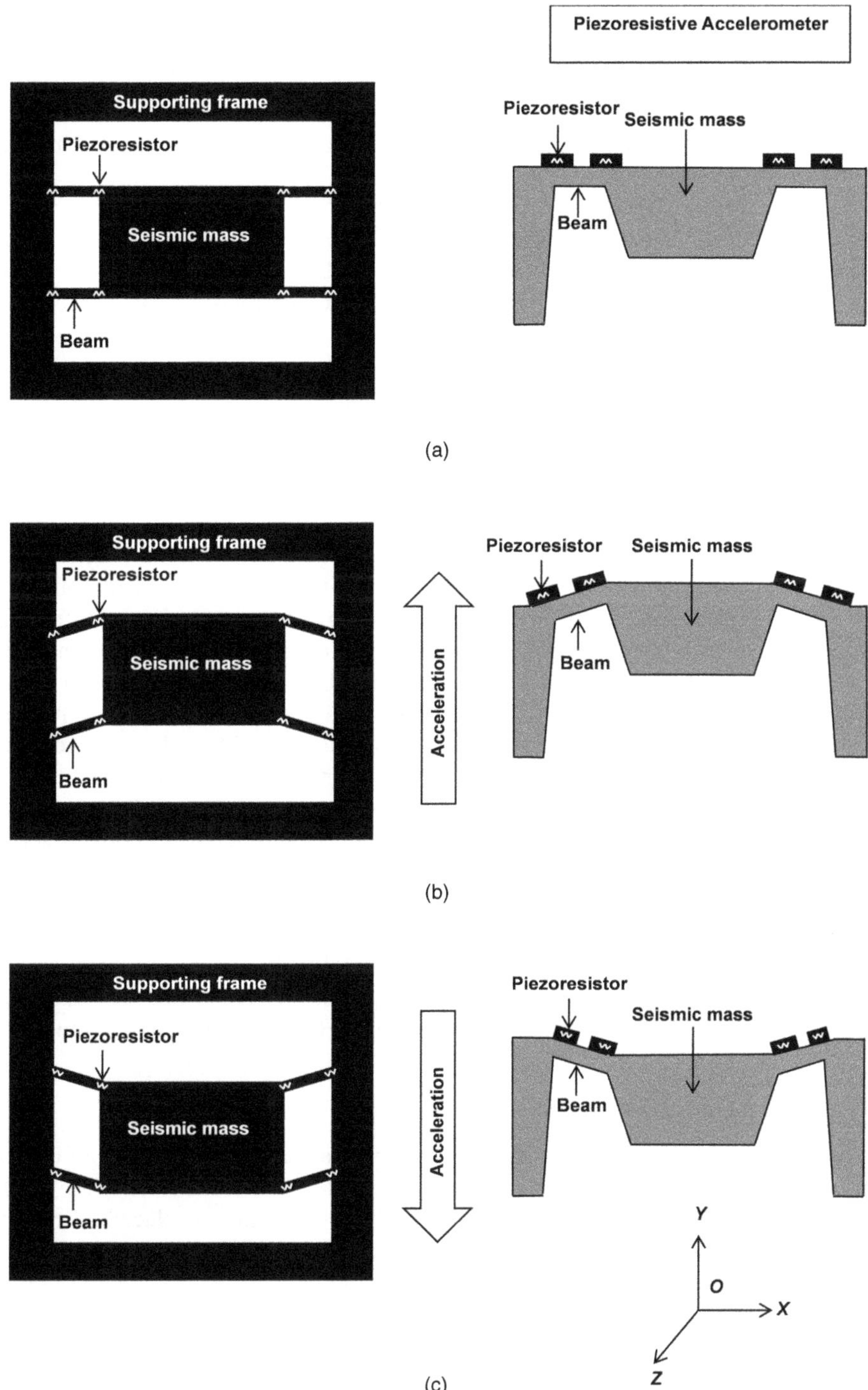

FIGURE 4.22 Piezoresistive accelerometer: (a) in a stable position and (b), (c) under acceleration. Parts (a), (b), and (c): Top view: The seismic mass is hanging from a supporting frame by four beams with four resistors at the seismic mass/beam and beam/supporting frame joints on one side and another four resistors on the other side. Cross section: A trapezium-shaped mass is hanging from beams over a cavity in silicon and is supported by pillars on two sides. Four resistors are seen at the seismic mass/beam and beam/supporting frame joints. *X*, *Y*, and *Z* coordinate axes are drawn. Part (a): The beams are straight. Part (b): The beams are bent upward. Part (c): The beams are bent downward.

beams with the frame as well as with the seismic mass (Figure 4.22(a)). Under an applied acceleration, the seismic mass experiences a force (Figure 4.22(b), (c)). It opposes this force by virtue of its
inertia. The opposition of seismic mass produces stresses on the flexures. The joints of the flexures
with the seismic mass/frame are the sites which suffer the maximum stress. Resistors made at these
joints change in values due to the stresses developed. The change in resistor values is an indicator of
the magnitude of acceleration.

Silicon undergoes plastic deformation above 500°C. Therefore, silicon accelerometers fail in high-
temperature environments. For these situations, 4H-SiC is used as the accelerometer material. The
4H-SiC accelerometer is fabricated by adopting the elastic beam-proof mass structure. The device
works with 5 V input. Its dynamic sensitivity is 0.21 mVg^{-1}. Linearity is 99.8% (Zhai et al. 2022).

4.8.3 Capacitive Accelerometer

In this accelerometer (Figure 4.23), the seismic mass is suspended at its four edges by flexible arms
or springs attached to anchor points on the substrate. The seismic mass hangs at a small height above
the substrate so that it can move freely. As the seismic mass can move along the X-axis, this accelerometer is sensitive to acceleration along this axis.

The seismic mass has electrodes on both sides which protrude outward like fingers. Because the
seismic mass can move, its electrodes move along with it. Hence, they are the movable electrodes.
Besides the movable electrodes, the accelerometer also has fixed electrodes on its substrate. The fixed
electrodes are made in such places that each movable electrode has fixed electrodes on its two sides.
When the seismic mass is displaced from its position, a movable electrode attached to it changes position in the space between two fixed electrodes. Its distance from the fixed electrodes changes, and so
also the capacitances of the two capacitors formed between the movable and fixed electrodes. These
changes in capacitance are correlated to the acceleration acting on the accelerometer.

The accelerometer shown in Figure 4.24 is identical in operation to that of Figure 4.23 except that
its seismic mass can move along the Y-axis. Hence, this accelerometer responds to accelerations
along the Y-axis in the same way that the accelerometer of Figure 4.23 worked for accelerations in
the X-direction.

A capacitive accelerometer is fabricated using an SOI (silicon-on-insulator) wafer with a similar
device and buried oxide layers on both sides (Zhou et al. 2015). The proof mass thickness is 560 μm.
The accelerometer has symmetrical double-sided H-shaped beams. It shows a resonant frequency of
2.24 kHz. Its sensitivity is 0.24 Vg^{-1}, and the nonlinearity is 0.29% over the 0–1 g range.

4.8.4 Piezoelectric Accelerometer

The vital component of this accelerometer (Figure 4.25) is a piezoelectric crystal of quartz or lead
zirconate titanate. Metal films are deposited on its opposite faces to act as electrodes. Wires bonded
to these electrodes are connected across a voltmeter. The seismic mass is placed over the top electrode of the piezoelectric crystal. The opposite side of the seismic mass is connected by a spring to
the housing of the accelerometer.

In the normal rest position of the accelerometer (Figure 4.25(a)), a small force acts on the piezoelectric crystal due to the weight of the seismic mass. So, a small output voltage is developed
between the electrodes.

Under an applied acceleration (Figure 4.25(b)), the seismic mass is displaced from its position,
stretching the spring above it and compressing the piezoelectric crystal below it. A voltage is produced across the crystal by the piezoelectric effect. The greater the acceleration, the more the compressive force on the crystal, and therefore higher the resulting voltage. The output voltage read
from the voltmeter is a measure of the acceleration.

A triaxial piezoelectric accelerometer having folded beam and fabricated using the Mo/AlN/
ScAlN/Mo stack as the sensing material shows a charge sensitivity ~1.07 pCg^{-1} along X-axis,
~0.66pC g^{-1} along Y-axis and ~3.35 pC g^{-1} in the Z-axis direction (Liu et al. 2021).

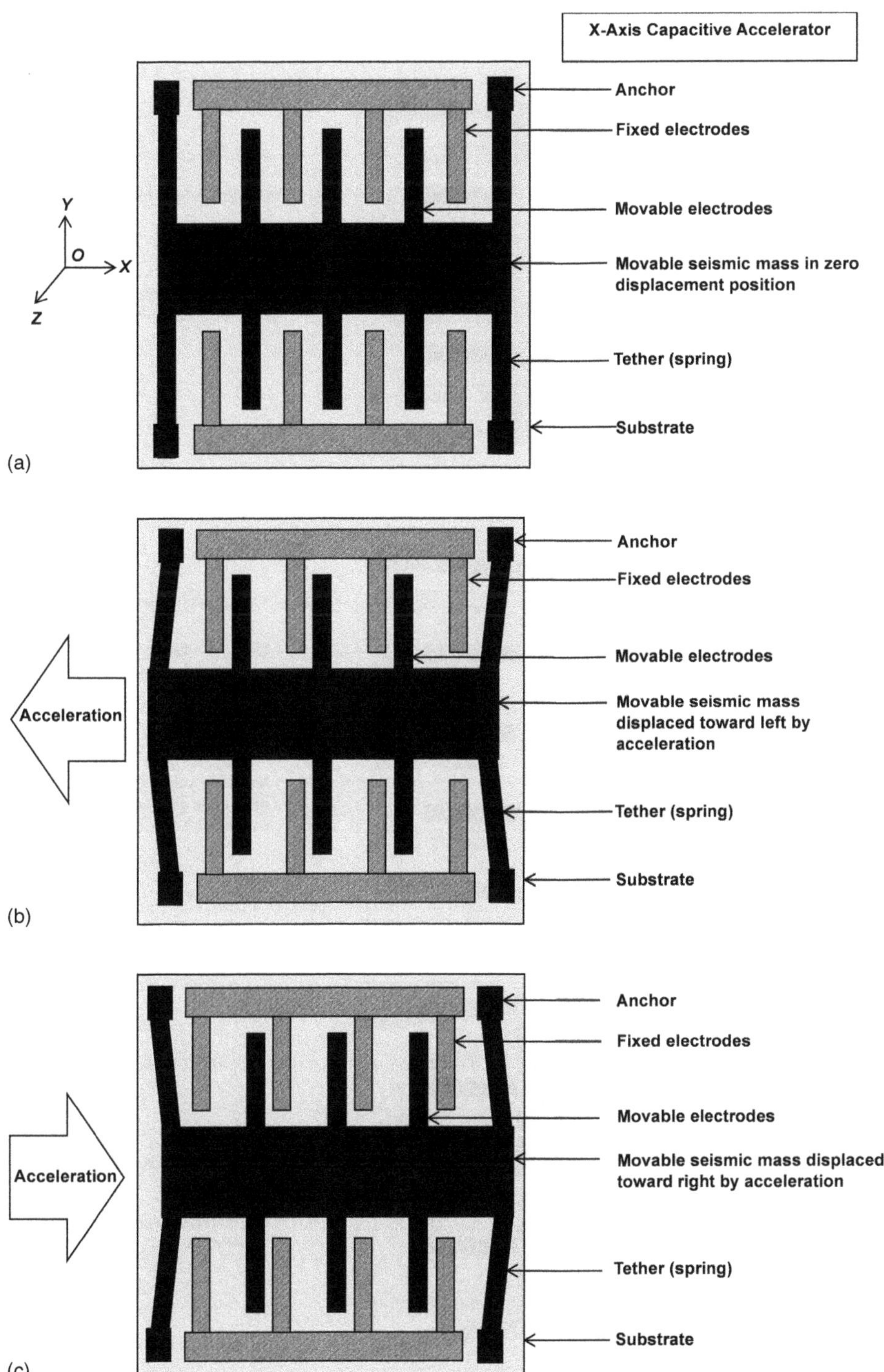

FIGURE 4.23 *X*-axis capacitive accelerometer: (a) at rest, (b) and (c) under acceleration. Parts (a), (b), and (c): Movable seismic mass (black portions) with attached movable electrodes in the vertical direction is hanging from tethers (springs) fixed at anchor points on a substrate. The mass is deflected horizontally leftward and rightward along the *X*-axis. The vertical movable electrodes are translated horizontally left and right along the *X*-axis between fixed vertical electrodes (hatched portions). Part (a): Movable seismic mass in zero displacement position. Part (b): Movable seismic mass displaced horizontally toward the left by acceleration. Part (c): Movable seismic mass displaced horizontally toward the right by acceleration.

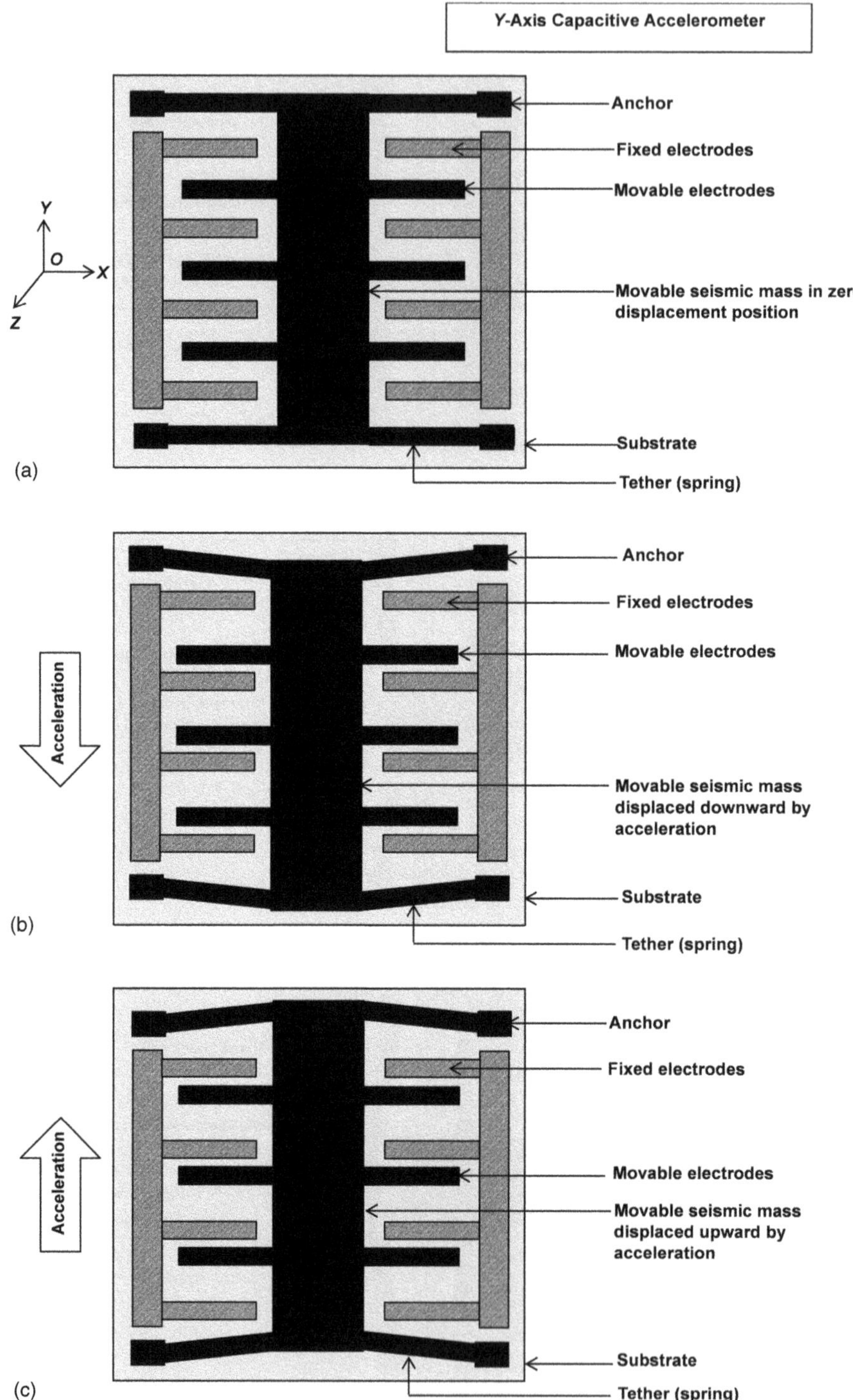

FIGURE 4.24 *Y*-axis capacitive accelerometer: (a) at rest, (b) and (c) under acceleration. Parts (a), (b), and (c): Movable seismic mass (black portions) with attached movable electrodes in the horizontal direction is hanging from tethers (springs) fixed at anchor points on a substrate. The mass is deflected vertically downward and upward along the *Y*-axis. The horizontal movable electrodes are translated vertically down and up along the *Y*-axis between fixed horizontal electrodes (hatched portions). Part (a): Movable seismic mass in zero displacement position. Part (b): Movable seismic mass displaced vertically downward by acceleration. Part (c): Movable seismic mass displaced vertically upward by acceleration.

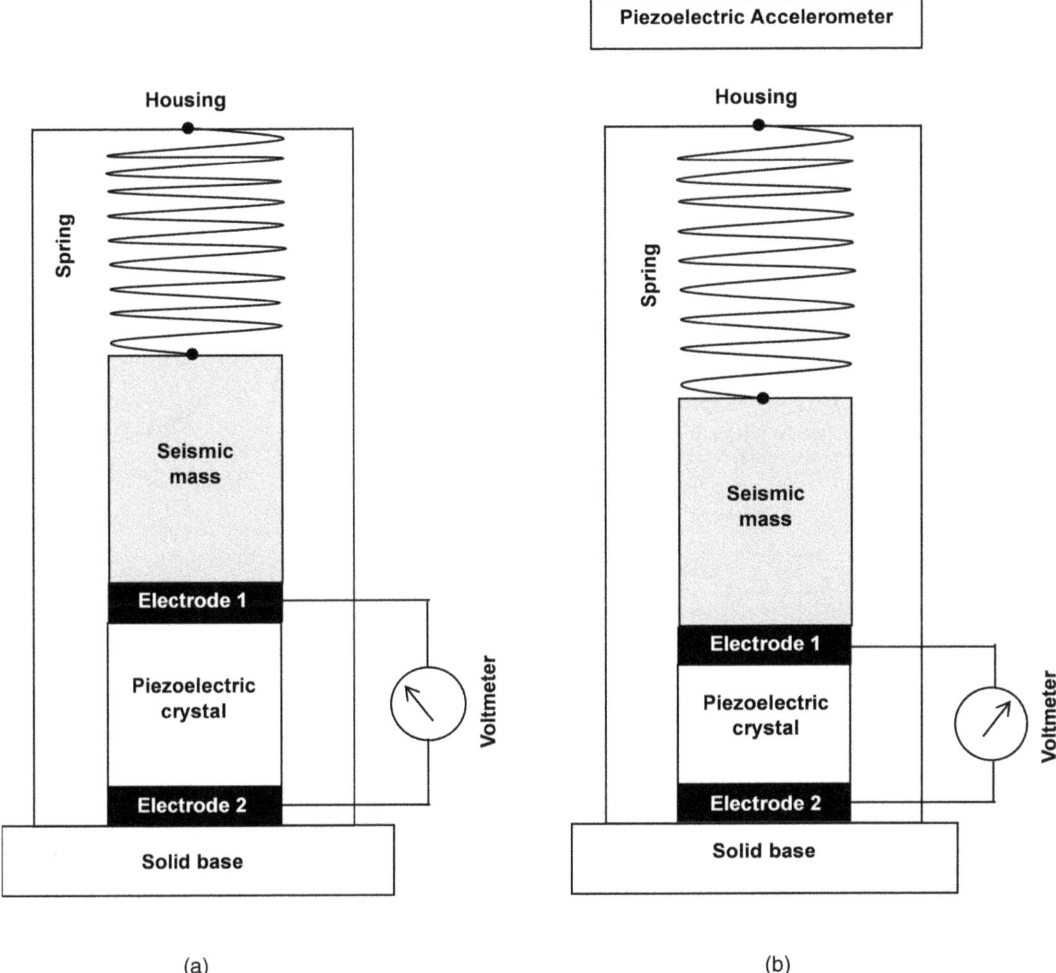

FIGURE 4.25 Piezoelectric accelerometer: (a) in normal position without acceleration when a small voltage is produced by seismic mass pressure and (b) under acceleration in the vertical direction when the seismic mass compresses the piezoelectric crystal, generating a larger voltage. Parts (a) and (b): A spring is attached to the housing roof. To the other end of the spring, the seismic mass is tied. Below the seismic mass is the electrode 1/ piezoelectric crystal/ electrode 2 structure, resting on a solid foundation The electrodes are connected by wires to a voltmeter. Part (a): The spring is in the normal position. A small voltage is seen on the voltmeter. Part (b): The spring is elongated pushing the seismic mass downward. The downward force produces a large piezoelectric voltage, as seen from the voltmeter deflection.

4.8.5 MEMS GYROSCOPE

The gyroscope is an angular rate sensor. It measures the rate of change of angular position of an object, i.e., its angular velocity. Most of the MEMS gyroscopes contain vibrating mechanical components, not rotating elements. So, they can be fabricated in miniaturized forms (Jeong et al. 2004, Patel and McCluskey 2012).

4.8.5.1 Coriolis Effect

Vibratory gyroscopes work on the Coriolis effect involving the transfer of energy between two modes of vibration of a microstructure (Xia et al. 2014). The Coriolis effect is the apparent deflection of an object in a frame of reference rotating with respect to an inertial frame, a frame in which

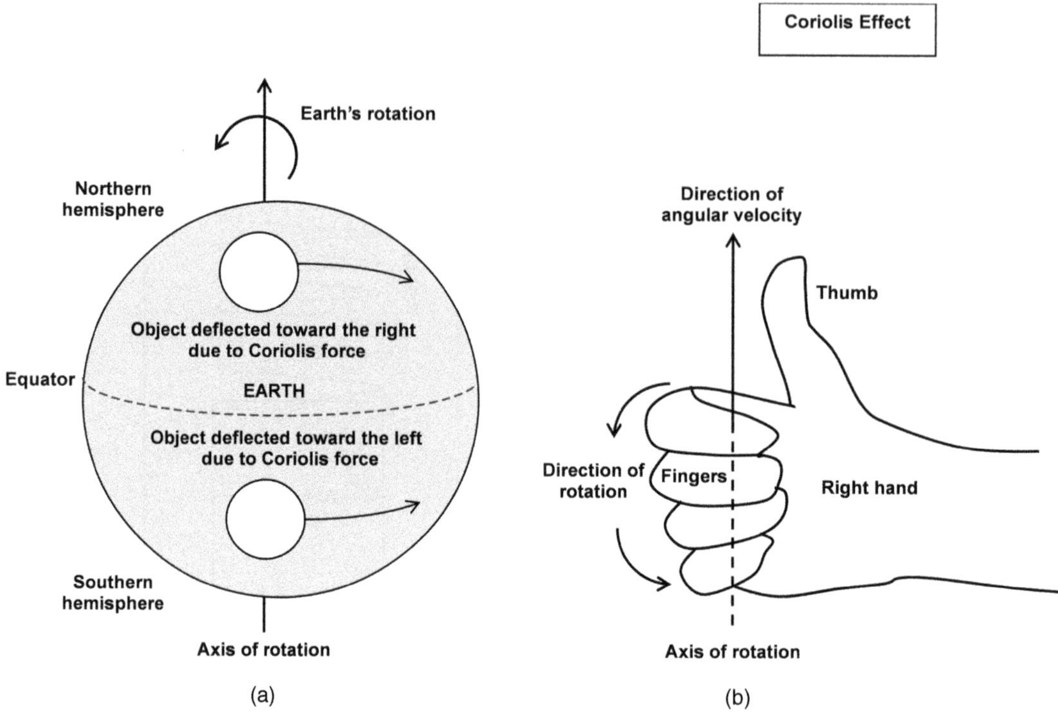

FIGURE 4.26 Coriolis force and angular velocity directions: (a) Coriolis force and (b) angular velocity. Part (a): Earth rotates from west to east which is in the counterclockwise direction as seen from the north pole. The axis of rotation, equator, and northern and southern hemispheres are marked. An object in the northern hemisphere is deflected toward the right direction due to the Coriolis force while an object in the southern hemisphere is deflected toward the left direction owing to the Coriolis force. Part (b): The right-hand rule according to which: When we curl the fingers of the right hand around the axis of rotation with fingers indicating the direction of rotation, and we extend the thumb, then the direction in which the thumb points is the direction of angular velocity.

Newton's laws of motion are valid. This deflection is caused by Coriolis force, a fictitious force acting toward the left of a moving object in a reference frame which is rotating clockwise with respect to the inertial frame and toward the right in a reference frame rotating in the anticlockwise direction relative to the inertial frame (Figure 4.26(a)). So, an object is deflected toward the right in the northern hemisphere whereas an object in the southern hemisphere is deflected leftward. The deflection of the object is said to be apparent because the object does not undergo any deviation from its path, rather it appears to do so because the frame of reference is rotating.

The Coriolis force acts in a direction that is orthogonal to the directions of angular velocity of the rotating frame of reference with respect to the inertial frame, and the velocity of the object with respect to the rotating frame of reference. The right-hand rule is applied to find the direction of the angular velocity of the frame. This rule is stated as: If one holds the axis of rotation of the frame in the right hand with fingers pointing along the direction of rotation, then the thumb indicates the direction of the angular velocity of the frame (Figure 4.26(b)).

4.8.5.2 Vibratory Gyroscope

At the heart of a vibratory gyroscope is the seismic mass (Figure 4.27). It has a perforated body. Perforations are made to reduce the damping of vibrations of the mass caused by opposition due to air film. The seismic mass is suspended at a small distance above the substrate so that it can move

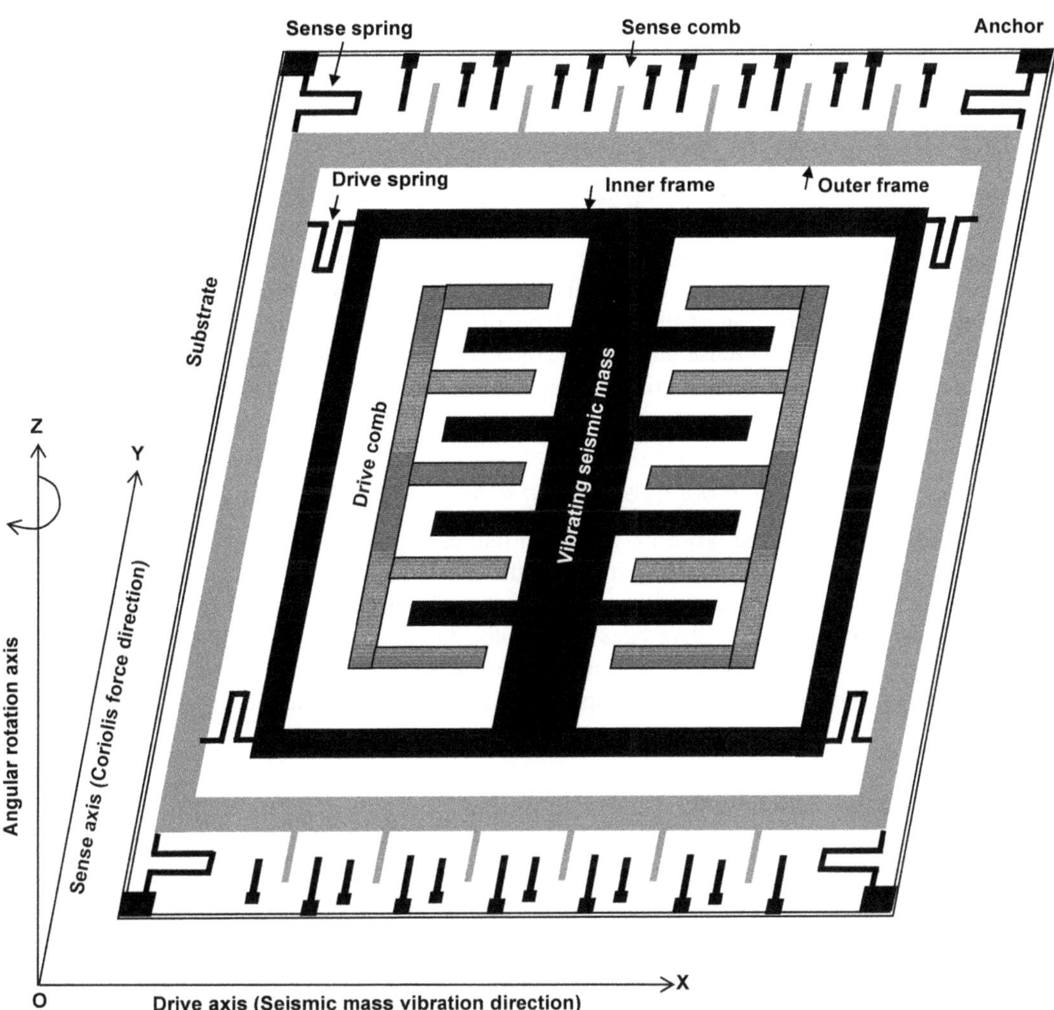

FIGURE 4.27   MEMS vibratory gyroscope showing that a seismic mass driven into vibration along the X-axis (driving axis) by applying a voltage to a drive comb experiences a Coriolis force-induced force along the Y-axis (sensing axis) when an angular rotation occurs around the Z-axis (rotational axis). The diagram shows a vibrating seismic mass (black portion) with black horizontal electrodes hanging from an inner frame by drive springs which are themselves hanging from an outer frame (gray portion). The black horizontal electrodes of seismic mass move between hatched horizontal electrodes, forming the drive comb. The outer frame with attached gray vertical electrodes is hanging from anchor points on a substrate by sense springs. Black vertical electrodes are fixed between gray vertical electrodes to form the sense comb. The seismic mass vibrates along the X-axis (drive axis). Angular rotation takes place about the Z-axis (rotation axis). The resulting Coriolis force acts along the Y-axis (sense axis).

freely. The suspension is done with the help of flexible beams which act as springs. Comb fingers are attached to the seismic mass. These fingers move with the proof mass when it moves. Hence, they are called movable fingers. Another set of comb fingers, the fixed or static fingers remain stationary. The pair of movable and static comb fingers constitute a comb drive.

4.8.5.2.1 Drive Motion, Drive Mode, and Drive Direction

When a voltage is applied between the movable and static fingers, an electrostatic force acts between them. Electrostatic comb driving is used to vibrate the proof mass in opposite directions along the X-axis at its resonance frequency. The axis (here, the X-axis) along which the proof mass is vibrating is referred to as the drive direction of the gyroscope. The motion is called drive motion, and the operation mode of the gyroscope is the drive mode.

4.8.5.2.2 Sense Motion and Sense Direction

Suppose the gyroscope undergoes an angular rotation along the Z-axis. Recalling the Coriolis effect, a Coriolis force starts acting on the proof mass. The direction of this Coriolis force is perpendicular to the angular velocity of the proof mass pointing along the axis of rotation, the Z-axis, and the direction of vibration of the proof mass, the X-axis. Hence, the rotation-induced Coriolis force on the proof mass acts in the direction of the Y-axis. This direction is called the sense direction, and the motion in this direction is the sense motion. The mode of operation of the gyroscope in sense direction is its sense mode. The movement of proof mass in the sense direction, the Y-axis, is proportional to its angular rotation.

4.8.5.2.3 Measurement of Proof Mass Motion in Sense Direction

This is done by translating the proof mass motion into capacitance changes between an interdigitated pair of electrodes. The supporting frame of the proof mass is suspended by springs and can move with it. One set of electrodes of the pair is fixed to the supporting frame of the proof mass so that it is movable. Another set of electrodes remains fixed on the substrate at their places. As the set of movable electrodes shifts from its position due to the Coriolis force, the capacitance between the movable and fixed electrodes varies. This capacitance change is related to the angular rotation of the gyroscope.

4.8.5.3 Gyroscope Parameters

The full-scale range of a gyroscope is the maximum angular velocity that can be measured by it. The full-scale range is specified in deg-s^{-1}. Its typical values are 450, 1000, and 2500 deg-s^{-1}. The sensitivity of a gyroscope is the change in output voltage for a given angular velocity, e.g., 30 mV $(\text{deg s}^{-1})^{-1}$. The bias offset of a gyroscope is the rotation shown when it is stationary. Bias instability is the instability of bias offset at constant temperature. Bias stability for a consumer-grade gyroscope may range between 30 and 1000 deg-hr^{-1}. The same for an industrial gyroscope is 1–30 deg-hr^{-1} but for a navigational gyroscope, it is 0.01–0.1 deg-hr^{-1}. (KVH white paper 2014). The frequency response of a gyroscope is represented by its bandwidth, the frequency at which its output decreases to 70% (~3 dB) of the magnitude of motion, e.g., a bandwidth of 120 Hz (Wu et al. 2017).

4.9 CONCLUDING REMARKS AND PREPARING FOR THE UPCOMING CHAPTER

The sensors considered in this chapter constitute a small segment of the team of sensors required for the efficient running of a factory. The second part of this chapter (Chapter 5) will give further glimpses of the large family of sensors on which the functioning of a factory relies.

REFERENCES

Huang D., D. Zhang, J.-P. Wang, and X. Wang 2022 Magnetization dynamics in synthetic antiferromagnets with perpendicular magnetic anisotropy, *Condensed Matter: Mesoscale and Nanoscale Physics*, arXiv: 2211.07744 [cond-mat.mes-hall], 24 pages.

Jeong C., S. Seok, B. Lee, H. Kim and K. Chun 2004 A study on resonant frequency and Q factor tunings for MEMS vibratory gyroscopes, *Journal of Micromechanics and Microengineering*, 14: pp. 1530–1536.

Khan M. A., J. Sun, B. Li, A. Przybysz and J. Kosel 2021 Magnetic sensors-A review and recent technologies, *Engineering Research Express*, 3:022005, pp. 1–22.

KVH White Paper 2014 Guide to Comparing Gyro and IMU Technologies https://caclase.co.uk/wp-content/uploads/2016/11/Guide-to-Comparing-Gyros-0914.pdf Accessed on 16th December 2022.

Liu Y., B. Hu, Y. Cai, W. Liu, A. Tovstopyat and C. Sun 2021 A novel tri-axial piezoelectric MEMS accelerometer with folded beams, *Sensors*, 21(2): 453, pp. 1–12.

Ono T., H. Naganuma, M. Oogane and Y. Ando 2010 Synthetic CoFeB/Ru/NiFe free layer on MgO barrier layer for spin transfer switching, *International Conference on Magnetism (ICM 2009) Journal of Physics: Conference Series*, 200:062019, pp. 1–4.

Patel C. and P. McCluskey 2012 Modeling and simulation of the MEMS vibratory gyroscope, *13th InterSociety Conference on Thermal and Thermomechanical Phenomena in Electronic Systems*, 30 May–01 June, San Diego, CA, 6pp.

Ranjbar S., M. Al-Mahdawi, M. Oogane and Y. Ando 2020 High-temperature magnetic tunnel junction magnetometers based on $L1_0$-PtMn pinned layer, *IEEE Sensors Letters*, 4(5): pp. 1–4. Article sequence no. 2500504.

Rohde & Schwarz 1MA182_5e Near Field Communication (NFC) Technology and Measurements White Paper, pp. 1–18.

Sankar A. R. and S. Das 2009 A high performance MEMS piezoresistive accelerometer with electroplated gold atop a thickness reduced proof mass, *8th IEEE International Sensors Conference, IEEE Sensors*, October 25–28, 2009, Christchurch, New Zealand, pp. 459–462.

Wu G.Q., G.L. Chua, and Y. D. Gu 2017 A dual-mass fully decoupled MEMS gyroscope with wide bandwidth and high linearity, *Sensors and Actuators A: Physical*, 259: pp. 50–56.

Xia D., C. Yu and L. Kong 2014 The development of micromachined gyroscope structure and circuitry technology, *Sensors*, 14: pp. 1394–1473.

Yang S. and J. Zhang 2021 Current progress of magnetoresistance sensors, *Chemosensors*, 9(8):211, pp. 1–23.

Yen C. T., W.-C. Chen, D.-Y. Wang, Y.-J. Lee, C.-T. Shen, S.-Y. Yang, C.-H. Tsai, et al. 2008 Reduction in critical current density for spin torque transfer switching with composite free layer, *Applied Physics Letters*, 93: pp. 092504–092504.

Zhai Y., H. Li, Z. Tao, X. Cao, C. Yang, Z. Che and T. Xu 2022 Design, fabrication and test of a bulk SiC MEMS accelerometer, *Microelectronic Engineering*, 260(C): pp. 111793.

Zhou X., L. Che, S. Liang, Y. Lin, X. Li and Y. Wang 2015 Design and fabrication of a MEMS capacitive accelerometer with fully symmetrical double-sided H-shaped beam structure, *Microelectronic Engineering*, 131: pp. 51–57.

5 IoT Sensors for Factory Automation-II

Temperature, Pressure, Vacuum, Force, Fluid Flow, and Level Measurements

5.1 INTRODUCTION TO CRITICAL PROCESS CONTROL PARAMETERS

Continuing from our discourse on factory sensors from the previous chapter, the present chapter will cast a look at some sensors that are critically vital to control industrial processes to get the desired outcomes. Among these sensors, mention may be made of the sensors for temperature, pressure, and vacuum level measurements, and force sensors, to name a few types. No less crucial is the part played by sensors that determine the flow rates of fluids and determine fluid levels. These sensors not only influence the quality of the finished product to provide quality assurance but they are also imperative for achieving the aimed manufacturing yields and for guaranteeing the safety of workers.

5.2 PREVENTION OF MACHINE OVERHEATING TO REDUCE DOWNTIME AND MAINTAINING CORRECT TEMPERATURES IN INDUSTRIES DEMANDING CRITICAL TEMPERATURE CONTROL VIZ., FOOD, MEDICAL, PHARMACEUTICAL, PETROCHEMICAL, AND SEMICONDUCTOR CHIP MANUFACTURING: TEMPERATURE SENSORS

Four types of temperature sensors are commonly used for different temperature ranges using dissimilar sensing principles for disparate applications (Kalsoom et al. 2020):

(i) Thermocouples (−250°C to +3,000°C; voltage change across a junction of different materials, e.g., copper/constantan; most widely used in an industry being rugged, reliable, and reasonably priced),

(ii) Resistance temperature detectors (RTDs; −50°C to +850°C; resistance change of a high-purity metal, e.g., Pt, Cu, Ni; used in HVAC, refrigerators, motors for overload protection, and automobiles, for air and oil temperature measurements),

(iii) Thermistors (−50°C to +250°C; resistance change of glass-coated oxides of Co, Mn, and Ni; used in automobiles to measure coolant temperature), and

(iv) Semiconductor IC chip-based sensors (−100°C to +400°C; change in a forward voltage drop of a PN junction diode; used for low-temperature measurements).

Small form factors, flexible and high-speed network communication, and remote configurability are desirable features of industrial temperature sensors.

5.2.1 THERMOCOUPLES

5.2.1.1 Principle and Governing Equation of Thermocouple

A thermocouple is a transducer that converts thermal energy into electrical energy. The thermocouple works on the thermoelectric effect known as the Seebeck effect (Morse 2022). This effect is

DOI: 10.1201/9781003374442-5

observed when a closed circuit is formed by joining two conductors, such as wires made of dissimilar metals. By joining the two conductors, two junctions are formed between them. On maintaining these two junctions at different temperatures, an electromotive force (EMF) is created in the circuit. As a result, a current begins to flow in the circuit. The magnitude of EMF is proportional to the temperature difference between the junctions. Furthermore, the values of the EMF are different for various combinations of conducting materials.

The junction placed at the point of unknown temperature is called the sensing junction while the junction at the point of known temperature, the room temperature, is known as the reference junction (Figure 5.1(a)). The EMF produced = voltage difference ΔV between the sensing junction at temperature T_{Sensing} and reference junction at temperature $T_{\text{Reference}}$ is given by

$$\Delta V = S\left(T_{\text{Sensing}} - T_{\text{Reference}}\right) = S\Delta T \tag{5.1}$$

where

$$\Delta T = T_{\text{Sensing}} - T_{\text{Reference}} \tag{5.2}$$

Also, S is the Seebeck coefficient of the thermocouple for the pair of materials used and is defined as the voltage difference generated per unit temperature difference between the sensing and reference junctions.

When

$$T_{\text{Sensing}} > T_{\text{Reference}} \tag{5.3}$$

ΔV is positive. In a situation where the sensing junction is at a temperature lower than that of the reference junction, ΔV is negative.

5.2.1.2 Measurement Circuit

Practically, a thermocouple is made by welding together two wires made from dissimilar metals at a point to form a junction. In the thermocouple, this junction is the point at which temperature is measured. It is the sensing junction. The opposite ends of the two wires at which the second junction is to be formed are not joined together but are kept separate. These ends constitute the reference junction. A voltmeter is connected between these ends (Figure 5.1(b)).

5.2.1.3 Construction

The wires of the thermocouple are enclosed in an Inconel sheath (Figure 5.1(c)). MgO powder is filled around the wires preventing them from vibration-induced damage and enhancing the transference of heat between the junction and surroundings.

5.2.1.4 Thermocouple Classes

Combinations of dissimilar conductors have been standardized and classified into eight types (Table 5.1) (Thermocouples-LibreTexts Engineering 2021, Thermocouples-madur.com, n.d.).

5.2.1.5 Advantages of Thermocouples

   (i) Simple and rugged in construction, compact in size, hence small inertia and thermal capacity,
  (ii) Ability to measure the temperature of a small localized area,
 (iii) A wide measurement range,
 (iv) Easy calibration, good reproducibility, reliability, and durability,
  (v) Fast response,
 (vi) Require no excitation power and hence no self-heating, and
(vii) Low cost for general-purpose thermocouples.

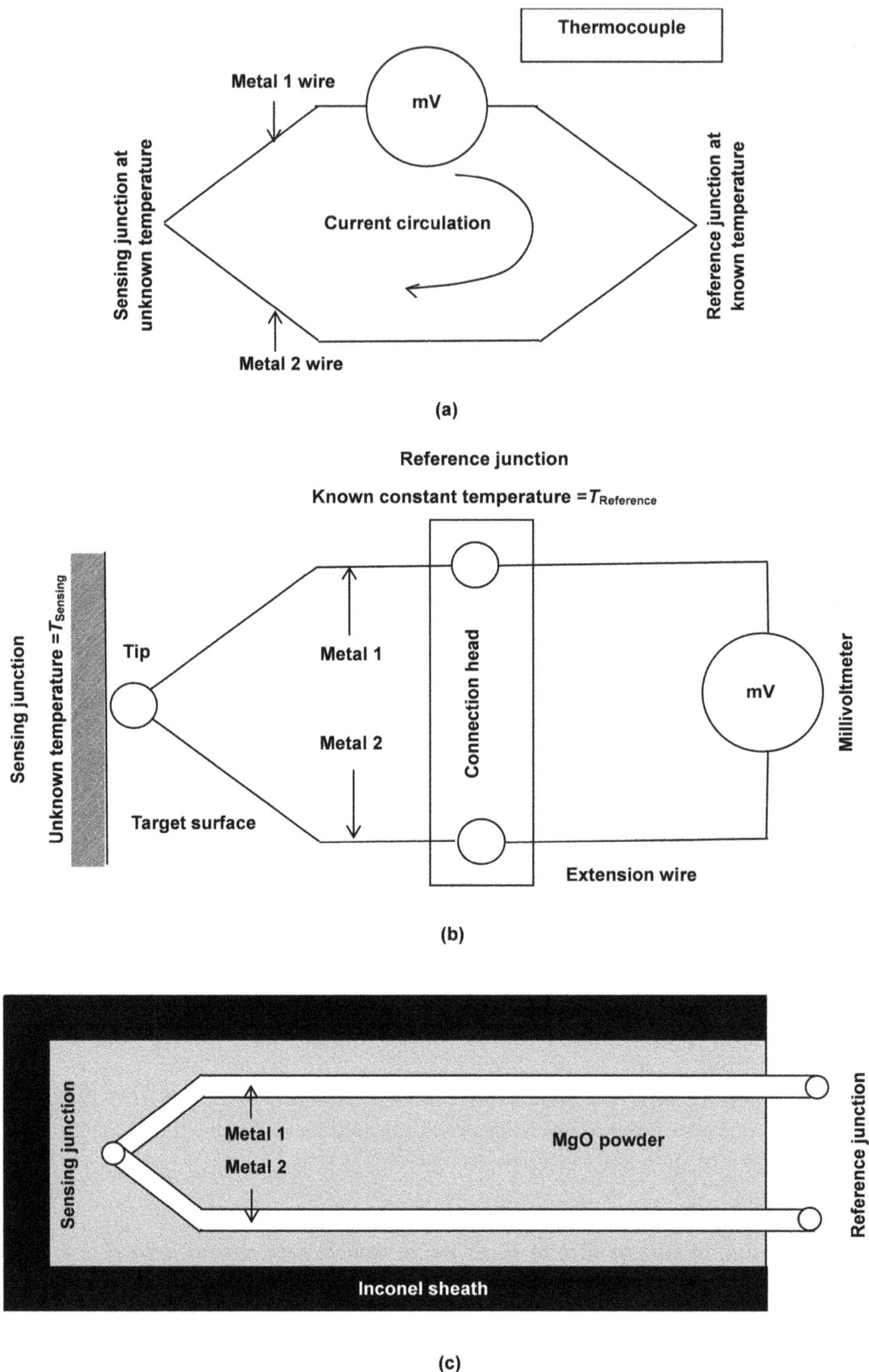

FIGURE 5.1 Thermocouple principle and use: (a) setting up of a current flow between hot and cold junctions of dissimilar metals, (b) appearance of a voltage when the circuit is opened, which is measured with a voltmeter, and (c) construction of the thermocouple. Part (a) shows metal 1 and metal 2 wires, a sensing junction at an unknown temperature, a reference junction at a known temperature, a millivoltmeter, and the direction of current circulation. Part (b) shows the sensing junction at an unknown temperature $T_{Sensing}$, target surface; metal 1 and metal 2 wires; connection head, extension wire, and millivoltmeter. Part (c) shows metal 1 and metal 2 wires, MgO powder, Inconel sheath, sensing, and reference junctions.

TABLE 5.1
Types of Thermocouples

Sl. No.	Type	Conductor Combinations	Seebeck Coefficient µVK⁻¹	Operating Range
1.	K	Ni-Cr/Ni-Al: Chromel/Alumel (Chromel = 90%Ni, 10% Cr; Alumel = 95% Ni, 2% Al, 2% Mn, and 1% Si alloy)	40	−200°C to +1260°C
2.	J	Fe/Constantan (Constantan = 55% Cu, 44% Ni, 1% Si alloy)	51	0°C to +760°C
3.	T	Cu/Constantan	40	−250°C to +370°C
4.	E	Chromel/Constantan	62	−200°C to +870°C
5.	N	Nisil/Nicrosil (Nisil = 95.5% Ni, 4.4%, Si, and 0.1% Mg alloy Nicrosil = 84.1% Ni, 14.4% Cr, 1.4% Si, and 0.1% Mg alloy)	27	−270°C to +1260°C
6.	S	(90%Pt, 10% Rhodium)/Pt	11	−50°C to +1480°C
7.	R	(87%Pt, 13% Rhodium)/Pt	12	−50°C to +1480°C
8.	B	(70% Pt, 30% Rhodium)/(94% Pt, 6% Rhodium)	8	0°C to +1700°C

5.2.1.6 Limitations of Thermocouples

   (i) Give differential temperature measurements requiring that the temperature of the reference junction be known to find the sensing junction temperature,
  (ii) Stray voltage and noise pickup,
 (iii) Susceptible to corrosion,
 (iv) The voltage produced for a small temperature difference is too low and requires amplification,
  (v) Unsuitable for high-accuracy measurements,
 (vi) Non-linearity of signal,
(vii) Need for compensation cables and complex signal conditioning, and
(viii) Expensive noble metal thermocouples.

5.2.2 RESISTANCE TEMPERATURE DETECTORS (RTDs)

5.2.2.1 Platinum RTD

RTD is a temperature sensor that works on the change in electrical resistance of a metal resistor with temperature. This property is governed by the temperature coefficient of resistance (TCR) for the metal. Platinum is the most commonly used metal for fabricating the resistor. It has a TCR of 3,850 ppm-K⁻¹. The value 100 Ω at 0°C is chosen for Pt resistor. It is called a PT100 RTD. Copper and nickel are also sometimes used.

5.2.2.2 Callender-Van Dusen Equation

The Callendar-Van Dusen equation relates the resistance R_{RTD} (T) of a platinum resistor with temperature T in °C. For the PT100 RTD for temperatures T below 0°C, the equation is (Texas Instruments 2018)

$$R_{\mathrm{RTD}}(T) = R(0)\left\{1 + AT + BT^2 + CT^3(T - 100)\right\}, -200°C < T < 0°C \qquad (5.4)$$

At temperatures $T > 0$°C, the equation is

$$R_{\mathrm{RTD}}(T) = R(0)\left(1 + AT + BT^2\right), 0°C \leq T < 661°C \qquad (5.5)$$

In these equations, the coefficients A, B, and C have the values: $A = 3.9083 \times 10^{-3}$, $B = -5.775 \times 10^{-7}$, and $C = -4.183 \times 10^{-12}$.

5.2.2.3 Configurations

The platinum resistor is made in two configurations:

(i) Wire wound on a ceramic or glass bobbin as an inner or outer coil; a resistor with an outer wound coil is shown in Figure 5.2(a),
(ii) Platinum thin film on a ceramic substrate in a meander pattern (Figure 5.2(b)) to increase the length, with resistance value adjusted by laser trimming. A thin film RTD is better than a wire wound RTD because it is smaller in size, cheaper in price, faster in response, and more vibration-tolerant than the wire-wound RTD.

5.2.2.4 Measurement Circuits

For determining the resistance of a Pt resistor, a small current is passed through the resistor from a constant-current source to avoid self-heating and the consequent undesirable change in resistor value. The voltage drop produced across the resistor by the small current is measured.

A Wheatstone bridge is used for measuring the resistance (Figure 5.2(c)). The simplest measurement circuit is a 2-wire circuit (Figure 5.2(d)) in which the RTD of resistance R_{RTD} is connected across an arm of the Wheatstone bridge using two lead wires 1 and 2 of resistances R_{L1} and R_{L2}, respectively. The circuit component R_{RTD} is the resistance of RTD and R_1, R_2, and R_3 are fixed-value resistors. As can be seen, the bridge is balanced against the resistance $(R_{RTD} + R_{L1} + R_{L2})$ so that the resistance of the RTD together with the resistances of lead wires are measured, and not the pure RTD resistance, as required. So, this uncompensated method is erroneous.

The 3-wire circuit shown in Figure 5.2(e) works better than the 2-wire circuit. In this circuit, lead wires 1 and 2 are the current leads. Lead wire 3 is the potential lead through which current flows when the bridge is balanced. On the left side of the balance condition, the resistance $(R_{L1} + R_{L3})$ appears in the numerator while $(R_{L3} + R_{L2})$ can be seen in the denominator. By making

$$R_{L1} = R_{L2} = R_{L3} = R_L \tag{5.6}$$

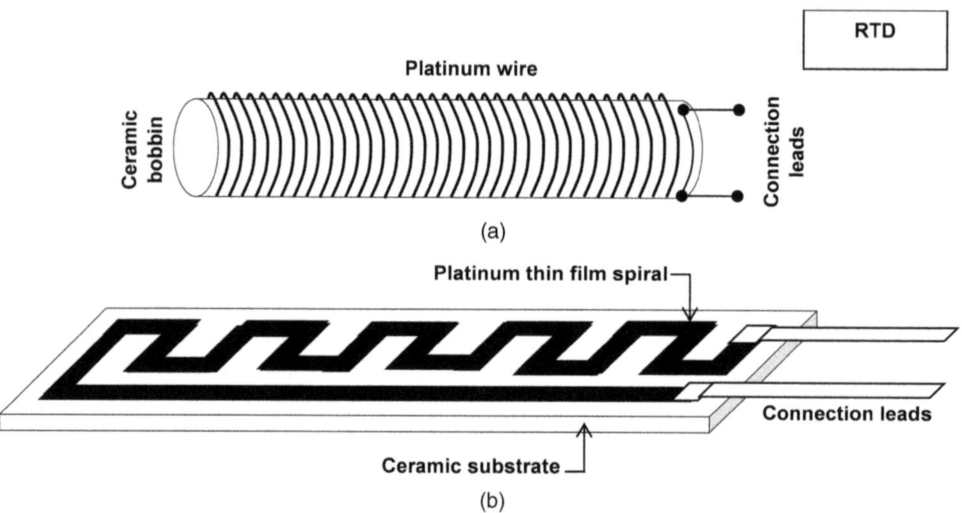

FIGURE 5.2 Resistance temperature detectors and their circuit diagrams: (a) wire wound, (b) thin film, (c) standard Wheatstone bridge, (d) 2-wire circuit, (e) 3-wire circuit, and (f) 4-wire circuit. Part (a) shows a ceramic bobbin, Pt wire wound around the bobbin, and connection leads. Part (b) shows a ceramic substrate, a Pt thin film spiral made on the substrate, and connection leads.

(Continued)

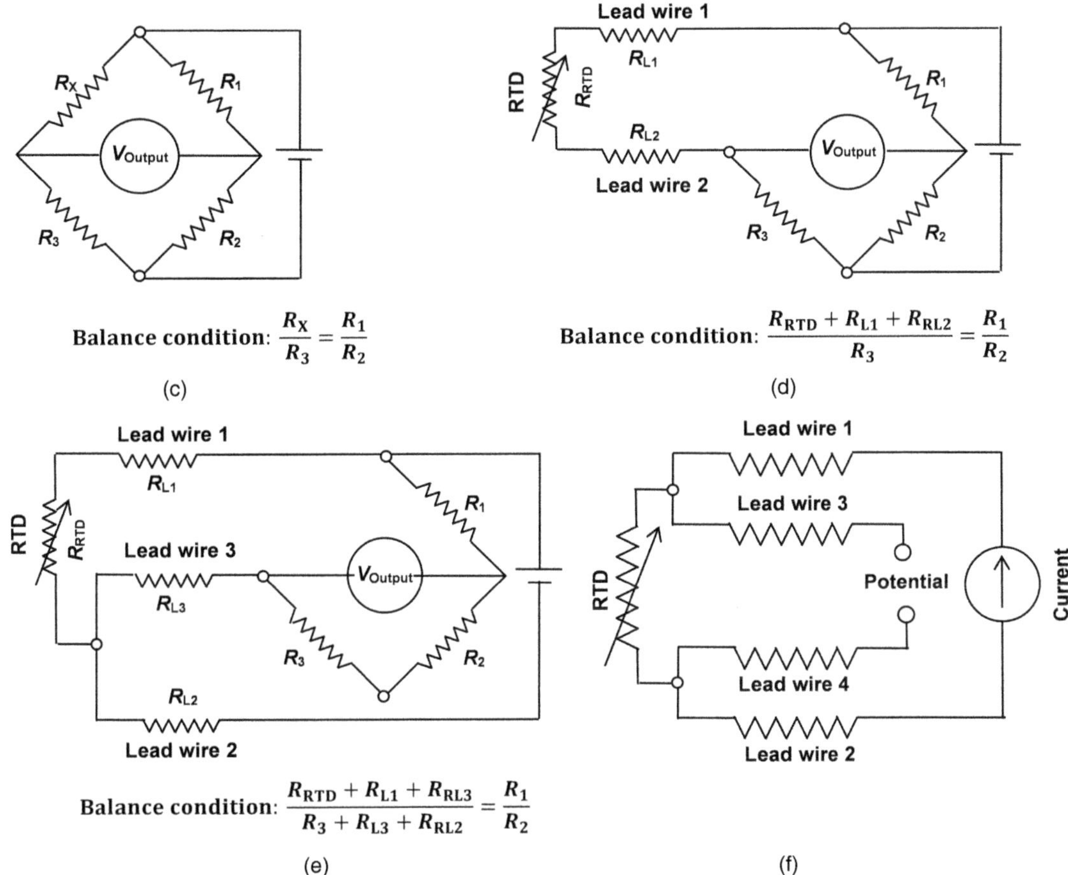

FIGURE 5.2 (CONTINUED) Part (c) shows the resistors R_1, R_2, R_3, and R_X, connected in a Wheatstone bridge circuit, and a battery supplying power to the bridge. The formula of the bridge balance condition is written. Part (d) shows resistors R_1, R_2, and R_3, as in part (c) but resistor R_X is replaced by a variable resistor R_{RTD} of the resistance temperature detector with resistances R_{L1} and R_{L2} of its lead wires 1, 2, respectively. A battery is connected to the circuit. The condition of the bridge balance changes from that given in part (c), and the modified equation is written. Part (e) shows resistors R_1, R_2, and R_3, as in part (c) but resistor R_X is replaced by a variable resistor R_{RTD} of the resistance temperature detector with resistances R_{L1}, R_{L2}, and R_{L3} of lead wires. A battery is connected to the circuit. The condition of the bridge balance changes from that given in part (d), and the modified equation is written. Part (f) shows the variable resistor R_{RTD} of the resistance temperature detector with lead wires 1 and 2 serving as current leads connected to a constant-current source, and lead wires 3 and 4 acting as potential leads.

the balance condition becomes

$$\frac{R_{RTD} + R_L + R_{RL}}{R_3 + R_L + R_{RL}} = \frac{R_1}{R_2} \tag{5.7}$$

or,

$$\frac{R_{RTD} + 2R_L}{R_3 + 2R_L} = \frac{R_1}{R_2} \tag{5.8}$$

so that both the numerator and denominator are equally affected by resistance $2R_L$, serving to weaken the influence of lead resistances. Hence the 3-wire circuit is more accurate than the 2-wire circuit. It can handle up to 30-m-long wire runs.

For most accurate measurements, a 4-wire circuit (Figure 5.2(f)) is used. It has two current leads delivering current to the resistor from a constant-current source. The voltage across the RTD is recorded by two potential leads which offer a high impedance to offset any voltage drop due to current flowing through the RTD. This circuit cancels and eliminates the effects of lead wire resistances. A 4-wire circuit is usable up to much longer distances than a 3-wire circuit.

RTDs show a good hysteresis behavior. They show stable long-term operation. They are good for accurate and precise measurements.

5.2.3 THERMISTORS

These are thermal resistors, i.e., temperature-sensitive resistors. They are nonmetallic resistors. The resistance of thermistors changes fast with temperature by more than 1,000 times.

5.2.3.1 Classification of Thermistors

Thermistors are subdivided into two classes: positive temperature coefficient (PTC) thermistors and negative temperature coefficient (NTC) thermistors.

(i) PTC Thermistors: The resistance of PTC thermistors increases with temperature. They are made from titanates of Ba, Pb, and Sr. The PTC thermistors are also made with ceramics and silicon. Switched PTC thermistors are used as self-regulating heaters with transition temperatures between 60 and 120°C, providing both the heat source and temperature control functions, and as resettable fuses for the protection of motors and transformers.

(ii) NTC Thermistors: The resistance of NTC thermistors decreases with temperature. The NTC thermistors consist of metal oxides, e.g., those of Cr, Ni, Mn, Co, and Fe. The metals are chemically oxidized, and the oxides produced are crushed to a fine powder, compressed, and heated to form different shapes such as discs, rods, or beads (Figure 5.3(a)–(c)), and sizes, and covered with an insulating glass or epoxy coating. Lead wires are fixed for electrical connection. The properties of thermistors are governed by the composition of oxides and the stabilizing agents used. They are used for measuring temperatures in the range $-100°C$ to $+300°C$.

5.2.3.2 NTC Thermistor Equation

NTC thermistors have nonlinear resistance-temperature characteristics described by the equation

$$R(T) = R_0 \exp\left\{\beta\left(\frac{1}{T} - \frac{1}{T_0}\right)\right\} \tag{5.9}$$

where $R(T)$ is the resistance of the thermistor at a temperature T, R_0 is its temperature at reference temperature T_0, and β is a material constant = 3,400–4,600 depending on the composition of the thermistor material; T, T_0 are in Kelvin scale. Taking the natural logarithm of both sides,

$$\ln R(T) = \ln R_0 + \beta\left(\frac{1}{T} - \frac{1}{T_0}\right) \tag{5.10}$$

or,

$$\beta\left(\frac{1}{T} - \frac{1}{T_0}\right) = \ln R(T) - \ln R_0 \tag{5.11}$$

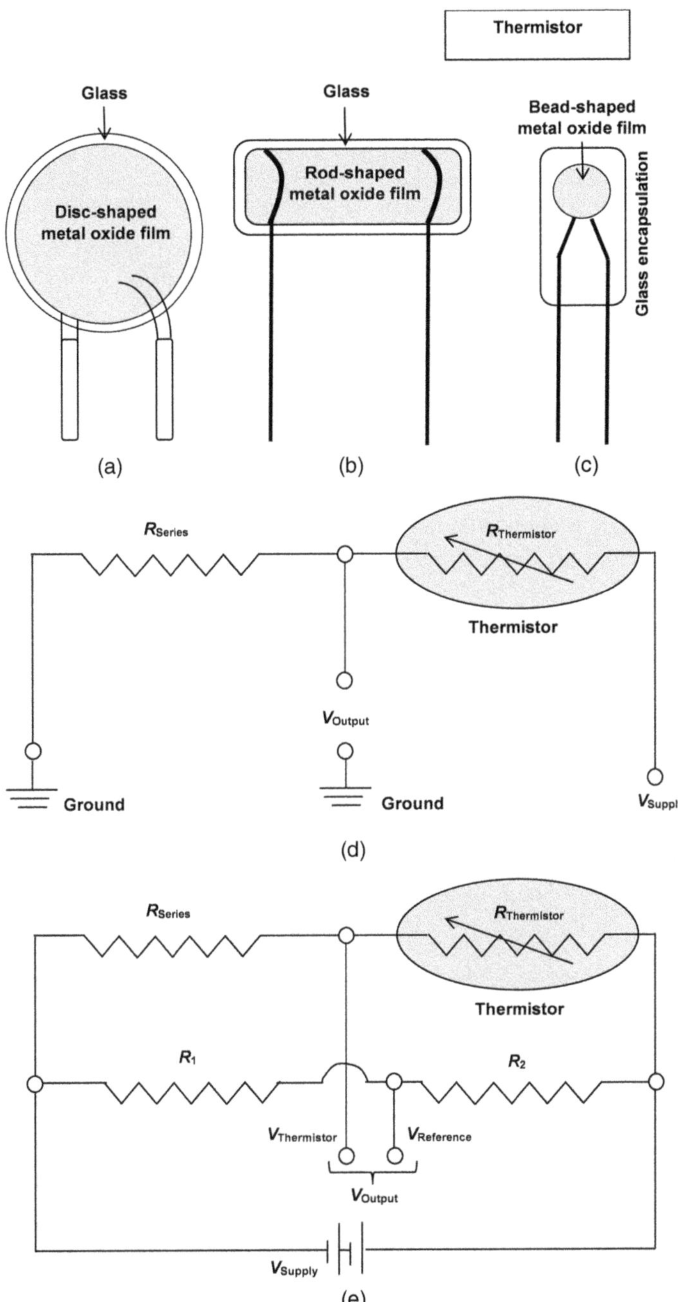

FIGURE 5.3 Thermistors: (a) disc, (b) rod, (c) bead, (d) voltage divider circuit, and (e) Wheatstone bridge circuit. Part (a): A disc-shaped metal oxide film is deposited on a glass substrate with one lead attached to its bottom surface and the other tied to the top surface of the disc. Part (b): A rod-shaped metal oxide film is deposited on a glass substrate with leads connected to the two edges of the film. Part (c): A bead-shaped metal oxide film is formed with leads connected from the two sides, and the metal oxide film is protected with a glass coating. Part (d): A thermistor with resistance $R_{Thermistor}$ is connected with a series resistor R_{Series} across the V_{Supply} and ground terminals. The output voltage V_{Output} is measured between the joining point of the $R_{Thermistor}$ and R_{Series} and ground. Part (e): A thermistor with resistance $R_{Thermistor}$ is connected with a series resistor R_{Series} across the negative and positive terminals of the V_{Supply} battery. Resistors R_1 and R_2 are also connected between the terminals of the V_{Supply} battery to form a bridge with resistors $R_{Thermistor}$ and R_{Series}. The output voltage V_{Output} is measured between the joining point of $R_{Thermistor}$ and R_{Series}, and the joining point of R_1 and R_2.

or,

$$\frac{\beta}{T} = \ln R(T) - \ln R_0 + \frac{\beta}{T_0} \tag{5.12}$$

or,

$$T = \frac{\beta}{\ln R(T) - \ln R_0 + \dfrac{\beta}{T_0}} = \frac{\beta T_0}{T_0 \ln\left\{\dfrac{R(T)}{R_0}\right\} + \beta} \tag{5.13}$$

5.2.3.3 Voltage Divider Circuit

The thermistor is a passive component. A power supply is connected across it for excitation. A potential divider circuit (Figure 5.3(d)) is a circuit which acts as a resistance change-to-voltage change converter, yielding voltage outputs in accordance with temperature variations. In this circuit, a constant voltage of V_{Supply} is applied across the thermistor $R_{\text{thermistor}}$, and a fixed series resistor R_{Series}, and the voltage drop across the thermistor is measured. This voltage V_{Output} is

$$V_{\text{Output}} = \left(\frac{R_{\text{Series}}}{R_{\text{Series}} + R_{\text{Thermistor}}}\right) V_{\text{Supply}} \tag{5.14}$$

5.2.3.4 Wheatstone Bridge Circuit

The thermistor is inserted in one arm of a Wheatstone bridge (Figure 5.3(e)). The values of resistors R_1 and R_2 are chosen to adjust the reference voltage $V_{\text{Reference}}$ to the required value. As soon as the temperature changes, the resistance of the thermistor ($R_{\text{Thermistor}}$) is altered affecting the voltage $V_{\text{Thermistor}}$ at the point between the thermistor and resistor R_{Series}. The difference between $V_{\text{Reference}}$ and $V_{\text{Thermistor}}$ is the output voltage V_{Output} for the given temperature:

$$V_{\text{Output}} = V_{\text{Thermistor}} - V_{\text{Reference}} = \left(\frac{R_{\text{Series}}}{R_{\text{Series}} + R_{\text{Thermistor}}} - \frac{R_1}{R_1 + R_2}\right) V_{\text{Supply}} \tag{5.15}$$

5.2.4 Semiconductor Diode Temperature Sensors

The forward voltage of a silicon diode has a linear temperature coefficient -2 mV°C^{-1}, making it useful as a temperature sensor (Figure 5.4(a)).

However, it is recommended that for temperature measurement, the emitter-base diode of a bipolar junction transistor should be used with the collector-base junction short-circuited (Figure 5.4(b)). The reason is that electrons flowing through the depletion region of an N-P junction diode recombine with holes leading to a recombination current. So, the forward current of the diode contains a component due to this recombination current, which makes the variation of the diode current with voltage and temperature non-linear. This recombination current is also present in the BJT but the difference is that it mostly flows into the base of the BJT. Therefore, its contribution to the collector current is reduced. As a consequence, the non-linearity of characteristics arising from this current is appreciably decreased (Bramble 2020).

The Ebbers–Moll model is a large-signal, steady-state representation of a bipolar transistor in terms of two diodes and two controlled current sources. At a temperature T, the collector current I_C of a transistor is related to the reverse saturation current I_S and the base-emitter voltage V_{BE} as

$$I_C = I_S \left\{ \exp\left(\frac{q V_{\text{BE}}}{k_B T}\right) - 1 \right\} \tag{5.16}$$

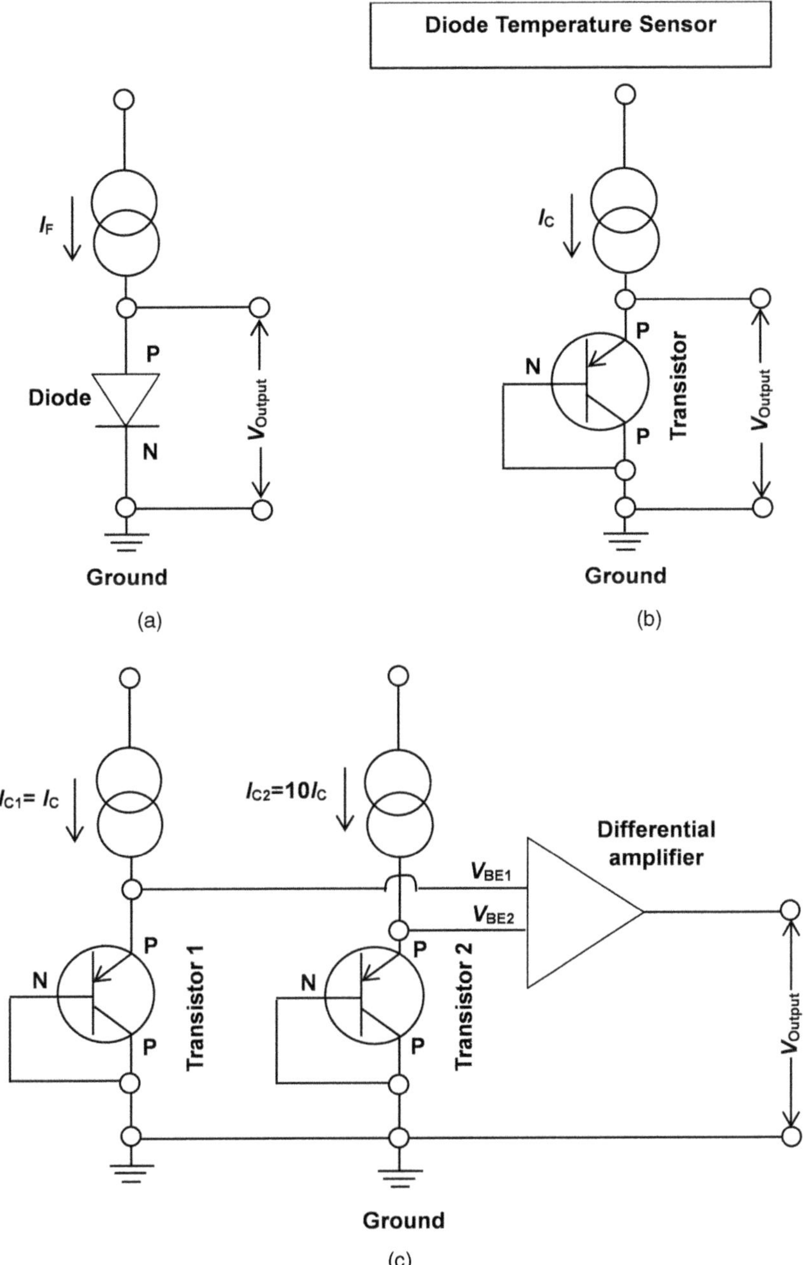

FIGURE 5.4 Semiconductor diode temperature sensors: (a) single diode sensor, (b) single transistor replacement for the diode sensor, and (c) two-transistor-based diode temperature sensor. Part (a): A P-N junction diode is connected between a constant-current source I_F and ground, and the output voltage V_{Output} is measured across the terminals of the diode. Part (b): A PNP transistor is connected with its emitter terminal joined to a constant-current source I_C, the collector terminal grounded, and the base terminal shorted with the collector terminal. The output voltage V_{Output} is measured between the emitter and collector terminals. Part (c): A PNP transistor 1 is connected with its emitter terminal joined to a constant-current source $I_{C1} = I_C$ and collector terminal connected to the ground. A PNP transistor 2 is connected with its emitter terminal joined to a constant-current source $I_{C2} = 10I_C$ and collector terminal to the ground. The base and collector terminals of transistor 1 are shorted together. Likewise, the base and collector terminals of transistor 2 are shorted together. The emitter terminals of transistors 1 and 2 at potentials V_{BE1} and V_{BE2}, respectively, are connected to a differential amplifier. The output voltage V_{Output} is measured between the output terminal of the differential amplifier and the ground.

where q is the electronic charge and k_B is the Boltzmann constant. Equation (5.16) can be rearranged as

$$I_C = I_S \exp\left(\frac{qV_{BE}}{k_B T}\right) - I_S \tag{5.17}$$

or,

$$I_S \exp\left(\frac{qV_{BE}}{k_B T}\right) = I_C + I_S \tag{5.18}$$

or,

$$\exp\left(\frac{qV_{BE}}{k_B T}\right) = \frac{I_C + I_S}{I_S} = \frac{I_C}{I_S} + 1 \tag{5.19}$$

or,

$$\frac{qV_{BE}}{k_B T} = \ln\left(\frac{I_C}{I_S} + 1\right) \tag{5.20}$$

$$\therefore V_{BE} = \frac{k_B T}{q} \ln\left(\frac{I_C}{I_S} + 1\right) \tag{5.21}$$

Since

$$I_C \gg I_S \tag{5.22}$$

or,

$$\frac{I_C}{I_S} \gg 1 \tag{5.23}$$

So, the second term in the bracket on the right-hand side of equation (5.21) is neglected, getting

$$V_{BE} = \frac{k_B T}{q} \ln\left(\frac{I_C}{I_S}\right) \tag{5.24}$$

In this equation k_B and q are constants. If I_C and I_S are constant then V_{BE} becomes proportional to temperature. It is possible to force a constant current I_C through the transistor. However, I_S depends on the geometry of the transistor and is also temperature sensitive, doubling for every 10°C rise in temperature, although the 'ln' function suppresses the change in $\ln(I_C/I_S)$. Moreover, the absolute value of V_{BE} exhibits device-to-device variation making calibration necessary. To avoid these problems, temperature sensors are made by

(i) Using two transistors 1 and 2 of similar geometry and electrical parameters, and
(ii) Scaling the current through the diodes of these transistors, e.g., forcing a current $I_{C1} = I_C$ through transistor 1 and upscaled current $I_{C2} = 10I_C$ through transistor 2 (Figure 5.4(c)).

Since the two transistors are at the same temperature T, the difference in voltages across their base-emitter diodes is expressed as

$$\Delta V_{BE} = V_{BE2} - V_{BE1} = \frac{k_B T}{q} \ln\left(\frac{10 I_C}{I_S}\right) - \frac{k_B T}{q} \ln\left(\frac{I_C}{I_S}\right) = \frac{k_B T}{q}\left\{\ln\left(\frac{10 I_C}{I_S}\right) - \ln\left(\frac{I_C}{I_S}\right)\right\}$$

$$= \frac{k_B T}{q}\left\{\ln\left(\frac{10 I_C}{I_S} \div \frac{I_C}{I_S}\right)\right\} = \frac{k_B T}{q}\left\{\ln\left(\frac{10 I_C}{I_S} \times \frac{I_S}{I_C}\right)\right\} = \frac{k_B T}{q} \ln(10)$$

$$= 2.3026 \frac{k_B T}{q} = \text{Constant} \times T \tag{5.25}$$

where

$$\text{Constant} = 2.3026 \frac{k_B}{q} \tag{5.26}$$

Hence,

$$\Delta V_{BE} \propto T \tag{5.27}$$

Thus, the effects of I_S along with its dependence on transistor geometries have been removed as well as those of different V_{BE} values of transistors. With the 10-fold collector current scaling, ΔV_{BE} shows a linear variation with the temperature at the rate of 198 μV°C^{-1} (Bramble 2020).

5.2.5 THERMOSTATS

The working of a thermostat is explained in Figure 5.5.

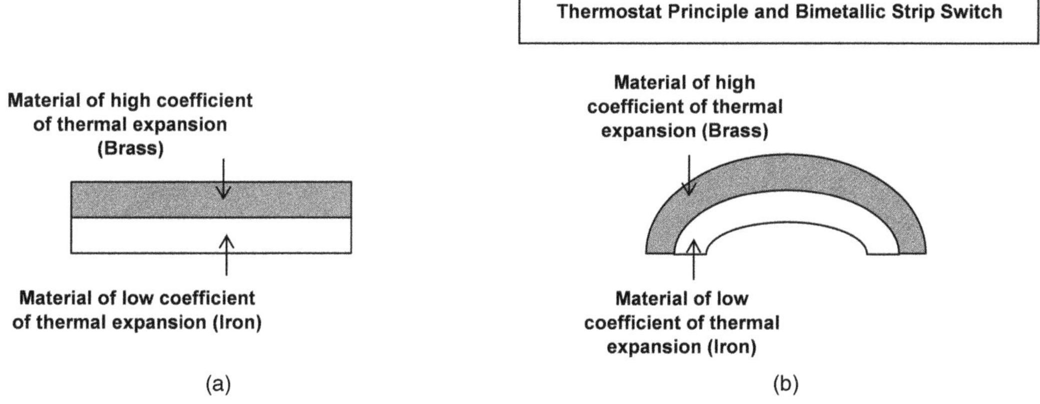

FIGURE 5.5 Thermostat principle and bimetallic strip-based switch: (a) straight bars at normal temperature, (b) straight bars upon heating, (c) switch closed at room temperature and (d) switch open on heating. Parts (a) and (b): Straight bars of brass (high-temperature coefficient = 1.92 × 10^{-5} °C^{-1}) and iron (low-temperature coefficient = 1.2 × 10^{-5} °C^{-1}) are joined together. Part (a): The condition of bars at room temperature is shown. Part (b): The bent shape of bars on heating is shown. Parts (b) and (c): A brass strip and an iron strip are welded or bolted together; they are clamped at one end to a rigid support but are free to move at the opposite end. Part (c): At room temperature or when cold, the two strips are fastened together as a straight bar touching and closing the contact point.
(Continued)

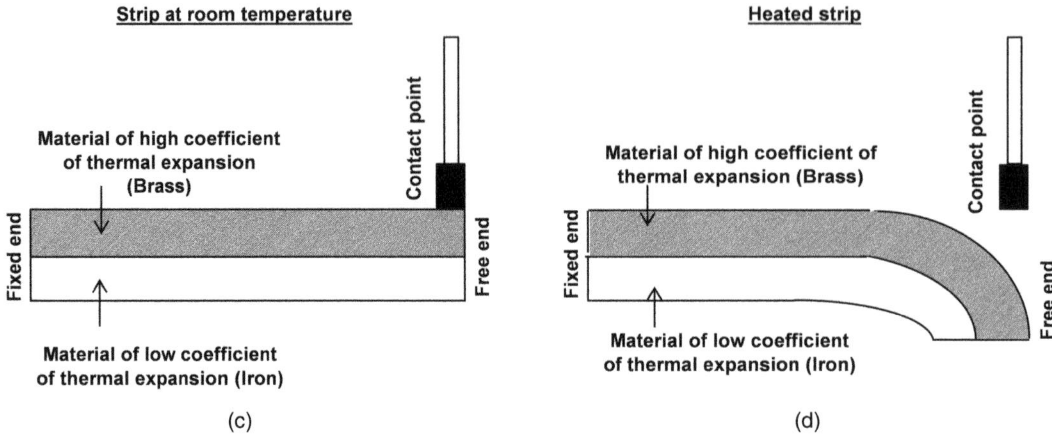

FIGURE 5.5 (CONTINUED)   Part (d): When heated equally, the two materials (brass and iron) expand by different amounts depending on their thermal expansion coefficients. The brass strip elongates more than the iron strip for a given rise of temperature so that the bimetallic strip bends with brass outside and iron inside. On bending, the strip no longer touches the contact point, and the circuit is broken.

5.3   GAS COMPOSITION MONITORING IN WELDING ENVIRONMENTS, E.G., WELDING OF TI: OXYGEN, HUMIDITY AND TEMPERATURE SENSORS

Titanium is extremely sensitive to small quantities of oxygen which markedly decreases its ductility (Chong et al. 2020). Due to the oxygen poisoning effect, titanium welding is critically dependent on argon shielding gas. It requires 100% Ar shielding with a minimum of 99.995% purity. Oxygen concentration should be less than 20 ppm. To ensure the correct shielding gas supply, the oxygen level is monitored using a zirconia sensor. It is an electrochemical galvanic cell which measures the oxygen concentration by the EMF produced between two porous electrodes across a zirconium oxide (ZrO_2) solid electrolyte disc heated to 750°C with a ceramic heater. The hygroscopic TiO_2 absorbs moisture from the atmosphere which adversely affects the welding of Ti. So, a dewpoint sensor is used to measure the water vapor content from +20°C to −100°C. Additionally a type K thermocouple (nickel–chromium/nickel–alumel, −270°C to +1260°C) is used for measuring temperature.

5.4   MEASURING FLUID PRESSURE IN PIPELINES TO FIND ANY IRREGULARITIES ALERTING FOR MAINTENANCE AND REPAIR: MEMS PRESSURE SENSORS

5.4.1   Types of Pressure Sensors

Two types of pressure sensors are widely used: piezoresistive and capacitive (Eaton and Smith 1997, Song et al. 2020). In both types, a thin membrane or diaphragm is deformed when the pressure is applied. In the piezoresistive sensor, the bending of the diaphragm causes changes in the resistances of resistors fabricated on the diaphragm surface. The resistance changes are utilized for pressure measurement. In the capacitive sensor, the bending of the diaphragm decreases the gap between it and a facing fixed plate, enabling the determination of pressure from capacitance changes. Advantages of piezoresistive sensors include linearity of characteristics and operational stability. Their temperature sensitivity is the main drawback. Capacitive sensors show higher sensitivity and insignificant temperature dependence.

5.4.2 PIEZORESISTIVE PRESSURE SENSORS

Pressure sensors of various specifications for different applications are made by microelectromechanical systems (MEMS) manufacturing processes (Bhat and Nayak 2013). Figure 5.6 shows two types of piezoresistive pressure sensors fabricated using MEMS technology: differential and absolute.

5.4.2.1 Differential Pressure Sensor

As shown in Figure 5.6(a), the sensor consists of a silicon base with a cavity on top of which a diaphragm is suspended. At the edges of the diaphragm where it joins with silicon, piezoresistors are formed by thermal diffusion. The silicon base is bonded to a glass plate. A hole is drilled in the glass plate from which a gas can be introduced. If the pressure acting on the upper surface of the diaphragm is P_1 and that on its lower surface is P_2, the differential pressure acting on the diaphragm is given by the equation

$$P_{\text{Differential}} = P_1 - P_2 \tag{5.28}$$

Often the pressure P_2 is the atmospheric pressure. A differential pressure sensor is used when a differential measurement is required such as with respect to atmospheric pressure or some other pressure defined in an experiment. However, for performing comparative measurements, it is necessary to fix the value of P_2 at a particular value. The atmospheric pressure being a varying pressure which changes every moment, cannot be used as a reference in such situations. Absolute pressure sensors are used for this purpose.

5.4.2.2 Absolute Pressure Sensor

An absolute pressure measurement is performed with reference to vacuum or zero pressure. Figure 5.6(b) shows an absolute pressure sensor. The absolute pressure sensor looks identical in structure to the differential pressure sensor. However, there is an important difference between the two types of sensors. A careful inspection of the diagram reveals that the main difference between this sensor and the differential pressure sensor is that the glass plate is bonded to the silicon wafer in such a manner as to maintain vacuum in the cavity. Hence, the pressure in the cavity is zero, $P_2 = 0$. So, the absolute pressure sensor measures pressure acting on the diaphragm with reference to vacuum. Because the pressure in the cavity is always kept constant at zero value, the pressure values recorded with this pressure sensor in different conditions can be compared among themselves allowing us to assess their relative magnitudes and perform a comparative analysis.

Figure 5.7(a) shows the 3D view of an absolute pressure sensor with a square diaphragm. On the diaphragm surface, four resistors are seen near the boundaries of the diaphragm. The resistors are made at these places because these are the locations which experience maximum stress when the diaphragm bends on applying pressure upon it. These resistors are connected together to form a Wheatstone bridge circuit. The power source is connected between the terminals V_{Input} and ground, as indicated on the diagram. Across the terminals designated as V_{Output}, the output voltage of the bridge is measured.

Comprehensive numerical simulations on pressure-induced longitudinal/transverse stresses in the diaphragm, positions of piezoresistors on the diaphragm, and diaphragm thickness are performed. By positioning two resistors on the center and two resistors on the edges, and using a diaphragm thickness of 2 μm, a high sensitivity of 37.79 mV-V^{-1} MPa^{-1} with a full-scale output of 472.33 mV is demonstrated on a MEMS pressure sensor fabricated with SOI wafer and glass wafer; the pressure range is 0.25–2.5 MPa (Meng et al. 2021).

Pressure sensors for harsh environments in industrial applications require close attention to packaging (Vierinen 2018). When a pressure sensor is to be used to measure the pressure of a liquid, it must be properly packaged in a hermetic enclosure in such a manner that its vulnerable parts and wires are protected from exposure to the liquid. At the same time, there should be a mechanism by

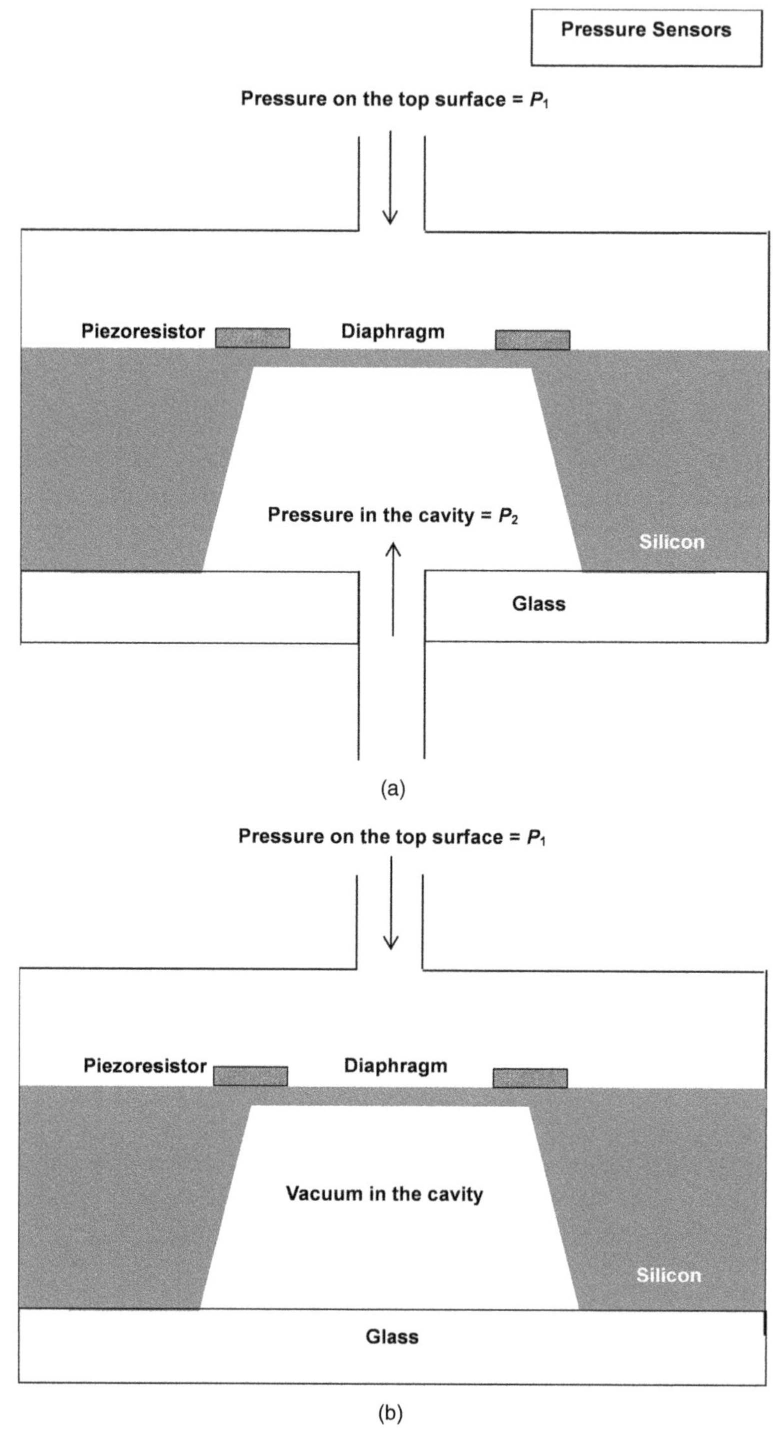

FIGURE 5.6 Pressure sensors: (a) differential sensor with a pressure P_1 acting on the top surface and P_2 on the bottom surface, (b) absolute sensor with the pressure P_1 acting on the top surface measured with reference to the bottom surface exposed to vacuum. Parts (a) and (b): A diaphragm is suspended over a cavity in silicon with resistors made at the joining points of the diaphragm with the supporting pillars. The bottom of the cavity is closed with a glass substrate. Pressure P_1 acts on the top surface of the diaphragm through a hole in the roof of the enclosure. Part (a): Pressure P_2 is acting on the bottom surface of the diaphragm through a hole in the glass substrate. Part (b): There is no hole in the glass substrate, and the cavity is vacuum-sealed.

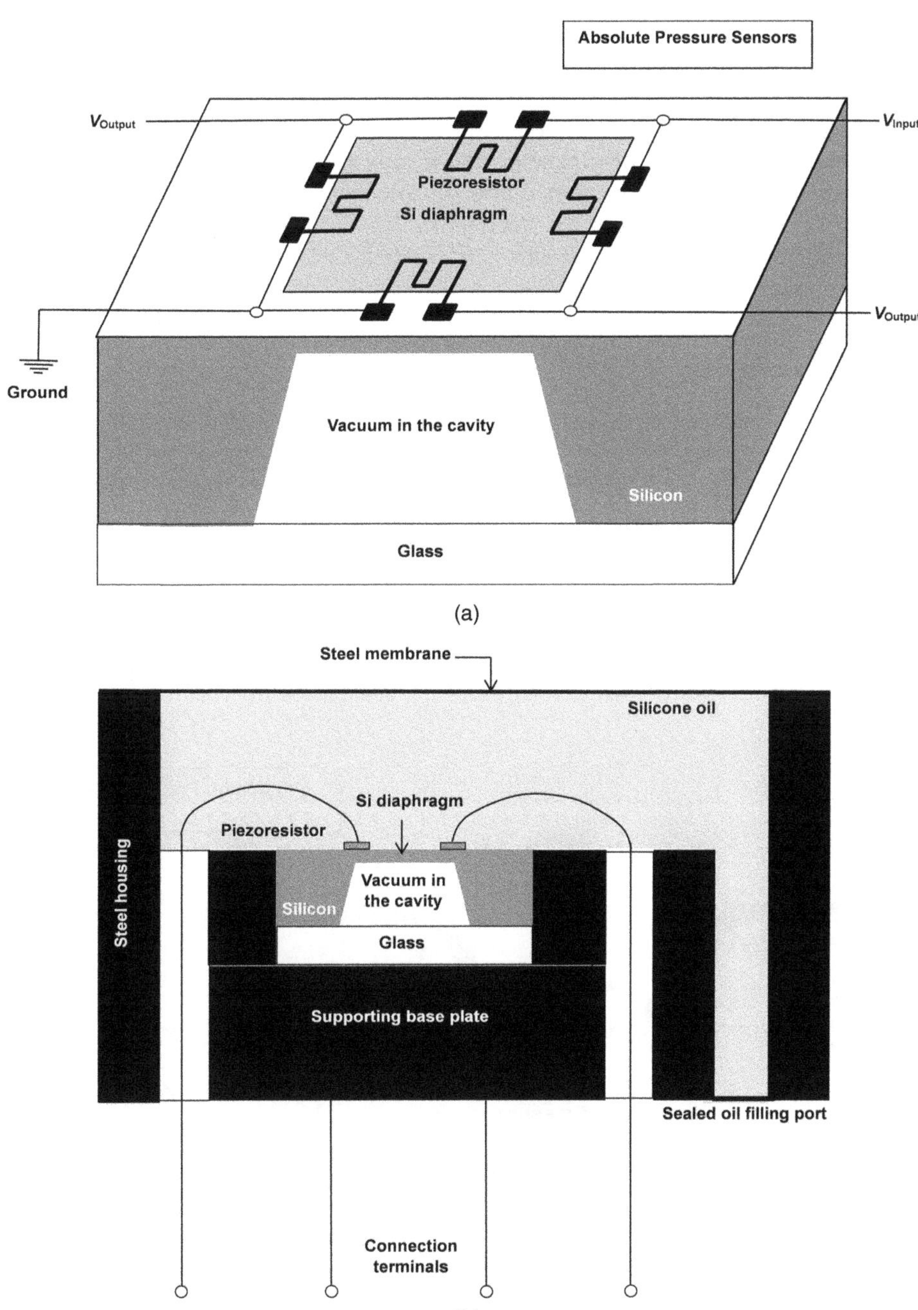

FIGURE 5.7 Absolute pressure sensors: (a) three-dimensional view and (b) hermetically packaged sensor in a stainless-steel housing for measurement of liquid pressure. The space over the silicon diaphragm is filled with silicone oil. The liquid pressure acting on the steel membrane is transmitted to the silicon diaphragm through the oil. Part (a): A rectangular Si diaphragm is suspended over a cavity in silicon with four spiral resistors formed at the boundaries of the diaphragm with the surrounding supporting silicon. These resistors are connected together to form a Wheatstone bridge. The cavity is closed on the bottom side with a glass substrate after evacuation to produce vacuum in the cavity. A power supply is connected across V_{Input}/ground terminals. The output voltage is measured across the two terminals marked as V_{Output}. Part (b): Cross-sectional view of the pressure sensor of Part (a) supported on a base plate and packaged in a steel housing filled with silicone oil through an oil filling port, which is sealed after filling oil. Wires are bonded to the piezoresistors, and the connection terminals are shown.

which the pressure of the liquid is faithfully transmitted to the diaphragm otherwise sensitivity of the pressure sensor will be impaired.

An absolute pressure sensor is packaged in a steel housing, as shown in Figure 5.7(b). The sensor chip is mounted on a steel base plate. Over the diaphragm, the package is filled with silicone oil. A steel membrane is fixed to cover the oil and protect it from spilling. The liquid pressure acts on the steel membrane from which it is transported through silicone oil and delivered to the diaphragm of the silicon chip. The oil filling port of the steel housing is closed by welding after packaging. The housing has openings through which the wires are inserted for connecting the piezoresistors to the connection terminals of the packaged device.

5.4.3 Capacitive Pressure Sensors

A capacitive pressure sensor (Figure 5.8) has one conductive layer on the diaphragm surface to serve as the top plate of the capacitor and another conductive layer on the base of the cavity to act as the bottom plate. The movable Al electrode of the sensor is made on a recessed silicon diaphragm formed by bulk micromachining of silicon (Chandra et al. 2014). The fixed Al electrode is made on a glass wafer. The glass plate with the metal pattern is fixed to the silicon wafer under vacuum by silicon-to-glass anodic bonding forming a press-on-contact. The sensor shows a sensitivity of 1 pF/kg-cm^{-2} over the pressure range of 0–0.6 kg-cm^{-2}.

The sensitivity of a capacitive pressure sensor can be increased by using a larger and thinner diaphragm and keeping a smaller sensing gap. However, these improvements are achieved at the

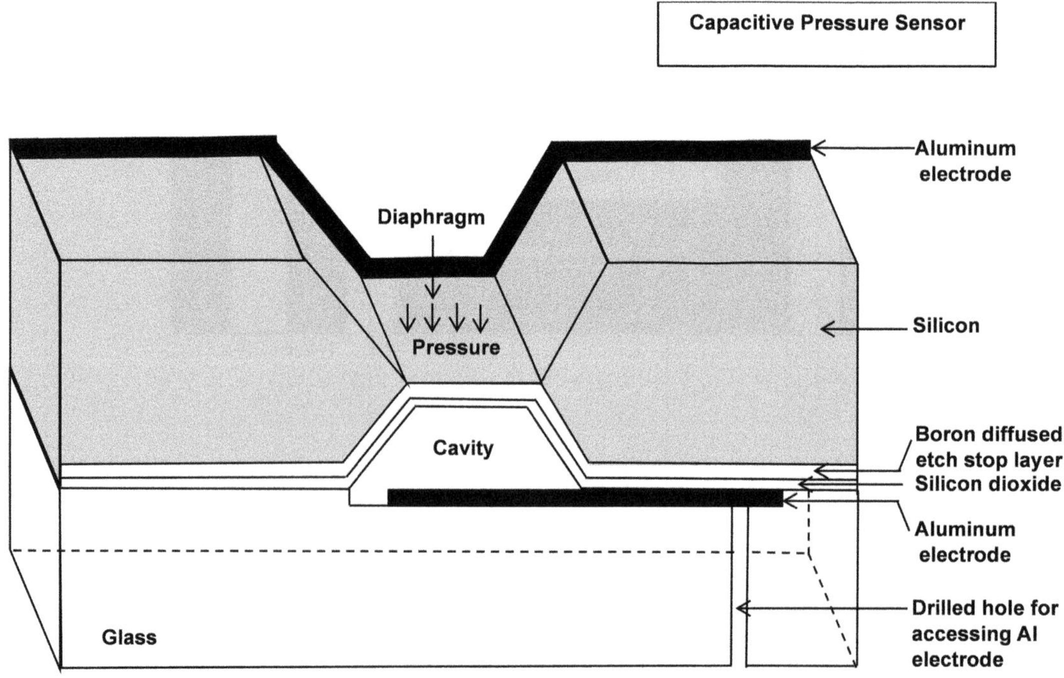

FIGURE 5.8 MEMS capacitive pressure sensor fabricated by anodic bonding between silicon and glass. A Si diaphragm having a boron-diffused etch stop layer covered with an SiO$_2$ film is hanging over a cavity under vacuum. An Al electrode passing over the Si diaphragm is used for one plate connection. The second plate connection is provided by an Al electrode made on a glass wafer, which is bonded to the above Si wafer to serve as the floor of the cavity. Contact to this Al electrode is established by drilling a hole in the glass wafer to reach aluminum. Pressure applied on the diaphragm is measured.

expense of increasing the size of the sensor and the non-linearity of its characteristics together with the imposition of limitations on its dynamic range (Zhang et al. 2011).

The change in capacitance with pressure is measured by using the capacitive pressure sensor as the changing capacitive component of an inductance–capacitance (LC) or resistance–capacitance (RC) oscillator circuit. The frequency of the oscillator circuit is altered when the capacitance of the pressure sensor changes with variation of pressure. Pressures are obtained from frequency values. Another way to determine the capacitance is to measure the time taken to charge the capacitance from a current source. In all cases, the measurement circuit must be integrated with the pressure sensor because it is essential to minimize the stray capacitances that come into play with long wiring distances.

5.5 MEASURING VACUUM PRESSURE IN VACUUM SYSTEM-BASED MACHINES: VACUUM SENSORS AND GAUGES

5.5.1 VACUUM SENSOR

A vacuum sensor is a pressure-measuring device used to measure the degree of vacuum in an enclosed space such as pressure in a vessel at sub-atmospheric pressure. Vacuum is divided into several categories: low vacuum ($10^5 - 10^2$ Pa), medium vacuum ($10^2 - 10^{-1}$ Pa), high vacuum ($10^{-1} - 10^{-5}$ Pa), ultrahigh vacuum ($10^{-5} - 10^{-8}$ Pa), and extremely high vacuum ($\leq 10^{-8}$ Pa). 1 Pa = 7.50062×10^{-3} Torr, 1 Torr = 133.32 Pa = 0.999 mm Hg = 1.33322 mbar, and 1 bar = 750.0617 mm Hg.

Vacuum sensors are used in factories for the measurement of vacuum levels to ensure the proper functioning of machines. They are widely used in paper milling, pharmaceutical, and off-shore drilling work, petroleum refineries, semiconductor IC manufacturing, and chemical plants. Broadly, vacuum sensors are divided into three classes: mechanical, thermal conductivity, and ionization gauges.

5.5.2 MECHANICAL GAUGES

5.5.2.1 Bourdon Vacuum Gauge

The gauge is shown in Figure 5.9. It consists of a C-shaped hollow tube of stainless steel or phosphor bronze with:

(i) A Free Moving End, closed at its edge and connected to a pivot and a pin attached to a sector gear to which is tied an indicator needle moving on a calibrated scale, and
(ii) A Fixed End connected to the fluid inlet pipe from which the fluid flows into the tube. This pipe is enclosed in a socket block.

When the fluid enters the C-tube, the pressure exerted by the fluid tends to straighten the C-shaped tube. The straightening movement is transferred by the pivot and pin to the sector gear which amplifies it so that a small pressure change causes a large deflection of the indicator needle. It is used to measure pressure in the range of 0.6–7,000 bar.

Advantages: High accuracy, easy to use, and low-cost gauge.
Disadvantages: Slow response, hysteresis, vibration, and shock sensitivity.
Applications: Pneumatic and hydraulic systems, power generation, oil and gas production, automotive and transportation, and HVAC systems.

5.5.2.2 Diaphragm Vacuum Gauge

Also called a seal pressure gauge, its sensitive component is a circular corrosion- and high-temperature-resistant stainless-steel diaphragm (Figure 5.10). The diaphragm is clamped in a housing between two flanges. The housing is connected at its lower end to the fluid inlet. Fluid pressure acting on the

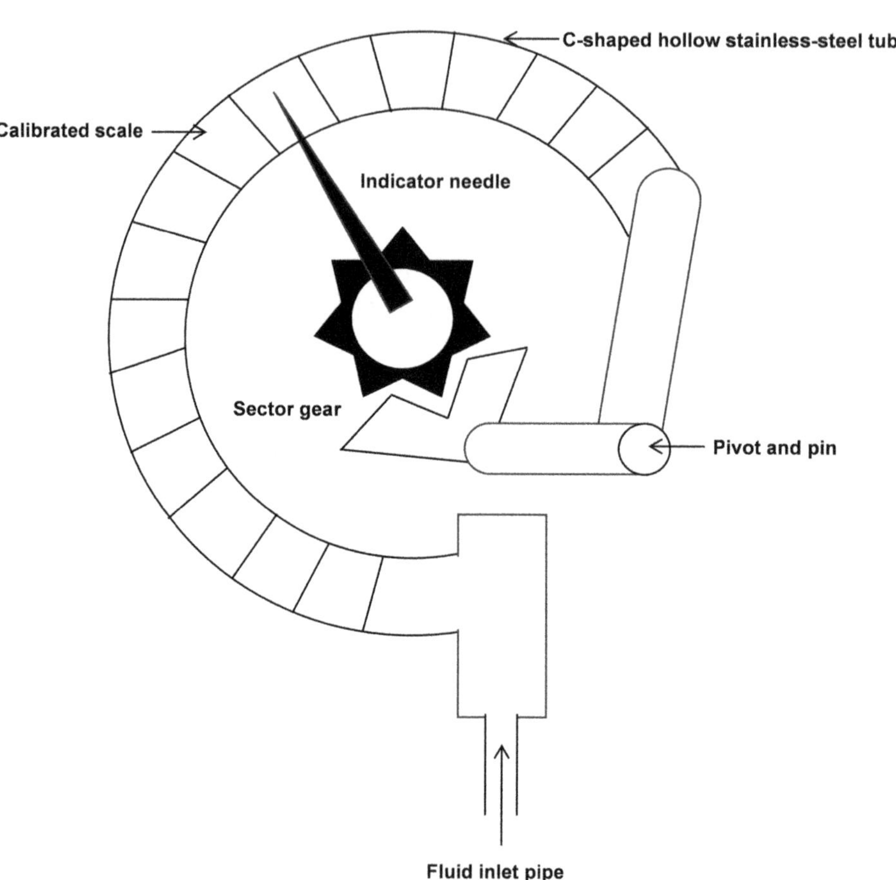

FIGURE 5.9 Bourdon vacuum gauge. The fluid inlet pipe is connected to one end of a C-shaped hollow stainless-steel tube with a calibrated scale. The opposite end of the C-shaped tube is fixed to a pivot and pin with an attached sector gear for actuating an indicator needle.

diaphragm deforms it. The elastic deformation of the diaphragm is proportional to the applied pressure, which is connected to links and gears for rotating the pointer on the readout dial proportionally to the pressure. It is used for measuring pressures from 1 mbar to 2,000 bar.

Advantages: Small size, affordable cost, linearity, and usable for viscous fluids.
Disadvantages: Poor seismic and impact resistance, and difficult maintenance.
Applications: Measurement of pressure in gas and liquid storage tanks, chemical and power plants, pneumatic and hydraulic systems, and HVAC boilers.

5.5.2.3 Capsule Vacuum Gauge

There are two types of capsule vacuum gauges used over the pressure span from 2.5 to 600 mbar (Figure 5.11).

(i) Capsule Absolute Pressure Gauge: A capsule is prepared from two thin corrugated diaphragms made of non-ferrous metal or 316i Ti, a Mo-containing, Ti-stabilized, Cr-Ni austenitic stainless steel (Figure 5.11(a)). It is evacuated and hermetically sealed by welding around the periphery. The sealed capsule, similar in appearance to a cylindrical bellows chamber, is enclosed in a housing with an

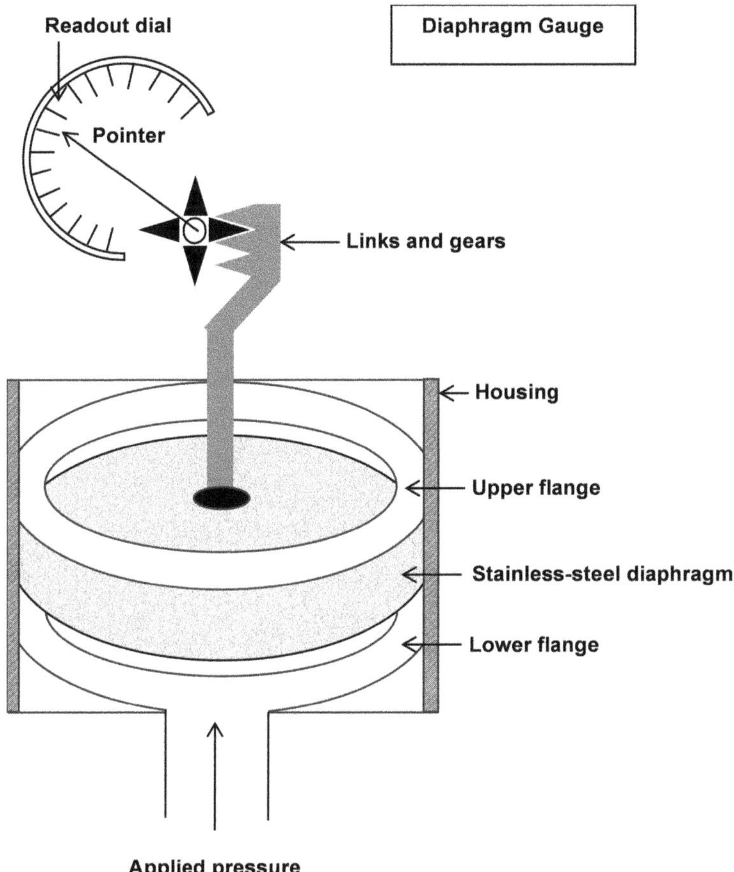

FIGURE 5.10 Diaphragm vacuum gauge. The pressure acting on a stainless-steel diaphragm fixed between a lower flange and an upper flange in a housing is transferred through links and gears to an indicator needle moving on a readout dial.

inlet pipe connected to the gas whose pressure is to be measured. One diaphragm of the capsule is coupled to a transmission mechanism, enabling the movement of a pointer.

When the pressure of the gas in the housing is less than the pressure inside the capsule, the capsule expands and bulges out. However, when the external pressure is larger than the pressure inside the capsule, the capsule contracts and squeezes in size. In both cases, the pointer moves showing the pressure of the gas. Since the pressure of the gas in the housing is measured with reference to vacuum or zero pressure inside the capsule, the pressure value obtained is absolute in nature.

(ii) Capsule Differential Pressure Gauge: Here, the capsule is formed by welding together two metal diaphragms, as in the absolute pressure gauge but the capsule is not evacuated (Figure 5.11(b)). Rather an opening is made at one end of the capsule to introduce the process gas and a transmission mechanism is coupled to its other end to convey its deflection to the pointer. Mechanical linkages translate the capsule deflection into rotary motion which makes the pointer move over the gauge dial.

The capsule expands as the pressure of the introduced gas exceeds the atmospheric pressure and contracts when it falls below it. The pressure measured with respect to atmospheric pressure is a differential pressure = the difference between the two pressures.

Advantages of Capsule Gauges: Rugged, vibration-resistant, linear readout, and independent-of-gas operation.

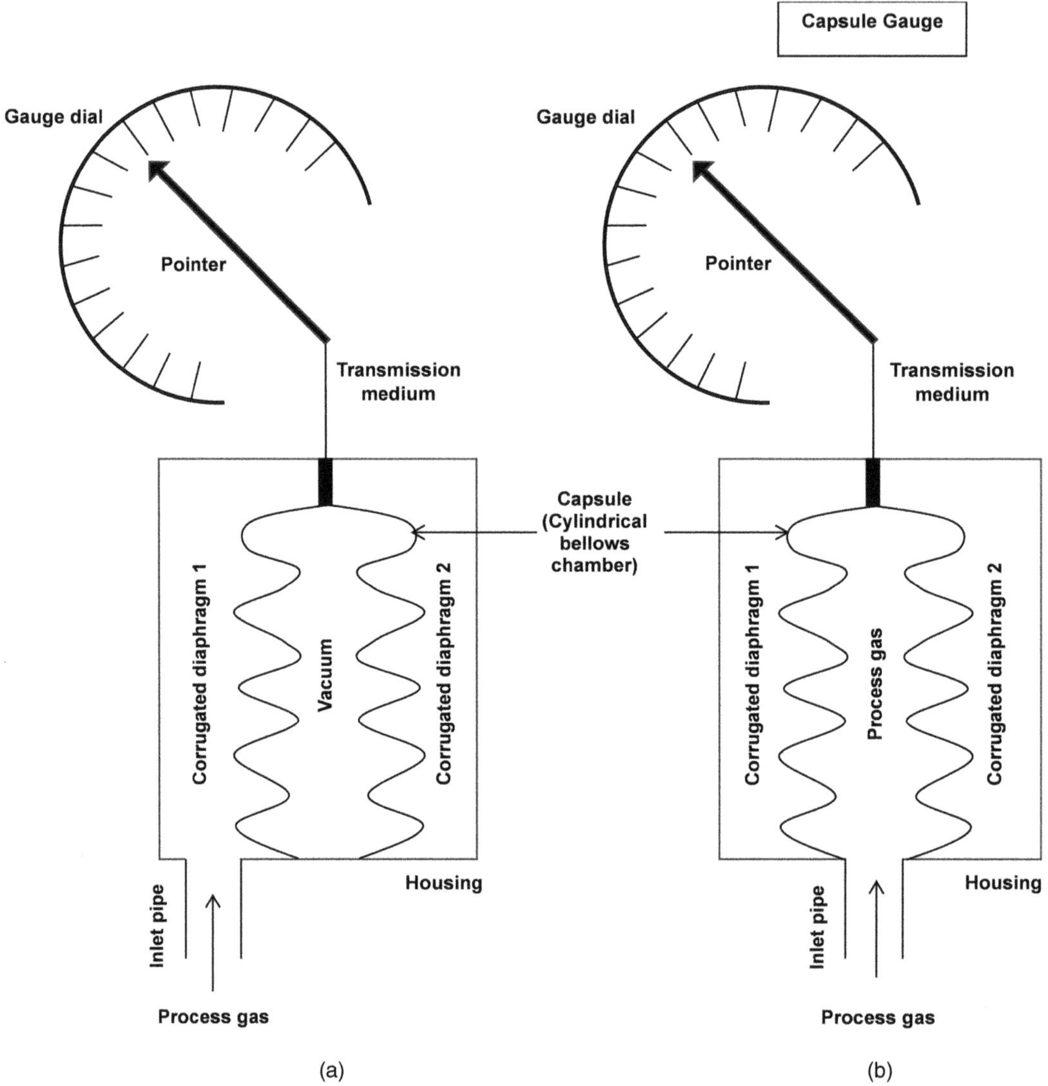

FIGURE 5.11 Capsule vacuum gauge: (a) absolute and (b) differential. Parts (a) and (b): Corrugated diaphragms 1 and 2 are fixed inside a housing to form a cylindrical bellows chamber known as a capsule. The top end of the capsule is connected through a transmission medium to a pointer moving on the gauge dial. Part (a): The capsule is sealed under vacuum, and the process gas is introduced through an inlet pipe in the housing outside the capsule. Part (b): The capsule is open, and process gas is introduced through an inlet pipe into the capsule.

Disadvantages of Capsule Gauges: Require frequent calibration, poor vibration, and impact resistance; and prone to failures by overpressure, mishandling, and shock.

Applications of Capsule Gauges: Food and beverages, pharmaceutical, and packaging industries, especially suitable for low-pressure measurements.

5.5.2.4 Capacitance Vacuum Gauge

It consists of a permanently sealed cavity with pressure $<1 \times 10^{-7}$ mbar (Figure 5.12). One side of the cavity is closed by an Inconel diaphragm on the opposite side of which lies the chamber containing the gas whose pressure is to be determined. By this arrangement, the diaphragm bends or relaxes as the pressure in the gas chamber becomes larger or smaller than the vacuum pressure in the cavity. The Inconel diaphragm serves as one electrode of the capacitor to which a wire is connected.

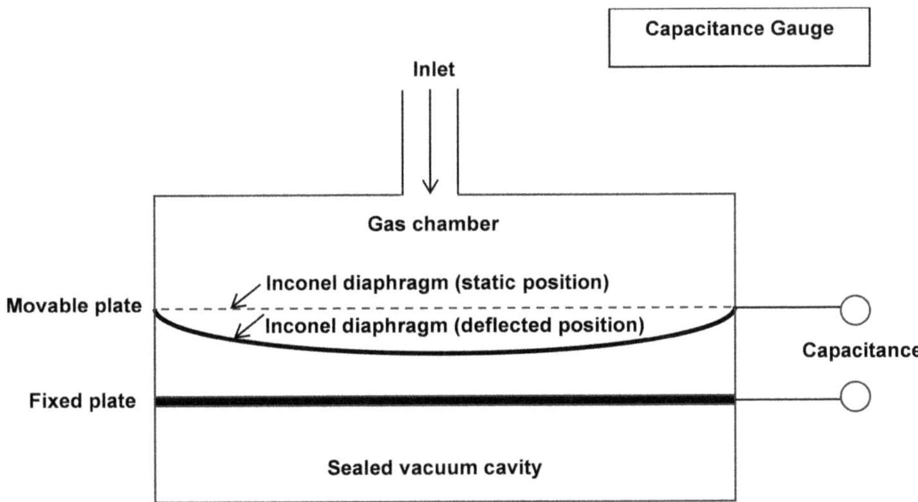

FIGURE 5.12 Capacitance vacuum gauge. The diagram shows a sealed vacuum cavity. The top surface of the cavity is closed by an Inconel diaphragm, serving as the movable plate of the capacitor. The fixed pate of the capacitor is a plate hanging in the cavity at a small distance below the diaphragm and firmly fastened to the walls of the diaphragm on its two sides. Above the Inconel diaphragm is the gas chamber with an opening. The chamber contains the gas whose pressure is to be measured. Capacitance is measured between the diaphragm and the fixed plate by connecting wires to them and taking these wires outside the chamber. Static and deflected positions of the diaphragm are marked.

The other electrode is a metal electrode mounted in the cavity near the diaphragm, and facing the diaphragm surface. A wire connected to this electrode is taken outside the cavity. The capacitance is measured across the wires connected to the diaphragm and the metal electrode.

The distance between the fixed metal electrode and the elastic diaphragm changes with the bending of the diaphragm in accordance with pressure in the gas chamber, and the capacitance of the gauge follows this variation. It is thus a gas pressure-dependent capacitor. The gas pressure obtained from the measured capacitance value is an absolute pressure as it is measured relative to almost zero pressure in the sealed cavity. This happens because one side of the diaphragm is exposed to vacuum in the cavity and the other side to gas in the chamber. The range of measurement of pressure of this gauge spans from 1.0×10^{-5} mbar to well above atmospheric pressure.

Advantages: High sensitivity and accuracy, good resolution, fast response, low hysteresis, low temperature-sensitivity, and low power consumption because of zero DC current flow.

Disadvantages: Costly, non-linear characteristics, and sensitive to vibrations.

Applications: Wearable devices, semiconductor wafer processing, switches and keyboards, and car tires.

5.5.2.5 Piezoresistive Pressure Gauge

A digital diaphragm pressure gauge has piezoresistors on the diaphragm whose resistance changes with applied pressure (Figure 5.13). The resistance change, measured by a Wheatstone bridge as a change in voltage, is converted by a microprocessor chip into a pressure value which is shown on a digital display. It has temperature compensation and data logging features together with internet connectivity. Its pressure measurement range is $1\text{--}10^4$ Torr.

Advantages: Robustness, good shock- and vibration resistance, stability, and high-speed and gas-composition-independent operation

Disadvantages: Temperature sensitivity (needing temperature compensation)

Applications: Semiconductor processing, mass spectrometers, and scanning electron microscopes.

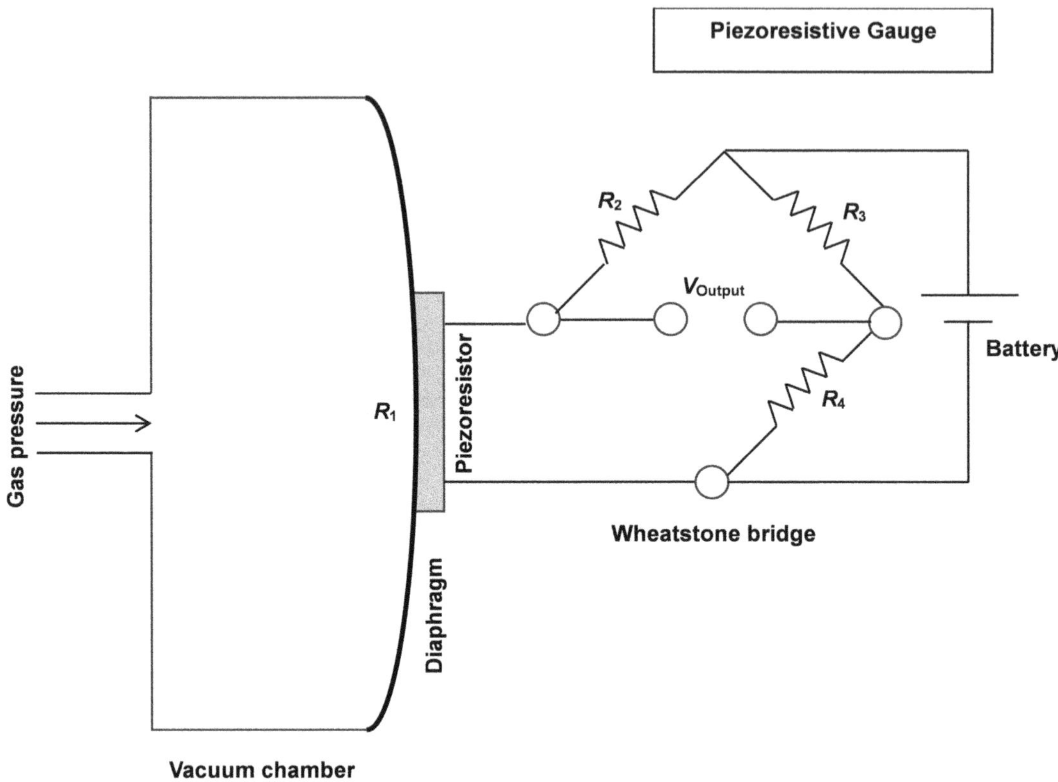

FIGURE 5.13 Piezoresistive pressure gauge. A diaphragm with a piezoresistor R_1 on its surface is bent by the applied gas pressure. The piezoresistor R_1 is placed in one arm of a Wheatstone bridge. The remaining resistors of the bridge are R_2, R_3, and R_4. A battery is connected between the indicated points. The output voltage is measured between V_{Output} terminals.

5.5.3 THERMAL CONDUCTIVITY GAUGES

They measure the degree of vacuum pressure from its effect on the ability of a gas to conduct heat, known as its thermal conductivity. These gauges used to measure low to medium vacuum pressures in the range of 10^{-4} mbar to atmospheric pressure, are of two types: Pirani and thermocouple gauges.

5.5.3.1 Pirani Gauge

The Pirani gauge has a sensing head and a readout circuit (Figure 5.14). The sensing head is a tube in which a filament wire is suspended. The sensing head is connected to the system whose pressure is to be measured. The filament in the sensing head is made of a W, Pt, or Ni wire and has a thickness of less than 25 μm. It is heated by passing an electric current through it. The filament forms one arm of the four-arm Wheatstone bridge.

The gauge works on the variation of thermal conductivity of a gas with its pressure which causes a variation in resistance of the filament with gas pressure.

At high pressure, the gas contains a high density of molecules whereby it can remove heat from the filament efficiently by the conduction mechanism. This happens because of more frequent collisions of gas molecules with the filament as their mean free path is shortened. The increased thermal conductivity of gas leads to a reduction of filament temperature. As the temperature of the filament is lowered, its resistance becomes smaller.

Similarly, at low pressure, the density of gas molecules is less so that its efficiency of removing heat from the filament falls. The collisions of gas molecules with the filament are now less frequent

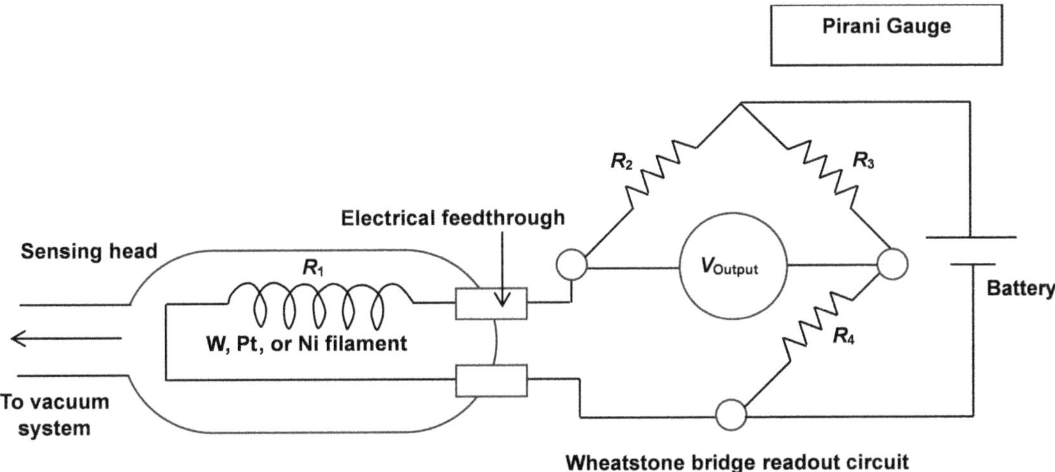

FIGURE 5.14 Pirani gauge. A tungsten, nickel, or platinum filament of resistance R_1 is mounted inside the sensing head connected to the vacuum system. Resistor R_1 is connected via feed-throughs to resistors R_2, R_3, and R_4 of a Wheatstone bridge circuit. Battery connection and V_{Output} points are marked.

as their mean free path is lengthened. The decreased thermal conductivity of gas causes an increase in filament temperature, which makes its resistance larger.

The gauge is operated in three modes: constant voltage, constant current, and constant temperature. The constant-temperature mode, also called the constant-resistance mode, is the preferred mode because it curtails the undesired signal drift caused by the variation in ambient temperature. In this mode, the voltage supply to the Wheatstone bridge is changed to keep the temperature at a fixed value. Hence, the pressure-dependent voltage is converted to the pressure value.

In the Pirani gauge fabricated using MEMS technology, a resistor made on an ultra-thin silicon membrane takes the place of the filament. The MEMS Pirani gauge displays superior performance to the filamentary vacuum gauge owing to its miniaturized size, low operating temperature, reproducible semiconductor processing, and improved temperature compensation.

5.5.3.2 Thermocouple Gauge

As in the Pirani gauge, a filament is electrically heated (Figure 5.15). Interactions of surrounding gas molecules with the filament cause transference of heat from the filament, thereby changing its temperature. A thermocouple attached to the filament records its temperature. The output voltage of the thermocouple is measured with a mill-voltmeter. When the pressure is high, the filament temperature is low and hence the output voltage is small. However, when the pressure reduces, the filament temperature increases, and the output voltage becomes large. The millivoltmeter is calibrated to give a direct readout of pressure.

5.5.3.3 Pirani vs. Thermocouple Gauge

Pirani gauges are often preferred over thermocouple gauges because they are 10-fold faster than them, besides responding better to pressure changes, and due to their usability over a wider pressure range and the linearity of the pressure–resistance relationship. Both the gauges are rugged and inexpensive but require frequent calibration.

5.5.3.4 Ionization Gauges

Ionization gauges, the most sensitive gauges for high vacuum ranges, measure the gas pressure indirectly from the ionization-induced electric current generated by the bombardment of gas molecules

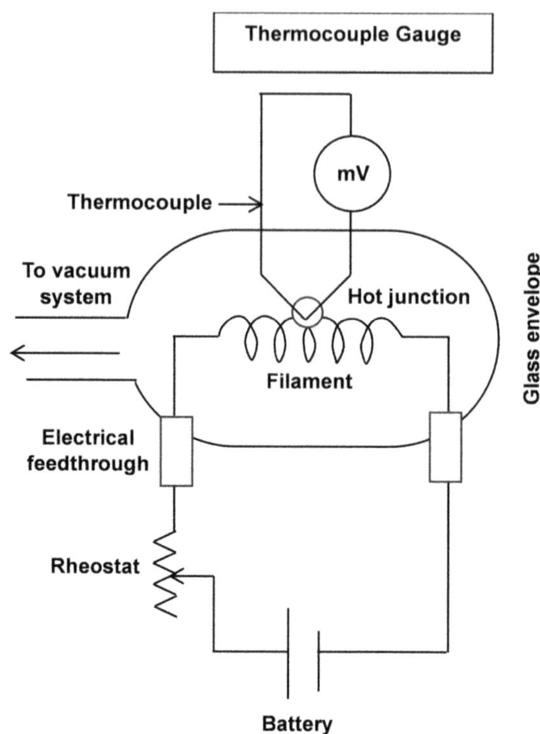

FIGURE 5.15 Thermocouple gauge. A filament inside a glass envelope is connected via feed-throughs to a battery through a rheostat. The glass envelope is connected to the vacuum system. The filament is heated and its temperature is measured by a thermocouple tied to it.

by accelerated electrons (IG1hotapp.pdf). They are divided into two groups: hot-cathode and cold-cathode gauges. In a hot-cathode gauge, the electron beam is produced by heating a filament while in a cold-cathode gauge, it is obtained by a high-voltage gas discharge. Both types of gauges have the same pressure measurement ranges: 10^{-2} and 10^{-10} Torr, encompassing 7 decades of dynamic range, and both give pressure readings dependent on the gas composition owing to differences in ionization efficiencies of gases. They are generally calibrated using nitrogen or argon.

5.5.3.4.1   Hot-Cathode Ionization Guage

The hot-cathode gauge (Figure 5.16(a)) has a triode construction with three active electrodes: a tungsten filament cathode, a grid electrode at a positive potential, and a fine collector wire electrode at a negative potential, all sealed inside a glass tube. The grid electrode has the shape of a cylindrical grid cage. It is maintained at a positive potential (+180–210 V) larger than that of the cathode (+30V). The electrons released by thermionic emission from the cathode are accelerated toward the grid. On their way to the anode, they collide against the gas molecules, ionizing them to form positive ions. These positive ions are attracted by the negatively biased collector wire (−30V). Amplification of the ionization current is done using a high-gain differential amplifier. An electrometer measures the ion current.

5.5.3.4.2   Cold-Cathode Ionization Guage

The cold cathode gauge (Figure 5.16(b)) is based on the field electron emission phenomenon, usually called field emission, which is caused by applying a high electric field between two electrodes, the cathode and the anode, enclosed in a glass envelope. Besides the anode and cathode, it also has

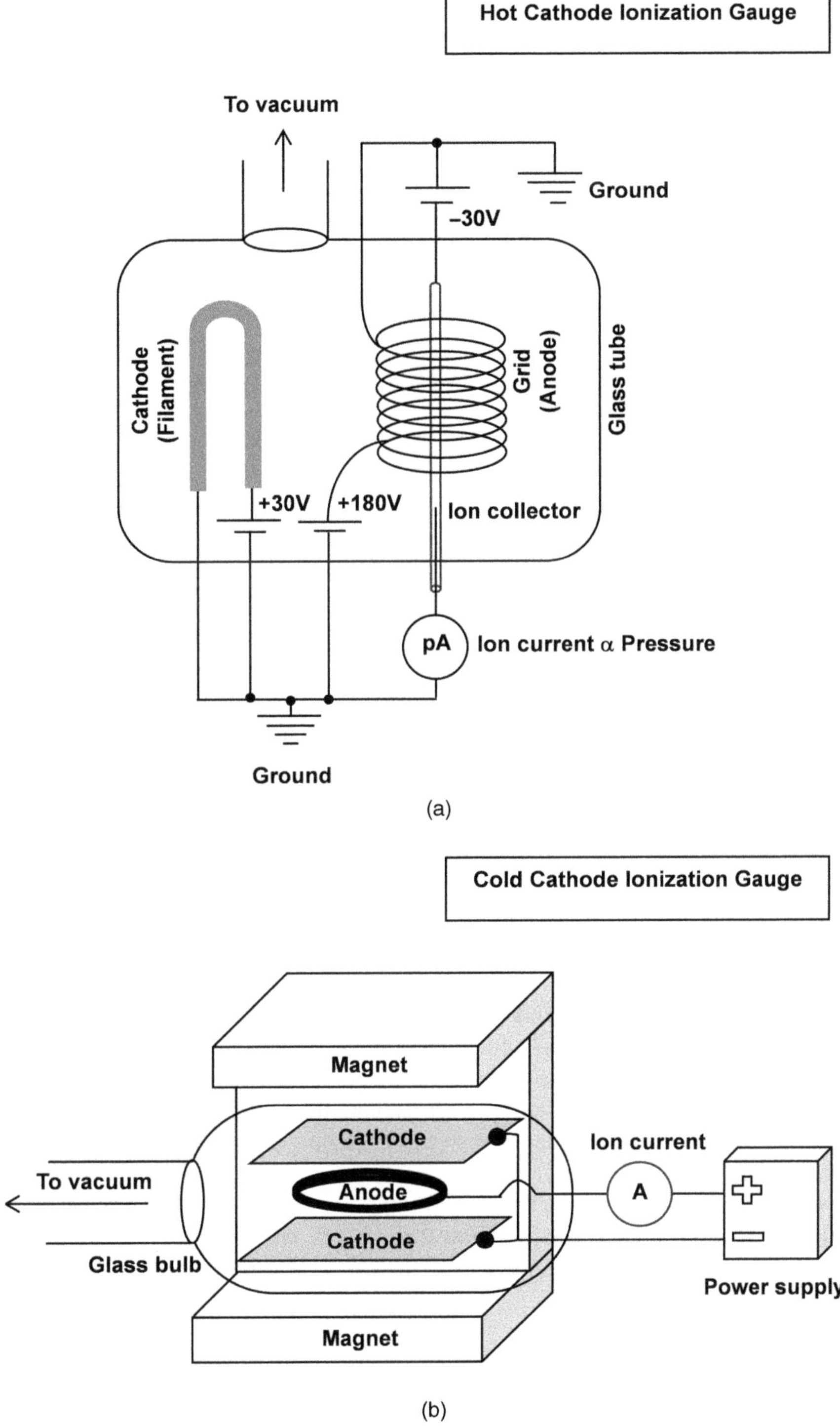

FIGURE 5.16 Ionization gauges: (a) hot cathode and (b) cold cathode. Part (a): A glass tube containing a cathode at +30V, and a grid electrode at −30V, surrounded by an ion collector coil at +180V, is connected to a vacuum enclosure in which vacuum is to be measured. Part (b): A glass bulb connected to a vacuum enclosure contains a ring anode with interconnected rectangular cathodes on both sides. It is placed between the pole pieces of a magnet. A power supply is connected between the anode and cathode, and an ammeter measures the ion current.

a permanent magnet. It uses a high electric field ~2–6 kV along with a magnetic field ~1–2 kG in a crossed configuration for electron entrapment. The combined electric and magnetic fields cause electrons to travel in long spiral trajectories, thus enhancing the probability of causing gas ionization. The generation of electron plasma known as a glow discharge is triggered by the random liberation of an electron at the cathode by a cosmic ray particle, a photon, or the decay product from a radioactive event smashing into a gas molecule. The discharge gradually builds up to the level at which no further electrons can enter the plasma because of electrostatic repulsion from those inside the plasma. The electrons encircle the anode in cycloidal loops and bombard the gas molecules causing impact ionization. The collected ion current varies with the gas pressure and provides an indication of its magnitude.

5.5.3.5　Advantages of HCGs

(i) Easy calibration due to the linearity of collector current-pressure characteristic with deviations $<\pm 25\%$ over the dynamic range.
(ii) Better accuracy, stability, and repeatability than CCGs.

5.5.3.6　Disadvantages of HCGs

(i) Thermal degassing from gauge components can influence low-pressure measurements if the gauge is not sufficiently degassed.
(ii) Measurement errors at 10^{-10} Torr caused by the X-ray limit (the lowest pressure at which all the output current comes from X-ray-induced photoelectric emission), electron-stimulated desorption (ESD), and outgassing, which can be avoided by proper gauge degassing and bakeout.
(iii) Inaccuracies due to reactions of gas molecules with the hot filament of the gauge.
(iv) Operating life is limited by filament degradation.
(v) Gauge stabilization time ranges from minutes to weeks.
(vi) Safety hazards due to burns from hot gauge walls, and possible glass implosions.

5.5.3.7　Advantages of CCGs

(i) Simpler and more economical electronics than HCGs.
(ii) Quick response, enabling faster stabilization of readings than HCGs.
(iii) The absence of hot filament eliminates burnout risks like those in HCGs.
(iv) Slow and calculable outgassing than HCGs.
(v) Degassing is not required because of low input power and the absence of internal heating.
(vi) Lower pumping speeds than HCGs, preventing errors in pressure measurement.
(vii) Provision of more accurate pressure measurements than HCGs in applications involving repeated pump-downs with no degassing option. Hence, it is favored over HCGs in material outgassing investigations.
(viii) No residual current issues due to freedom from X-ray and ESD effects.
(ix) Less sensitivity to external magnetic fields than HCGs. The sensitivity is further decreased by using sleeves for shielding.

5.5.3.8　Disadvantages of CCGs

(i) Lower accuracy than HCGs makes them inappropriate for use as standards.
(ii) Complicated calibration than HCGs due to non-linearity of current-pressure characteristic.
(iii) Difficulty in starting because the discharge is not initiated immediately on application of a high voltage. Starting delay increases from seconds at 10^{-6} Torr to several hours at 10^{-10} Torr.
(iv) Electron current and energy depend on the gauge construction and operating parameters, hence they are not under user control, unlike HCGs where most parameters can be acquired by the user from the controller.

5.5.3.9 Applications of Ionization Gauges

Medium to ultra-high vacuum measurements in vacuum furnaces, calibration equipment, and vacuum physics laboratories.

5.6 MEASURING COMPRESSIVE/STRETCHING FORCES OR RATE OF CHANGE OF AN APPLIED FORCE: FORCE SENSORS

Force sensors convert forces into electrical signals reflecting the magnitude of forces applied. These signals are fed to computers serving as inputs to control machines. Depending on the type of force under consideration, a suitable force sensor is chosen from the various types available.

5.6.1 LOAD CELL

It is a force transducer that converts a mechanical force, such as due to weight, compression, tension, pressure, or torque, into a measurable pneumatic, hydraulic, piezoelectric, inductive, capacitive, magnetostrictive, or mechanical strain signal.

> Advantages: Small size, compact, sensitive, accurate, low hysteresis, low-cost, durable high-strength aluminum and steel devices with no moving parts; temperature-compensated operation is possible; high-frequency response for dynamic loading, highly configurable, readily available in a variety of shapes and geometries, and in several and specially customized mounting versions; and can easily fit in any industrial application due to wide choice of mounting options.
>
> Disadvantages: Slow response; clean, dry air requirement, non-linear devices needing frequent calibration, subject to temperature variations, measure strain in one direction only (uniaxial gauge); being passive transducers, they need a voltage source for excitation which limits their use in areas with power scarcity or with ignition risks.
>
> Applications: Scales (bathroom, truck, railway), vehicle weighbridges, crane load monitors; medical, gaming, and robotic devices; dynamometers; automotive, rail transport (stress measurement in railway lines) and aerospace (monitoring wing deflection/deformation during flight); bag and bottle-filling machines; architecture and construction engineering; and real-time monitoring of strain in bridges to avoid accidents.

5.6.2 CLASSIFICATION OF LOAD CELLS

According to the type and nature of the signal produced, the load cells are categorized, as given in the following sub-sections.

5.6.2.1 Pneumatic Load cell

It is a load cell that works using compressed air pressure. It consists of a steel chamber filled with pressurized air or gas through an air supply valve and a pressure regulator (Figure 5.17). The chamber has a bleed valve to release the gas, if required. On the sidewall of the chamber, a pressure gauge is connected to measure the pressure of gas.

The top surface of the chamber is an elastic diaphragm over which there is a load platform. When the force to be measured is applied by placing a weight on the load platform, the gas below the elastic diaphragm is squeezed, and its pressure increases. The air pressure reading of the pressure gauge is proportional to the force. Thus, the force is read as an air pressure which balances out the weight of the object examined.

> Advantages: Intrinsically safe and explosion-proof, less sensitive to temperature variations; and not affected by humidity and environmental factors.

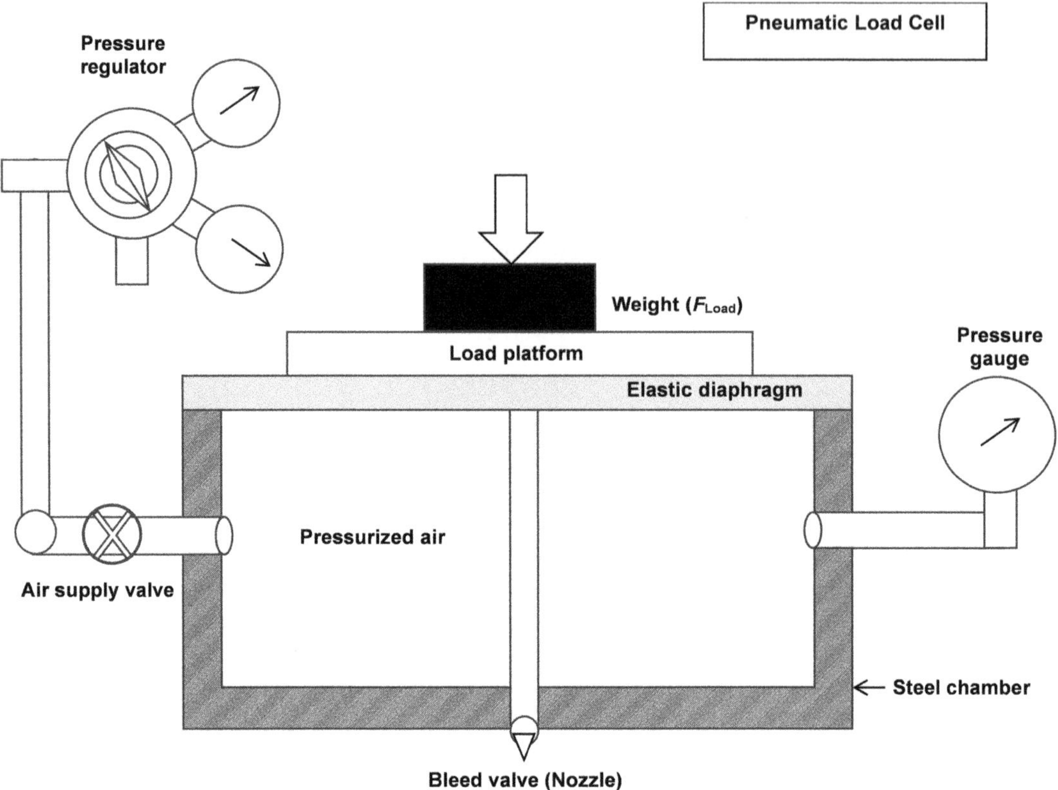

FIGURE 5.17 Pneumatic load cell. A steel chamber is filled with pressurized air from an air supply valve fitted with a pressure regulator. The weight F_{Load} is placed over a load platform fixed to an elastic diaphragm serving as the roof of a steel chamber. The diaphragm is connected to a bleed valve. The pressure inside the steel chamber is measured by a pressure gauge.

Disadvantages: Bulky devices with rigid construction; require an amplification circuit and electrical power for output display.

Applications: Industries in which safety and cleanliness are primary apprehensions, conveyor belts, and weighing machines.

5.6.2.2 Hydraulic Load Cell

It is a load cell working with oil, water, or any liquid under pressure. It consists of a steel housing filled with a hydraulic fluid, frequently oil-filled up to a certain height (Figure 5.18). At the bottom of the steel housing, there is a hole through which a tube comes out and bends toward the side. At the end of this tube, a Bourdon gauge is fitted to measure the oil pressure. The surface of the oil is covered with an elastic membrane. Over this membrane there is a piston, and on the top surface of the piston lies the load platform.

This load cell is similar in operation to the pneumatic load cell. When a weight is placed on the platform, the piston is pushed downward whereby the oil below it is pressurized. The oil pressure is converted into weight put on the platform. Hence, the force is measured as an oil pressure balancing the weight of the object.

The pneumatic and hydraulic load cells share the common feature that they do not use an electrical power supply, which makes them particularly suited for environments using dangerous substances at the risk of fire and explosion. Nevertheless, their meter readings can be converted into electrical signals representing the applied forces, if necessary.

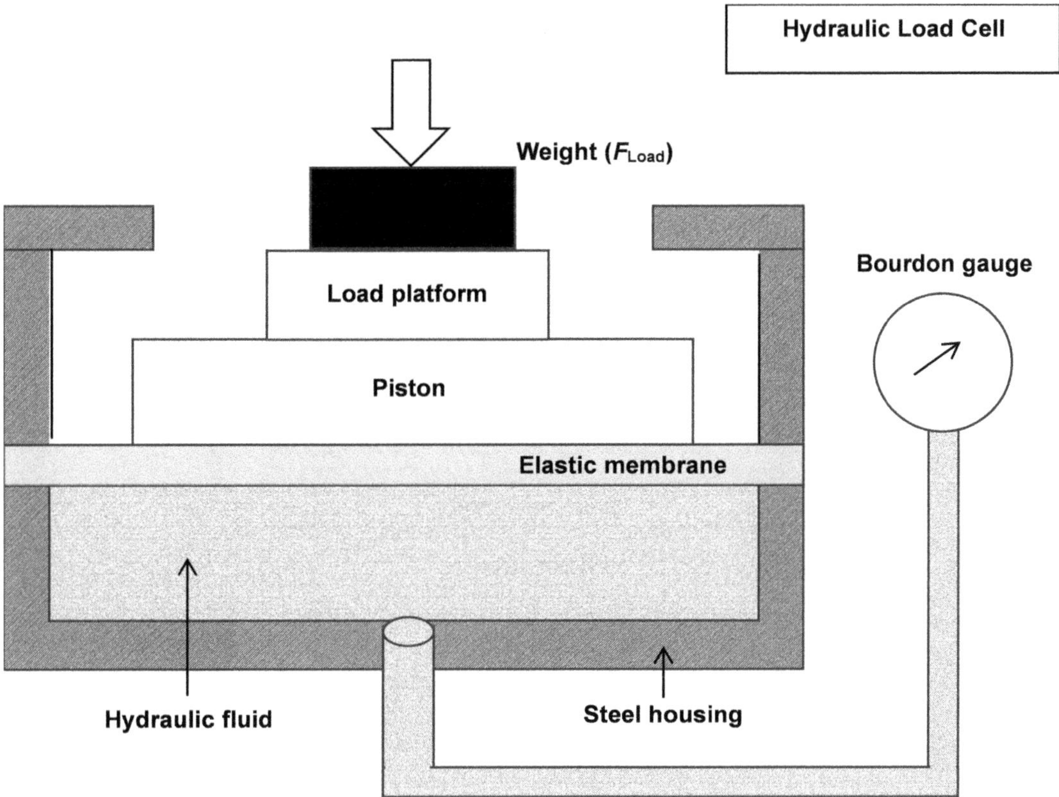

FIGURE 5.18 Hydraulic load cell. A steel housing is filled with a hydraulic fluid. The weight F_{Load} is placed over a load platform fixed to a piston mounted on an elastic diaphragm and pressed against the hydraulic fluid. A Bourdon gauge connected by a tube through a hole at the bottom of the steel housing measures the fluid pressure in the housing.

Advantages: Good electrical damage immunity and environment resistance, can be used in hazardous areas and extreme temperatures, low maintenance expenditure, and longevity.

Disadvantages: Expensive, difficult calibration, tedious mounting, damaged by overloading, impaired by chemicals and humidity, and cannot withstand electrical surges.

Applications: Hazardous areas with explosive materials such as inflammable oil leaking from an automobile, heavy-duty weighing machines, e.g., truck and railway scales; mechanical testing, monitoring, and control systems.

5.6.2.3 Piezoelectric Load Cell

It is a load cell involving the use of electric polarization of a material by mechanical stress called the piezoelectric effect. This effect is observed in materials such as quartz, Rochelle salt, lead zirconate titanate (PZT), aluminum nitride, and lithium niobate. When mechanically stressed, these materials get deformed, resulting in a shift of orientation of internal dipoles. A consequent change in the center of symmetry of the electrical charges takes place. It causes the appearance of a charge difference or voltage between electrodes on the opposite faces of the material. The charge difference depends on the magnitude of the force and the nature of the force, whether compressive or tensile. It serves as the basis of the operation of the piezoelectric load cell.

As its name implies, the piezoelectric load cell consists of a piezoelectric crystal sandwiched between metal electrodes (Figure 5.19). The electrical connections are made by attaching wires to the electrodes. The assembly is mounted on a base plate. It is covered with another plate on the

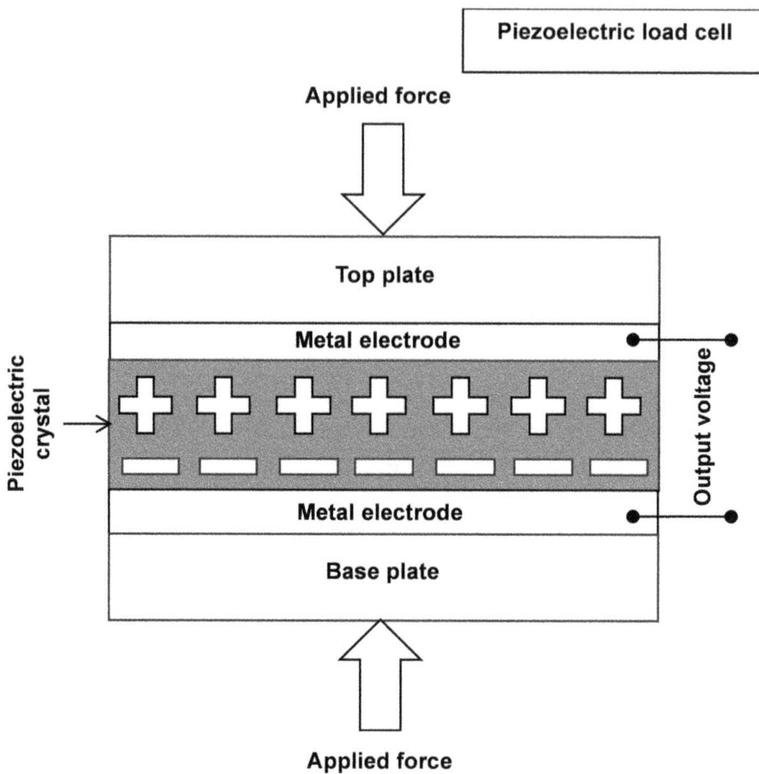

FIGURE 5.19 Piezoelectric load cell. A piezoelectric crystal has metal electrodes on both sides. It is fixed between a base plate and a top plate. When forces are applied on the top and bottom plates, positive and negative charges appear on opposite sides of the crystal. The output voltage is measured to find the force.

upper surface to form a housing. This housing is designed in such a way that loading forces can be applied from opposite sides. The output piezoelectric voltage is a measure of the sum-total loading forces.

It must be noted that the piezoelectric effect is a dynamic phenomenon and, therefore, the electrical output of a piezoelectric load cell is observed as an impulse function when the force is changing and becomes zero under static conditions. Hence, this load cell is useful only for measuring transient forces. It must be further noted that the piezoelectric load cell does not require an external power supply making it the appropriate choice in applications where provision of power source is difficult. So, it is an active transducer.

   Advantages: Self-powered, enables accurate measurements for dynamic and quasi-static processes, small size; better frequency response; robust, rugged, and more durable than other load cells.

   Disadvantages: Cannot be used for static force measurements; affected by vibration, acceleration, and temperature; also problematic in humid conditions due to water-solubility and dissolution of some crystals.

   Applications: Forces rapidly changing with time, harmonic forces varying sinusoidally with time; vibration pickups in rockets and heavy machinery, and seismographs.

5.6.2.4 Inductive Load Cell

It is a load cell involving electrical or magnetic induction. It consists of a ferromagnetic core moving inside a current-carrying solenoid coil (Figure 5.20). It works on the change in inductance of the

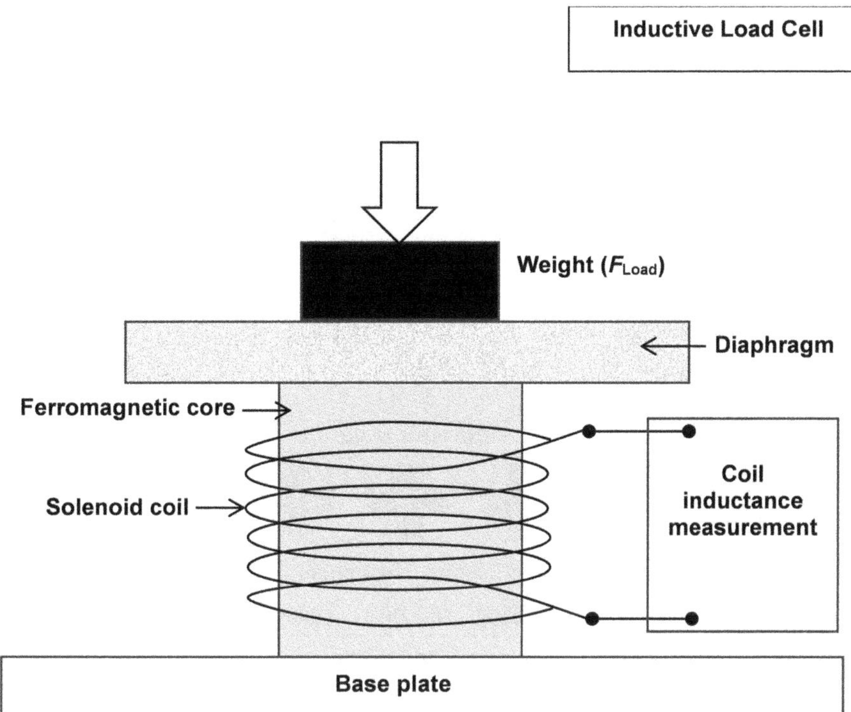

FIGURE 5.20 Inductive load cell. A solenoid coil wound on a ferromagnetic core is mounted on a base plate. The force is applied on a diaphragm placed over the core by putting a weight F_{Load} on the diaphragm. The core position varies with the weight, changing the inductance of the coil, which is measured for different weights.

solenoid coil arising from the change in position of its ferromagnetic core caused by different weights put on the load diaphragm. The inductance of the coil changes proportionally with a change in core position, and the greater the weight put on the platform, the larger the change in core position. The load cell is calibrated by plotting the change in coil inductance with the quantity of weight put on the diaphragm, and hence the force acting on the load cell. Thus, the inductive load cell responds to the force-induced core displacement. The change in inductance is usually converted into a voltage variation.

Advantages: High accuracy, resistance to environmental degradation, enables touchless measurement without any moving parts, reliable and long life.

Disadvantages: More expensive than other load cells and need complex signal-readout circuitry.

Applications: Crane load monitoring, vehicle weighing, weighing machines in construction sites, industrial scales, and laboratory balances.

5.6.2.5 Capacitive Load Cell

It is concerned with a load cell working on the capability of collecting and storing electrical charge. It consists of two parallel metal plates with connecting wires (Figure 5.21). The plates are separated by an air gap or an elastic dielectric polymer such as an insulating elastomer to form a capacitor. They are enclosed in a housing with a loading platform. One plate is movable, and the other plate is fixed. A rigid rod connects the loading platform with the movable plate.

A power source is connected between the plates to charge the capacitor. After a stable positive charge forms on one plate and a negative charge on the other, the force to be measured is applied on the platform. The applied force is conveyed through the rigid rod to the movable plate, which moves

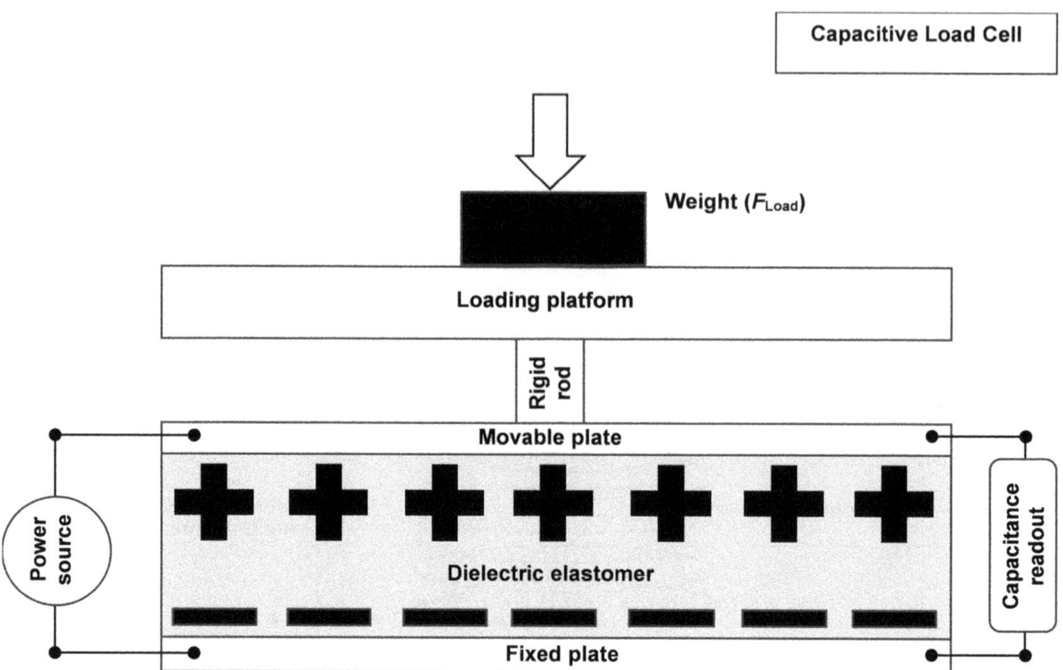

FIGURE 5.21 Capacitive load cell. A capacitor consists of a dielectric elastomer film crammed between movable and fixed plates. It is fed by a power source producing positive and negative charges on its plates. Its capacitance is measured after putting a weight F_{Load} on a loading platform fixed by a rigid rod to the movable plate.

downward decreasing the gap between the plates, and thereby causing an increase in the capacitance of the load cell. The increase in capacitance indicates the intensity of the force.

>Advantages: High sensitivity and accuracy, wide force range from small to large forces, simple design, low power consumption, cost-effective, durable; and can be hermetically sealed making it suitable for hygiene-critical medical applications.
>Disadvantages: Unsuitable for environments containing flammable and combustible substances; also the performance is affected if the dielectric used is temperature-sensitive.
>Applications: Industrial weighing, process automation, and control; medical devices, mechanical and automotive testing.

5.6.2.6 Magnetoelastic Load Cell

It relates to a load cell utilizing the magnetoelastic effect, also called the Villari effect. It is the inverse magnetostrictive effect describing the change in magnetic permeability of a ferromagnetic material when it is mechanically stressed. Under the influence of a mechanical force, the magnetic moments of the domains of a ferromagnetic material are altered resulting in a change in the magnetic properties of the material in the direction of the force.

The magnetoelastic load cell (Figure 5.22), also called the pressductor load cell, translates the force to a change in the permeability of a ferromagnetic material. It consists of a steel core comprising laminated layers with four holes through which two perpendicular coils are wound (Nowicki 2018). The primary winding is named as the magnetizing coil and the secondary winding as the sensing coil. The operation of the load cell under two different conditions is described further.

When No Load Is Applied to the Core: When an alternating current flows in the magnetizing coil, the magnetic flux induced in the magnetizing coil does not cross the plane of the sensing coil for a

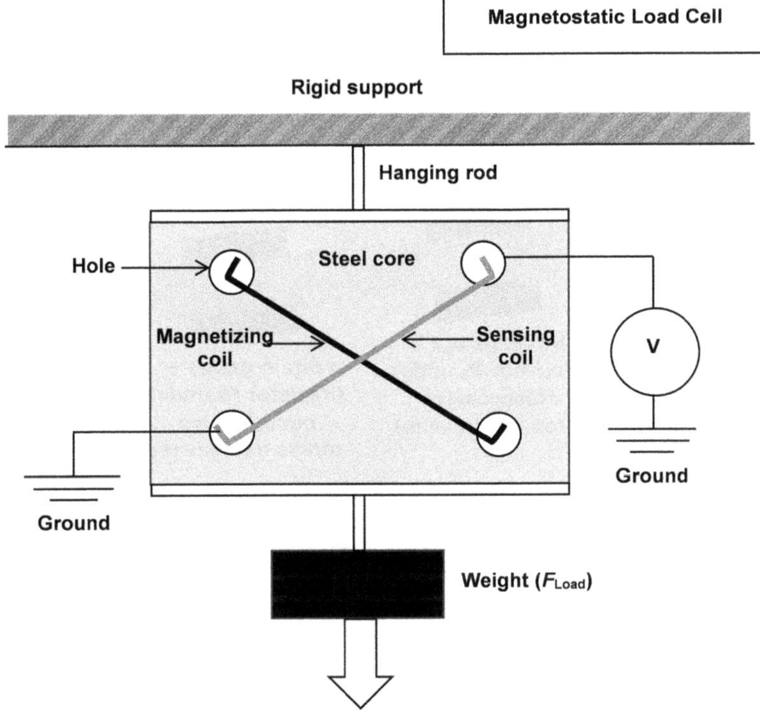

FIGURE 5.22 Magnetoelastic load cell. A steel core is taken on which the magnetizing and sensing coils are wound in perpendicular directions through four holes. The coil-wound core is hung from a rigid support. A weight F_{Load} attached to its lower end induces a voltage in the sensing coil.

symmetrical system assuming that the magnetic material is perfectly isotropic. Therefore, no voltage is induced in the perpendicular sensing coil.

When a Load Is Applied to the Core: The permeability of the ferromagnetic core changes. The stress-induced anisotropy of permeability changes the orientation of flux lines, directing some flux lines to cut the sensing coil as the system loses its symmetry. Now, a voltage is induced in the sensing coil from which an electrical signal is generated enabling the force to be measured.

    Advantages: Easily machined non-contact sensors, mechanical wear-free, long life, and generate high output signals.

    Disadvantages: Slow in response, requires clean, dry air environment; useful for dynamic strains only, static values not measured, and temperature-sensitive.

    Applications: Bolted joint constructions for monitoring bolt loads to improve joint safety and dependability.

5.6.2.7 Strain Gauge Load Cell

It refers to a load cell using a strain gauge as the force sensor. The strain gauge is a device that converts the mechanical deformation of a body, caused by an applied force, pressure, tension, or weight, into a change in its electrical resistance. Hence, it is an electrical component whose resistance varies with applied force. The 'mechanical deformation' means a relative change in the shape and size of the body compared to a reference state. It represents the displacement of particles in the body.

    Construction of Strain-Gauge Load Cell: It consists of a loading platform on which a weight is placed to apply the force. In a single-beam load cell (Figure 5.23), one end of the beam is fixed, and

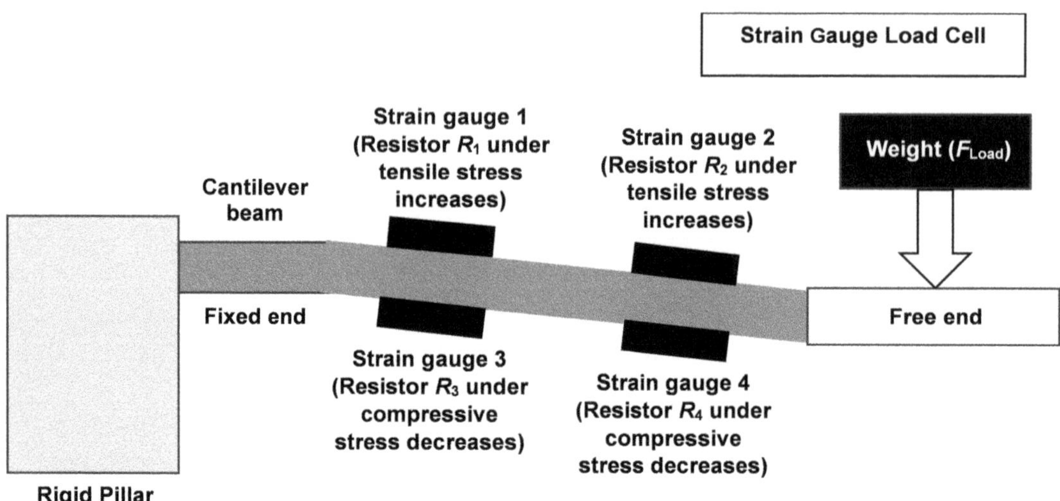

FIGURE 5.23 Strain gauge load cell. A weight F_{Load} is put on the free end of a cantilever beam which is clamped to a rigid pillar at the other end, the fixed end. Strain gauge 1 (resistor R_1) and strain gauge 2 (resistor R_2) glued on the upper surface of the cantilever beam experience tensile stress, and therefore increase in their values. Strain gauge 3 (resistor R_3) and strain gauge 4 (resistor R_4) glued on the lower surface of the cantilever beam undergo compressive stress, and therefore decrease in their values.

the other end is free for loading. It is a cantilever beam clamped at one end. The beam bends when a force is applied. Four strain gauges are bonded on the beam. The two strain gauges fastened to the top surface of the beam undergo tensile strain while the two strain gauges fitted on its bottom surface experience compressive strain. The strain gauge is excited by a power supply and the output signal is measured across its terminals. The strain gauge most widely used in load cells is known as the bonded strain gauge.

5.7 BONDED STRAIN GAUGE

5.7.1 CONSTRUCTION AND PRINCIPLE

A bonded strain gauge (Figure 5.24(a)) consists of a conductive zig-zag-shaped pattern of metal laid out in a back-and-forth style on an insulating substrate with large pads at both ends for electrical connections. The metal pattern acts as a resistor made of loops of metallic wire. When a force is applied to this resistor, the length L and cross-sectional area R of the constituent loops of wire vary. When the force F is tensile (Figure 5.24(b)), the length increases and the cross-sectional area A decreases. If the metal has a resistivity ρ, the resistance R of the metallic pattern is given by

$$R = \frac{\rho L}{A} \tag{5.29}$$

increases. For a compressive force F (Figure 5.24(c)), the reverse happens and the resistance R decreases. The ratio of change in length (ΔL) of the metal pattern when the force is applied to its original length L before the application of force, is known as mechanical strain ε, which is expressed as

$$\varepsilon = \frac{\Delta L}{L} \tag{5.30}$$

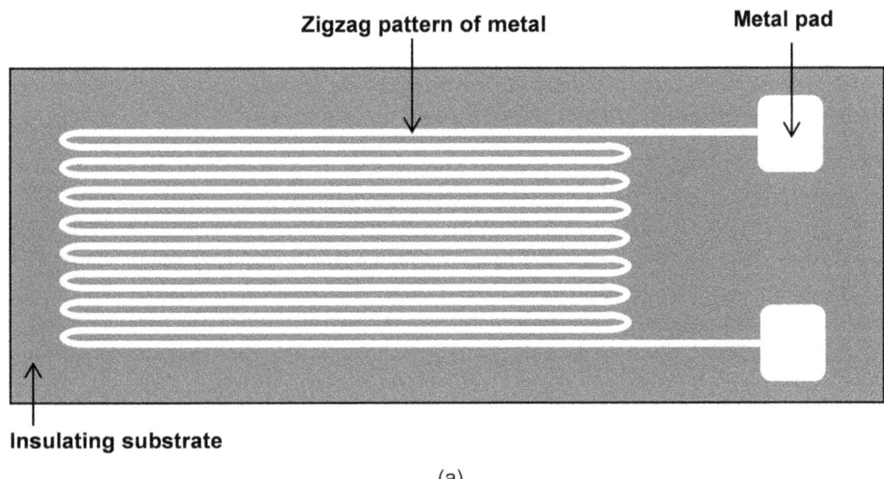

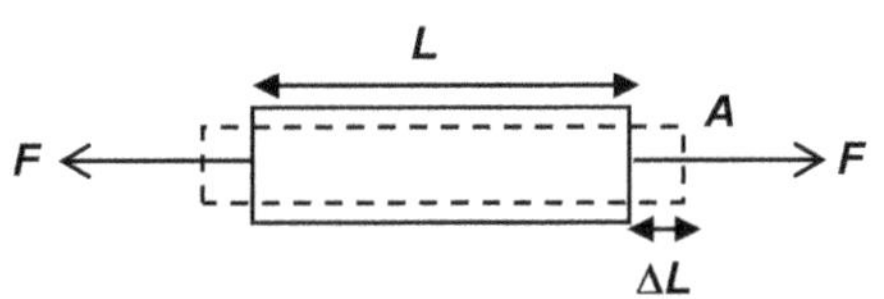

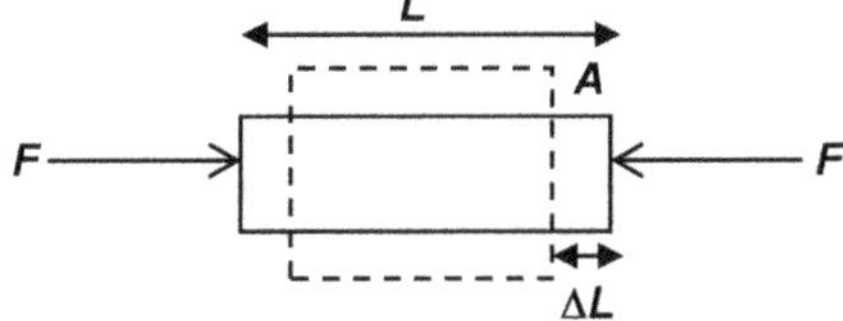

(b) **Tensile F, R increases**   (c) **Comprehensive F, R decreases**

FIGURE 5.24   Bonded strain gauge load cell: (a) the resistor pattern of the load cell, (b) load cell under a tensile force F, and (c) load cell under a compressive force F. Part (a): A zigzag pattern is etched in a metal film coated on an insulating substrate. The pattern has metal pads at its two ends. Part (b): Application of a tensile force F on the load cell increases the resistor length L by ΔL and decreases its cross-sectional area A. So, its resistance increases. Part (c): Application of a compressive force F on the load cell decreases the resistor length L by ΔL and increases its cross-sectional area A. So, its resistance decreases.

Confining the operation of the strain gauge within the elastic limit of the metal, the relationship between the relative change in resistance of the metal pattern to the mechanical strain producing it is approximately linear, and is called the gauge factor denoted by G:

$$G = \frac{\dfrac{\Delta R}{R}}{\varepsilon} = \frac{\dfrac{\Delta R}{R}}{\dfrac{\Delta L}{L}} \tag{5.31}$$

For steel, the elastic limit occurs at a strain of 0.12–0.2% = 0.0012–0.002. The gauge factor of most metals is ~2, so that for a strain gauge having a resistance = 125 Ω in unloaded condition, the maximum resistance change achievable, is from eq. (5.31),

$$\Delta R = RG\varepsilon = 125 \times 2 \times 0.0012 = 0.3\,\Omega \tag{5.32}$$

assuming the strain at the elastic limit to be $\varepsilon = 0.0012$. Since this resistance change is too small to be measured with an ohm-meter, the strain gauge load cell uses a diamond-shaped circuit known as a Wheatstone bridge circuit for this purpose. The Wheatstone bridge arrangement consists of two voltage divider circuits connected in the parallel arms of a circuit with a common power supply (Figure 5.25(a)).

5.7.2 REASONS FOR USING A WHEATSTONE BRIDGE

(i) It compares the voltage drop across the strain gauge against the voltage across a similar resistor, enabling accurate measurement of relative resistance changes $\sim 10^{-4}\text{--}10^{-2}$.
(ii) The bridge is highly sensitive to imbalances. It has the capability to detect small resistance changes, even in milliohms. It is called an ohm-meter due to its ability to give precise resistance results.
(iii) The resistance is measured by the null approach. At the null point, no current flows through the galvanometer. So, there is no voltage drop across it, and its resistance does not affect the measurements.
(iv) It can be easily interfaced with different circuit combinations.

5.7.3 WHEATSTONE BRIDGE CONFIGURATIONS

Three different Wheatstone bridge configurations are used, namely, quarter, half, and full bridges (Figure 5.25(b)–(e)). A quarter bridge uses only one strain gauge as a resistive element in one arm and it is so called because 1/4th the total number of 4 resistors in the bridge = 1 resistor is the strain gauge resistor. A half bridge uses two strain gauges in two arms and is called by this name because ½ the total number of 4 resistors in the bridge = 2 resistors are strain gauge resistors. Similarly, a full bridge uses four strain gauges in all four arms and derives its name from the fact that all 4 resistors in the bridge are strain gauge resistors.

5.7.3.1 Quarter Wheatstone Bridge

Resistors (R_1, R_3) compose one voltage divider and resistors $(R_2 + \Delta R, R_4)$ make up another voltage divider (Figure 5.25(b)). Suppose V_{IN} is the excitation voltage of the bridge. The strain gauge is represented by the resistance $R_2 + \Delta R$. Let the resistors R_1, R_2, R_3, and R_4 in the bridge be chosen such that

$$R_1 = R_2 = R_3 = R_4 = R \tag{5.33}$$

(a) Zero Load Condition: Here, no force is acting on the load cell, hence $\Delta R = 0$. The voltages at the output terminals V_{OUT+} and V_{OUT-} are equal giving

$$V_{OUT+} = V_{OUT-} \tag{5.34}$$

so that the output voltage across the bridge is

$$V_{OUT} = V_{OUT+} - V_{OUT-} = 0 \tag{5.35}$$

(b) Loaded Condition: In this case, a force is applied on the load cell as a weight, so $\Delta R \neq 0$. By Ohm's law, we get

$$V_{OUT+} = V_{IN} \frac{R_2 + \Delta R}{R_2 + \Delta R + R_4} \tag{5.36}$$

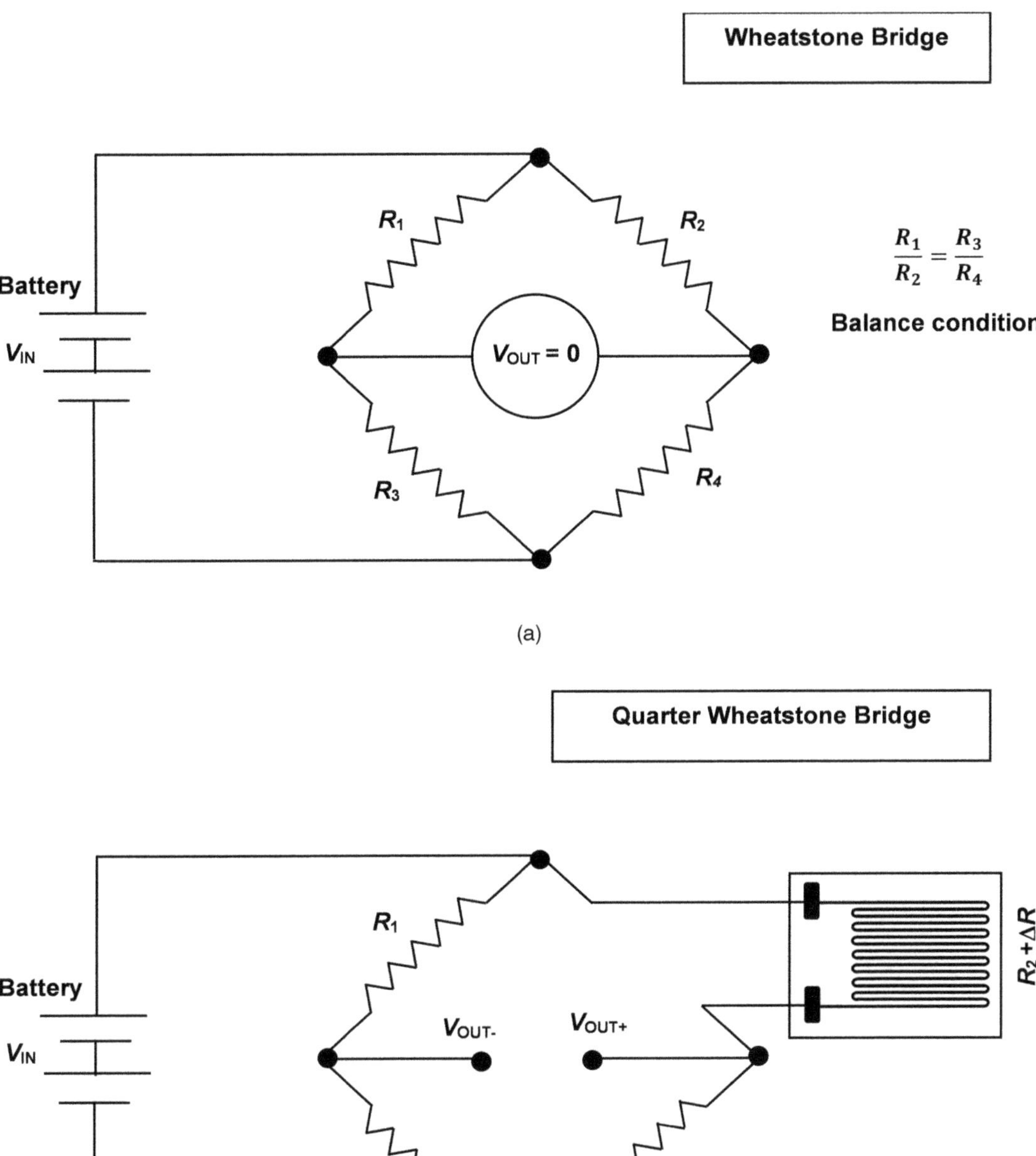

FIGURE 5.25 Wheatstone bridge circuits: (a) general circuit, (b) quarter bridge, (c) half bridge (case I), (d) half bridge (case II), and (e) full bridge. Part (a): Resistors R_1, R_2, R_3, and R_4 are connected together to form a Wheatstone bridge. A battery V_{IN} is joined across one pair of diagonally opposite points. The output voltage V_{OUT} is measured between the other pair of diagonally opposite points. At the balance condition of the bridge, $V_{OUT} = 0$, and $R_1/R_2 = R_3/R_4$. Part (b): Resistors R_1, $R_2 + \Delta R$, R_3, and R_4 are connected together to form a Wheatstone bridge. A battery V_{IN} is joined across one pair of diagonally opposite points. The output voltage is measured between the other pair of diagonally opposite points V_{OUT+} and V_{OUT-}. The resistor $R_2 + \Delta R_2$ is a strain gauge under tensile stress.

(Continued)

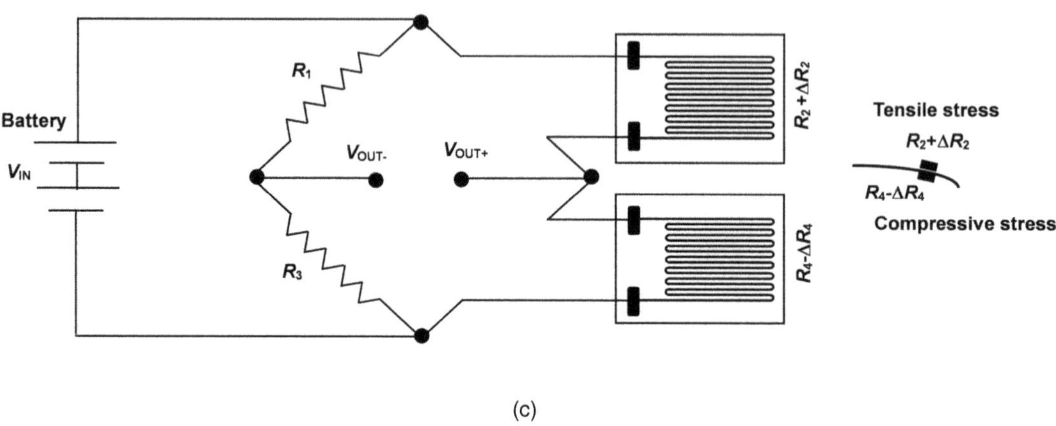

(c)

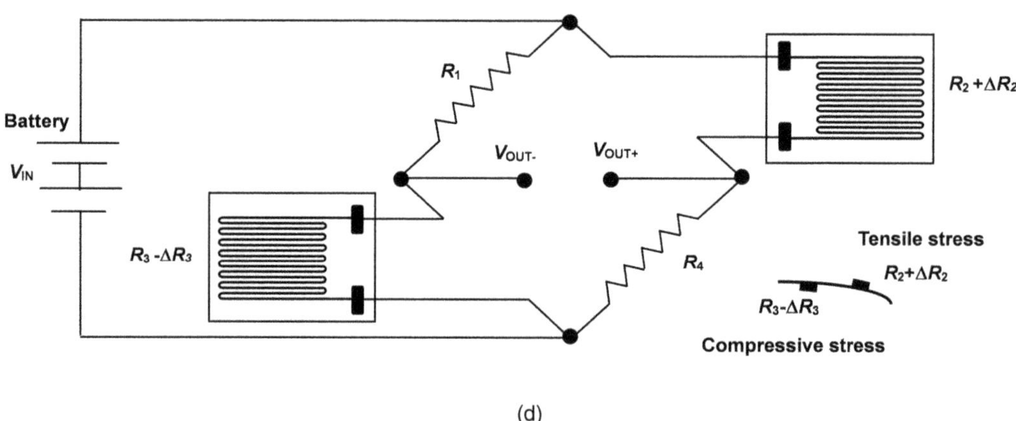

(d)

FIGURE 5.25 (CONTINUED) Part (c): Same as part (b) except that strain gauge resistor $R_4 - \Delta R_4$ (under compressive stress) is inserted in place of resistor R_4. Part (d): Same as part (b) except that strain gauge resistor $R_3 - \Delta R_3$ (under compressive stress) is inserted in place of resistor R_3.

(Continued)

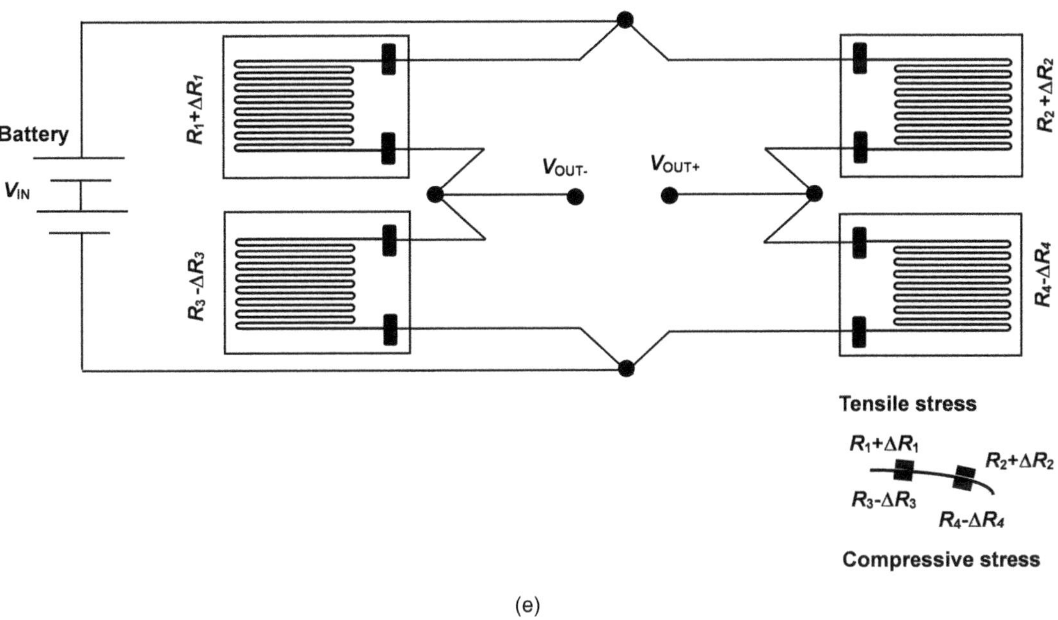

(e)

FIGURE 5.25 (CONTINUED) Part (e): Same as part (b) with some differences. The differences from part (b) are that: Two strain gauge resistors $R_1 + \Delta R_1$ (under tensile stress) and $R_4 - \Delta R_4$ (under compressive stress) are inserted in place of resistors R_1 and R_4, and strain gauge resistor $R_3 - \Delta R_3$ (under compressive stress) is inserted in place of resistor R_3. The resistor $R_2 + \Delta R_2$ remains as before, i.e., a strain gauge under tensile stress.

and

$$V_{\text{OUT}-} = V_{\text{IN}}\,\frac{R_3}{R_1 + R_3} \tag{5.37}$$

$$\therefore V_{\text{OUT}}\,|_{\text{Quarter Bridge}} = V_{\text{OUT}+} - V_{\text{OUT}-} = V_{\text{IN}}\,\frac{R_2 + \Delta R}{R_2 + \Delta R + R_4} - V_{\text{IN}}\,\frac{R_3}{R_1 + R_3}$$

$$= V_{\text{IN}}\,\frac{\left(R_2 + \Delta R\right)\left(R_1 + R_3\right) - R_3\left(R_2 + \Delta R + R_4\right)}{\left(R_2 + \Delta R + R_4\right)\left(R_1 + R_3\right)}$$

$$= V_{\text{IN}}\,\frac{R_2 R_1 + R_2 R_3 + R_1\Delta R + R_3\Delta R - R_3 R_2 - R_3\Delta R - R_3 R_4}{\left(R_2 + \Delta R + R_4\right)\left(R_1 + R_3\right)}$$

$$= V_{\text{IN}}\,\frac{R_2 R_1 + R_1\Delta R - R_3 R_4}{\left(R_2 + \Delta R + R_4\right)\left(R_1 + R_3\right)} = V_{\text{IN}}\,\frac{\left(R_2 + \Delta R\right)R_1 - R_3 R_4}{\left(R_2 + \Delta R + R_4\right)\left(R_1 + R_3\right)} \tag{5.38}$$

which simplifies to

$$V_{\text{OUT}}\,|_{\text{Quarter Bridge}} = V_{\text{IN}}\,\frac{\left(R + \Delta R\right)R - R^2}{\left(2R + \Delta R\right)\left(2R\right)} = V_{\text{IN}}\,\frac{R\Delta R}{\left(4R + 2\Delta R\right)R} = V_{\text{IN}}\,\frac{\Delta R}{4R + 2\Delta R} \tag{5.39}$$

on applying eq. (5.33). Noting that

$$4R \gg 2\Delta R, \tag{5.40}$$

Eq. (5.39) reduces to

$$V_{OUT}\big|_{Quarter\ Bridge} = \frac{V_{IN}}{4}\left(\frac{\Delta R}{R}\right) \tag{5.41}$$

Since from eq. (5.31),

$$\frac{\Delta R}{R} = G\varepsilon \tag{5.42}$$

eq. (5.41) becomes

$$V_{OUT}\big|_{Quarter\ Bridge} = \frac{V_{IN}}{4}G\varepsilon \tag{5.43}$$

Difficulty Faced with the Quarter Bridge: The distance between the strain gauge resistor and the remaining three resistors in the bridge is unknown. The wire resistance influences the measurements by contributing to the output voltage. So, the resistance measured is not that of the strain gauge resistor only.

5.7.3.2 Half Wheatstone Bridge

Case I: Strain Gauges Connected in Adjoining Arms of the Bridge

Due to the elongation of the top-surface strain gauge and contraction of the bottom-surface strain gauge (Figure 5.25(c)), the output voltage of the bridge is obtained from eq. (5.41) as

$$V_{OUT}\big|_{Half\ Bridge} = \frac{V_{IN}}{4}\left(\frac{\Delta R}{R}\right) = \frac{V_{IN}}{4}\left(\frac{\Delta R_2 - \Delta R_4}{R}\right) \tag{5.44}$$

where $+\Delta R_2$ and $-\Delta R_4$ are the changes in the values of resistors R_2 and R_4, respectively, one change is positive (tensile stress in R_2) and the other change is negative (compressive stress in R_4).

Applying eq. (5.43), the equation for output voltage is written in terms of strains $+\varepsilon_2$ (tensile strain) in R_2 and $-\varepsilon_4$ (compressive strain) in R_4 in the form

$$V_{OUT}\big|_{Half\ Bridge} = \frac{V_{IN}}{4}G\varepsilon = \frac{V_{IN}}{4}G(\varepsilon_2 - \varepsilon_4) = \frac{V_{IN}}{4}G(\varepsilon + \epsilon) = \frac{V_{IN}}{4}G(2\varepsilon) = \frac{V_{IN}}{2}G\varepsilon \tag{5.45}$$

because

$$\varepsilon_2 = -\varepsilon_4 = \varepsilon \tag{5.46}$$

Eq. (5.45) gives

$$V_{OUT}\big|_{Half\ Bridge} = \frac{V_{IN}}{2}G\varepsilon = 2 \times \frac{V_{IN}}{4}G\varepsilon = 2 \times V_{OUT}\big|_{Quarter\ Bridge} \tag{5.47}$$

where eq. (5.43) is applied.

Case II: Strain Gauges Connected in Diagonally Opposite Arms of the Bridge

Considering the changes $+\Delta R_2$ and $-\Delta R_3$ in the values of resistors R_2 and R_3 (Figure 5.25(d)),

$$V_{OUT}|_{\text{Half Bridge}} = \frac{V_{IN}}{4}\left(\frac{\Delta R_2 - \Delta R_3}{R}\right) \tag{5.48}$$

If the corresponding strains are $+\varepsilon_3$ and $-\varepsilon_2$, we have

$$V_{OUT}|_{\text{Half Bridge}} = \frac{V_{IN}}{4}G\left(\varepsilon_2 - \varepsilon_3\right) = \frac{V_{IN}}{4}G\left(\varepsilon + \epsilon\right) = \frac{V_{IN}}{2}G\varepsilon \tag{5.49}$$

because

$$\varepsilon_2 = -\varepsilon_3 = \varepsilon \tag{5.50}$$

Similar to eq. (5.45), eq. (5.49) also shows that

$$V_{OUT}\big|_{\text{Half Bridge}} = 2 \times V_{OUT}\big|_{\text{Quarter Bridge}} \tag{5.51}$$

Thus, in both cases I and II, the output voltage of the half bridge is double that of the quarter bridge.

Advantages of Half Bridge:

(i) For a given strain, the half-bridge circuit gives twice the output voltage than that of the quarter-bridge circuit, resulting in sensitivity enhancement.
(ii) The half-bridge circuit provides temperature compensation because strain changes induced by temperature on the two strain gauges, being equal and opposite, cancel out. Other environmental effects are likewise mitigated.

Disadvantages of Half Bridge: Two complementary strain gauges are required in place of one strain gauge in the quarter-bridge circuit. Such a gauge may not be readily available to the experimenter.

5.7.3.3 Full Wheatstone Bridge

The expression for the output voltage of the full bridge (Figure 5.25(e)) is

$$V_{OUT}|_{\text{Full Bridge}} = \frac{V_{IN}}{4}\left(\frac{\Delta R}{R}\right) = \frac{V_{IN}}{4}\left(\frac{\Delta R_1 + \Delta R_2 - \Delta R_3 - \Delta R_4}{R}\right) = \frac{V_{IN}}{4}G\left(\varepsilon_1 + \varepsilon_2 - \varepsilon_3 - \varepsilon_4\right)$$

$$= \frac{V_{IN}}{4}G\left(\varepsilon + \varepsilon + \varepsilon + \varepsilon\right) = \frac{V_{IN}}{4}G\left(4\varepsilon\right) = V_{IN}G\varepsilon \tag{5.52}$$

From eqs. (5.52) and (5.43),

$$V_{OUT}\big|_{\text{Full Bridge}} = 4 \times \frac{V_{IN}}{4}G\varepsilon = 4 \times V_{OUT}\big|_{\text{Quarter Bridge}} \tag{5.53}$$

meaning that the output voltage of a full bridge is four times that of the quarter bridge. In the same way, eqs. (5.52), (5.45), and (5.49) lead to

$$V_{OUT}\Big|_{\text{Full Bridge}} = 2 \times \frac{V_{IN}}{2} G\varepsilon = 2 \times V_{OUT}\Big|_{\text{Half Bridge}} \tag{5.54}$$

implying that the output voltage of a full bridge is twice that of the half bridge.
Advantages of Full Bridge

(i) For a given strain, the full-bridge circuit gives four-fold the output voltage of the quarter bridge, and two-fold the output voltage of the half-bridge circuit, resulting in better sensitivity to both quarter and half-bridge circuits.
(ii) The full-bridge circuit provides temperature compensation.

Disadvantages of Full Bridge: Two pairs of complementary strain gauges are required in place of one strain gauge in the quarter-bridge circuit and a single complementary pair in the half bridge. Such pairs of gauges are not always available. So, one suitable Wheatstone bridge configuration is adopted from the three available choices, depending on the application and mounting constraints. The quarter-bridge circuit is often preferred.

5.8 FORCE SENSING RESISTOR (FSR)

5.8.1 PRINCIPLE AND MECHANISMS

It is a resistor made of a material such as a polymer sheet or screen-printable ink whose resistance changes when subjected to a mechanical force. It works on the piezoresistive principle showing a high resistance in mega ohms when unloaded, which decreases to kiloohms when pressed. The polymer contains sub-micron size conducting and non-conducting particles dispersed in a matrix. When a force is applied on the surface of the polymeric resistive film, e.g., by squeezing, pressing, or loading, the particles come closer together touching the conducting electrodes so that the resistance value of the sensor decreases in proportion to the force magnitude. The FSR works by two mechanisms: percolation and quantum-mechanical tunneling. Both phenomena can occur simultaneously or one phenomenon may dominate over the other depending on the concentration of particles in the FSR. An FSR is fabricated in two technological versions (Figure 5.26), as described in the subsections below.

5.8.2 SHUNT-MODE FSR

It is used for wide-range force sensing (up to 5 kg) (Ohmite 2018). It consists of two PET layers (Figure 5.26(a)). On one PET layer, a textured resistive carbon ink is coated. On the other PET layer, interdigitated silver electrodes are printed. The two PET layers are laminated together using an adhesive with an intermediate perimeter spacer layer between the layers.

5.8.3 THROUGH-MODE FSR

It is used for narrow-range force sensing (<0.5–1 kg). Here, both PET layers are first printed with silver and then resistive carbon film (Figure 5.26(b)). The two layers are attached using an adhesive with a perimeter spacer layer between them.

Advantages: Easy-to-use, thin, flexible, lightweight, available in various shapes and sizes, durable, cheap, shock-resistant, and low power consumption

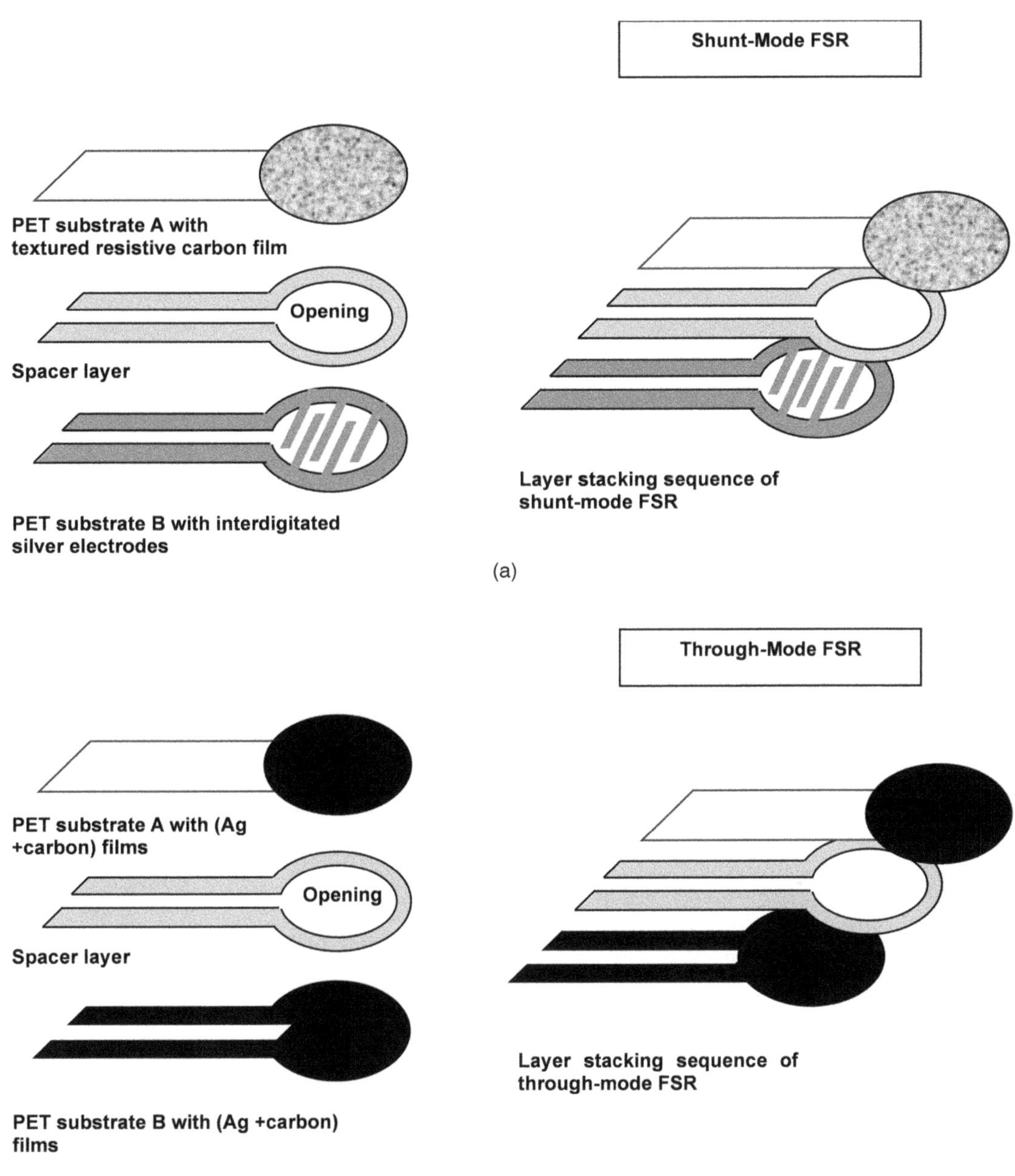

FIGURE 5.26 Force sensing resistors: (a) shunt mode and (b) through-mode. Part (a): Left side: PET substrate A with textured resistive carbon film, spacer layer with opening, and PET substrate B with interdigitated silver electrodes. Right side: Layer stacking sequence of shunt-mode FSR. Part (b): Left side: PET substrate A with (Ag + carbon) films, spacer layer with opening, and PET substrate B with (Ag + carbon) films. Right side: Layer stacking sequence of through-mode FSR.

Disadvantages: Low precision and repeatability with sensor-to-sensor variability ~10%
Applications: Pressure sensing buttons, keys or pads in mobile phones, musical instruments, and gaming devices; robot fingertips, car occupancy sensors, infusion pumps, medical evaluation of prosthetic sockets, foot pronation systems, and inventory monitoring

5.9 MONITORING FLOW OF FLUIDS, LIQUIDS, AND GASES: FLOW SENSORS

5.9.1 ROLE IN FACTORY

Commonly called a flowmeter, a flow sensor is a device for measuring the flow rate of a fluid (liquid or gas) through a pipe or conduit, in terms of the volume or mass of the fluid passing through it per unit of time, giving volumetric flow rate in liters per minute (L/min) or mass flow rate in kilogram per second (kg/s).

Flow sensors play crucial roles:

(i) In manufacturing automation by providing real-time measurements of flow rate to control valves for process control and optimization,
(ii) In resource management by accurate supervision over water, oil, or gas consumption,
(iii) In quality control by maintaining correct flow rates,
(iv) In energy efficiency by ensuring optimal flow rates in pumps and compressors to minimize power consumption,
(v) In protecting machines from damage by triggering alarms to alert about improper flow conditions, and
(vi) In safety and compliance with environmental regulations by detecting leaks and blockages to prevent accidents.

5.9.2 CLASSES OF FLOW SENSORS

Flow sensors are divided into three categories, as explained in the ensuing subsections.

5.9.2.1 Positive Displacement Flow Sensor

This sensor is shown in Figure 5.27. It measures the flow of a fluid directly by performing repeated, sequential cycles of:

(i) Entrapping discrete pockets of fluid of known volume with mechanical seals between its rotating parts housed in a metering chamber, and
(ii) Fully emptying these pockets downstream before allowing entry to more fluid.

Note that the movement of rotating parts of the sensor is caused by the flowing fluid itself. Since the velocity of rotation of rotating parts is proportional to the rate of flow of fluid, the volume of fluid flowing per unit time is obtained by counting the number of rotations per unit time.

Standard accuracy is 0.5% and linearity is 0.02–0.5%. The accuracy of recorded value is only limited by bypass or slippage, which is the fluid managing to pass through the device without being measured.

Flow rates of fluids in a wide viscosity range of less than 1 cP to greater than 5,00,000 cP can be measured, e.g., domestic water or gas supply, oil, and gasoline (Omega Engineering Inc. 2003–2023). High viscosity fluids show less slippage giving high metering accuracy. Slippage is reduced by allowing minimal clearance between sealing faces.

This flow sensor requires low maintenance, requiring servicing and calibration after several years of operation, 10–20 years. However, its installation is expensive due to its high-precision moving parts.

5.9.2.2 Mass flow Sensors

5.9.2.2.1 *Hot-Wire Anemometer*

This mass flow sensor works on the dependence of heat loss from a hot surface such as an electric current-heated wire on the flow rate of the fluid stream in which it is positioned (Figure 5.28).

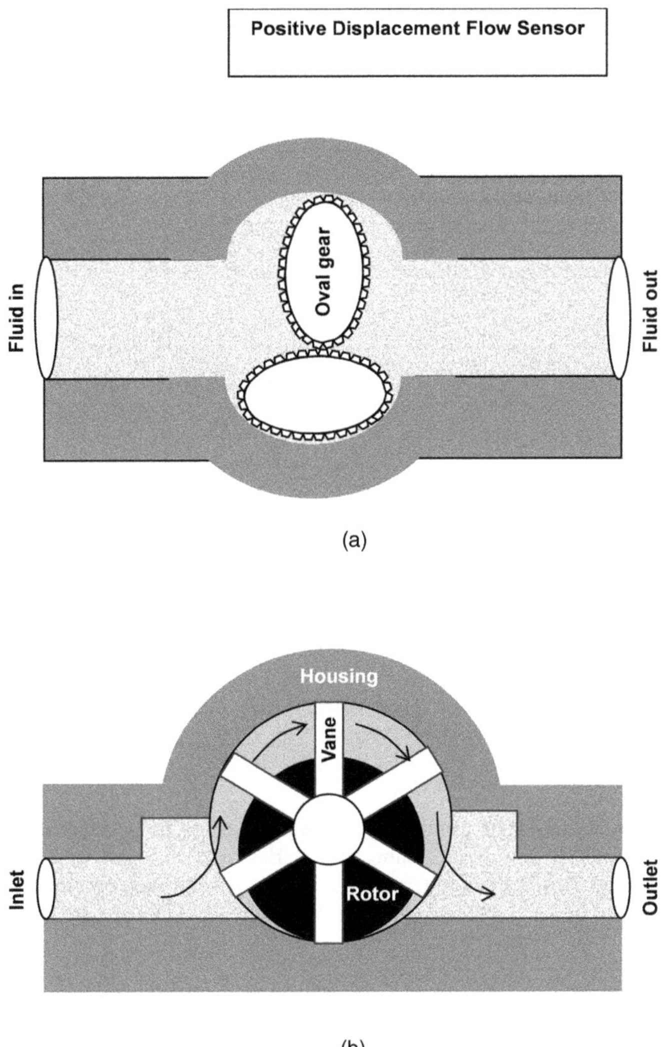

FIGURE 5.27 Positive displacement flow sensors of (a) oval gear and (b) rotor and vane types. Part (a): Oval gears are moved by fluid flow transferring the fluid from the inlet to the outlet. Part (b): Vanes fitted on an eccentrically mounted rotor inside a housing are moved by the flowing fluid, displacing the fluid from the inlet to the outlet.

The fluid stream cools it by convective heat transfer, and the cooling rate gives the fluid flow rate. The faster the fluid movement, the more rapidly the wire cools. The sensor is operated in one of the following three modes.

(i) Constant-Current Mode: In this mode, a constant current is passed through the wire. The fluid flow alters the temperature of the wire and thereby changes its resistance. The hot wire is connected in one arm of a Wheatstone bridge, and the change in its resistance caused by fluid flow is accurately measured. The fluid flow rate is derived from this resistance variation.

(ii) Constant-Voltage Mode: In this mode, a constant voltage is applied across the wire. The temperature of the hot wire is measured with a thermocouple. The change in temperature of the hot wire with fluid flow gives its flow rate.

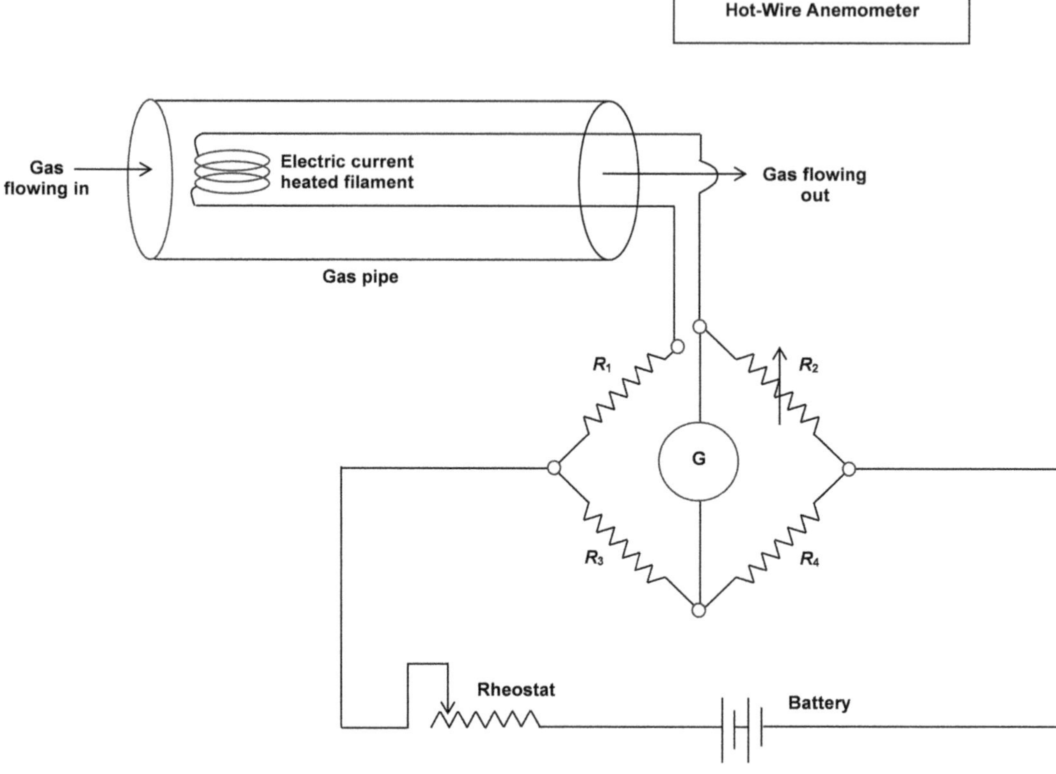

FIGURE 5.28 Hot-wire anemometer. A filament heated by an electric current is installed inside a pipe through which gas is flowing. The filament is connected in series with resistor R_1 of a Wheatstone bridge. Other resistors in the bridge are R_2, R_3, and R_4. A battery with a rheostat is connected between two diagonally opposite points. A galvanometer G is connected between other two diagonally opposite points.

(iii) Constant-Temperature Mode: In this mode, the temperature of the wire is maintained constant by increasing the current. The sensor is calibrated by measuring the change in current required to keep the temperature constant for different flow rates. Hence, the change in current indicates the flow rate.

Applications are in the following industries:

(i) Automotive industry to measure the mass of air entering the internal combustion engine from which the electronic control system decides the quantity of fuel to be supplied to the engine for keeping the proper fuel–air ratio for ignition to save fuel and reduce environmental pollution.

(ii) Pharmaceutical and semiconductor chip manufacturing industries to monitor airflow in chemical fume hoods and laminar flow hoods.

(iii) Metallurgical industry for controlling air flow in blast furnaces.

(iv) Chemical industry for ammonia gas flow measurement in fertilizer plants.

(v) Oil and gas industry for fuel gas metering.

(vi) Power plants for determining gas flow in boilers and furnaces.

(vii) Heating, ventilation, and air-conditioning (HVAC) systems for detecting changes in airflow in critical zones.

5.9.2.2.2 *Coriolis Flow Meter*

Also called an inertial flow meter, it applies the Coriolis effect for flow measurement. According to the Coriolis effect (Sec. 4.8.5.1), any body or object moving on earth's surface such as an air or ocean current is deflected in the sideways direction from its path because of earth's rotation. The process fluid is split into two halves and forced to flow through two parallel tubes (Figure 5.29). These tubes are excited by a magnetic drive coil to oscillate opposite to each other at their resonant frequency. Assemblies containing magnets and coils are mounted on both tubes at their inlets and outlets to act as sensors. As the coils move through the magnetic fields created by the magnets, a sine wave is produced. Thus, the motions of the tubes with respect to time are converted into sine waves.

In the absence of any fluid in the tubes, both the sensors show the same deflection of the two tubes at any given instant. Then inlet and outlet sine waves are in phase. This indicates that the tubes are in synchronized motion.

But when a fluid flows through the tubes, the tubes begin to twist due to the mass of fluid in the tubes. This happens in the same way as the sideways drift occurring due to the rotation of the earth, which is known as the Coriolis effect. As a consequence, the sine waves show a phase difference portraying the asynchronous motion of tubes. This phase shift between sine waves is proportional to the mass flow rate of the fluid. The mass flow rate is obtained from the time difference Δt between the sine waves.

While the phase shift is a measure of mass flow rate, the frequency of vibration of the tubes is an indicator of the fluid density. A denser fluid vibrates slower than a less dense one. The fluid density varies inversely to the square of the frequency of vibration of the tubes. The density of fluid is extracted from the frequency of sine waves. Division of the mass flow rate with density gives the volumetric flow rate of the fluid.

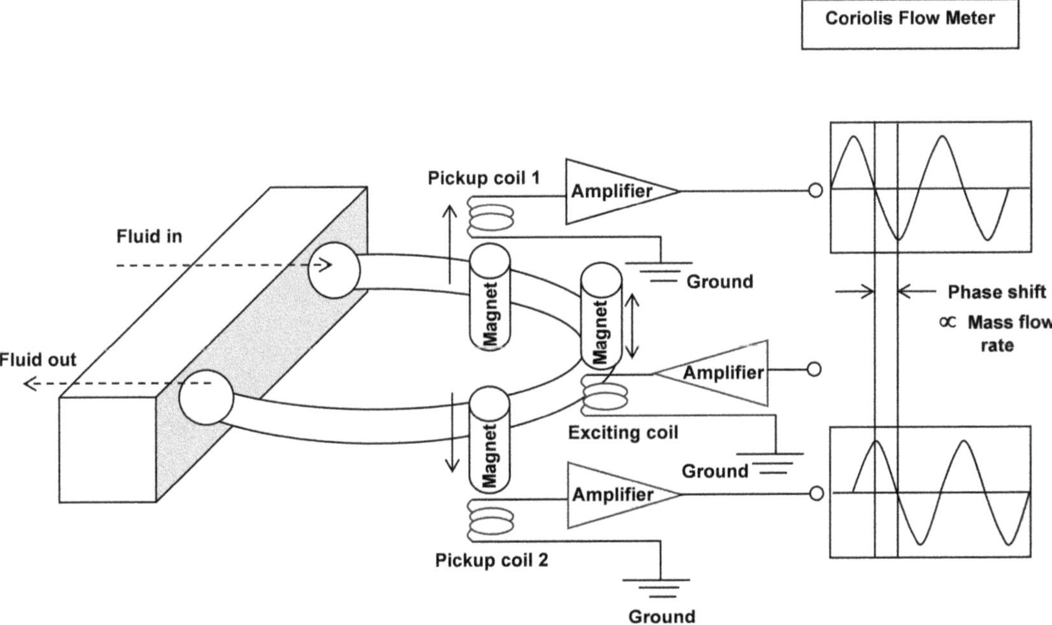

FIGURE 5.29 Coriolis flow meter. A fluid is flowing through a U-shaped tube. Three magnets are fixed on the three sides of the U-shaped tube. Coils are placed near each magnet. The exciting coil is supplied current from an amplifier circuit causing up/down vibrations of the U-shaped tube. The pickup coils 1 and 2 collect sinusoidal current signals, which are amplified and compared to find the phase difference between them. This phase shift depends on the fluid flow rate. Arrows indicate the directions of forces acting on the three sides of the U-tube. The side near the pickup coil 1 moves up while the side near the pickup coil 2 moves down creating a twist in the U-shaped tube.

Advantage of the Coriolis Flow Meter over the Hot-Wire Anemometer: The Coriolis flow meter provides a direct measurement of mass flow rate so that any errors from physical properties of fluid do not creep into the measurements. The hot-wire anemometer is a thermal mass flow sensor providing indirect flow measurement through the heat capacity of the fluid.

Applications: Measurement of flow of liquids, steam, and high-pressure gas in oil, gas, water, petrochemicals, power, chemical, food, and beverage industries.

5.9.2.3 Velocity Flow Sensor

A velocity flow sensor calculates the flow rate of a fluid from a measurement of the velocity of the flowing fluid.

(i) Mechanical Paddle Wheel Flow Sensor: This sensor is a paddle wheel consisting of a number of paddles around the periphery of a wheel (Figure 5.30). When it is placed in the fluid stream, it rotates with the flowing fluid. Through this rotation, the linear motion of the fluid is converted into the rotary motion of the wheel. The flow rate of the fluid is determined from its proportionality relationship with the speed of rotation of the paddle wheel measured in the number of revolutions per minute (RPM). The faster the fluid flows, the greater the spinning speed of the wheel. For non-contact measurement of wheel speed, a magnet is fixed on a paddle of the wheel. An output voltage pulse is produced across a Hall element (placed outside the fluid and hermetically separated from the fluid) when a magnetic paddle of the wheel inside the fluid passes near the Hall element outside. The fluid flow rate is obtained from the pulse frequency. This sensor performs best in water and low-viscosity liquids (Djalilov et al. 2023).

Advantages: Cost-effective, compact, and usable for a variety of fluids because the sensor operation is not dependent on fluid conductivity.

Disadvantages: Subject to wear and tear because of moving parts; also prone to contamination effects; and the requirement of a minimum flow rate to move the wheel.

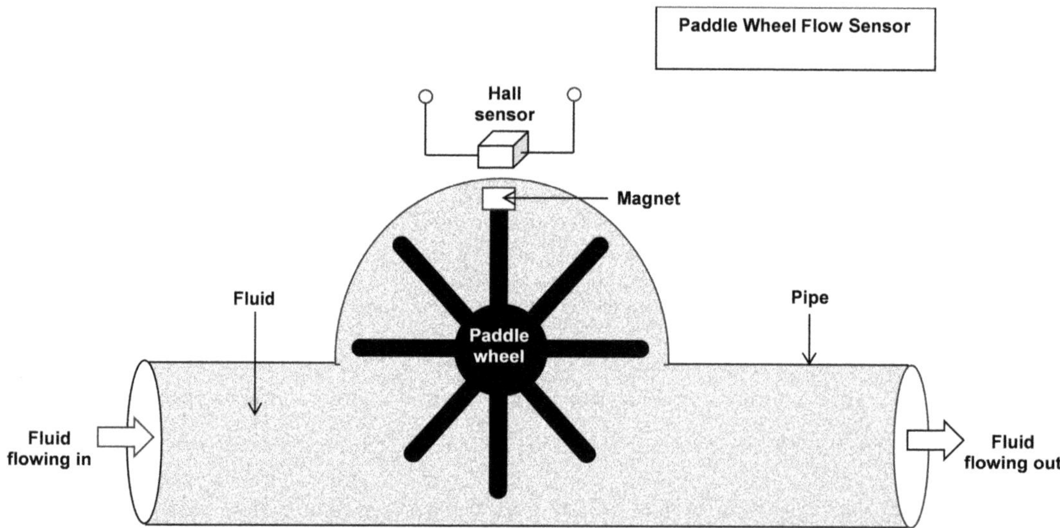

FIGURE 5.30 Mechanical paddle wheel flow sensor. A paddle wheel with outward protruding rods is moved by the fluid flow in the pipe. A magnet is fixed at the end of one rod. During each revolution of the paddle wheel, this magnet fixed at the end of the rod inside the pipe passes near a Hall element placed outside the pipe. When this happens, an output signal is produced in the Hall element.

Applications: Cooling towers and chillers, fume scrubbers, reverse osmosis; chemical metering, dispensing systems, medical equipment, wastewater treatment plants, HVAC, and irrigation systems for alerting users when the flow rate exceeds the programmed limit.

(ii) Ultrasonic Flow Sensor: Sections 3.11.1 and 11.12.1

(iii) Electromagnetic Flow Sensor: Sections 3.11.1 and 11.12.2

5.10 RECORDING LIQUID LEVELS IN CONTAINERS AND TANKS: FLUID LEVEL SENSORS

5.10.1 TYPES OF LEVEL SENSORS AND THEIR USE IN INDUSTRY

Level sensors are level-indicating devices used to measure the levels of materials such as liquids, fluidized and granular solids, or slurries kept in containers or storage tanks. They are used in industrial manufacturing to check levels of chemicals, in food and beverage industries to monitor liquids and powdered eatables, in HVAC to control oil or refrigerant levels, and in homes and offices for water level inspection to prevent overflow. There are two types of level sensors: point level sensors showing whether the level is below or above a specified point, and continuous level sensors giving continuous level readings as the level increases or decreases.

5.10.2 CAPACITIVE FLUID LEVEL SENSOR

This sensor works on the change in capacitance with the level of fluid. The capacitor is formed with one insulated electrode called the conductive probe immersed inside the fluid and the metallic wall of the fluid storage tank as the other electrode (Figure 5.31(a)). If the tank wall is an insulator, a reference electrode is inserted inside the fluid and taken as the second electrode. Taking the example of a water tank, the capacitor for an empty tank has air as dielectric (relative permittivity = 1) while that for a water-filled tank has water as dielectric (relative permittivity = 78.39). Obviously, the empty tank has a lower capacitance than the water-filled tank, and the capacitance increases as the tank is filled to a greater height with more water. The capacitance is measured by applying an RF signal between the conductive probe and the wall of the tank (for a metallic wall) or between the conductive probe and a reference electrode (in the case of an insulating tank wall) (Figure 5.31(b)), producing a small current flow between the electrodes. The change in water level is converted by the internal circuit of the sensor into a change in the relay state of a level switch for a point level detector, or by a processor chip into scaled analog output in case of a continuous level detector.

Advantages: Small, inexpensive devices with no moving parts requiring minimal maintenance, easily installed, and having a large measurement range from a few cm to 100 m.

Disadvantages: Invasive, and difficult to work with air-filled low-density materials having low dielectric constants.

Applications: Liquids and solids in powdered form, molten metal at a very high temperature, liquid air at an extremely low temperature, and corrosive acids, e.g., HCl.

5.10.3 OPTICAL FLUID LEVEL SENSOR

It consists of a light source such as an infrared LED, a cone-shaped prism, and a photodetector such as a photoresistor, a photodiode, or a phototransistor (Figure 5.32). When the prism is above the liquid surface (Figure 5.32(a)), the light from the LED is reflected back to the photodetector. However, when it is submerged inside the liquid (Figure 5.32(b)), the light is mostly transmitted into the liquid with very little reflected light received by the photodetector. Hence, the received light intensity decreases.

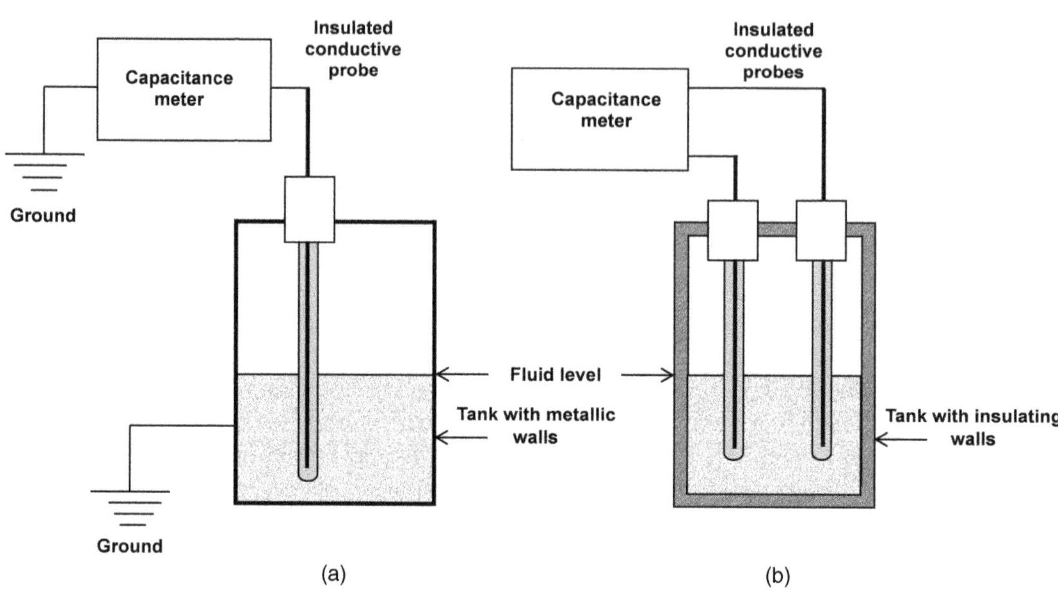

FIGURE 5.31 Capacitive fluid level sensor for a tank with (a) metallic walls and (b) insulating walls. Part (a): A tank with metallic walls contains a fluid into which a single insulated conductive probe is immersed. The tank wall is grounded. A capacitance meter is connected between the probe and the ground terminal. Part (b): A tank with insulating walls contains a fluid into which two insulated conductive probes are immersed. The capacitance meter is connected between the two probes.

Then the photodetector activates a relay switch for sounding an alarm or for executing the necessary control function.

> Advantages: A small device with no moving parts, can be used at high temperatures and pressures and in explosive environments; and is immune to electromagnetic interference.
>
> Disadvantages: Invasive, degradable by corrosion, and long reading time due to bubbles or coating.
>
> Applications: Detection of levels of aqueous or organic liquids or liquids with suspended solids in tanks in medical, food, and beverage industries; off-road adventure vehicles used for steep or uneven terrain and aircraft; leak detection such as rainwater leakage in outdoor equipment; lubricant leakage detection in robotic automation systems, etc.

5.10.4 Conductivity Fluid Level Sensor

This sensor is used for conductive liquids with a minimum conductivity of $10~\mu Scm^{-1}$. Water, acids, and alkali metal hydroxides fall in this class of liquids. Pure solvents have poor conductivity and are therefore excluded.

Two electrodes are fitted in the liquid container, one longer electrode to always remain dipped in the liquid, and one shorter electrode placed at the height at which the liquid filling must be stopped to prevent overflow. A DC-free RF alternating current is passed through the electrodes to avoid electrolytic dissociation of the liquid (Figure 5.33). As long as the liquid level is below the height at which it does not touch the shorter electrode, the electrical circuit is broken because of the air layer

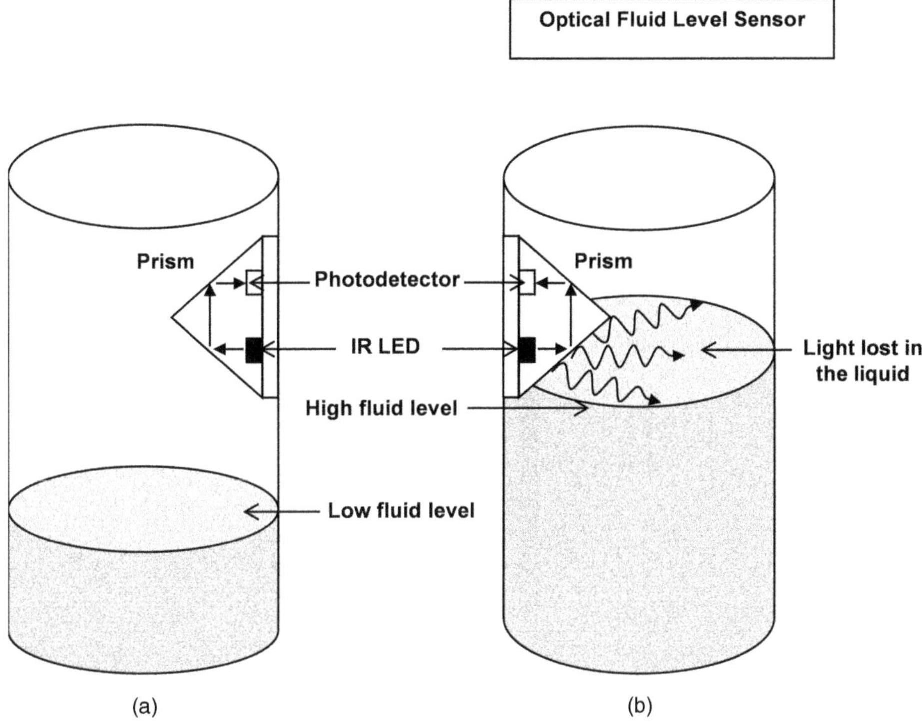

FIGURE 5.32 Optical fluid level sensor: (a) above fluid level and (b) under fluid level. Parts (a) and (b): Light from an infrared LED is falling on one face of a cone-shaped prism. It is reflected from the opposite face of the prism into a photodetector. Part (a): The prism is in air. The light received by the photodetector has a high intensity. Part (b): The prism is dipped under the fluid. The received light has lower intensity because some fraction of light is lost by transmission and scattering into the liquid.

between the electrodes. As soon as the liquid is filled in the container to a height at which it touches the shorter electrode, the electrical circuit is completed, and the current starts flowing from one electrode through the conductive liquid to the other electrode. The moment this happens, the HIGH alarm is triggered, and a switch is actuated to close the valve from which the liquid is entering the container. Similarly, when the liquid level drops below the height at which it loses contact with the shorter electrode, a LOW alarm is triggered, and a switch is actuated to open the valve for starting the filling of the container.

Advantages: Easy-to-use, low-cost device with no moving parts.
Disadvantages: Invasive, corrodes with time, and useful for conductive liquids only.
Applications: Generating high- and low-level alarms and alerts.

5.10.5 Vibrating Tuning Fork Fluid Level Sensor

A stainless-steel tuning fork is used for high-temperature and pressure applications (Figure 5.34). It is set into vibration at its natural frequency by applying an alternating voltage to the attached piezoelectric crystal. The frequency and amplitude of its vibration are measured by an electronic circuit. Both these parameters change when the tuning fork is vibrating in the air medium of the empty tank (Figure 5.34(a)), and the liquid medium of the filled tank (Figure 5.34(b)). The amplitude of vibration of the tuning fork is dampened and the frequency also decreases in the liquid medium as compared to the air medium.

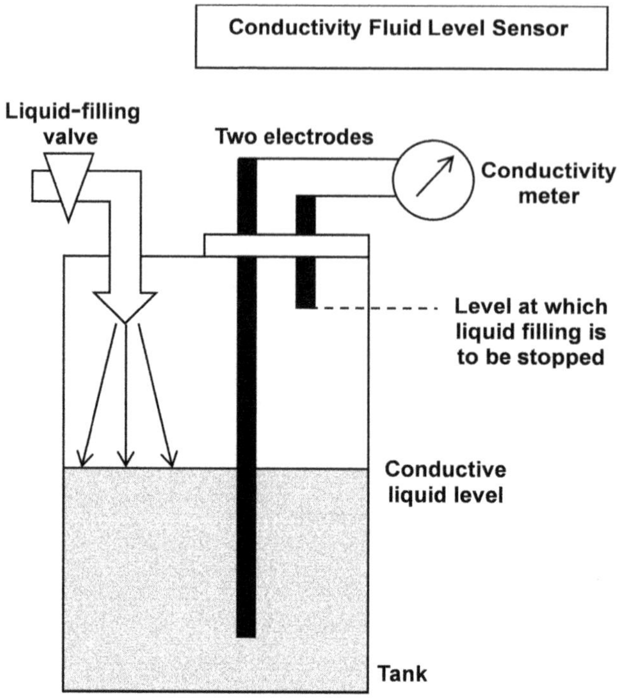

Conductivity Fluid Level Sensor

Liquid-filling valve

Two electrodes

Conductivity meter

Level at which liquid filling is to be stopped

Conductive liquid level

Tank

FIGURE 5.33 Conductivity fluid level sensor. The sensor has two electrodes, one longer electrode always immersed in the fluid and one shorter electrode placed at a height above the fluid level. As soon as the conductive fluid fills in the tank up to the height of the shorter electrode, both electrodes are dipped in the fluid completing the circuit for current flow and triggering the switch to stop further fluid flow.

When the amplitude or frequency of the vibrating tuning fork diminishes below the adjusted threshold value, the relay is prompted to do the needful.

> Advantages: Cheap, compact, and easily installed sensor providing maintenance-free operation.
> Disadvantages: Invasive, does not give continuous level measurements, and limited in usage to overfilling or running dry alerts.
> Applications: Mining, chemical processing, food and beverage industries.

5.10.6 Float Switch Fluid Level Sensor

It works on the buoyancy principle, with the float ascending and descending as the liquid level rises and falls. The sensing arrangement comprises a reed switch encased in a hermetically sealed tube (Figure 5.35). This tube is surrounded by a float. The float contains a magnet which moves up and down as the float moves with the changes in the liquid level. When the float raises the magnet to the height at which it is aligned with the reed switch (Figure 5.35(b)), the switch closes so that the pump stops filling or a water-level alarm is sounded. Likewise, when the liquid level falls (Figure 5.35(a)), the magnet moves down with the float losing alignment with the reed switch. The switch is opened to start refilling, thus serving as an automatic water level control system.

- Advantages: Inexpensive device giving a direct indication of fluid level.
- Disadvantages: Invasive, large size, has moving parts exposed to corrosion and fouling, and is often stuck by viscous liquids ceasing actuation.
- Applications: Liquid storage tanks requiring full/empty warning.

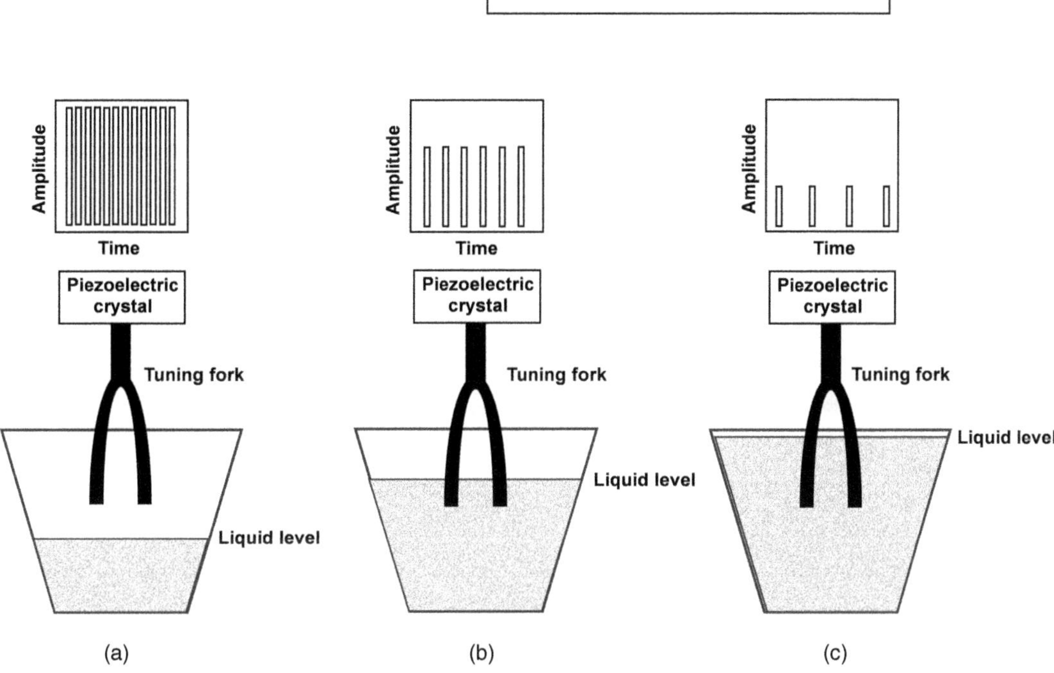

FIGURE 5.34 Vibrating tuning fork fluid level sensor when the fork is: (a) above the fluid level in air, (b) inside the fluid up to a small depth, and (c) inside the fluid up to a greater depth. Diagrams (a), (b), and (c) show a tuning fork set into vibration by actuating a piezoelectric crystal attached to it. Part (a): The amplitude and frequency of tuning fork vibration are high because the fork is vibrating in air. Impedance to fork vibration due to air is low. Part (b): The amplitude and frequency of tuning fork vibration decrease because some portion of the fork is under the fluid, say water, and its vibrations are slightly impeded by the fluid. The vibrations of the submerged portion of the fork are opposed by the fluid. Part (c): The amplitude and frequency of tuning fork vibration decrease still further because a greater portion of the fork is under the fluid, and its vibrations are impeded beyond that in part (b).

5.10.7 ULTRASONIC FLUID LEVEL SENSOR

It is based on the time-of-flight principle. An ultrasonic transducer acting as a transmitter/receiver is mounted on the top of the fluid tank facing downward (Figure 5.36). It sends ultrasonic pulses toward the fluid surface. The ultrasonic pulses propagate through the air space above the fluid surface and strike the fluid surface where they suffer reflection back toward the ultrasonic transducer. The time taken by an ultrasonic pulse to travel from the ultrasonic transducer to the fluid surface and return from the fluid surface to the ultrasonic transducer is proportional to the distance between the ultrasonic transducer and the surface of the fluid. The speed c of the ultrasonic pulse, its traveling time t, and the transmission distance s are related as

$$s = \frac{ct}{2} \tag{5.55}$$

The fluid level is determined from this time by applying knowledge of storage tank geometry and dimensions.

- Advantages: Non-contact measurements of levels of fluids, sludges, pastes, and fine-to-coarse powdery materials; maintenance-free operation; neither contaminates the fluid nor

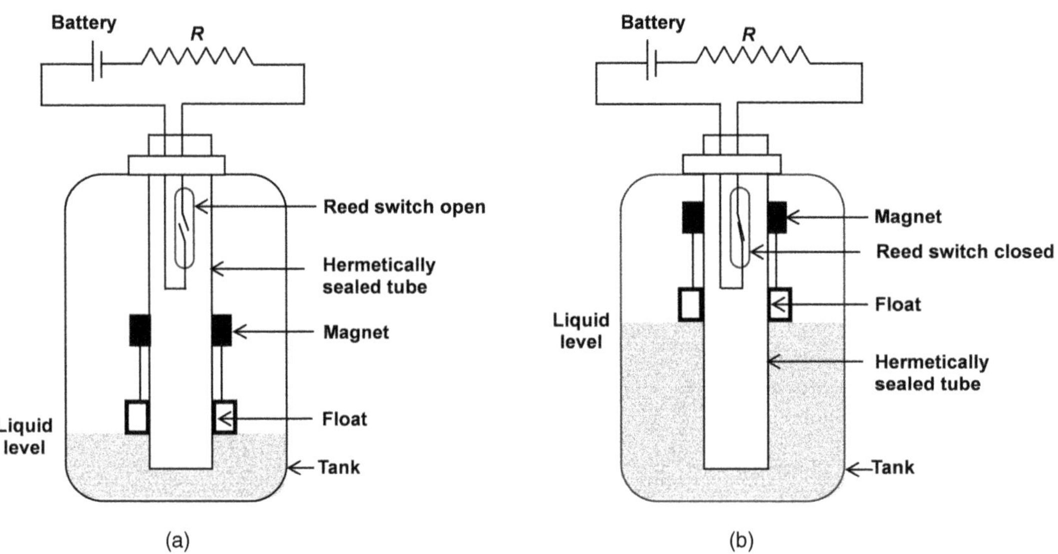

FIGURE 5.35 Float switch fluid level sensor when: (a) the fluid level is low and (b) the fluid level is high. Diagrams (a) and (b) show a reed switch inside a hermetically sealed tube outside which a float with an attached magnet is resting over the liquid surface. The switch is connected to a battery through a resistor. Part (a): The float is much below the level of the reed switch, hence also the magnet so that the reed switch is OFF. Part (b): The float rises as the liquid level increases. Then the magnet is aligned to and is at the same height as the reed switch, hence turning the reed switch ON.

damaged by corrosive fluids; uninfluenced by dielectric constant, conductivity, or density of fluid or by ambient humidity; high resolution (0.1 mm) and long range of detection (6–12 m); and self-cleaned by vibrations.

- Disadvantages: Performance impairment by fluid foaming, agitation, splashing, and turbulence; and only top mounting can be done in a tank, with no side or bottom mounting flexibility.
- Applications: Measuring the level of non-foaming water and chemicals at temperatures up to 150°C and pressures up to 3 bar; liquid level measurement in food and beverage, oil and petroleum industries, and in wastewater plants; especially useful for abrasive, rough, and aggressive fluids.

5.10.8 FMCW RADAR FLUID LEVEL SENSOR

It works on the principle of difference between the frequency of reflected microwave from a fluid surface and the present frequency of the microwave transmitter. In this sensing arrangement, a frequency-modulated continuous wave (FMCW) signal is transmitted toward the fluid (Figure 5.37). This signal moves at the velocity of light. Its frequency ranges between 76 and 81 GHz. An important feature of the transmitted signal is that its frequency continuously varies at a fixed rate. So, a signal transmitted at a later time instant t_2 has a different frequency f_2 than the frequency f_1 of a signal transmitted at an earlier instant t_1. By this reasoning, the echo signal received by reflection from the liquid surface, as detected by the microwave receiver, has the frequency f_1 because it is the reflected signal corresponding to the signal transmitted at an earlier instant t_1. The signal transmitted

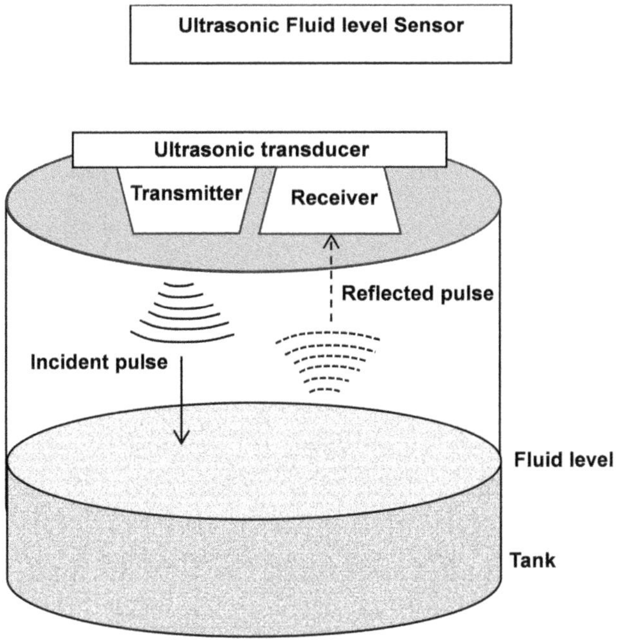

FIGURE 5.36 Ultrasonic fluid level sensor. An ultrasound signal is sent by an ultrasound transmitter toward the fluid. The incident pulse is reflected from the fluid surface. The reflected pulse is detected by the ultrasound receiver. It is used to measure the fluid level.

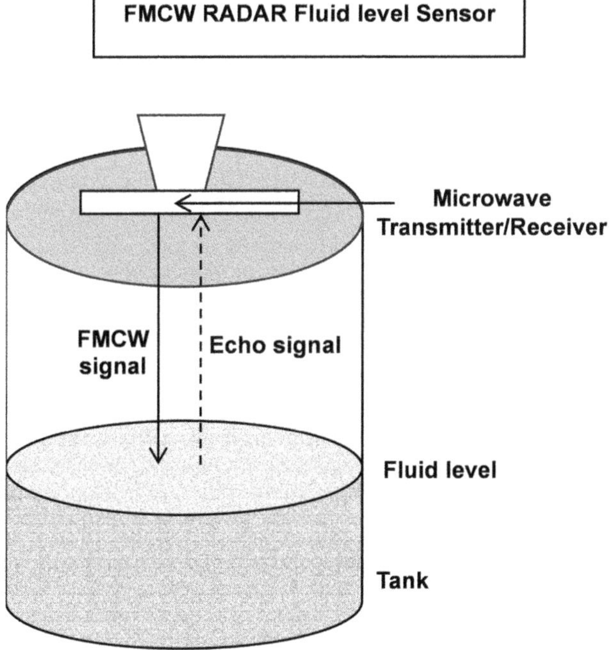

FIGURE 5.37 FMCW RADAR fluid level sensor. The FMCW signal is sent by a microwave transmitter toward the fluid. The echo signal from the fluid surface is detected by the microwave receiver. It is used to determine the fluid level.

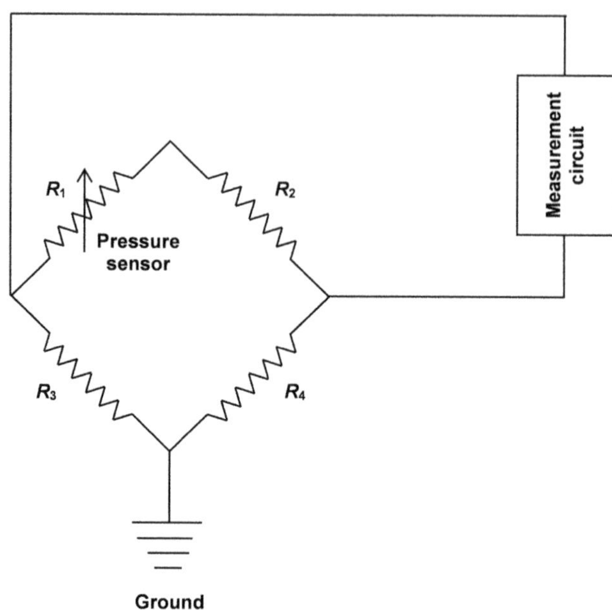

FIGURE 5.38 Hydrostatic fluid level sensor. A pressure sensor of resistor R_1 is placed below the fluid surface at the bottom of the tank. This resistor constitutes one arm of a Wheatstone bridge made from resistors R_1, R_2, R_3, and R_4. As the pressure exerted by a fluid column at a point is proportional to the depth of this point from the fluid surface, the measurement circuit extracts the fluid level from the pressure values given by the pressure sensor.

at the time of reception of the echo signal at instant t_2 has frequency f_2. The two signals are mixed together, and their difference signal is determined. The difference signal is an intermediate frequency signal having a frequency = the difference between the frequencies of mixed signals. From the known rate of frequency variation ($\Delta f / \Delta t$), and the measured frequency difference $\Delta f = f_2 - f_1$, the time taken Δt by the wave to travel from the microwave transmitter antenna to the fluid surface and back, at the velocity c of light, is measured, and hence the distance $\Delta s = c\Delta t/2$ between the transmission antenna and the fluid surface is calculated with high precision.

Advantages: Non-contacting measurement prevents the sensor from corrosion making it suitable for abrasive, gluey, and high-viscosity fluids; highly adaptive to harsh environments; immune to dust particles, dirt, and vapors of measured fluid; neither affected by high temperature and pressure nor by density and conductivity of fluid; high accuracy, and simple installation.

Disadvantages: Expensive, affected by the shape of the container, and limited range of detection.

Applications: Oil and water tanks, acid and caustic tanks, and perfectly suited for hot liquid containers.

5.10.9 Hydrostatic Fluid Level Sensor

The hydrostatic liquid level sensor (Figure 5.38) consists of a submersible piezoresistive MEMS pressure sensor. This sensor is housed in a hermetic package to protect it from the liquid. It is configured in a Wheatstone bridge arrangement and installed at the bottom of the liquid tank (Yuliza et al. 2016). At any particular liquid level, the hydrostatic pressure exerted by the liquid column on the diaphragm of the pressure sensor, is given by the product of the height of the superincumbent liquid (h) at that level, its density (ρ), and acceleration due to gravity (g). Utilizing this relationship, the height of the liquid above the installation point of the pressure sensor is calculated from the measured hydrostatic pressure.

- Advantages: Highly accurate and reliable; neither affected by disturbances due to foam, dust, vapors, or contamination nor by physical properties of liquid, e.g., density, viscosity, and dielectric constant; and not influenced by container geometry.
- Disadvantages: Limited to liquid media only, calibration changes with specific gravity of liquid; narrow dynamic range, costly maintenance, and thermal instability.
- Applications: Fluid level measurement in liquid storage tanks, boreholes, reservoirs, sewage plants, waterways, tidal water, or seawater; food, chemical plants, medical and metallurgical industries, hydrological explorations, etc.

5.11 CONCLUDING REMARKS AND PREPARING FOR THE UPCOMING CHAPTER

We have thoroughly examined the applications of sensors in industrial manufacturing in this chapter and its predecessor and appreciated that sensors form a part and parcel of factory operations, playing an essential role in almost all industrial processes. In the forthcoming chapter, we try to see how sensors help us to commute safely and conveniently from our homes to our places of work by controlling the traffic on the roads connecting them by preventing traffic congestions, jams, and accidents..

REFERENCES

Bhat K. N. and M. M. Nayak 2013 MEMS pressure sensors: An overview of challenges in technology and packaging, *Journal of ISSS*, 2(1):39–71.

Bramble S. 2020 Silicon temperature sensing with precision - An autobiographical look at measuring temperature to ±0.1°C, Analog Devices, pp. 1–3, https://www.analog.com/en/technical-articles/silicon-temperature-sensing-with-precision.html

Chandra S., R. Tiwari and C. Parthiban 2014 A novel technique for fabrication of MEMS based capacitive pressure sensor using press-on-contact in anodic bonding, *2014 Symposium on Design, Test, Integration and Packaging of MEMS/MOEMS (DTIP)*, 1–4 April, Cannes, France, ©EDA Publishing/DTIP, pp. 1–5.

Chong Y., M. Poschmann, R. Zhang, S. Zhao, M. S. Hooshmand, E. Rothchild, D. L. Olmsted, J. W. Morris Jr, D. C. Chrzan, M. Asta and A. M. Minor 2020 Mechanistic basis of oxygen sensitivity in titanium, *Science Advances*, 6(43), pp.1–10.

Djalilov A., O. Nazarov, E. Sobirov, U. Tasheva, J. Abdunabiyev, and S. Urolov 2023 Research of water flow measuring device based on Arduino platform, In: D. Bazarov (Ed.), *E3S Web of Conferences, V International Scientific Conference "Construction Mechanics, Hydraulics and Water Resources Engineering" (CONMECHYDRO – 2023)*, Tashkent, Uzbekistan, April 26–28, 401, 04039, pp. 1–10.

Eaton W. P. and J. H. Smith 1997 Micromachined pressure sensors: review and recent developments, *Smart Materials and Structures*, 6: 530, 12pp.

IG1hotapp.pdf: Hot vs Cold Ionization Gauge, SRS Stanford Research Systems. https://www.thinksrs.com/downloads/pdfs/applicationnotes/IG1hotapp.pdf

Kalsoom T., N. Ramzan, S. Ahmed and M. Ur-Rehman 2020 Advances in sensor technologies in the era of smart factory and industry 4.0, *Sensors*, 20, 6783, pp. 1–22.

Meng Q., Y. Lu, J. Wang, D. Chen and J. Chen 2021 A piezoresistive pressure sensor with optimized positions and thickness of piezoresistors, *Micromachines*, 12, 1095, pp. 1–13.

Morse C. 2022 Temperature Sensors: Choosing between thermocouples, RTDs, and thermistors, MADGETECH, https://www.madgetech.com/posts/blogs/temperature-sensors-choosing-between-thermocouples-rtds-and-thermistors/

Nowicki M. 2018 Tensductor—Amorphous Alloy Based Magnetoelastic Tensile Force Sensor, *Sensors*, 18, 4420, pp. 1–11.

Ohmite FSR Series Integration Guide: Force Sensing Resistor, v1.0 Jan 2018 pp. 1–10, https://www.mouser. com/pdfdocs/Ohmite-FSR-Integration-Guide-V1-0_11-01-18.pdf

Omega Engineering Inc. 2003–2023 Lessons about Positive Displacement Flow Meters, https://www.omega. co.uk/technical-learning/positive-displacement-flow-meter.html#:~:text=Mechanical%20positive%20 displacement%20flow%20sensors,pipe%20runs%20for%20their%20installation

Song P., Z. Ma, J. Ma, L. Yang, J. Wai, Y. Zhao, et al. 2020 Recent progress of miniature MEMS pressure sensors. *Micromachines*, 11(1):56, pp. 1–38.

Texas Instruments 2018, Application Report SBAA275, A basic guide to RTD measurements, pp. 1–42. https:// www.newport.com/medias/sys_master/images/images/h97/h74/9163083317278/TN-RTD-1-Callendar-Van-Dusen-Equation-and-RTD-Temperature-Sensors.pdf

Thermocouples-LibreTexts Engineering 2021. https://eng.libretexts.org/Bookshelves/Materials_Science/ Supplemental_Modules_(Materials_Science)/Electronic_Properties/Thermocouples

Thermocouples-madur.com n.d. https://www.madur.com/pdf/article/en/Thermocouples_EN.pdf

Vierinen K. 2018 Microsystems, MEMS-applications, manufacturing methods for MEMS. https://www. researchgate.net/publication/323069589_Microsystems_MEMS-applications_manufacturing_methods_ for_MEMS

Yuliza E., R. A. Salam, I. Amri, E. D. Atmajati, D. A. Hapidin, I. Meilano, M. M. Munir, M. Abdullah, and Khairurrijal 2016 Characterization of a water level measurement system developed using a commercial submersible pressure transducer, *International Conference on Instrumentation, Control and Automation (ICA)*, Bandung, Indonesia, 29–31 August, pp. 99–102.

Zhang Y., R. Howver, B. Gogoi and N. Yazdi 2011 A high-sensitive ultra-thin MEMS capacitive pressure sensor, *2011 16th International Solid-State Sensors, Actuators and Microsystems Conference*, June 5–9, Beijing, Chaina, IEEE, NY, USA, pp. 112–115.

6 IoT Sensors for Traffic Control and Car Parking

6.1 INTRODUCTION

Between homes and offices/institutes/factories/markets/other workplaces, one has to commute regularly on some form of transport. The aggregation of pedestrians and vehicles (cars, trucks, buses plying on the roads; the boats and ships on rivers and seas; and the airplanes in the air) in a particular locality constitute the surface, water, and aerial traffic in that area. Here, we shall be mostly concerned with road traffic, and see how sensors can be used to divert traffic to vacant routes to prevent overcrowding, jams, and accidents on roads, especially during peak rush hours. Indeed, sensors have proved to be incredibly useful in the optimization of road utilization and traffic control.

6.2 THE INTELLIGENT TRAFFIC MANAGEMENT SYSTEM

6.2.1 MEANING AND GOALS OF THE SYSTEM

Smart or intelligent traffic lights are a vehicular traffic control system in which conventional traffic lights are augmented with sensors and the Internet of Things to collect real-time traffic data on different roads of a city or highway (Nellore and Hancke 2016). The traffic data are processed by artificial intelligence algorithms to guide traffic to less crowded routes, thereby preventing traffic jams.

The traffic management system has the following principal goals:

(i) To avoid traffic overcrowding and congestion on the roads,
(ii) To decrease the average waiting times of vehicles at traffic lights,
(iii) To allow quick passage of emergency vehicles such as ambulances,
(iv) To ensure safe traffic flow, and
(v) To improve air quality by reducing traffic-induced environmental pollution.

6.2.2 SUBSYSTEMS IN THE INTELLIGENT TRAFFIC MANAGEMENT SYSTEM

It comprises three subsystems (Figure 6.1).

(i) Traffic Sensor Subsystem: Its purpose is to collect real-time information about the traffic including:
   (a) detection of the presence of vehicles at distinctive points,
   (b) counting the number of vehicles passing through specific points,
   (c) measuring the speed and direction of motion of the vehicles, and
   (d) transmission of gathered information to the next subsystem.
(ii) Wireless Traffic Control Subsystem: This subsystem acts as a vehicle information and communication system center (VICS Center). It processes the information received from the traffic sensor subsystem by applying adaptive algorithms. It tries to reduce traffic congestion for regular and emergency vehicles at the points of intersection of roads.
(iii) Traffic Safety Subsystem: It provides security to the wireless control subsystem to prevent traffic jams and to apprehend traffic rule violators.

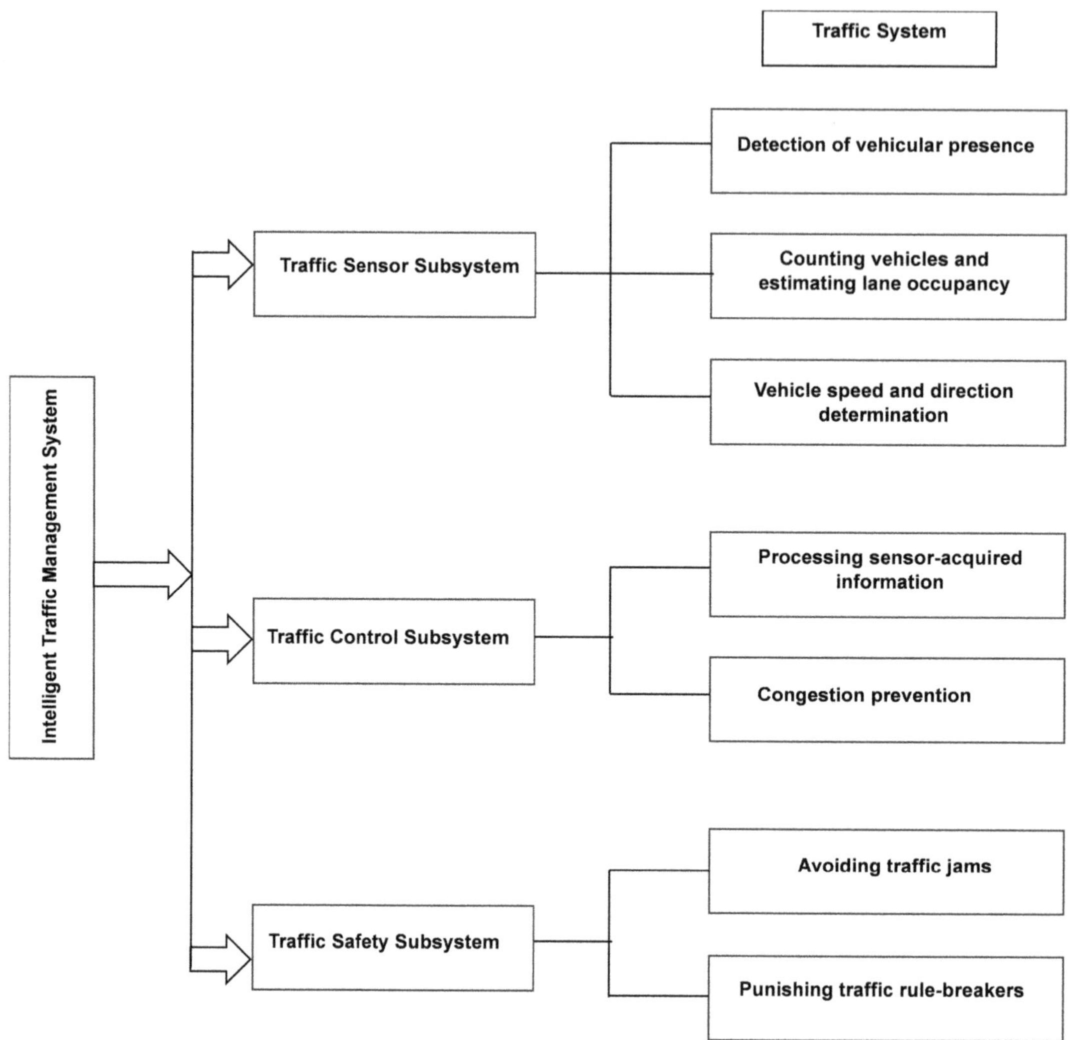

FIGURE 6.1 Traffic management system consisting of three subsystems concerned with traffic sensors, control, and safety: the sensors and accompanying electronics for detecting vehicles, counting them, and acquiring their speed and direction data; the information processing component to prevent congestions; and the safety part for dealing with jams and for grabbing offenders.

6.3 ROADWAY INDUCTIVE LOOP SENSOR FOR VEHICLE DETECTION

The inductive loop sensor consists of one or more coils of conducting wire (Klein et al. 2006). The road pavement is cut, and the coils are laid down below the surface (Figure 6.2). A high-frequency magnetic field (10–200 kHz) is set up in the coil. When a vehicle passes over or stops over the area influenced by the loop, the vehicle experiences the changing magnetic field in that area, and eddy currents are induced in its metallic body. The induced eddy currents flowing in the vehicle's body set up their own magnetic field, opposing the magnetic field of the loop coil. The net effect is that the inductance of the loop sensor decreases. Hence, the measurement of the inductance of the loop sensor reveals the passage or presence of a vehicle over or near the loop. The same phenomenon occurs when any metallic object moves over the loop. So, the loop acts as a metal detector.

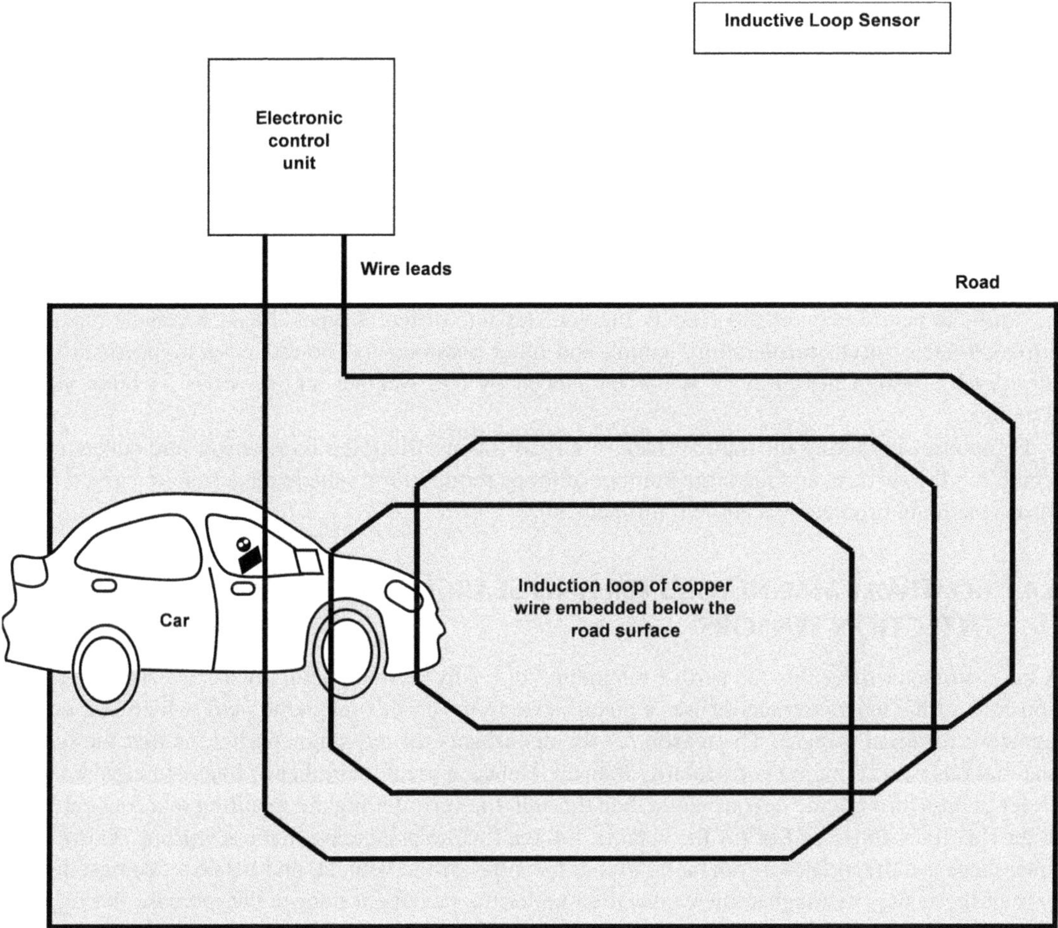

FIGURE 6.2 Inductive loop sensor. The diagram shows an induction loop comprising three turns of copper wire embedded in the road pavement, and wire leads connecting the copper coil to the cabinet housing the electronic equipment by the roadside called the electronic control unit. A car is passing over the coil.

If L_V is the loop inductance when the vehicle is moving over the loop and L_{WV} is the loop inductance when there is no vehicle over the loop, the change in inductance caused by the vehicle is

$$\Delta L = L_{WV} - L_V \tag{6.1}$$

The sensitivity of the sensor is defined as

$$S = \frac{\text{Change in inductance of the sensor caused by the vehicle passing over or standing over the loop's detection area}}{\text{Original inductance of the sensor in the absence of the vehicle}}$$

$$= \frac{\Delta L}{L_{WV}} = \frac{L_{WV} - L_V}{L_{WV}} \tag{6.2}$$

The electronic circuit consisting of a tuning circuit and an oscillator converts the decrease in loop inductance into a decrease in frequency. The inductance–capacitance combination formed by the loop and connecting cable inductance L and the capacitor C in the electronic unit decides the oscillator frequency f by the relation

$$f = \frac{1}{2\pi\sqrt{LC}} \qquad (6.3)$$

The output signal from this circuit triggers a relay, sending a pulse to the controller unit, informing that a vehicle is detected. Another option is the optically isolated solid-state output, which assures greater reliability because of the absence of moving parts. Advanced versions of the electronic units use artificial neural networks to classify the vehicles into different types. Besides vehicle type, they display vehicle speed, acceleration, length, and other parameters. The inductive loop sensor offers considerable design flexibility. It is not influenced by bad weather and provides accurate vehicle counting.

Difficulties in cutting the hard surface of a road for installing the loop sensor and subsequently rebuilding the surface, and the large number of loops required for vehicle detection in a given locality are the main problems faced with this sensor.

6.4 ROADWAY MAGNETOMETER AND SEARCH COIL VEHICLE DETECTION SENSORS

A magnetometer measures the earth's magnetic field. Any ferrous metal object entering the detection zone of the magnetometer causes a disturbance in the earth's magnetic field, which is used as a signature for identifying it. The reason for the disturbance of the magnetic field is that the ferrous material has a much higher permeability than air. Hence, a greater number of lines of magnetic force prefer to pass through the ferrous metal than through the surrounding air, resulting in a concentration of the flux lines above and below the vehicle. So, the flux lines increase in these regions. At the same time, there is a divergence of magnetic flux at the sides of the vehicle, and hence a decrease in flux lines at these sites. A magnetometer installed under the pavement notices the increase in magnetic flux below the vehicle. Thus, it detects the presence of a vehicle from the magnetic shadow cast by it due to the distortion of the flux lines by the vehicle (Figure 6.3).

A two-axis flux gate magnetometer can detect vehicles at rest and in motion while the search coil magnetometer can detect only moving vehicles. A single-axis magnetometer which is sensitive only along vertical direction, is not useful near equatorial regions because of the horizontal lines of terrestrial magnetism at the equator. When a 3-axis magnetometer is used, one need not worry about its orientation during installation. The forthcoming subsections will describe the operation of the different magnetometers and the search coil.

6.4.1 SINGLE-AXIS FLUXGATE MAGNETOMETER WITH TWO BAR-SHAPED CORES

A single-axis, two-core fluxgate magnetometer is shown in Figure 6.4. It consists of two exactly similar ferromagnetic bar-shaped cores 1 and 2 placed in close proximity to each other (In et al. 2004). Each core is wound with an excitation coil. The windings of the two cores are done in such a manner that the electric current flows in opposite directions around the two cores.

6.4.1.1 Without an External Magnetic Field

Suppose there is no external magnetic field, and the coils are supplied current from a source of alternating current. Consider one half-cycle of current. When the current flows in the coils, the cores are magnetized. Both the cores reach saturation at the same time. The cores being similar, the induced magnetic fields produced in them, are equal. Further, the directions of magnetic fields induced in the two cores are opposite to each other. These oppositely directed fields are induced because the current is flowing in opposite directions in the coils of the cores. Since the induced

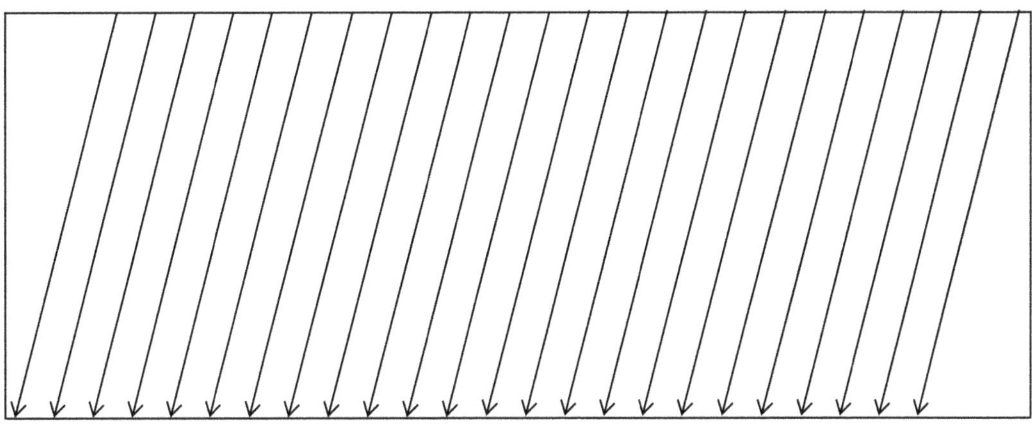

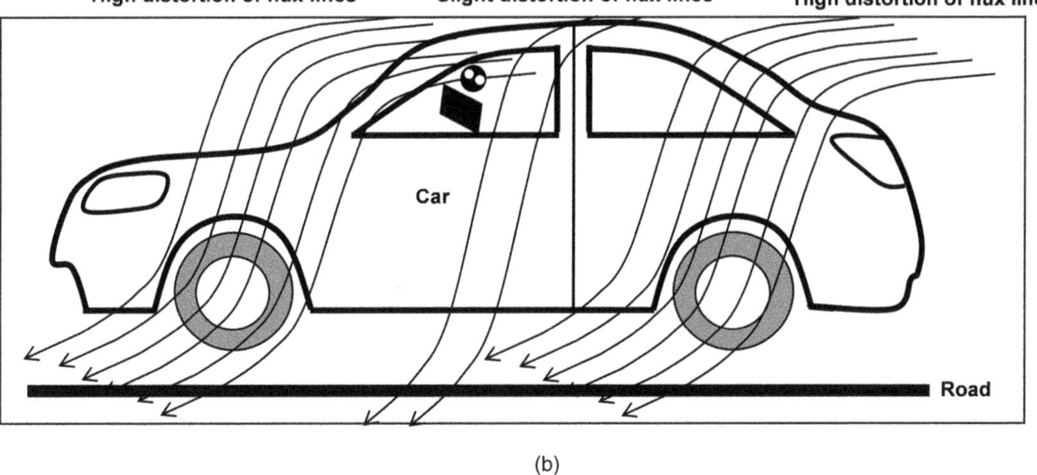

FIGURE 6.3 Terrestrial magnetic flux lines: (a) without a car on the road and (b) when a car is on the road. Part (a) shows that the flux lines are parallel, uniform, and spaced at equal distances apart. Part (b) shows that the flux lines passing through different parts of an ongoing car experience varying degrees of distortion, ranging from slight to high distortion.

magnetic fields are equal in magnitude and opposite in direction, they cancel out and the net magnetic field is zero.

In the next half-cycle, the cores come out of saturation together and are magnetized in the reverse direction. During this half-cycle also, the induced magnetic fields in the two cores are equal and opposite, and the total magnetic field is zero.

6.4.1.2   In the Presence of an External Magnetic Field, and during the First Half-Cycle of Excitation Current

Next, consider the situation where the magnetometer is placed in an external magnetic field. Then the magnetic field around one core, say core 1, will be in the same direction as the external field.

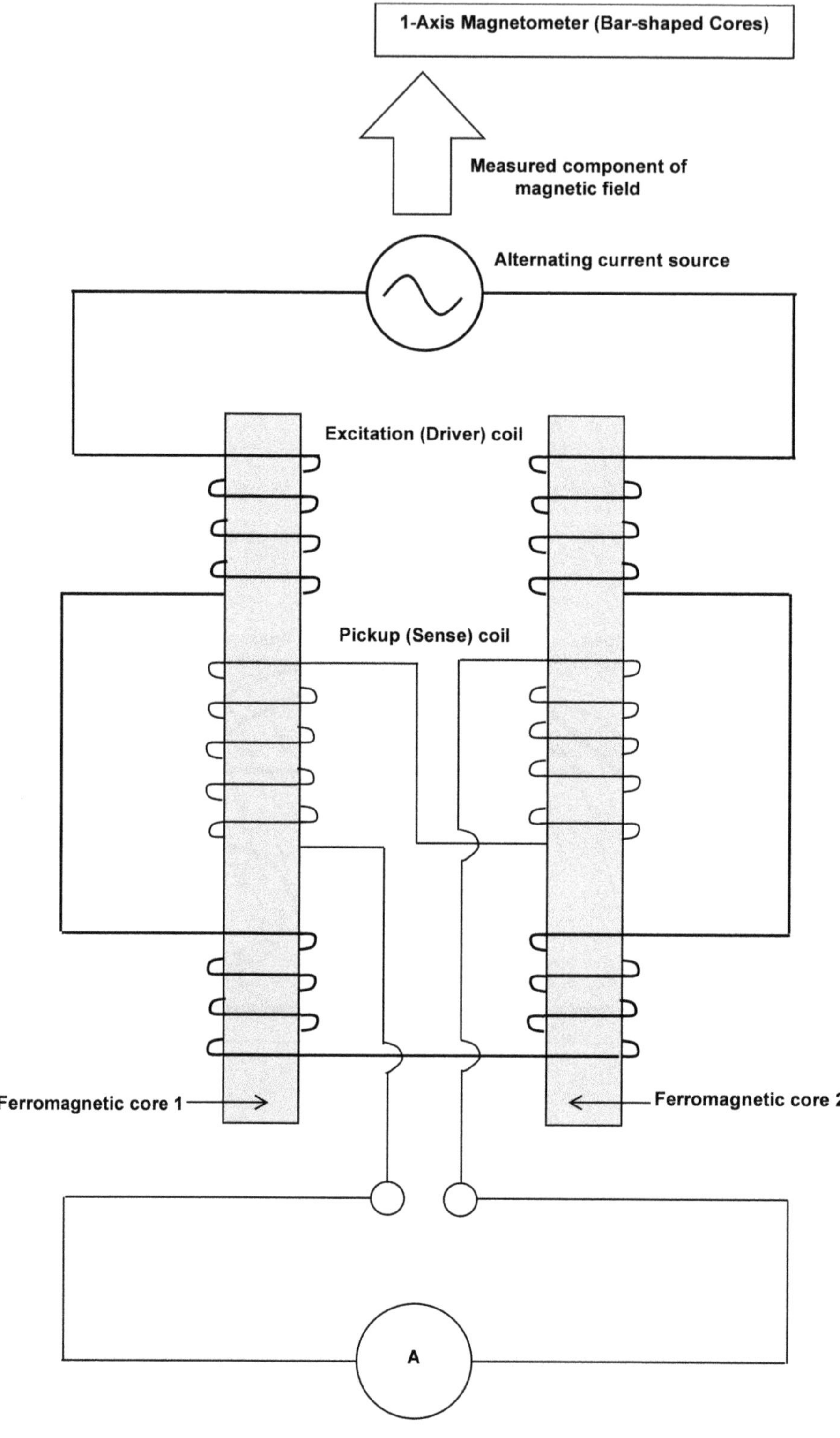

FIGURE 6.4 Single-axis flux-gate magnetometer made with two bar-shaped cores. The diagram shows two identical ferromagnetic cores: core 1 and core 2, around which are wound two coils: an excitation or driver coil fed by an AC source, and a pickup or sense coil with an ammeter for measuring current. The arrow indicates the direction of the measured component of the external magnetic field.

So, the two fields will add up, and the magnetic field of core 1 will be strengthened. Also, the magnetic field around the other core, i.e., core 2 will be in opposite direction to that of the external field. So, the external field will be subtracted from the magnetic field of core 2, and the magnetic field of core 2 will be weakened.

The difference between the magnetic fields of the two cores will be clearly evident when the two cores are driven into saturation. Core 1 with a stronger magnetic field (because of the external field acting in the same direction as its field) will go into saturation sooner than core 2 with a weaker magnetic field (because of the external field acting in the opposite direction to its field). The overall result is that during this period of going into saturation, the magnetic fields around the two cores will not cancel out.

This difference in magnetic fields causes a net change in flux across the pickup coil. This is the stage at which the pickup coil comes into play. According to Faraday's law of electromagnetic induction, a voltage will be induced in the pickup coil. As the external magnetic field is responsible for the production of this voltage, this induced voltage is an indicator of the magnitude and direction of the external field.

6.4.1.3 In the Presence of an External Magnetic Field, and during the Second Half-Cycle of Excitation Current

The difference between the magnetic fields of the two cores will also be evident when the two cores come out of saturation in the succeeding half-cycle. Core 2 with a weaker magnetic field (because of the external field acting in the opposite direction to its field) will come out of saturation sooner than core 1 with a stronger magnetic field (because of the external field acting in the same direction as its field). The result is that during this period of coming out of saturation, the magnetic fields around the two cores will not annul each other. This difference causes a net change in flux across the pickup coil. According to Lenz's law, the current induced in a conductor by a changing magnetic field flows in such a direction that the magnetic field produced by the induced current hampers the changes in the original magnetic field. Hence, a voltage will be induced in the pickup coil in the opposite direction to that during core saturation. As in the previous case, the external magnetic field is responsible for the production of this voltage. So, the magnitude and direction of the external field can be obtained from this voltage.

Thus, the fluxgate magnetometer detects and measures an external magnetic field from the potential difference generated in the pickup coil at the time of reaching saturation or at the time of coming out of saturation.

6.4.2 Single-Axis Fluxgate Magnetometer with a Single Ring Core

A ring-core fluxgate magnetometer consists of a ring-shaped core around which two coils are wound: the drive coil and the sense coil (Figure 6.5). The operation of the ring core magnetometer is similar to that of the magnetometer containing two separate bar-shaped cores and can be understood on the same lines. Instead of the two bar-shaped cores, there are two half-cores corresponding to the two halves of the circular ring.

6.4.2.1 When the External Magnetic Field Is Not Applied

Consider the situation when no external magnetic field is present. When current flows in the excitation coil, magnetic fields are produced around the two half-cores. The magnetic fields around the two half-cores are equal in magnitude and opposite in sign. They counterbalance each other, leading to zero total magnetic field.

The two half-cores go into saturation at the same instant. As a result, their magnetic fields cancel out. They also come out of saturation at the same instant with the consequent cancellation of their magnetic fields. In both cases, the combined magnetic field of two half-cores is zero.

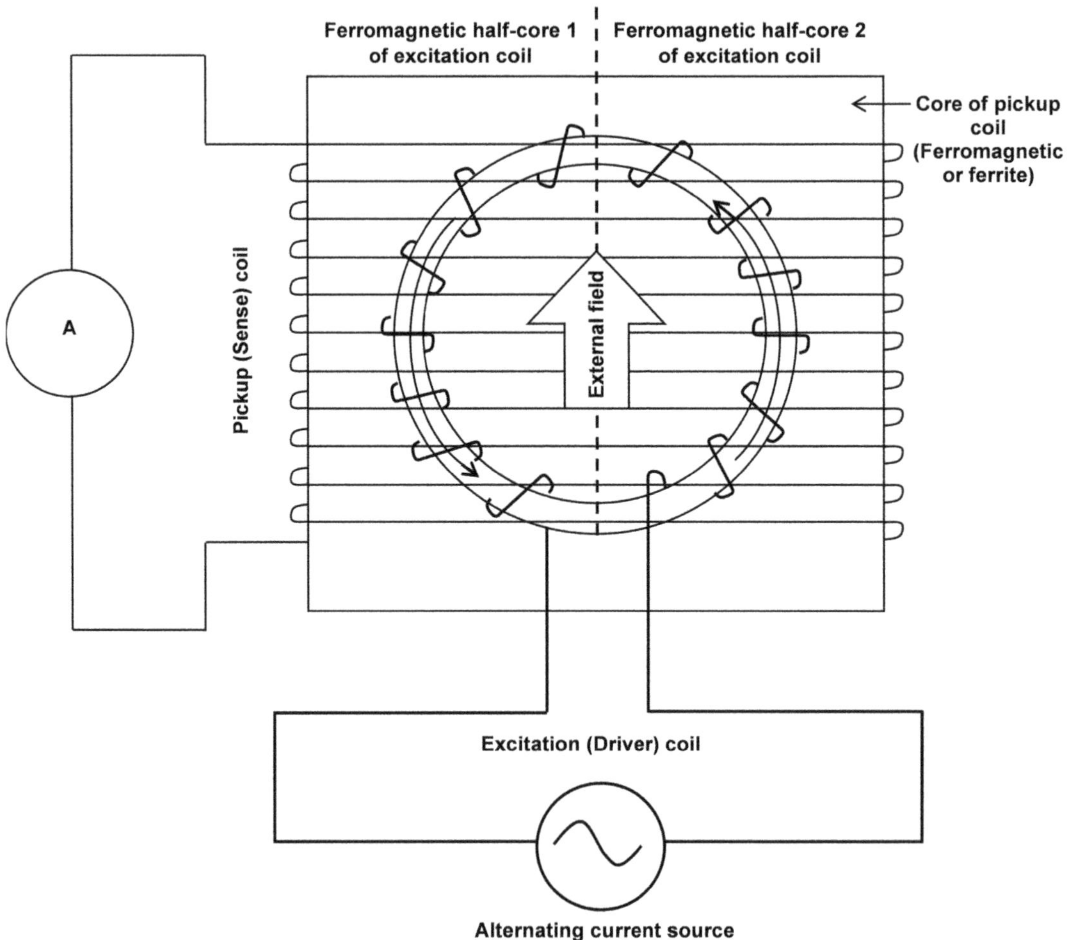

FIGURE 6.5 Single-axis ring-core fluxgate magnetometer. The diagram shows a ferromagnetic core having the shape of a circular ring. The circular ring core is imagined as made of two half-cores labeled as ferromagnetic half-core 1 and ferromagnetic half-core 2. An excitation or driver coil driven by an AC source energizes the excitation coil. Around the circular core, there is a rectangular core. The coil wound around this core is the pickup or sense coil. An ammeter is connected to the pickup coil for current measurement. The direction of the external field is shown by the arrow.

6.4.2.2 After the Application of the External Magnetic Field

Suppose the magnetic field of one half-core 1 is in the same direction as the external field. Hence the external field reinforces the magnetic field of this half-core.

The magnetic field of the other half-core 2 is in the opposite direction to the external field. So, the external field debilitates the magnetic field of this half-core.

Therefore, core 1 will go into saturation sooner than core 2. Also, it will come out of saturation later than core 2. According to Faraday's law, the change in flux during saturation will induce a voltage in the pickup coil. By Lenz's law, the flux change at the time of coming out from saturation will also induce a voltage in the pickup coil but in the opposite direction to that during saturation. The magnitude and direction of the external magnetic field are determined from the size and phase of the induced voltages.

6.4.3 Two-Axis and Three-Axis Fluxgate Magnetometers using Bar-Shaped Cores

A two-axis fluxgate magnetometer is constructed by using two single-axis fluxgate magnetometers with the axes of their sense or pickup coils aligned along the X- and Y-directions. Similarly, a three-axis fluxgate magnetometer comprises three single-axis fluxgate magnetometers. The axes of the pickup coils of these three single-axis magnetometers are aligned parallel to the X-, Y-, and Z-axes of a rectangular coordinate system (Figure 6.6).

6.4.4 Search Coil Magnetometer

The search coil magnetometer works on Faraday's law of electromagnetic induction according to which a change in magnetic flux linked with a coil produces an electromagnetic force across the ends of the coil causing the flow of an induced current through the coil (Figure 6.7). The change in flux takes place whenever a coil moves through a uniform magnetic field or rotates within this field, or the coil is subjected to a time-varying magnetic field. So, a search coil magnetometer is used to

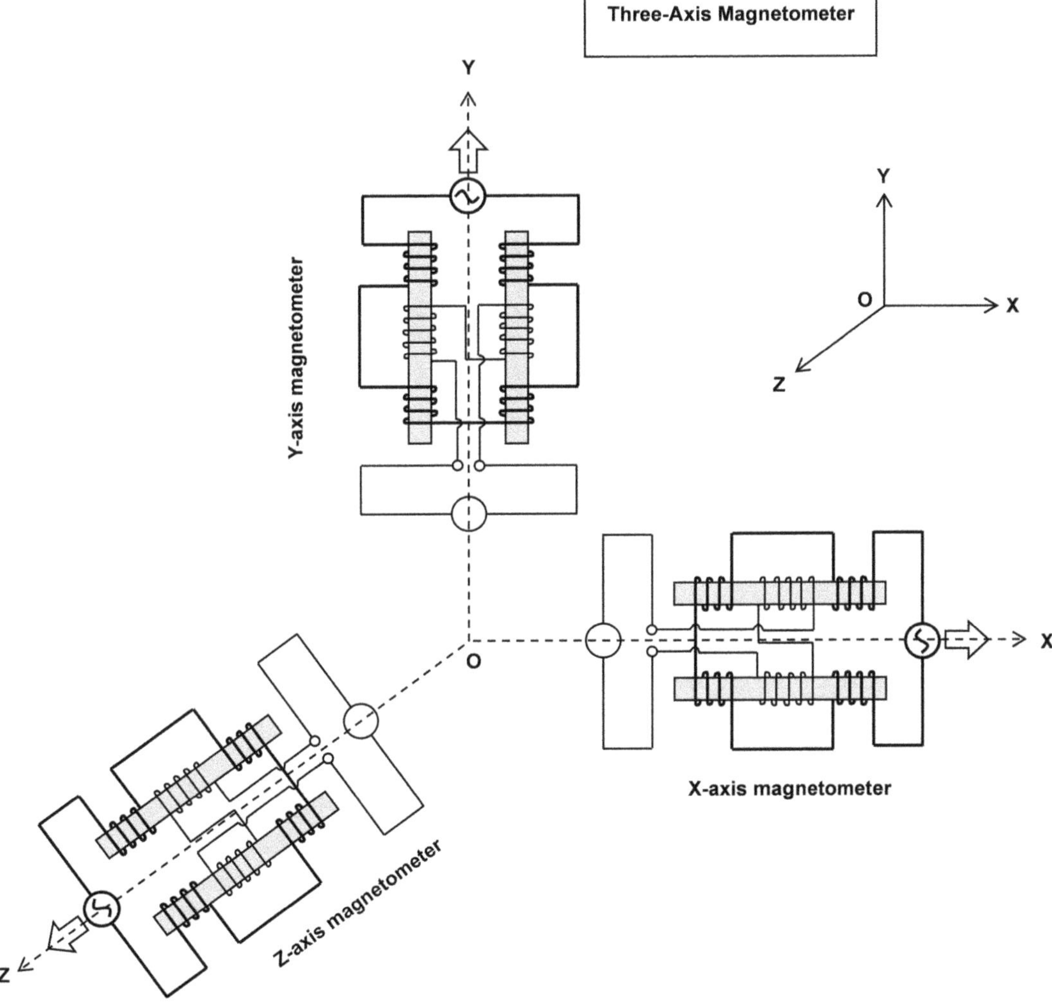

FIGURE 6.6 Three-axis fluxgate magnetometer. The diagram shows three fluxgate magnetometers with the axes of their sense or pickup coil aligned along the X-, Y-, and Z-axis directions of a rectangular Cartesian coordinate system. Accordingly, the three magnetometers are named as X-axis magnetometer, Y-axis magnetometer, and Z-axis magnetometer.

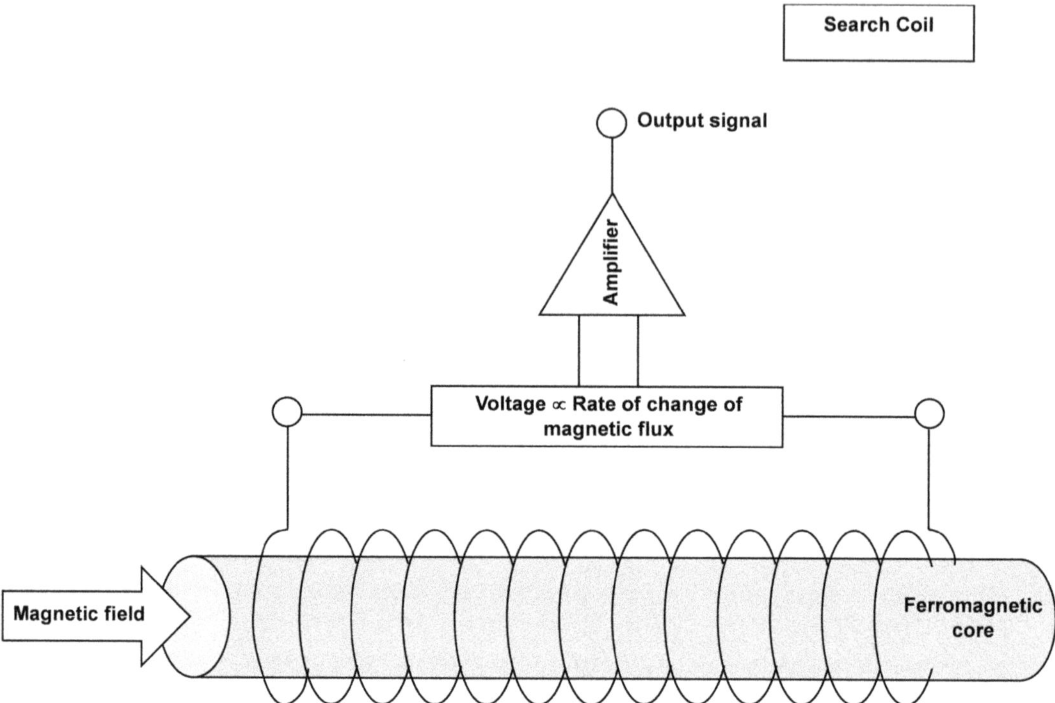

FIGURE 6.7 Search-coil magnetometer. The diagram shows a ferromagnetic core around which a coil is wound. The coil is exposed to a changing magnetic field. A voltage is induced across the terminals of the coil. Its magnitude is proportional to the rate of change of flux. The feeble voltage produced is fed to an amplifier to give a measurable output signal.

measure the strength of the magnetic field in such situations. The magnetometer is made by winding a coil around a rod of high-permeability ferromagnetic material. This rod acts as the core of the coil, and its purpose is to increase the flux density to efficiently collect the magnetic field surrounding and in the immediate neighborhood of the coil.

Besides the rate at which magnetic flux changes, the main parameters determining the magnetic field detection signal induced in the search coil are the area of the coil, the number of turns of wire used in making the coil, and the permeability of the magnetic material around which the coil has been wound. Search coils can measure fairly weak magnetic fields ~ 20 fT. Sizes of search coils range from 5 cm to 1.3 m. Their typical operating frequency range is 1 Hz to 1 MHz. The ratio of coil inductance to its resistance decides the maximum working frequency (Lenz and Edelstein 2006).

6.5 MICROWAVE RADAR SENSOR

6.5.1 SPECIAL FEATURES

As its name indicates, the microwave radar (radio detection and ranging) sensor is a radar-based sensor. The special merits of this sensor are its non-intrusive nature, together with easy installation, maintenance, and removal without disrupting the traffic. It can be mounted on a pole, an overpass bridge, or a mast arm, i.e., the bracket attachment to a lamp post for suspension of a luminaire (Figure 6.8). It transmits signals over the roads in the detection region and records the echo signals from the vehicles on the roads. From these measurements, it estimates the occupancy of the lanes to find the number of vehicles present per lane as well as the velocities and directions of vehicle movement (Ho and Chung 2016).

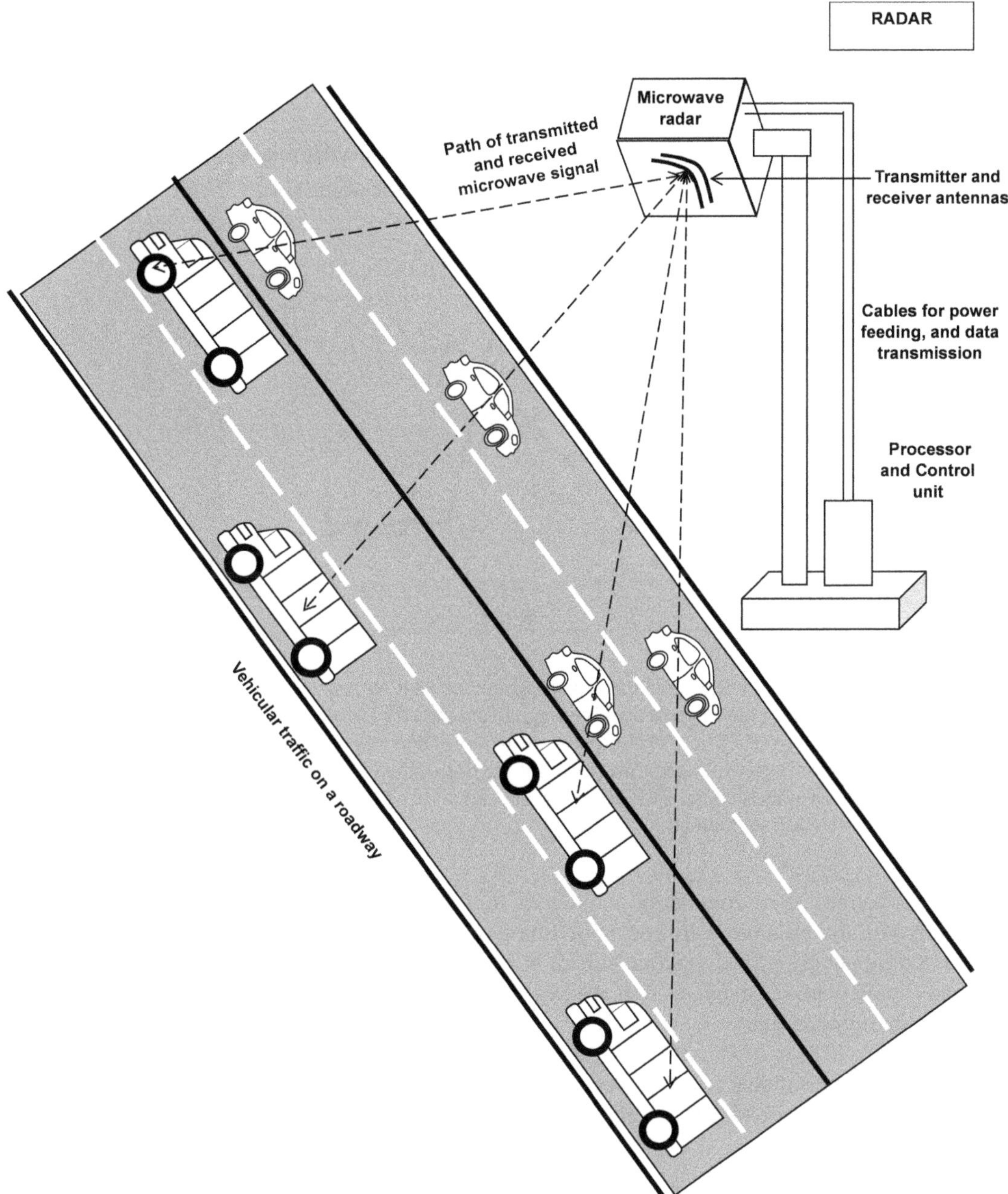

FIGURE 6.8 Remote traffic microwave radar sensor. The diagram shows cars and trucks plying in their lanes on a roadway. A microwave radar sensor is mounted on a pole and incident/reflected microwave signals are transmitted/received by its antenna. Paths of these signals are shown by dashed arrows. Also shown are the power and data cables, and the processor and controlling cabinet.

6.5.2 Principle of the Frequency-Modulated Continuous-Wave Radar

The radar sensor uses a frequency-modulated continuous-wave (FMCW) radar (Pongthavornkamol et al. 2021). In this radar, the frequency of the transmitted signal is continuously increased, at a fixed rate until it attains a maximum value (Figure 6.9). Then the frequency drops to the initial value and

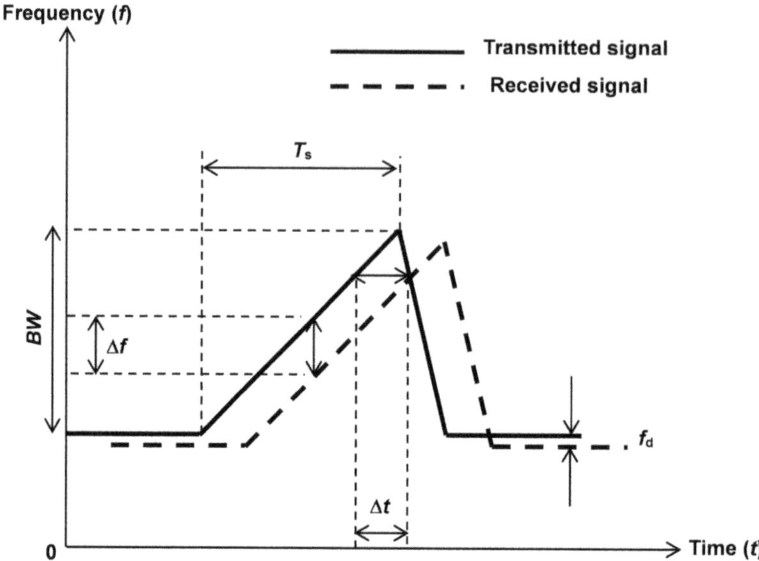

FIGURE 6.9 Transmitted and received sawtooth signals of FMCW radar. Two graphs are drawn between signal frequency and time, one for the transmitted signal and the other for the received signal. The graphs have the shapes of sawtooth waves. The transmitted signal is represented by full lines while the received signal is shown by dashed lines. Bandwidth and time period of the transmitted signal, frequency difference Δf between the instantaneous transmitted signal and received echo signal, time difference Δt between these signals, and Doppler frequency shift f_d are marked.

again starts increasing to commence another cycle. During each cycle, the received frequency is compared with the instantaneous transmitted frequency, and the instantaneous difference between transmitted signal and echo signal frequencies is measured. This difference arises because the signal received at a particular instant was sent at an earlier time when the transmission frequency was low. Under the assumption that

   (i) The object is stationary with respect to the receiving antenna, and
  (ii) The transmitted frequency increases linearly with time.

The time Δt taken by the radar signal to reach the target and return to the radar is related to the frequency difference Δf by the equation

$$|\Delta t| = \frac{\text{Frequency difference}}{\text{Rate of change of frequency with time}} = \frac{|\Delta f|}{\dfrac{df}{dt}} \tag{6.4}$$

If R is the distance between the reflecting object and radar, and c is the velocity of electromagnetic waves,

$$R = \frac{c|\Delta t|}{2} \tag{6.5}$$

Substituting for $|\Delta t|$ from eq. (6.4) in eq. (6.5) we get

$$R = \frac{c|\Delta t|}{2} = c\frac{|\Delta f|}{2\left(\dfrac{df}{dt}\right)} \tag{6.6}$$

Thus, by measuring the frequency difference $|\Delta f|$ between transmitted and received signals and the rate of frequency shift per unit time (df/dt), the distance R can be calculated since $c = 3 \times 10^8$ ms^{-1}.

The frequency difference is determined by mixing the received signal with the transmitted signal in the mixer. The frequency difference is obtained from the beat frequency f_b. Hence

$$|\Delta f| = f_b \tag{6.7}$$

Suppose the transmitted signal is a sawtooth wave of period T_s and bandwidth BW. Then,

$$\frac{df}{dt} = \frac{BW}{T_s} \tag{6.8}$$

Applying eqs. (6.7) and (6.8) to eq. (6.6), we have

$$R = c\frac{|\Delta f|}{2\left(\dfrac{df}{dt}\right)} = \frac{cf_b}{2\left(\dfrac{BW}{T_s}\right)} = \frac{cf_b T_s}{2BW} \tag{6.9}$$

If the reflecting object is moving with respect to the receiving object, then the frequency difference Δf is not only due to the frequency shift during the traveling of waves from the transmitter to the object and back called the run time but also due to the Doppler effect. When the reflecting object is moving toward or receding away from the radar, the frequency of the received signal undergoes an increment or decrement by the Doppler frequency f_D. In a sawtooth waveform, the echo signal moves to the right by the run time and downward by the Doppler effect. This Doppler frequency f_D is applied to determine the velocity of the object.

The total distance traveled by the wave between the radar and the object is $R + R = 2R$ because the distance is R from the radar to the object and R from the object to the radar. If λ is the wavelength of the wave, then the number of wavelengths in the two-way communication path is

$$N = \frac{2R}{\lambda} \tag{6.10}$$

Since one wavelength represents 2π radians, N wavelengths correspond to a phase angle

$$\phi = 2\pi N = 2\pi \times \frac{2R}{\lambda} = \frac{4\pi R}{\lambda} \text{ radians} \tag{6.11}$$

The object being in motion, R changes with time. So, the phase angle ϕ will also change with time. The change of phase angle ϕ with time is the angular frequency ω given by

$$\omega = \frac{d\phi}{dt} \tag{6.12}$$

Also,

$$\omega = 2\pi f_D \tag{6.13}$$

Equations (6.12) and (6.13) give

$$2\pi f_D = \frac{d\phi}{dt} \tag{6.14}$$

or,

$$f_D = \frac{1}{2\pi}\left(\frac{d\phi}{dt}\right) \tag{6.15}$$

Substituting for ϕ from eq. (6.11) in eq. (6.15), we have

$$f_D = \frac{1}{2\pi}\frac{d}{dt}(\phi) = \frac{1}{2\pi}\frac{d}{dt}\left(\frac{4\pi R}{\lambda}\right) = \frac{1}{2\pi} \times \frac{4\pi}{\lambda} \times \frac{dR}{dt} = \frac{2}{\lambda}\left(\frac{dR}{dt}\right) \tag{6.16}$$

If the object is moving with a velocity v_0 along a direction subtending an angle θ with the line joining the radar, then it has a radial velocity $v_r = v_0 \cos\theta$. The relative velocity v_r is expressed as

$$v_r = \frac{dR}{dt} \tag{6.17}$$

Hence, from eqs. (6.16) and (6.17),

$$f_D = \frac{2}{\lambda}(v_r) = \frac{2v_r}{\lambda} \tag{6.18}$$

But

$$\lambda = \frac{c}{f_t} \tag{6.19}$$

where f_t is the transmitted signal frequency.

Therefore, the Doppler frequency shift f_D and the transmitted signal frequency f_t are related by the formula

$$f_D = \frac{2v_r}{\lambda} = \frac{2v_r}{\dfrac{c}{f_t}} = \frac{2v_r f_t}{c} \tag{6.20}$$

from which the velocity v_r is calculated as

$$v_r = \frac{c f_D}{2 f_t} \tag{6.21}$$

6.5.3 Block Diagram of the Microwave Radar Sensor

Figure 6.10 shows the block diagram of the microwave radar sensor. The functions of the various blocks in the diagram are explained further.

6.5.3.1 Transmitter Side

(i) The control voltage from a microprocessor is supplied to the voltage-controlled oscillator through a digital-to-analog converter.

(ii) The VCO is an oscillator whose output frequency can be varied by changing the input DC voltage; hence the name 'voltage-controlled oscillator'. Typical operating frequencies of microwave radar sensors lie in the range of 0.3–40 GHz. Commonly used frequencies are 6, 10, and 26 GHz. Sensors working at 80 GHz are also commercially available (Vogt 2018).

(iii) The bandpass filter restricts the transmission frequencies to the allocated band, thereby limiting the bandwidth of the signal. It screens out frequencies that are lower or higher than those in the allowed band, together with any spikes or harmonics.

(iv) The power divider subdivides the power into two parts, feeding one part to the next stage of the power amplifier in the transmitter circuit, and the other part to the mixer in the

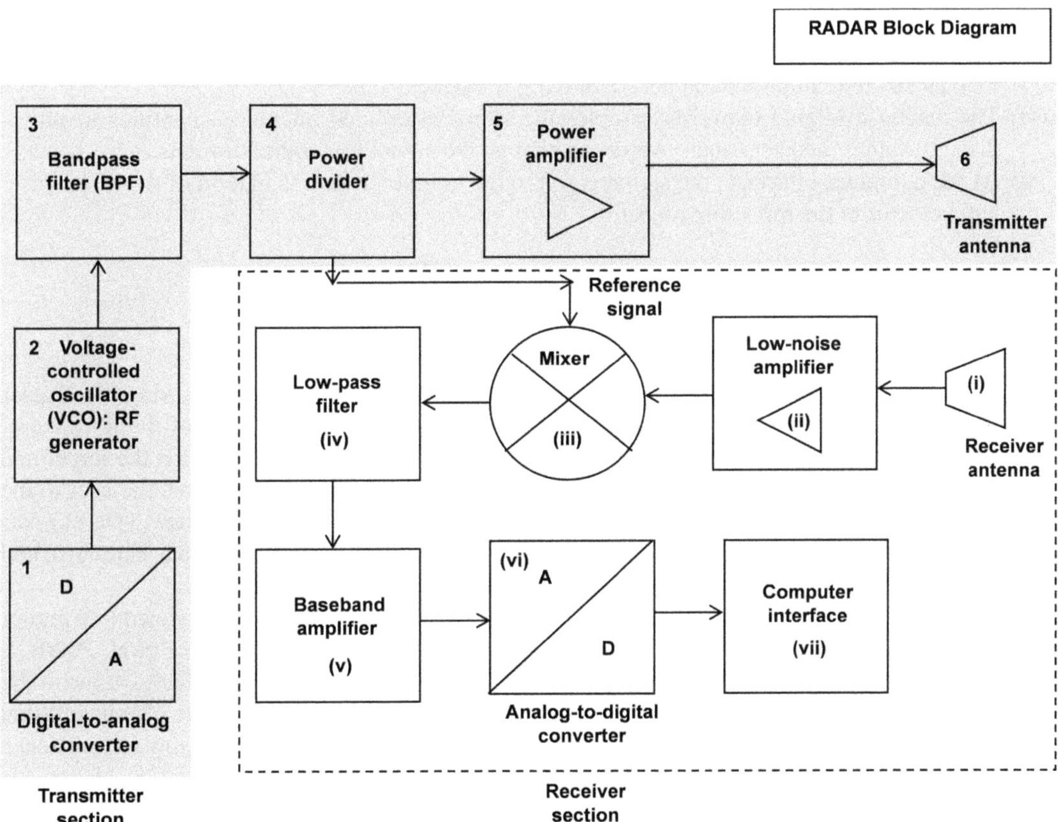

FIGURE 6.10 Block diagram of the remote traffic frequency-modulated continuous-wave (FMCW) microwave radar sensor. The transmitter section consists of (i) digital-to-analog converter, (ii) voltage-controlled oscillator (VCO): RF generator, (iii) bandpass filter (BPF), (iv) power divider, (v) power amplifier, and (vi) transmitter antenna. The receiver section consists of (i) receiver antenna, (ii) low-noise amplifier, (iii) mixer, (iv) low-pass filter, (v) baseband amplifier, (vi) analog-to-digital converter, and (vii) computer interface.

receiver circuit. It is here that the reference transmission frequency signal is fed to the mixer for comparison with the echo signal to meet the objective of determining the difference between the two frequencies.

(v) The power amplifier raises the power level of the signal. It converts the low-power signal into a high-power signal for transmission.

(vi) The transmission antenna couples the energy from the power amplifier into free space or air by radiating electromagnetic waves.

6.5.3.2 Receiver Side

(i) The receiver antenna produces an RF voltage from the incident electromagnetic waves.

(ii) The low-noise amplifier strengthens the weak received signal without any degradation in its signal-to-noise ratio. It boosts the signal to a satisfactory level above the noise floor in order that it can be subjected to further processing.

(iii) The mixer or frequency mixer produces new signals from the two signals applied to it. These new signals have frequencies equal to the sum and difference between the frequencies of the mixed signals. Since one input signal to the mixer is the transmission frequency at the particular instant of receipt of the echo signal, the difference between this instantaneous transmission frequency and the echo signal frequency gives the necessary beat frequency $f_b = \Delta f$, required for calculating the distance of the object from the radar.

(iv) The low-pass filter removes undesired mixed frequencies.

(v) The bandpass amplifier strengthens the weak echo signal in its original unmodulated form. Modulation had shifted the signal to much higher frequencies. The echo signal needs as high as 10^4-fold amplification before moving to the next stage.

(vi) The analog-to-digital converter samples the signal at defined intervals, quantifies it into discrete values, and sets binary values to change the signal into digital format.

(vii) At the computer interface, the signal is fed to the computer for calculation of the distance and velocity of the reflecting object.

6.6 ROADSIDE LIDAR SENSOR

6.6.1 Vehicle Detection Principle

LIDAR, the acronym for light detection and ranging, signifies an optical device working like a radar for determining the distance of a target object from the user. It works on the time-of-flight measurement principle. A laser pulse is transmitted by the user. The pulse is reflected from the target and returns to the user's instrument. The time taken by the laser pulse to propagate from the user to the target, and from the target back to the user is measured by a time-to-digital converter (TDC) (Texas Instruments 2016). TDC works as a stopwatch to measure the time elapsed between a start pulse and a stop pulse. The distance between the target and the user is found from this time.

A LIDAR sensor is shown in Figure 6.11. The LIDAR sensor consists of a co-located transmitter and receiver along with the TDC and data acquisition /processing electronics (Lee et al. 2020). It uses a laser diode to transmit laser pulses of a short width in the range of nanoseconds toward the target, detects the returning pulses by an avalanche photodiode, measures the time of transmission and return of pulses, and finally calculates and displays the target distance. The formula for distance calculation is

$$D = \frac{c\Delta t}{2} \tag{6.22}$$

where D is the distance of the target, c is the velocity of light and Δt is the total time taken by light for forward and reverse journeys.

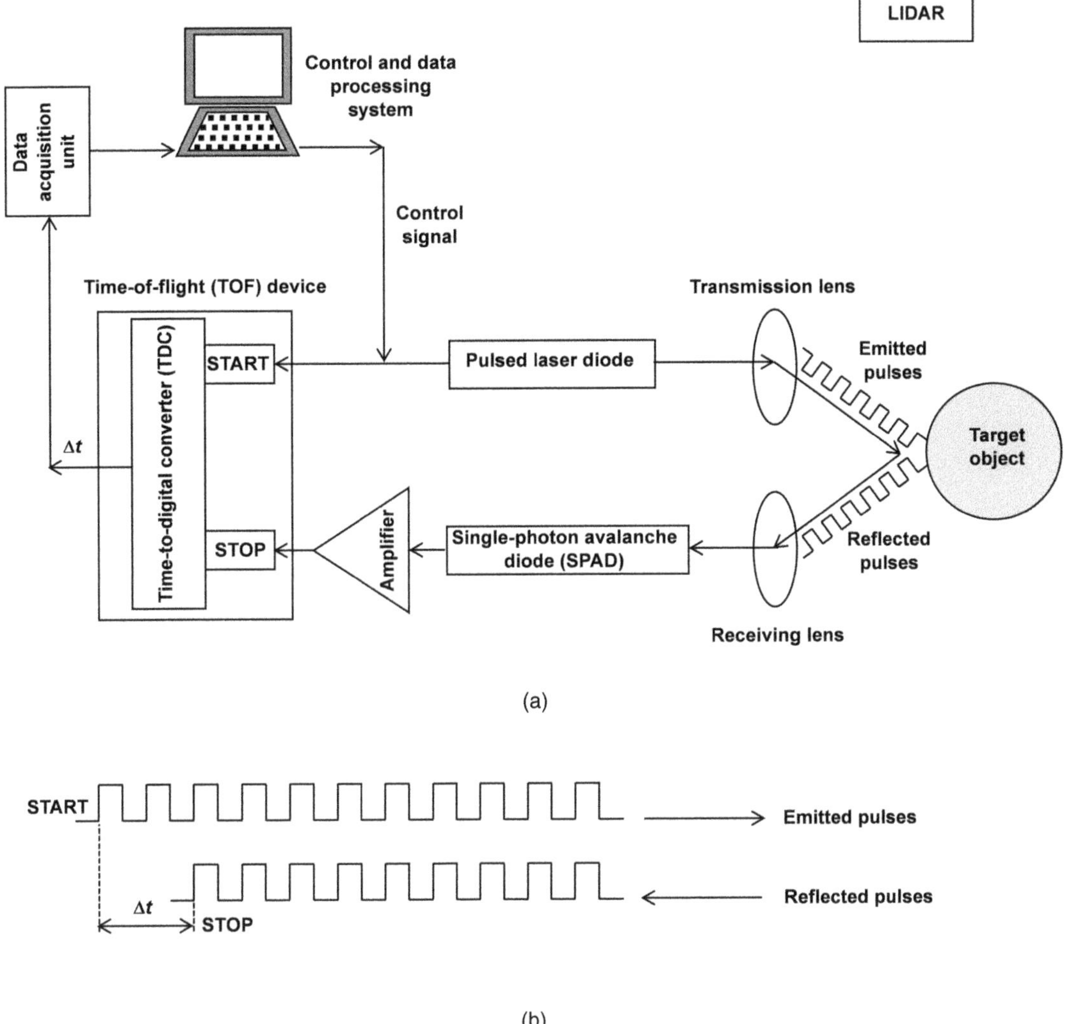

FIGURE 6.11 Direct time-of-flight LIDAR sensor: (a) block diagram of the sensor, and (b) the pulse trains. Part (a) shows the control and data processing system sending a control signal simultaneously to the pulsed laser diode and the START switch of the time-to-digital converter (TDC). The pulses fall on the transmission lens. The emitted pulses are directed toward the target object. The reflected pulses from the target object are incident on the receiving lens, and are directed to the single-photon avalanche diode. The electrical signal from the diode is amplified and applied to the stop switch of the TDC. The TDC with its starting and stopping switches constitutes a time-of-flight device. The output signal from the TDC is the time interval Δt between the actuation of START and STOP switches. This signal is fed to the data acquisition unit. From this unit, the data enters the control and processing system, which determines the distance of the target from the user, and displays it on the screen. Part (b) shows the emitted and reflected pulse trains, and the time difference Δt between the START and STOP switch triggering.

6.6.2 Single-Photon Avalanche Diode (SPAD)

The use of a single-photon avalanche diode in a LIDAR sensor provides a better resolution in time than an avalanche diode. Like an avalanche photodiode, an SPAD detects light by the avalanche current set off by a photon in a P-N junction diode in the reverse-biased mode (Figure 6.12). The primary difference between SPAD and the avalanche photodiode is that SPAD is operated at a reverse

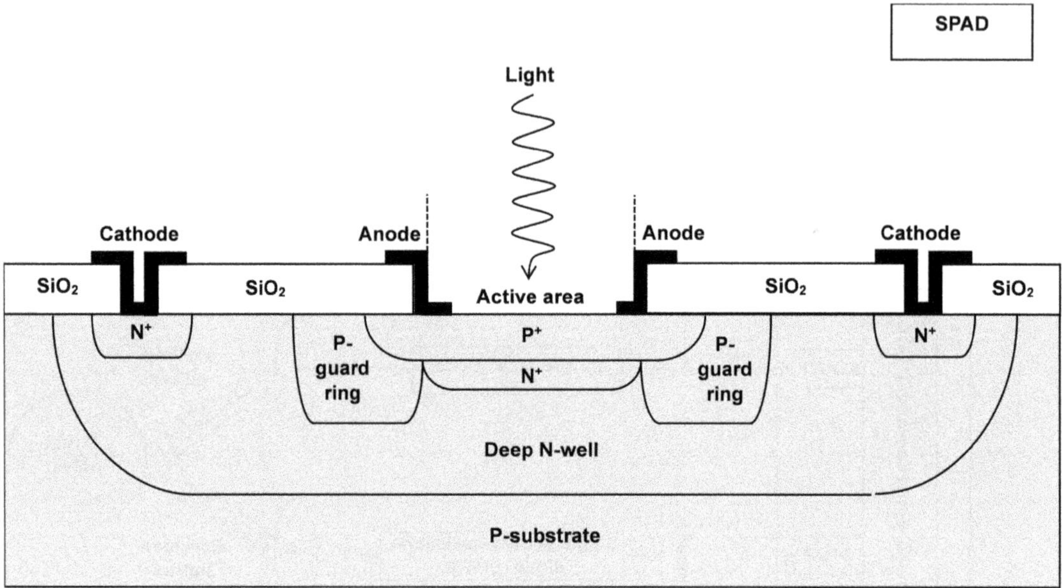

FIGURE 6.12 Cross-sectional diagram of a single-photon avalanche diode. The diagram shows a P-substrate, a P⁺/N⁺/deep N-well junction, P-guard rings, active area, anode contacts, and N⁺ regions on both sides of the P-guard rings with cathode metal contacts made above them. Light falls on the active area of the SPAD.

voltage much above its breakdown voltage called the Geiger mode, whereas the avalanche photodiode is operated at a voltage less than its breakdown voltage. The reason for deciding on such a high reverse voltage is that the associated large electric field ~ 3×10^5 V-cm⁻¹ accelerates the electrons to such high velocities that the impact of a single photon on the semiconductor is adequate to cause an avalanche, resulting in appreciable amplification of the current so that a measurable output current is produced even when the diode is exposed to very low-intensity light. Furthermore, the avalanche leading to amplification occurs extremely quickly providing a temporal resolution in picoseconds. This makes the SPADs suitable for direct time-of-flight measurements for the LIDAR sensor.

The avalanche ceases when the applied reverse voltage is decreased to the breakdown voltage value of the diode or lower than it. The circuit used to stop the avalanche is known as a quenching circuit. In a passive quenching circuit, a large ballast resistor is connected in series with the SPAD. In an active quenching circuit, the onset of an avalanche is sensed, and a pulse is generated in synchrony to avalanche buildup to lower the voltage across the diode.

The SPAD made on a P-substrate is a P⁺/N+/deep N-well structure surrounded by low P-doped guard rings to suppress abrupt doping profiles and corner effects, thereby lowering the electric potential gradients that cause premature breakdown (Niclass et al. 2005).

6.6.3 ADVANTAGES AND LIMITATIONS

LIDAR sensors have better resolution than RADAR sensors. Roadside LiDAR sensors provide high-resolution micro-level traffic data (HRMTD) to assist intelligent traffic applications, such as traffic signal optimization and control to minimize the waiting time of cyclists and pedestrians and also to increase the safety level of users at road intersections (Lin et al. 2023). Although LIDAR sensors are high-precision sensors for 3D visualization of the traffic, vehicle tracking is often difficult in complex traffic scenarios, primarily due to partial or complete occlusion of the target vehicle by other vehicles or objects on the way. The use of LIDAR sensors for vehicle motion detection becomes especially expedient under low-light conditions because they can see the surrounding objects during daytime as well as at night (Zhang and Jin 2022).

6.7 ROADWAY VIDEO IMAGE PROCESSOR SYSTEM

6.7.1 CONSTRUCTION AND WORKING

The video image processor (VIP) consists of a single or multiple cameras. The resolution of the cameras, their field of view, and the height at which they are installed are the determining parameters of the distance up to which they can track the vehicles. Other vital components of the processor system include a digital computer loaded with software for analysis of the black-and-white or colored images recorded by the cameras for a segment of the roadway. The algorithms used for analysis are designed to minimize the errors introduced by different weather conditions, the period of image recording, day or night, and other likely variable conditions without affecting the images of cars, trucks, motorcycles, or scooters plying on the road. After proper interpretation of the images, the data of traffic flow on the chosen road segment is generated. Vehicles are classified into different groups according to their lengths. Information on each group of vehicles present on a particular lane is extracted. The speeds of the vehicles are also calculated. Thus, lane occupancy is mapped.

6.7.2 TYPES OF ROADWAY VIDEO IMAGE PROCESSOR SYSTEMS

These systems are made in three formats in accordance with their approaches for identifying vehicles.

(i) Tripline Systems: In these systems, the user selects a given number of linear detection zones within the viewing range of the video camera. The vehicle entering any zone alters the properties of pixels in the camera. The difference between the properties of the pixels without and with the vehicle present is used for its recognition.

(ii) Closed-Loop Tracking Systems: These systems perform a continuous detection and tracking of vehicles moving on a segment of the roadway. The presence of vehicles is validated from these multiple detections. Vehicle speed values are also estimated.

(iii) Data Association Tracking Systems: Unique connected portions of pixels are searched for vehicle identification. Objects are distinguished using markers originating from morphologies and gradients. Then these portions of pixels are tracked from one frame to another to produce the tracking data for a particular vehicle or collection of vehicles.

6.8 ROADWAY ULTRASONIC VEHICLE DETECTION SENSORS

6.8.1 DETECTION OF VEHICLE

These sensors detect vehicular presence by transmitting sound pressure waves, usually as pulse waveforms, at frequencies ~25–50 kHz. The pulses strike the surface of the road at spots unoccupied by vehicles but they strike the vehicles whenever they are present on the road (Figure 6.13). The reflected pulses received from the road surface enable measurement of the distance between the sensor and the road surface whereas those received from the vehicle surface give the distance between the sensor and the vehicle surface. In each case, the distance is found by dividing the product of the velocity of ultrasound and the time of travel of ultrasound to the road surface or vehicle surface and back to the sensor by 2. Thus, the distance between the sensor and the surface of the road is different from the distance between the sensor and the surface of the vehicle, which is applied to detect the vehicle. To clarify, the sensor works by sending ultrasound signals toward the road/vehicle surface at different time instants from a region defined by the beamwidth of the ultrasound transmitter, calculating the corresponding distances, and looking for any differences between distance values obtained for ultrasound signals sent at different time instants. Whenever the measured distance differs from the distance between the sensor and the road surface, the presence of a vehicle is inferred.

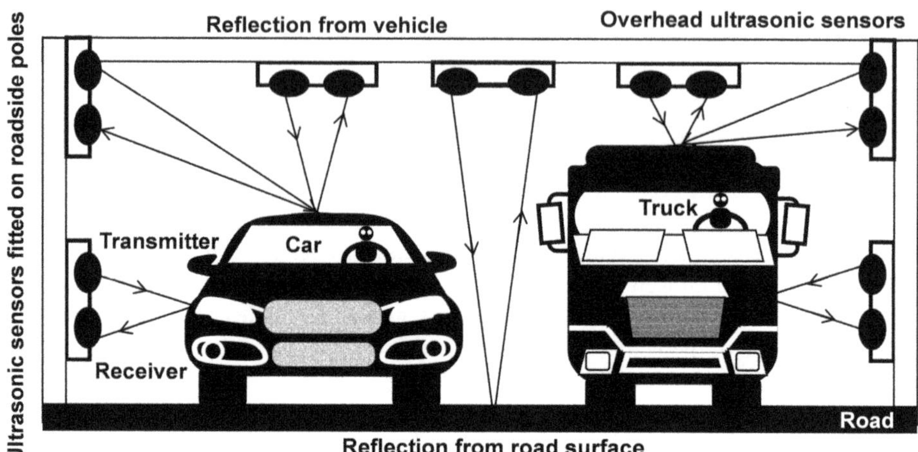

FIGURE 6.13 Ultrasonic sensors fitted overhead or on poles by the roadside to detect vehicles. The sensors consist of transmitter and receiver devices in a single package. Reflection of ultrasonic beams from vehicle rooftops, sides, and surface of the road are shown.

6.8.2 Determining Vehicle Speed

Two ultrasonic sensors are used for emitting ultrasonic pulses and receiving the reflected pulses. They are placed at a fixed, known distance apart. This distance is smaller than the length of the vehicle, e.g., 1 m, so that the reflected signals received by both sensors are from the same vehicle, not from two or more vehicles. The time instant t_1 at which a vehicle enters the detection zone of sensor 1 and time instant t_2 at which it enters the detection zone of sensor 2 are recorded. The vehicle speed v is given by the equation

$$v = \frac{\text{Separation distance between sensor 2 and sensor 1}}{\text{Time instant } t_2 \text{ at which vehicle reaches sensor 2} - \text{Time instant } t_1 \text{ at which the vehicle reaches sensor 1}} \tag{6.23}$$

6.8.3 Doppler Ultrasonic Sensor

It works on the Doppler effect according to which the frequency of ultrasound received by an ultrasonic sensor from a moving vehicle varies with the speed of the vehicle from which it is reflected. The speed of the vehicle is obtained from the frequency difference between transmitted and reflected ultrasound waves.

In the ensuing subsection, we describe the ultrasonic transducers fabricated by different methods, first piezoelectric and then capacitive.

6.8.4 Piezoelectric MEMS Ultrasonic Transducers (PMUTs)

In MEMS technology, PMUTs are made by different processes (Wang et al. 2020). Let us study the PMUT structures fabricated by various MEMS processes These processes differ in the way the membrane is made. The PMUT structures formed by three main processes are illustrated in Figure 6.14.

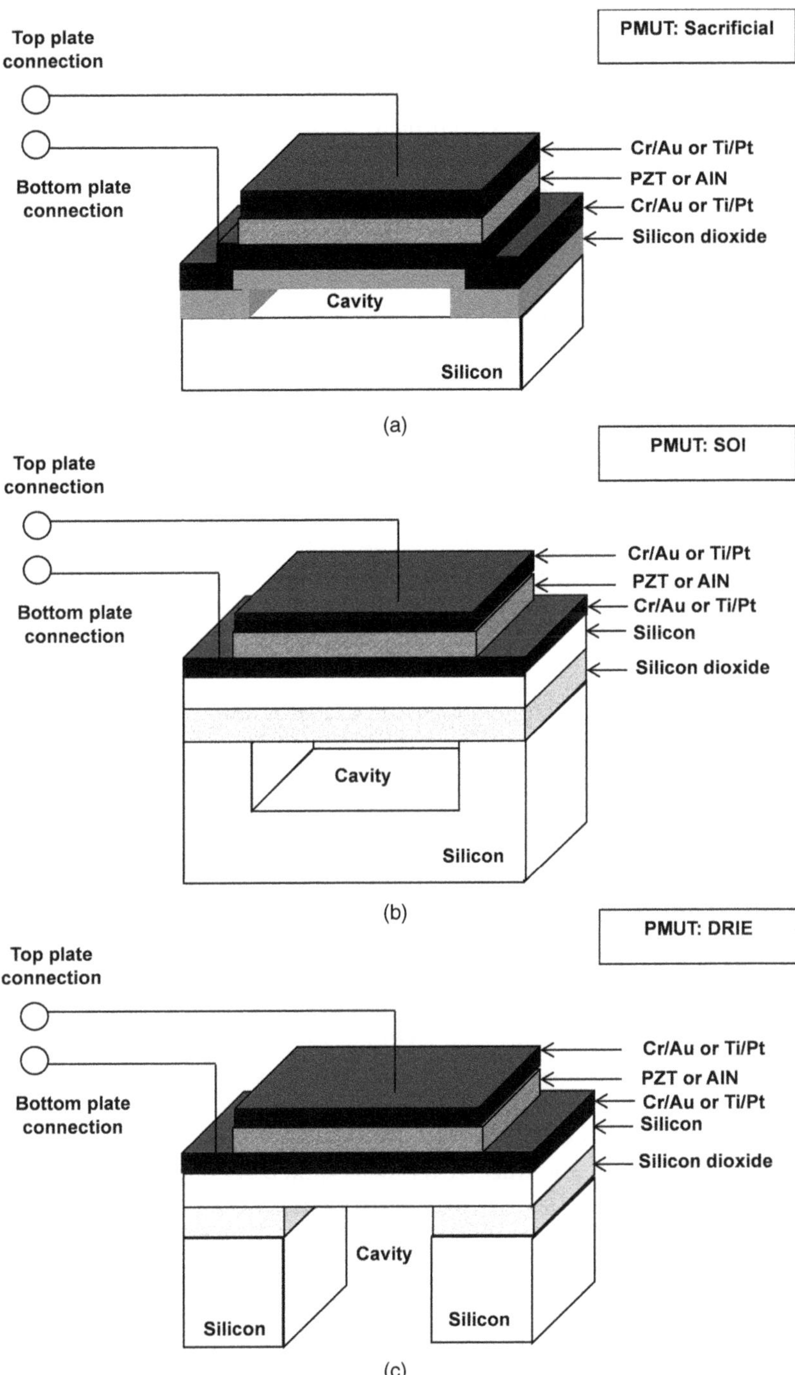

FIGURE 6.14 Piezoelectric MEMS ultrasonic transducers (PMUTs) with (a) membrane made by etching the sacrificial layer, (b) membrane made using silicon-on-insulator wafer and wafer bonding, and (c) membrane made by deep reactive ion etching. In all parts (a), (b), and (c), the piezoelectric material (PZT or AlN) is sandwiched between the bottom and top plates made of Cr/Au or Ti/Pt. All these PMUTs are made on a silicon substrate. The top and bottom plate connections and the cavities are shown. In part (a), the membrane is made of silicon dioxide. In part (b), the membrane is made of silicon dioxide and silicon. In part (c), the membrane is made of silicon. In parts (a) and (b), the silicon is partially etched to form cavities, and a layer of silicon is left at the bottom of the cavity. In part (c), the silicon is completely etched to form the cavity with no silicon left below the cavity.

6.8.4.1 Sacrificial-Etching Realized PMUT

In this process (Figure 6.14(a)), the membrane is formed over a sacrificial layer. At the end of the process, the sacrificial layer is removed with the etchant entering through holes or tunnels. As the sacrificial layer is etched away, the membrane is left behind hanging over anchoring supports. The method is suitable for making dense arrays containing a large number of closely spaced PMUT cells because it avoids the large silicon undercutting which happens when deep silicon anisotropic etching is performed.

6.8.4.2 PMUT Made by Wafer-Bonding Technique

Stress and surface tension effects cause the breakage of several membranes during etching for the removal of the sacrificial layer to release the membrane. This decreases the manufacturing yield of the process. To prevent the loss of chips due to broken membranes, the membrane and the cavity are formed on two separate wafers. Finally, these wafers are bonded together after careful alignment (Figure 6.14(b)). High accuracy in lithography along with proper surface cleaning and preparation of the bonding wafers, are critical requirements for the success of this process.

6.8.4.3 DRIE-Made PMUT

In this method (Figure 6.14(c)), deep silicon etching is done from the backside of the wafer using the deep reactive ion etching (DRIE) technique. Generally, silicon-on-insulator wafers are used in this process because the buried oxide layer of this wafer acts as an etch-stop layer for termination of the etching process.

6.8.5 CMUT STRUCTURES REALIZED BY VARIOUS TECHNIQUES

MEMS technology offers a diversity of process techniques for CMUT fabrication, a few of which are explained further (Figure 6.15).

6.8.5.1 CMUT Made by Glass-to-Silicon Anodic Bonding

After cleaning the glass substrate with piranha, it is coated with photoresist, and the regions for cavities are defined (Yamaner et al. 2014). The cavities are formed in the glass substrate by wet or dry etching (Figure 6.15(a)). Without removing the photoresist, Cr/Au is deposited and the metal pattern is obtained after removing the photoresist. This is a lift-off process. Separately, the device layer of a silicon-on-insulator wafer is covered with PECVD silicon nitride. This layer serves a two-fold purpose: it acts as an intermediate bonding film, and as an insulating layer to prevent electrical shorting between the two plates of the capacitor in case the membrane touches the underlying metal during vibration. The SOI wafer is bonded to the glass wafer with the silicon nitride layer in contact with the glass. The handle layer of the SOI wafer is removed by grinding, and the exposed buried oxide is etched. The silicon over the bottom electrode pad is etched, and the gas is evacuated from the cavities. Then the etched silicon regions are sealed with PECVD silicon nitride. The Cr/Au film is deposited over the bottom contact pad and silicon layer after the removal of silicon nitride from these regions while the location of sealing is protected.

Anodic bonding offers several advantages over other bonding techniques, viz., a greater tolerance to surface roughness and particle contamination together with the provision of a high bonding strength. These advantages make the method less prone to failures due to bonding.

6.8.5.2 CMUT by Wafer-Bonding Technology

The membrane is made on a silicon-on-insulator (SOI) wafer (Huang et al. 2003). The cavity is formed on a prime wafer, a high-quality polished silicon wafer (Figure 6.15(b)). The two wafers are bonded together in a vacuum environment. The cleanliness and surface finishing of the wafers is of paramount importance to prevent bonding failures. Before bonding, the surface of the wafer is activated in piranha, and then in 50:1 HF. After RCA1 cleaning, wafer bonding is done in a vacuum of 10^{-5} mbar at 150°C followed by annealing at 1,100°C for 2 h. The membrane is formed by grinding and etching up to the buried oxide layer of the SOI wafer. The buried oxide is etched to make the silicon membrane.

The process yields better control over the cavity depth, shortens the time of process run, and enables repeatability of fabrication of CMUTs with desired parameters.

6.8.5.3 CMUT with Through-Glass-Via Interconnections

Channels are made through the glass wafers by laser drilling (Zhang et al. 2015). Copper paste is filled in the channels. Cu paste-filled glass wafers are sintered and planarized by polishing (Figure 6.15(c)). Cavity locations are defined on the surface of the glass wafer by photolithography, and the cavities are formed by DRIE. The Cr/Au bottom electrodes are made. Separately, PECVD silicon nitride is deposited on the device layer of an SOI wafer. Then anodic bonding is done between the glass and SOI wafer in vacuum at 350°C at 700 V. Covering the backside of the bonded (glass + SOI wafer) structure with a protective coating, the handle layer of the SOI wafer is etched, and the buried oxide is removed to release the silicon membrane. After removal of the protective coating, the

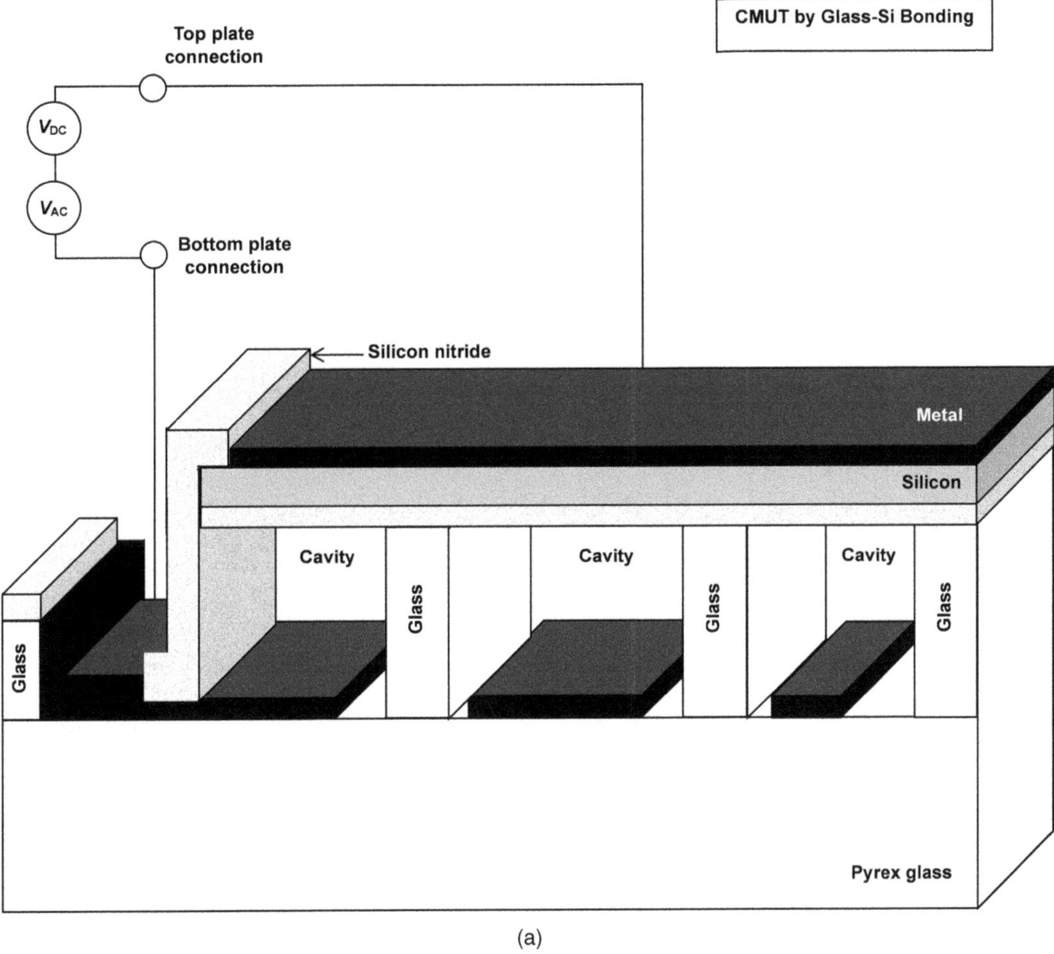

(a)

FIGURE 6.15 Capacitive MEMs ultrasonic transducers (CMUTs): (a) made by glass-to-silicon anodic bonding, (b) made by wafer bonding, and (c) made on a through-glass via (TGV) substrate. In all parts (a), (b), and (c), the top plate and bottom plate connections and cavities are shown. A DC source and an AC source are connected between the top and bottom plates. Part (a) shows a Pyrex glass wafer in which rectangular pits are etched, and Cr/Au is deposited in the pits. A membrane comprising silicon and silicon dioxide layers rests on two silicon pillars standing on the glass plate to form a cavity between the membrane and the glass plate.

(Continued)

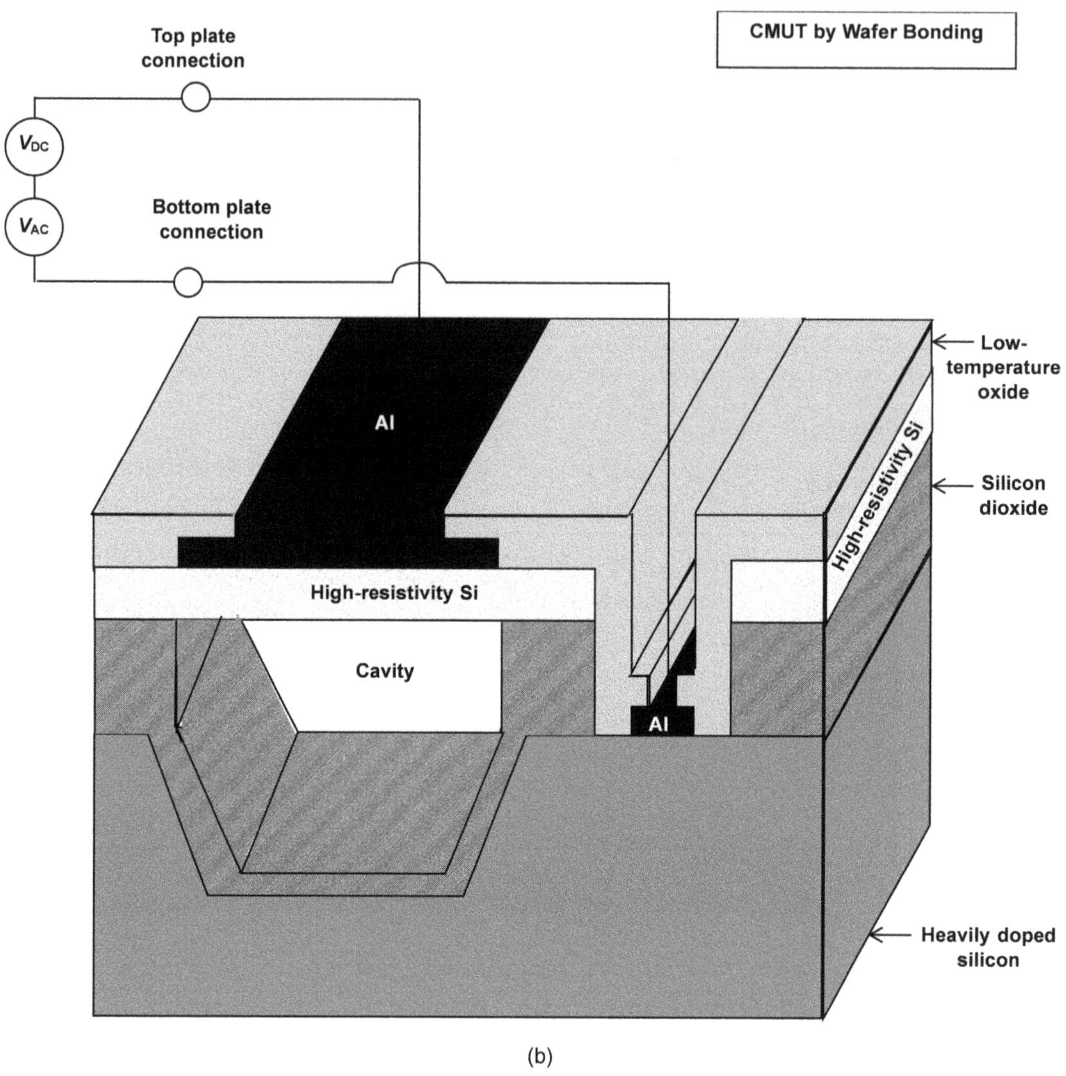

(b)

FIGURE 6.15 (CONTINUED)　Part (b) shows a heavily doped silicon substrate in which a cavity is etched. A silicon dioxide layer covers the surface of the cavity. The membrane is made of a high-resistivity silicon film. The aluminum film is deposited over the high-resistivity silicon membrane for the top plate contact. Access to the bottom plate is provided by the aluminum contact formed on the heavily doped silicon substrate at the bottom of a trench from the upper surface of the device to the silicon substrate. Then a low-temperature oxide is deposited. It is patterned to surround the aluminum contact regions.

(Continued)

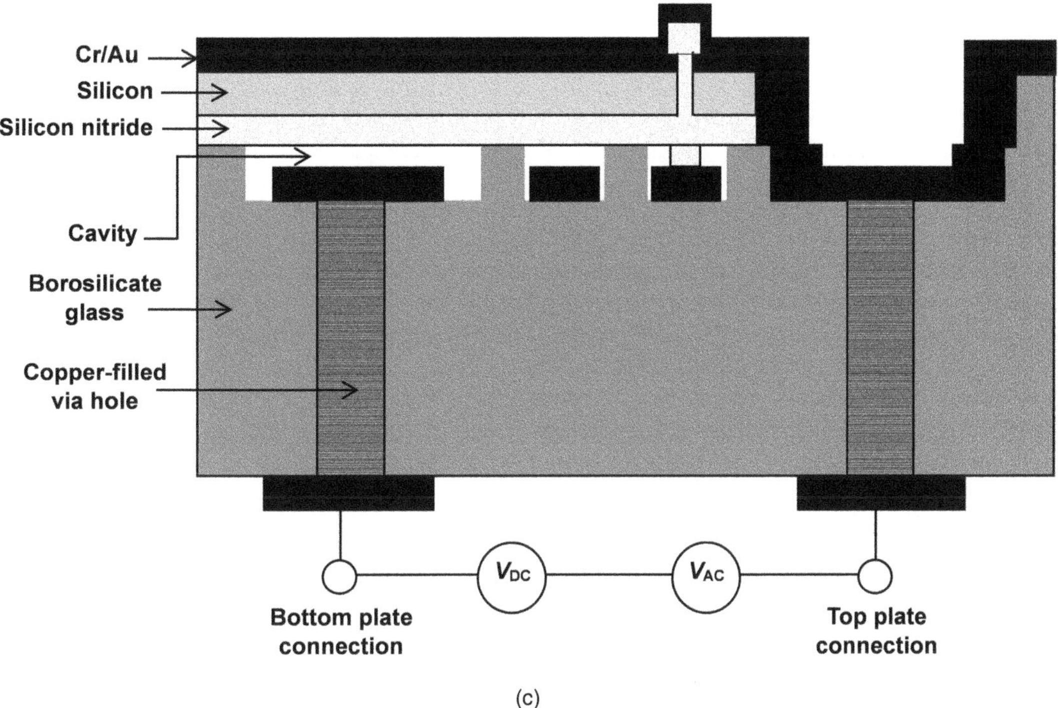

FIGURE 6.15 (CONTINUED)   Part (c) shows a through-glass-via substrate. In this substrate, the via holes are filled with copper to provide a conduction path across the substrate from one side to the opposite side. On the upper surface of the substrate, four pits are seen. The pit on the right-hand corner is completely covered with Cr/Au. Over the three pits on the left side, the membrane is suspended. It is made of Si and Si_3N_4 films. The Cr/Au film in the first pit from the left lies over the copper path extending to the bottom side of the substrate. This copper path is covered with Cr/Au on the lower surface of the substrate. The Cr/Au pad is the bottom plate connection. The second pit from the left has a Cr/Au film on its bottom and is a smaller cell of the CMUT working like the bigger cell constituted by the first pit. In the third pit from the left, a silicon nitride rod is seen piercing from the upper surface of the membrane to Cr/Au film. This rod-shaped structure is formed to plug the hole which was used for creating vacuum in the cavity during the process.

silicon/silicon nitride is etched to evacuate the trapped gas and reach the vias for accessing the top electrode. Vacuum sealing of the wafer with PECVD silicon nitride is done. To make the top electrode, silicon nitride is etched from the concerned area. Then top electrode is deposited. Pads are defined on the backside by photolithography followed by back metallization.

Besides process simplification, parasitic resistance and capacitance of interconnects is decreased. The process is useful for making 2D CMUT arrays.

6.9   ROADWAY INFRARED VEHICLE DETECTION SENSORS

Infrared sensors provide assistance in traffic signal control by estimating the volume of traffic on lanes and speeds of vehicles, and their classification into different types.

6.9.1 Passive Infrared Sensors

These sensors, mounted in the overhead or side-looking arrangement, do not emit any infrared radiation. Rather, they detect infrared energy falling on them from all infrared sources in their field of view. These infrared sources include the following:

(i) the direct infrared energy emitted from vehicles on the road, the road surfaces, and the objects on the road together with the infrared energy from the atmosphere, the cosmic rays, and the distant galaxies in outer space;
(ii) the indirect infrared energy falling from the atmosphere and outer space on the vehicles, road surfaces, and objects on the road, and reflected from them on the sensor.

The infrared energy received by the sensor is different when a vehicle is present on the road from the infrared energy when there is no vehicle on the road. In both cases, the received infrared energy is focused by an optical system upon the sensor placed in the focal plane of the system, where it is converted into an electrical signal. Assuming equality of surface temperatures of the road and the vehicle, the signal produced due to a vehicle's entry in the field of view of the infrared sensor is proportional to the product of two terms, an emissivity term corresponding to the different emissivity of the road and vehicle surfaces, and a temperature difference term representing the difference between the temperature of the road surface and the temperature resulting from the atmospheric, cosmic and galactic effects. By finding the emitted infrared energies from the signals in the two situations of vehicle presence and its absence, a decision is made about the presence or absence of the vehicle through real-time processing of the signals.

6.9.2 Active Infrared Sensors

These sensors are made in two configurations, each of which has its own advantages/disadvantages depending on the circumstances.

6.9.2.1 Two-Component Device

This device consists of two separate components, an infrared emitter and an infrared receiver (Figure 6.16). The infrared emitter is mounted on one side of the road and the infrared receiver is fixed on the opposite side exactly facing the emitter. The IR beam transmitted by the emitter is detected by the receiver when there is no obstacle, such as a vehicle, on the road to intercept the beam. When a vehicle moving on the road comes into the region where the monitoring emitter-receiver assembly is installed, its body interrupts the IR beam. Due to blockage of the IR beam by the vehicle, the receiver's output signal changes, registering the presence of the vehicle.

6.9.2.2 One-Component Device

The road is illuminated with low-power infrared radiation from an infrared laser diode serving as the infrared emitter. The infrared receiver is placed adjoining the emitter in the same package and facing the same direction. Thus, it is a single-component device in which the infrared emitter and receiver are placed side-by-side in close proximity.

The infrared beam from the emitter returns after striking the retroreflectors fitted under the pavement when there is no vehicle on the road. But when a vehicle enters the detection zone of the sensor, the incident infrared beam bounces off from the vehicle's surface, not from the retroreflector. So, the intensity of IR radiation changes with the entry of the vehicle, which is an indicator of its presence.

The single-component device performs the same function as the two-component device without the need for a separate remotely placed infrared receiver.

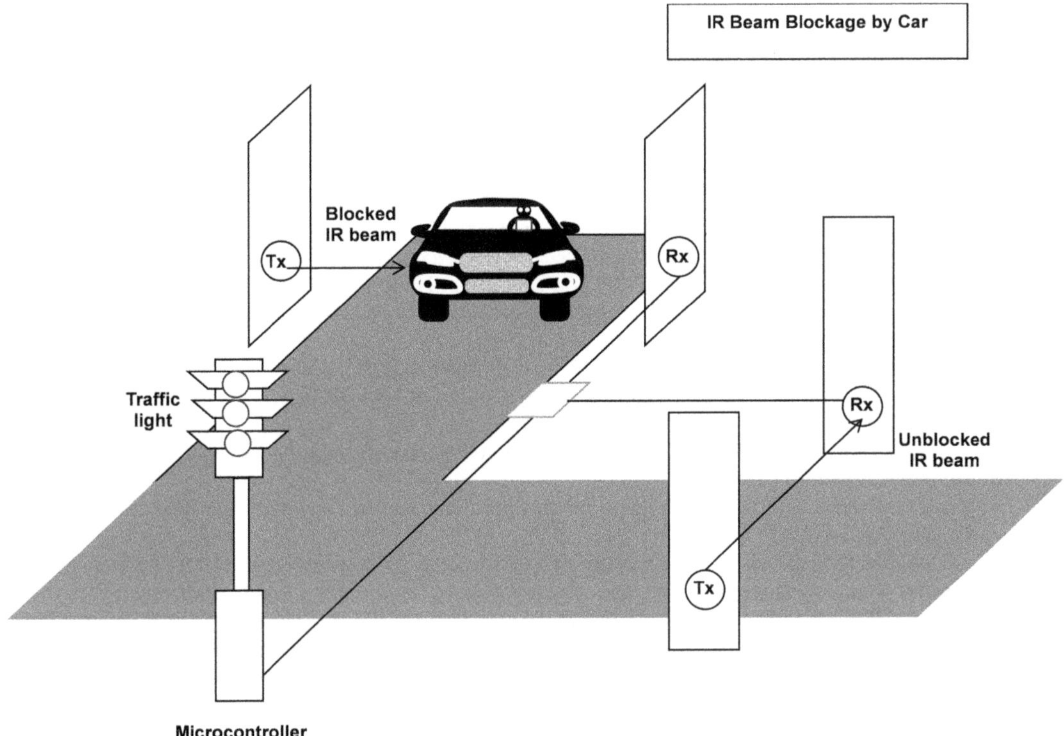

FIGURE 6.16 Blockage of the path of the infrared beam by a car plying on the road. The sensor signal is transmitted to a microcontroller which actuates a traffic light signal.

6.10 SMART VEHICLE PARKING SYSTEMS

This system keeps a watch on parking spaces in a locality with the help of properly mounted sensors. The occupancy or vacancy status of a parking space is determined by detecting whether a vehicle is present/absent in the parking space. The occupancy/vacancy data are supplied to a cloud platform readily accessible to drivers through a smartphone app. The drivers can move to the nearest available parking space in accordance with the size and type of vehicle. Thus, the drivers can easily find vacant parking spaces in nearby locations.

For indoor parking, ultrasonic sensors mounted on the ceiling of a parking garage detect the presence or absence of a vehicle at a parking spot. Wireless indicator lights guide cars to available spots. The wireless system can be quickly deployed with minimal downtime.

6.11 CONCLUDING REMARKS AND PREPARING FOR THE UPCOMING CHAPTER

Sensors are essential for managing the flow of traffic, keeping an eye on irresponsible unsafe driving, and providing road safety. At this point, it is necessary to look at the source of energy that is running our homes, offices, and factories, as well as the controlling of traffic en route, namely, the electrical power. The use of sensors in the proper distribution of power will form the subject of the next chapter.

REFERENCES

Ho T.-J. and M.-J. Chung 2016 Information-aided smart schemes for vehicle flow detection enhancements of traffic microwave radar detectors, *Applied Sciences*, 6:196, pp. 1–18.

Huang Y., A. S. Ergun, E. Hæggström, M. H. Badi, and B. T. Khuri-Yakub 2003 Fabricating capacitive micromachined ultrasonic transducers with wafer-bonding technology, *Journal of Microelectromechanical Systems*, 12(2):128–133.

In V., A. Bulsara, A. Kho, A. Palacios, P. Longhini, J. Acebron, S. Aglio and B. Ando 2004 Self-induced oscillations in electronically-coupled fluxgate magnetometers, CP742, Boccaletti S. et al. (Ed.) *Experimental Chaos: 8th Experimental Chaos Conference © 2004 American Institute of Physics*, pp. 57–62.

Klein L.A., M. K. Mills, and D. R.P. Gibson 2006 Traffic Detector Handbook: Third Edition—Volume I, Report No. FHWA-HRT-06-108, Chapter 2: Sensor technology, US Department of Transportation, Federal Highway Administration, McLean, VA, pp.2–1 to 2–95.

Lenz J. and A. S. Edelstein 2006 Magnetic sensors and their applications, *IEEE Sensors Journal*, 6(3): pp. 631–649.

Lee S., D. Lee, P. Choi and D. Park 2020 Accuracy–power controllable LiDAR sensor system with 3D object recognition for autonomous vehicle, *Sensors*, 20:5706, pp. 1–20.

Lin C., Y. Wang, B. Gong, and H. Liu, 2023 Vehicle detection and tracking using low-channel roadside LiDAR, *Measurement*, 218: p. 113159.

Nellore K. and G. P. Hancke 2016 A Survey on urban traffic management system using wireless sensor networks, *Sensors*, 16(2):157, pp. 1–25.

Niclass C., A. Rochas, P.-A. Besse, and E. Charbon 2005 Design and characterization of a CMOS 3-D image sensor based on single photon avalanche diodes, *IEEE Journal of Solid-State Circuits*, 40(9): pp. 1847–1854.

Pongthavornkamol T., A. Worasutr, D. Worasawate, La-or Kovaisaruch, and K. Kaemarungsi 2021 X-band front-end module of FMCW radar for collision avoidance application, *Engineering Journal*, 25(5): pp. 61–70.

Texas Instruments 2016 TDC7201 Time-to-Digital Converter for Time-of-Flight Applications in LIDAR, Range Finders, and ADAS, SNAS686 –MAY 2016, 50 pages.

Vogt M. 2018 Radar sensors (24 and 80 GHz Range) for level measurement in industrial processes, *2018 IEEE MTT-S International Conference on Microwaves for Intelligent Mobility (ICMIM)*, 15–17 April, Munich, Germany, pp. 1–4.

Wang H., Y. Ma, H. Yang, H. Jiang, Y. Ding and H. Xie 2020 MEMS ultrasound transducers for endoscopic photoacoustic imaging applications, *Micromachines*, 11:928, pp. 1–42.

Yamaner F. Y., X. Zhang, and O. Oralkan 2014 Fabrication of anodically bonded capacitive micromachined ultrasonic transducers with vacuum-sealed cavities, *2014 IEEE International Ultrasonics Symposium Proceedings*, 3–6 September, Chicago, IL, pp. 604–607.

Zhang T. and P. J. Jin 2022 Roadside LiDAR vehicle detection and tracking using range and intensity background subtraction, *Journal of Advanced Transportation*, 2022: pp. 1–14. Article ID 2771085.

Zhang X., F. Y. Yamaner, and O. Oralkan 2015 Fabrication of capacitive micromachined ultrasonic transducers with through-glass-via interconnects, *2015 IEEE International Ultrasonics Symposium (IUS) Proceedings*, 21–24 October, Taipei, Taiwan, pp. 1–4.

7 IoT Sensors for Energy-Efficient Power Grid

7.1 INTRODUCTION

Power travels from power plants to our homes, offices, businesses, street lamps and traffic lights, factories, hospitals, and other end users through a power grid, which is a complex network of power generation, transmission, and distribution lines comprising poles, wires, transformers, switchgear, circuit breakers, and protection circuitry. This infrastructure always keeps the voltage and current at safe customer levels, avoiding dangerous shocks. While this grid works silently and is always at our service, we often fail to realize its importance. But the whole life comes to a standstill, the moment the grid fails. We become helpless in this modern technological world where not only the lights, fans, and air conditioners but also the radio communication services, internet, and all online transactions come to a halt in case of a power breakdown. In this chapter, let us dive into the different ways in which sensors are utilized for real-time efficient power distribution, maintenance of services, line-fault location pinpointing, rapid repairing of disrupted connections, and so forth.

7.2 THE CONVENTIONAL AND SMART ELECTRICAL GRIDS

7.2.1 THE CONVENTIONAL ELECTRICAL GRID

The electrical grid is an electrical power distribution network connecting the power generating plant with consumers through substations, transmission lines, transformers, and various protection components to supply electricity to homes, offices, industries, and businesses to meet their energy requirements.

7.2.2 THE SMART ELECTRICAL GRID

A smart electrical grid is an improved version of the conventional power distribution network modernized with renewable energy sources and storage; enhanced with sensors, computers, software, and optical fiber broadband internet connectivity; and supported by digital technology to provide two-way communication and energy transfer between the power plant and consumers for quickly responding to the changing power demands, either by adjusting electricity flow in real time or restricting the load to match with power generation.

Re-emphasizing the capabilities of a smart power grid, it is clarified that:

(i) The smart grid is an automated power transmission and distribution system.
(ii) It is able to provide a two-way give-and-take mechanism, i.e., to-and-fro flow of electricity and information between the power plants and the consumers.
(iii) The to-and-fro energy and data flows are also applicable to all points between power plants/ consumers for monitoring, protecting, and controlling the operations of the grid.
(iv) For obtaining real-time data regarding bidirectional energy flow and status of the grid, the smart grid is equipped with several sensors.
(v) The main sensors in the smart grid are current and voltage sensors, smart energy meters, phasor measurement units (PMUs), temperature sensors, power failure sensors, transformer monitoring, and dynamic line rating (DLR) sensors.

DOI: 10.1201/9781003374442-7

(vi) The smart grid makes the power distribution network more capable of dealing with power outage emergencies during heavy rains, storms, and cyclones to fast restore the power supply by prompt detection, isolation, and repair of outage zones before they spread into wide-area blackouts.

(vii) It also reduces the network maintenance and operation costs and hence reduces the electricity bills of consumers.

Thus, the principal building blocks of a smart grid are sensing, communication, computing, visualization, and control units.

7.3 MONITORING TRANSFORMER PARAMETERS: CURRENT SENSOR, VOLTAGE SENSOR, TEMPERATURE SENSOR, AND TRANSFORMER OIL-LEVEL SENSOR

7.3.1 CURRENT SENSOR

Current sensing implies the creation of a voltage signal representative of the current flowing through the measured path. Hence, a current sensor is basically a current-to-voltage converter.

For current sensing, a current sense transformer is used. It is an instrument transformer. It usually consists of a single-turn primary winding, a silicon steel core, and a secondary coil made of several turns (Figure 7.1(a), (b)). The transmission line in which the current flow is to be measured is connected in series with the primary winding of the transformer. This is essential to confirm that the current flowing through the primary winding is the full line current. During the flow of an alternating current in the primary winding, a time-varying magnetic field is produced owing to the periodically reversing current. This time-varying magnetic field influences the core of the transformer, inducing a current in the secondary winding of the transformer. If I_P is the primary current and the secondary coil has N turns, the secondary current I_S is given by

$$I_S = \frac{1}{N} I_P \tag{7.1}$$

so that it is a current step-down, voltage step-up transformer and produces a smaller current from the large current flowing in the power grid wire.

7.3.2 VOLTAGE SENSOR

Like the current sense transformer, which transforms a high current to a lower safer value for measurement, a voltage sense transformer (Figure 7.1(c), (d)) converts a high voltage to a lower manageable level. It is a voltage step-down, current step-up transformer. Its core is made of high-quality steel. This steel works at low flux density. It is connected in parallel with the transmission line so that full line voltage appears across it. If V_P, V_S denote the primary and secondary voltages and N_P, N_S the number of turns in the primary and secondary coils,

$$V_S = \frac{N_S}{N_P} V_P \tag{7.2}$$

where

$$N_S < N_P, \text{ and } \frac{N_S}{N_P} < 1 \tag{7.3}$$

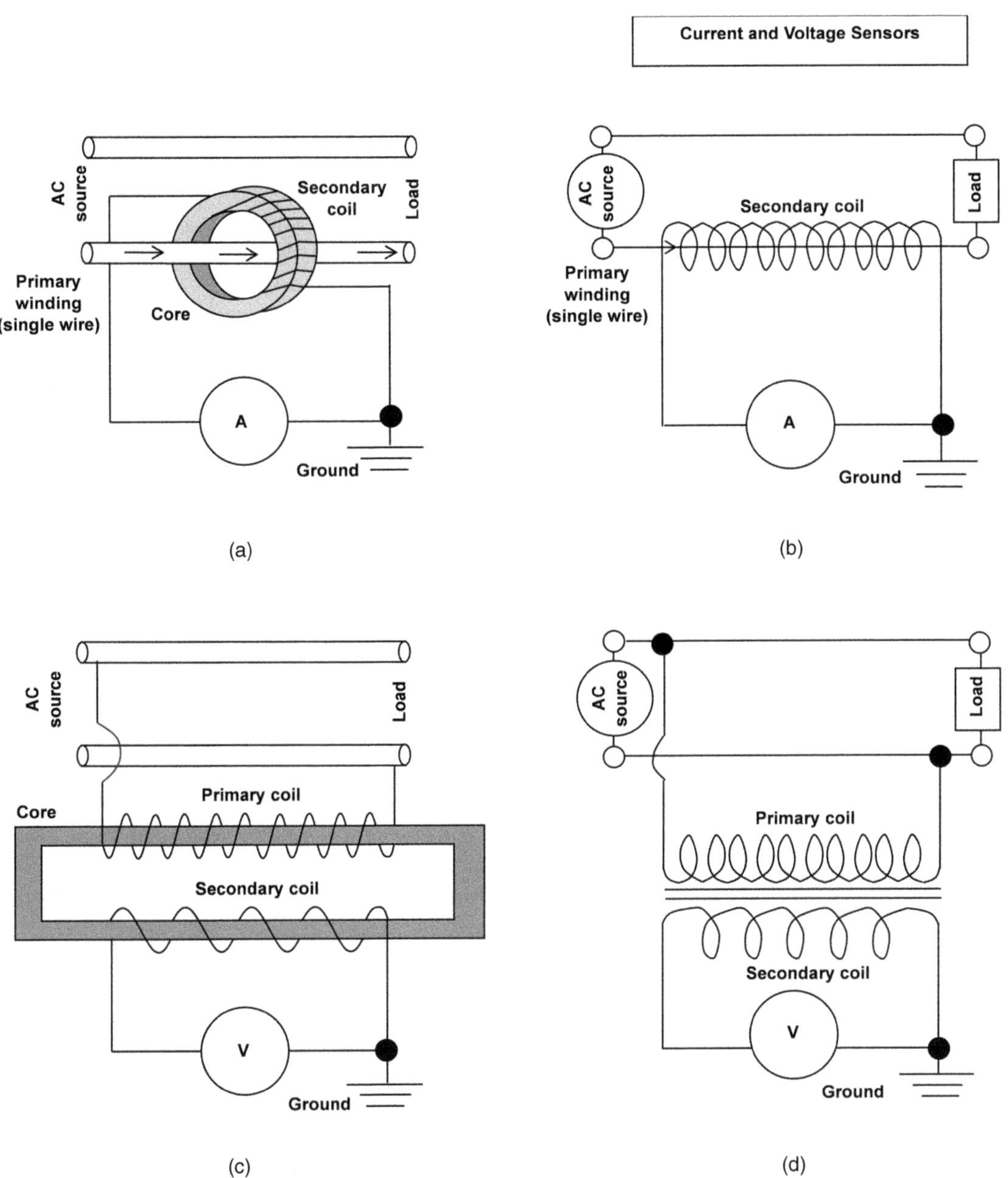

FIGURE 7.1 Current and voltage sensors: (a) Current sensor connected in series across supply line. The diagram shows the AC source, the load circuit, the primary winding (single wire), the secondary coil, the core, an ampere meter, and the grounded terminal. (b) Circuit diagram of a current sensor connected across the supply line. The diagram shows the AC source, the load circuit, the primary winding (single wire), the secondary coil, an ampere meter, and the grounded terminal. (c) Voltage sensor connected in parallel across supply line. The diagram shows the AC source, the load circuit, the primary coil, the secondary coil, the core, a voltmeter, and the grounded terminal. (d) Circuit diagram of a voltage sensor connected across the supply line. The diagram shows the AC source, the load circuit, the primary coil, the secondary coil, a voltmeter, and the grounded terminal.

7.3.3 Temperature Sensors

The reliable operation of a transformer depends on the maintenance of correct temperatures of its windings and the oil. Temperature control prevents transformer breakdown and increases its lifespan. Two temperature sensors, one for windings and the other for oil temperature measurements are used for monitoring transformer temperature continuously. Their output signals are digitized, and fed to a microcontroller. The preset and real-time recorded temperature values as well as the historical maximum and minimum temperature data are displayed on an LCD. An alarm rings when the temperature exceeds the permissible safe limit. The data are also sent by wireless transmission to a PC or a laptop where they are constantly watched.

7.3.3.1 Transformer Winding Temperature Sensors

Resistance temperature devices (RTDs), e.g., Pt100 sensors, have found widespread application in dry and cast resin transformers. High accuracy and cost-effectiveness are their main benefits. Fiber optic sensors incorporated into transformer windings during their manufacturing are also used. An increase and decrease in winding temperature cause expansion and contraction of the optical fiber, resulting in changes in its refractive index and other optical properties, from which temperature values are acquired.

7.3.3.2 Transformer Oil Temperature Sensors

7.3.3.2.1 Remote Oil Temperature Indicator

Generally, the top oil of the transformer is at the highest temperature. For large transformers, it is difficult to view the temperature dial from the ground level due to the height. A remote oil temperature indicator is useful in such cases. It consists of a sensing bulb linked by a capillary tube to a Bourdon tube or a bellows mechanism (Figure 7.2). The sensing bulb and the capillary tube are filled with a liquid. The sensing bulb is immersed in the top hottest region of the oil while the temperature reading mechanism and dial are positioned outside at eye level by taking proper flexible capillary length. Hence, the inconvenience of reading the dial due to height is avoided. Temperature measurement is done through the expansion of a liquid inside the bulb caused by the rise of temperature. The liquid volume changes are conveyed to the Bourdon tube or the bellows mechanism to drive a pointer moving on a dial.

7.3.3.2.2 Direct Mount Oil Temperature Indicator

These indicators are selected for measuring the oil temperature of small transformers. The temperature sensing bulb is directly connected to the Bourdon tube or the bellows mechanism without any connecting capillary because easy reading at eye level is possible. If bimetallic temperature sensors are used, they are directly submerged in the insulating transformer oil. Dissimilar expansion of the two metals actuates the indicating pointer on the dial.

7.3.4 Transformer Oil-Level Sensors

Every oil-immersed transformer has an expansion vessel known as the conservator which is a cylindrical tank mounted on the roof of the transformer. It accommodates the increase in the oil level caused by thermal expansion by providing sufficient space for the increased volume of the expanded oil. A minimum oil level in the conservator must be maintained even at the lowest expected temperature.

7.3.4.1 Transformer Oil-Level Float Mechanism Sensor

Also called a magnetic oil gauge, it consists of a float, either a hollow ball or drum, floating on the transformer oil (Figure 7.3). It is fixed to a long float arm. On the opposite side of the float arm, one

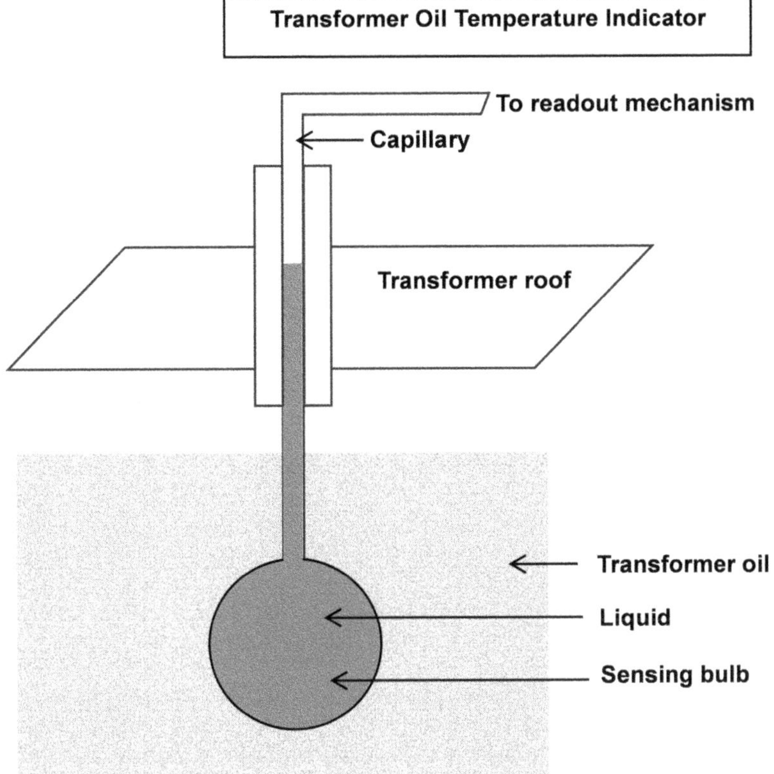

FIGURE 7.2 Transformer oil temperature indicator. A sensing bulb filled with a liquid is immersed in the transformer oil. A capillary tube is attached to the upper end of the bulb. It extends up to a height above the transformer roof so that it can be conveniently read by a person from a distance.

unit of a bevel gear is attached while the other unit of the bevel gear is magnetically coupled to a pointer moving on a dial. The bevel gears have conically shaped tooth-bearing faces. They are used for transmitting motion between intersecting shafts. The components fitted inside the transformer conservator tank are the float, the float arm, and the bevel gears, while the components outside the tank are the dial and the pointer. When the oil level rises, the float moves up, and when it falls, the float moves down. As a result, the alignment of the float arm changes, causing a rotation of the bevel gear. The magnet inside the tank rotates and along with it the magnet outside the tank also rotates moving the pointer on a graduated scale. The center of the scale is marked at 25°C. The high and low oil levels are also marked.

7.3.4.2 Transformer Oil-Level Infrared Sensor

The surface temperature of an object is related to the infrared energy emitted by it. The oil functions as a coolant taking away the heat from the windings. At night, the electrical load is high. So, the oil is at a high temperature. The outside atmospheric temperature is low. So, by measuring the infrared red emissions from the oil and the gas above it, the boundary of the oil level in the conservator is defined (Liu et al. 2021).

7.3.4.3 Transformer Oil-Level Fiber Optic Sensor

This sensing element is a multimode optical fiber (Figure 7.4) in which the cladding is removed from a selected length, either continuously (Figure 7.4(a)) or in discrete separate portions (Figure 7.4(b))

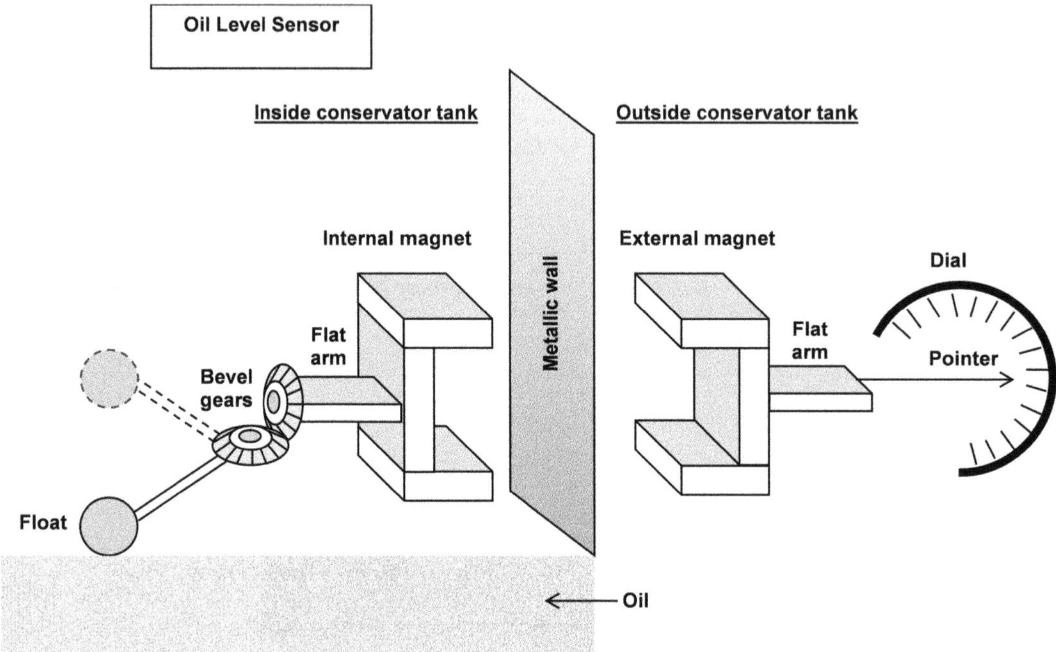

FIGURE 7.3 Transformer oil-level float mechanism sensor. The parts inside the conservator tank on the left side of the metallic transformer wall are: Transformer oil is seen. A ball is floating on the surface of oil. The float is carried by an arm whose opposite end terminates in the horizontal bevel gear. The horizontal bevel gear is coupled to a vertical bevel gear. A magnet is attached to the flat arm of the vertical bevel gear. This magnet interacts with an external magnet placed on the opposite side of the metallic transformer wall. The parts outside the conservator tank on the right side of the metallic wall are as follows: a magnet, called the external magnet, is mounted exactly at the same height and within interacting distance of the internal magnet. Its flat arm is connected to a pointer moving on a circular scale.

over the length (Mahanta and Laskar 2015). The uncladded portion is immersed in the oil. For this portion, the cladding medium is:

   (i) transformer oil when the oil level is high,
  (ii) air when the oil level is low, or
 (iii) partially oil and partially air at intermediate oil levels.

So, the cladding medium changes as the level varies. As the relative length of air and oil in contact with the uncladded portion of the sensing element changes, the refractive index of the cladding medium is altered. Variations of the refractive index of the cladding medium result in changes in the intensity of output light because of the different degrees of power loss incurred by the light beam propagating through the fiber in these situations.

The measurements are carried out in a differential configuration using two identical fibers, the sensing fiber, and the reference fiber, to cancel out environmental effects. The optical assembly comprises a He-Ne laser source, a sensing optical fiber, a reference optical fiber, beam splitters, a light-dependent resistor, a filter circuit, a data acquisition card, and a computer, as shown in Figure 7.4(c). The light outputs from the two fibers are fed via the light-dependent resistor-based voltage dividers to convert them into equivalent voltages. These voltage signals are filtered to remove high-frequency noise, and supplied to a data acquisition card and then to a computer for analysis.

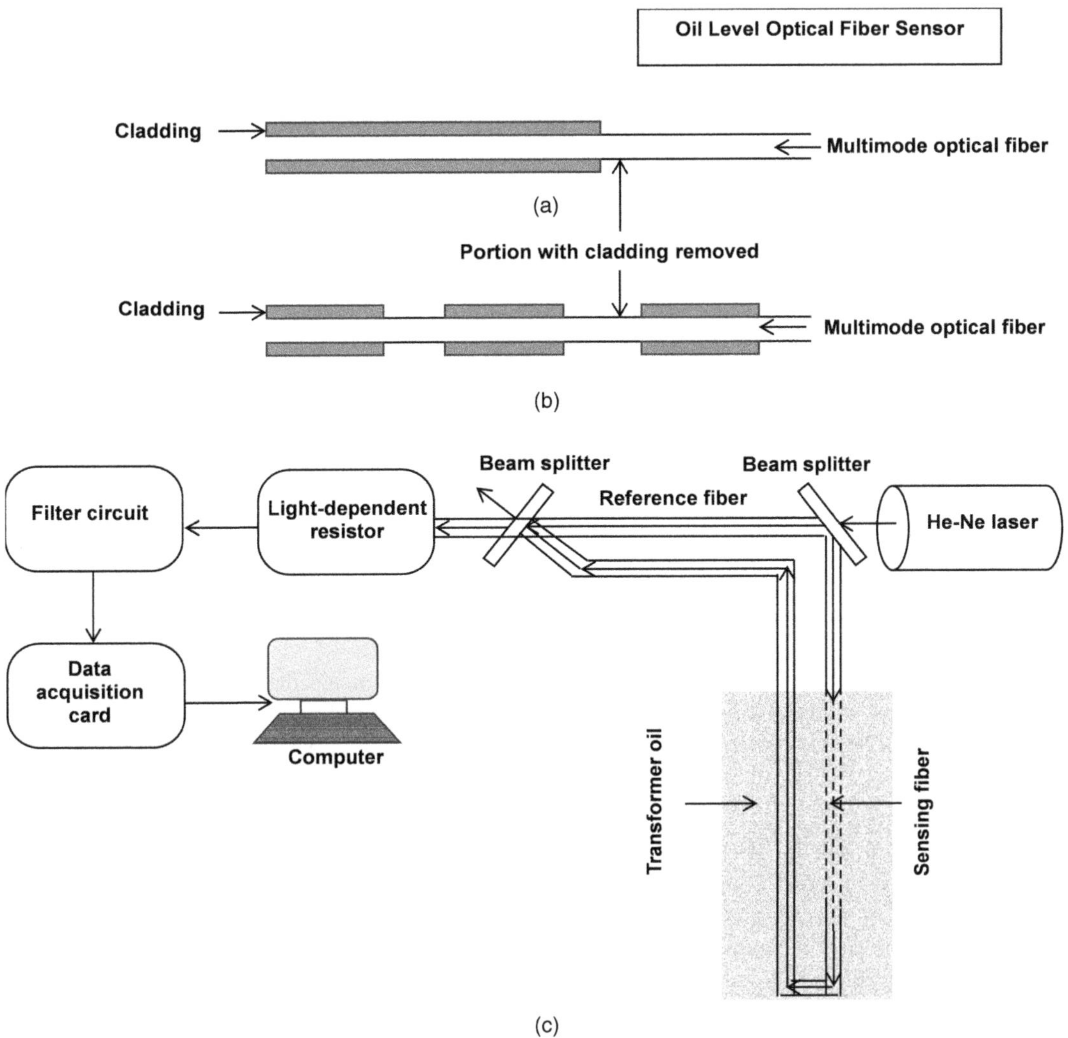

FIGURE 7.4 Transformer oil-level optical fiber sensor: (a) optical fiber with cladding removed from a portion, (b) optical fiber with cladding removed from discrete separate portions, and (c) the setup of optical fiber sensor. Part (a): The diagram shows a multimode optical fiber, which is covered with cladding in the left-hand portion but the cladding is removed from its right-hand portion. Part (b): The diagram shows a multimode optical fiber, which has several cladding-covered portions separated by portions without cladding. Part (c): Light from a He-Ne laser falls on a beam splitter where it is broken down into two beams. One beam travels along a straight line to a reference fiber, then a beam splitter, a light-dependent resistor, a filter circuit, a data acquisition card, and finally a computer. The second beam is bent by 90° toward the sensing fiber, immersed in a transformer coil. It first moves down the fiber, then bends leftward along the fiber, and subsequently bends upward along the same fiber. This beam also falls on the beam splitter on which the reference fiber beam was incident. Its onward path is the same as that of the reference fiber beam.

7.4 ELECTRONIC ENERGY MEASUREMENT: SMART METERING

7.4.1 ANALOG VOLTAGE-TO-DIGITAL CONVERSION

Let us start from the stage where analog voltage signals have been obtained corresponding to the current and voltage signals of the transmission line using the current and voltage sensors. These analog signals are converted into digital formats for further processing. So, let us understand how this is done.

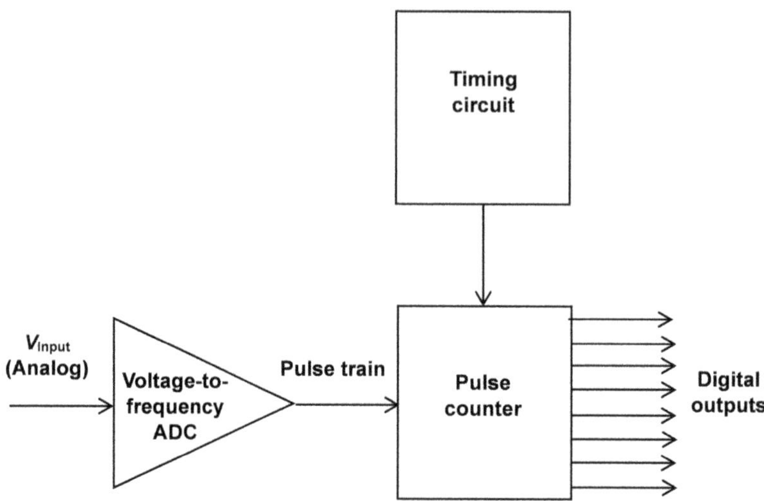

FIGURE 7.5 Analog voltage-to-digital output conversion circuit. The circuit contains a voltage-to-frequency ADC, timing circuit, and pulse counter. The analog signal V_{Input} is supplied to the ADC. The outgoing pulse train from the ADC moves to the pulse counter which also receives signals from the timing circuit delivering digital outputs.

The analog voltages are supplied to a voltage-to-frequency ADC (analog-to-digital converter) such as a sigma-delta ADC. This ADC produces a train of pulses having a frequency proportional to input voltage amplitudes (Figure 7.5). The frequency of the pulse train is determined by counting the pulses over a fixed period of time with the help of a pulse counter. This frequency is proportional to the input voltage magnitude, e.g., the delta-sigma ADC produces a stream of pulses whose frequency f is given by

$$f = kV_{\text{Input}} \tag{7.4}$$

where k is a constant for the particular circuit implemented and V_{Input} is the input analog voltage.

The sigma-delta ADC is an integrating ADC containing an integrator, a 1-bit DAC (digital-to-analog converter), a comparator, and a summing junction. When the integration time of the ADC input matches with a multiple of the AC period, the effects of noise pickup at line frequency are drastically reduced. A digital filter is included in the IC to enable working at a high sampling rate. The use of oversampling techniques allows to achieve a high resolution.

7.4.2 Electronic Energy Meter

The energy metering can be performed electronically using dedicated energy metering integrated circuits such as the AD775x from Analog ICs (Analog ICs 1999, Daigle 1999) specially designed for this purpose (Figure 7.6). These ICs take the input voltages from voltage and current sensors and carry out the necessary signal processing to display the power consumption of a domestic or industrial premise. The successive series of operations executed in these chips are analog-to-digital conversion, phase correction and high-pass filtering, multiplication, low-pass filtering, digital-to-frequency

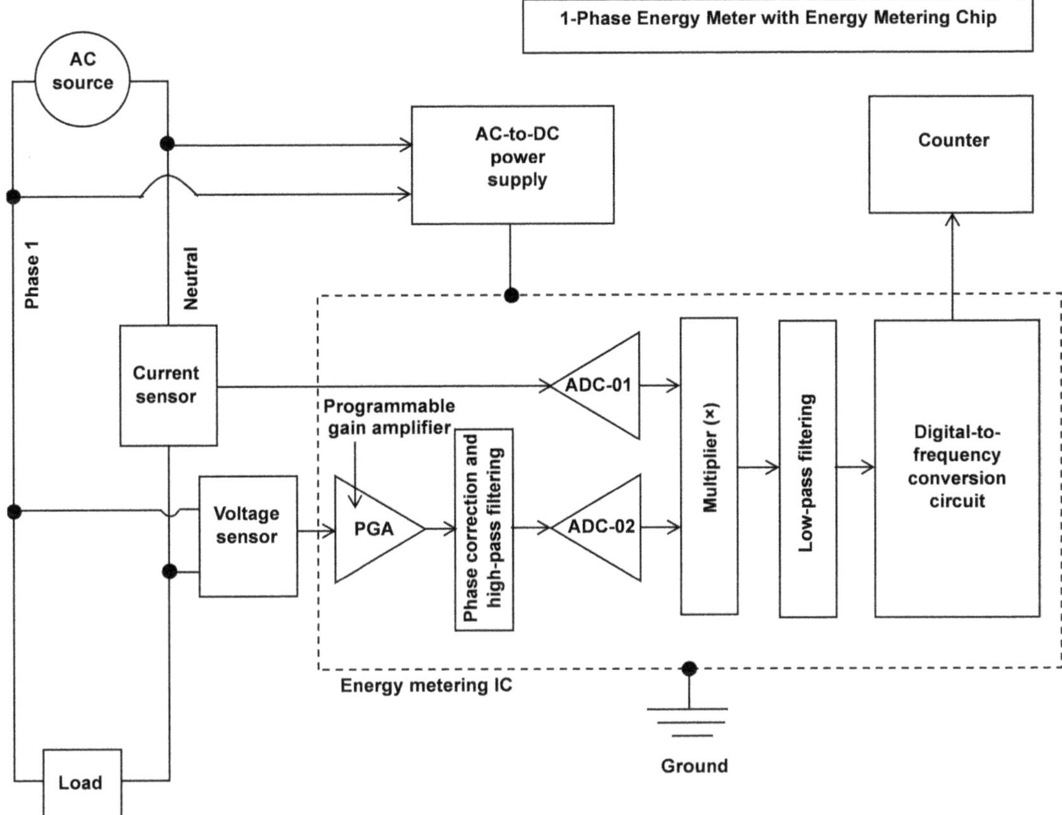

FIGURE 7.6 Single-phase energy meter using energy metering chip. The diagram shows the AC source, the load, and the circuit components: current sensor, voltage sensor, AC-to-DC power supply, PGA (programmable gain amplifier), ADC-01, ADC-02, phase corrector and high-pass filter circuits, multiplier, low-pass filter, digital-to-frequency converter, and counter. The components of the energy metering IC are enclosed inside a dashed rectangle. This IC is grounded. Phase 1 and neutral lines are indicated.

transformation, and counter activation for display. The voltage sensor signal is fed to ADC1 and the current sensor signal to ADC2. The digital signals from these ADCs flow to the multiplier, and the multiplication output signal flows to the voltage-to-frequency conversion block from which the output is delivered to the 3–8 encoder, Finally, the pulse counter is driven to show the energy consumed in kilowatt hour (kWh). The relay is used to interrupt the power if the power company so desires, such as during non-payment of bill.

7.4.3 Smart Energy Meter

A smart meter is an electronic meter that measures the voltage, current, power factor, and energy consumption. It sends the measured information instantly to the power supplier in order that the supplier can issue commands to operate the grid components like distribution switches to maintain reliable throughput to the consumer; this information is also used for customer billing. The throughput to the consumer can be varied remotely. It can be cut off in an extreme situation. The information is conveyed either by wires through a power line carrier (PLC) or in wireless mode. No employee from the power company needs to come to the consumer's house to record the meter reading.

The meter also provides the above information to the consumer to know about the consumption patterns and trends. It can be read locally as well as remotely, if required.

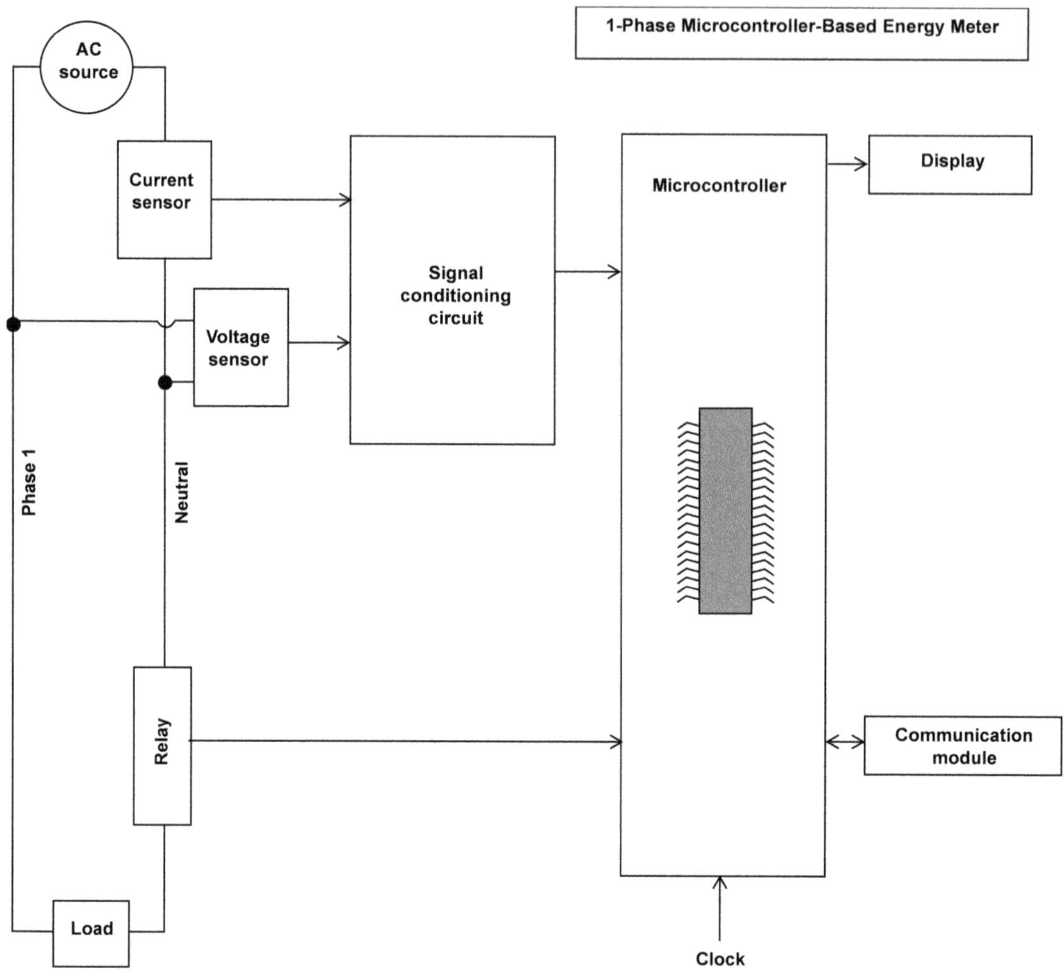

FIGURE 7.7 Single-phase microcontroller-based energy meter. The diagram shows the AC source, the load, and the circuit components: current sensor, voltage sensor, load, relay, signal conditioning circuit; the microcontroller, its clock input, and the LCD display and communication modules connected to it. Phase 1 and neutral lines are indicated.

7.4.3.1 Microcontroller-Based Smart Energy Meter

Figure 7.7 shows the functional block diagram of a single-phase energy meter in which the microcontroller takes care of several steps. Voltage and current sensors connected to the mains line feed the signal conditioning circuit made with OP-AMP integrated circuits. The signal conditioner matches the signal level to the required level for processing. In addition, it contains a multiplexer. The job of the multiplexer is to sequentially switch the signals from the voltage and current sensors to the analog input of the microcontroller. In the microcontroller, the first stage is the digitization of the received input signals by high-resolution sigma-delta ADCs converting them into pulse trains whose frequency is proportional to the amplitudes of the voltages. The next stage in the microprocessor is the multiplication of the digitalized signals from the voltage and current sensors to calculate the product. This value is proportional to power. The microcontroller is programmed to execute this function. The microcontroller is also programmed for integration of the power values over a pre-stimulated period of time, yielding the number of kWh units of electricity consumed over that period. The final stage is driving a counter to display the power consumption. Thus, computation of the product of

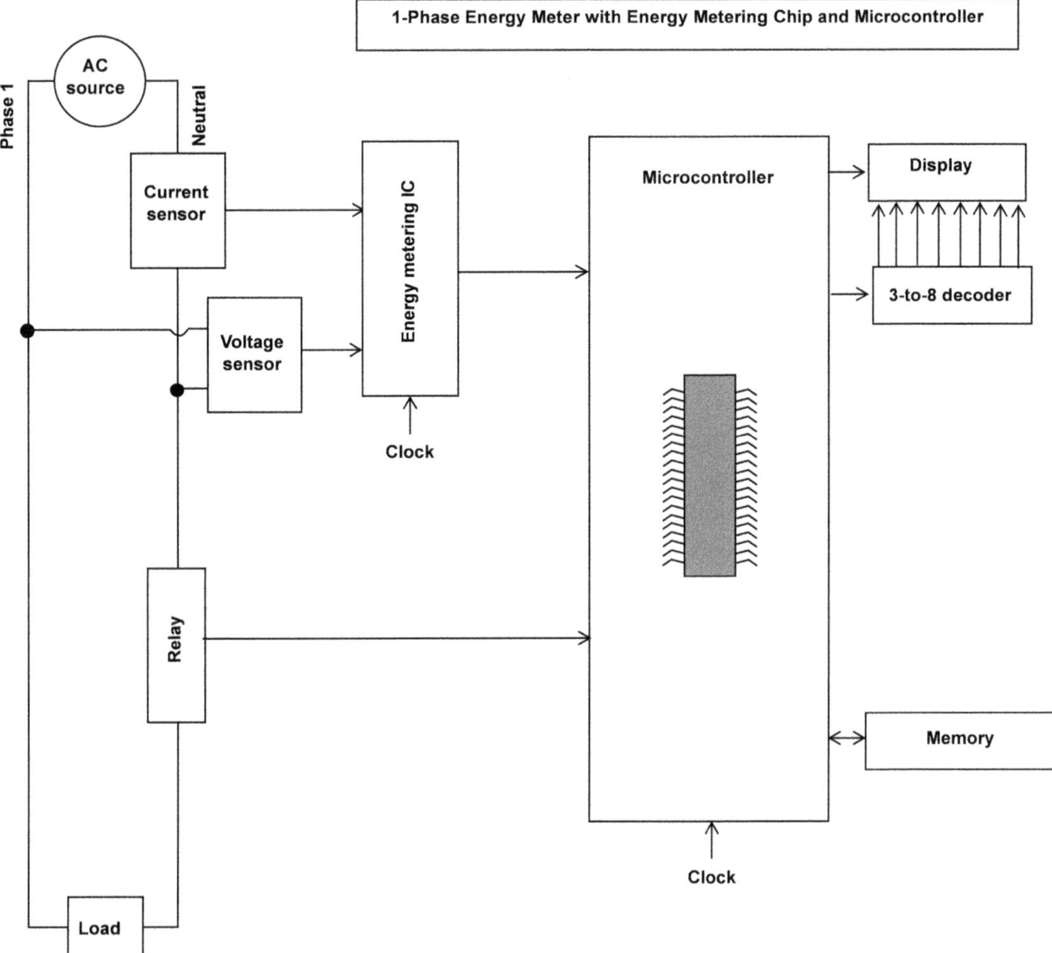

FIGURE 7.8 Single-phase energy meter using the energy metering chip and microcontroller. The diagram shows the AC source, the load, and the circuit components: current sensor, voltage sensor, load, relay, energy metering chip, microcontroller, a 3-to-8 decoder, display, and memory. Phase 1 and neutral lines are indicated.

instantaneous values of voltage and current gives power in kW. The determination of the integral of power over time yields the energy in KWh. A communication module is connected to the microcontroller for data transmission. The microcontroller is properly selected keeping in view the desired number of bits of the ADC, and ease of programming.

7.4.3.2 Smart Energy Meter Using Energy Metering Chip and Microcontroller

It is convenient to partition the work into two parts:

(i) energy metering and
(ii) overall supervision and control.

The first part can be looked after by the energy metering chip, and the second part by the microcontroller (Markow 1999). The microcontroller stores and retrieves data from EEPROM (electrically erasable programmable read-only memory). It operates the energy meter to view the display and helps the consumer select the desired service or utility. Functions such as data encryption, and intelligent services, viz., fault detection, load management, remote disconnection, and configuration of particular meters are possible. Thus, the microcontroller acts as the brain of the system whereas all signal processing operations for determination of energy units are done by the energy metering chip. Such combined energy metering chip-cum microcontroller circuits are shown in Figures 7.8 and 7.9 for single-phase and three-phase mains supply respectively. Figure 7.9 contains a power supply so that the circuit will continue working in a situation in which supply in any one phase fails. Both these circuits contain relays to disconnect the power if the power company wants to do so (Mohammad et al. 2013).

7.5 PHASORS, SYNCHROPHASORS, AND SYNCHROPHASOR MEASUREMENTS

7.5.1 PHASOR

7.5.1.1 Phasor Definition

Three parameters, namely, the amplitude, the frequency, and the phase angle completely define a sinusoidal signal. In some situations, the frequency is fixed. Then the signal is represented by only two parameters, the amplitude and the phase or phase angle. The phase of a sinusoidal signal is the angular separation of the peak value of the signal with respect to a stipulated reference point. This leads to the phasor depiction of signals, and it is essential that all phasors in a phasor diagram relate to the same frequency.

A phasor (Figure 7.10) is a shorthand vectorial representation of an alternating quantity (voltage V or current I) which varies sinusoidally with time as

$$V(t) = V_0 \sin(\omega t) \tag{7.5}$$

$$I(t) = I_0 \sin(\omega t + \phi) \tag{7.6}$$

This vector is a directed line segment signified by an arrow rotating in the counterclockwise direction with an angular velocity $\omega = 2\pi f$, about a fixed point (the origin) in a plane where f is the linear frequency. The length of the arrow is proportional to the peak value (amplitude V_0 or I_0) of the quantity, and the angle ϕ between this arrow and a reference line is the phase of the quantity. Thus, a phasor is a combination of amplitude V_0 or I_0 and phase ϕ. The projection of the phasor on the Y-axis gives the instantaneous value of the alternating quantity $V(t)$ or $I(t)$ at the instant t.

7.5.1.2 Mathematical Notation for Phasors

The voltage signal of eq. (7.5) is expressed in phasor form as a complex number

$$\mathbf{V} = V_0 \exp(j\phi) = V_0(\cos\phi + j\sin\phi) \tag{7.7}$$

The frequency ω is excluded because it is implicitly implied in the definition of phasor. In terms of magnitude V_0 and phase ϕ, the phasor is described by the equation

$$\mathbf{V} = V_0 \angle \phi \tag{7.8}$$

A similar equation is written for the current signal of eq. (7.6).

Thus, an AC waveform (voltage or current signal) can be written as

$$x(t) = X_0 \sin(\omega t + \phi) \tag{7.9}$$

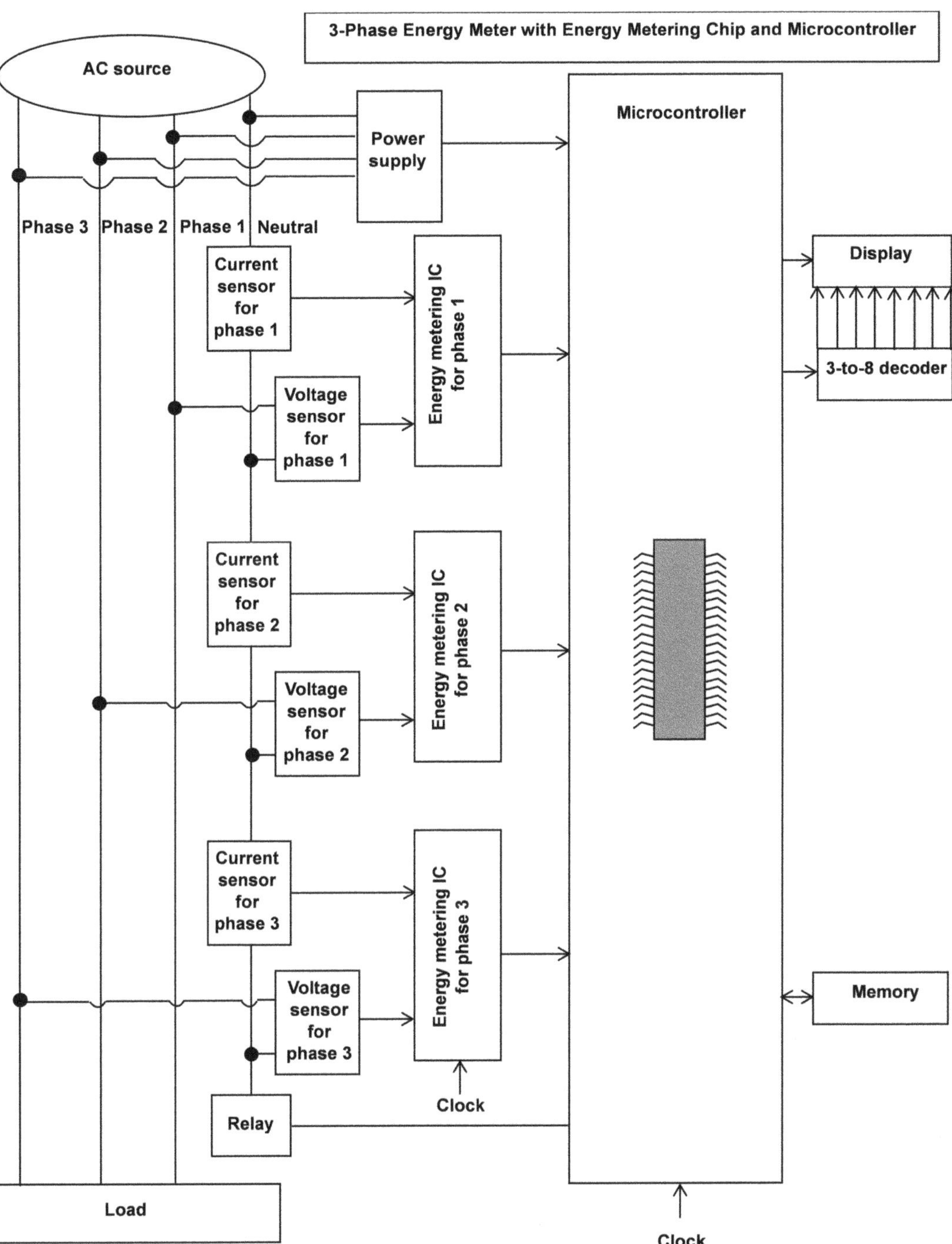

FIGURE 7.9 Three-phase energy meter using the energy metering chip and microcontroller. The diagram shows the AC source, the load, and the circuit components: three current sensors for phases 1–3, three voltage sensors for phases 1–3, load, a relay, three energy metering chips for phases 1–3, one microcontroller, a 3-to-8 decoder, display, and a memory unit. Also included is a power supply for the microcontroller, which is used in case of one-phase power failure. Phases 1, 2, and 3 and neutral lines are indicated.

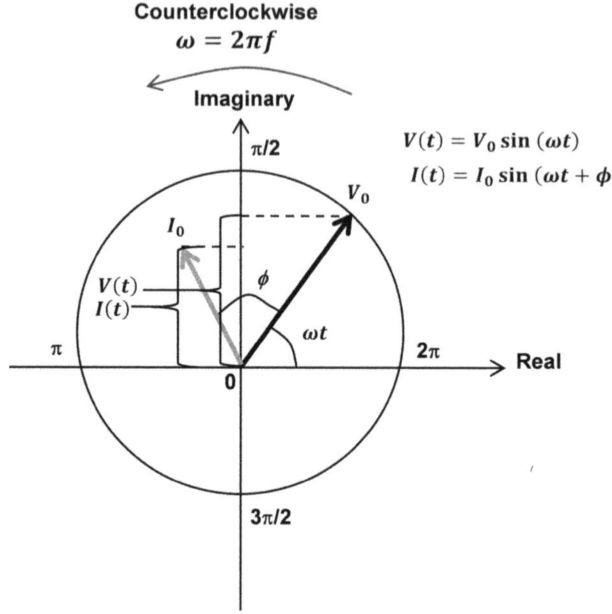

FIGURE 7.10 Definition of a phasor. Sinusoidal voltage and current waveforms are represented by the equations.

$$V(t) = V_0 \sin(\omega t)$$

$$I(t) = I_0 \sin(\omega t + \phi)$$

where $V(t)$ is the instantaneous voltage at time t, $I(t)$ is the instantaneous current at time t, $V_0(t)$ is the peak value or amplitude of voltage, $I_0(t)$ is the peak value or amplitude of current, and ω is the angular frequency of variation of voltage or current, related to its linear frequency f by the equation

$$\omega = 2\pi f$$

and ϕ is the phase angle. Here, the current leads the voltage by a phase angle ϕ. The X-axis is marked as the real axis and the Y-axis as the imaginary axis. The angle is measured by the rotation of a line in the counterclockwise direction, starting from the X-axis along the right direction. The zero angle (0°), $\pi/2$ angle (90°), π angle (180°), $3\pi/2$ angle (270°), and 2π angle (360°) are specified along the X-axis toward the right, along the Y-axis upward, along the X-axis toward the left, along the Y-axis downward and along the X-axis toward the right, respectively. The voltage wave is represented vectorially. This is accomplished by drawing a directed line segment. The magnitude of the voltage vector is taken to be equal to the peak value of the voltage wave (V_0). The direction of the voltage vector is shown by pointing it along a line subtending an angle ωt with the X-axis. The projection of this voltage vector on the Y-axis is the instantaneous value of the voltage at time t, i.e., $V(t)$. In analogy to voltage, the current wave is represented by drawing a vector of length equal to the peak value of current (I_0) in a direction subtending an angle ($\omega t + \phi$) with the X-axis. This angle is ($\omega t + \phi$) because the current leads the voltage by a phase angle ϕ in the given example. The projection of the current vector on the Y-axis gives the instantaneous value of current at time t, i.e., $I(t)$.

where $x = V$ or I, $X_0 = V_0$ or I_0. In the phasor format, we write

$$\mathbf{X} = X_0 \exp(j\phi) = X_0(\cos\phi + j\sin\phi) \tag{7.10}$$

and

$$\mathbf{X} = X_0 \angle \phi \tag{7.11}$$

Considering root-mean-square values, a scale factor of $1/\sqrt{2}$ is applied to give

$$\mathbf{X} = \frac{X_0}{\sqrt{2}}\exp(j\phi) = \frac{X_0}{\sqrt{2}}(\cos\phi + j\sin\phi) \tag{7.12}$$

and

$$\mathbf{X} = \frac{X_0}{\sqrt{2}} \angle \phi \tag{7.13}$$

from eqs. (7.10) and (7.11).

7.5.1.3 Drawing Phasor Diagrams

7.5.1.3.1 Resistive Circuit

Figure 7.11 depicts the phasor diagram of a resistive circuit.

7.5.1.3.2 Capacitive Circuit

Figure 7.12 illustrates the phasor diagram of a resistive circuit.

7.5.1.3.3 Inductive circuit

The phasor diagram of an inductive circuit is drawn in Figure 7.13.

7.5.2 Synchrophasor

Synchrophasors are time-synchronized phasors, i.e., accurately measured phasor values, representing the magnitude and phase angle of electrical sine waves, occurring at the same time at different locations. A synchrophasor is a phasor determined by universal time coordinated (UTC), which is the atomic clocks-based global standard. It is the standard with reference to which all the clocks across the world are regulated. The UTC time is also called Greenwich mean time (Figure 7.14). The phase angle is 0° when the positive maximum value of the sine wave is at the zero point, the start of UTC second (Figure 7.14(a)). It is −90° when the positive zero crossing of the sine wave lies at the start of UTC second (Figure 7.14(b)).

Phasor measurements from a wide region are synchronized (time-aligned) against the UTC time using the global positioning system (GPS) as the time source because GPS provides continuous global coverage for military and civil applications.

$$\text{GPS time} = \text{UTC time} + 18\text{s} \tag{7.14}$$

By time-tagged and synchronized phasor measurements from geographically dispersed locations of the grid, which may be hundreds to thousands of kilometers apart, it is possible to compare and

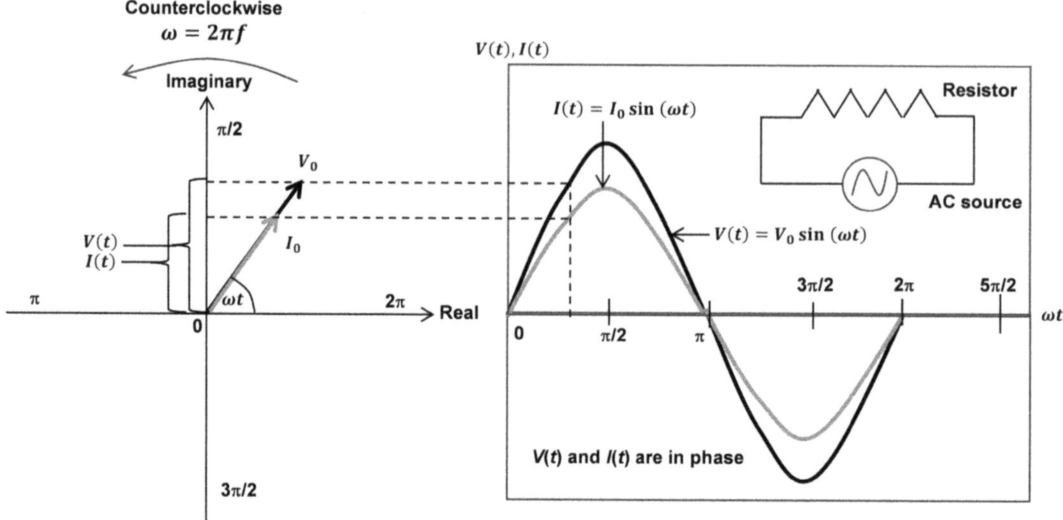

FIGURE 7.11 Phasor diagram of a resistor circuit with AC source. Right side: A resistor is connected across an AC source (inset). Instantaneous voltage $V(t)$ and instantaneous current $I(t)$ are plotted along the Y-axis. The angle ωt is plotted along the X-axis. The 0, $\pi/2$, π, $3\pi/2$, 2π, and $5\pi/2$ angles are marked. The voltage and the current are in phase, and the phase angle $\phi = 0$. The variation of voltage and current with angle are represented by drawing the sine waves

$$V\left(t\right) = V_0 \sin\left(\omega t\right)$$

$$I\left(t\right) = I_0 \sin\left(\omega t\right)$$

Left side: The X-axis (real axis), Y-axis (imaginary axis), and the angles 0, $\pi/2$, π, $3\pi/2$, 2π are marked. The counterclockwise-directed arrow shows the rotational direction of the phasor. The angular frequency

$$\omega = 2\pi f$$

is written. The voltage wave is represented by drawing a vector of length equal to the peak value of voltage (V_0) in a direction subtending an angle ωt with the X-axis. The projection of this voltage vector on the Y-axis is the instantaneous value of the voltage at time t, i.e., $V(t)$. Similarly, the current wave is represented by drawing a vector of length equal to the peak value of current (I_0) in a direction subtending an angle ωt with the X-axis because the current and the voltage are in phase, and the phase angle $\phi = 0$. The projection of the current vector on the Y-axis gives the instantaneous value of current at time t, i.e., $I(t)$.

analyze phasors from different places on the same phasor diagram. Hence, synchrophasor measurements are prime enablers for real-time acquisition of the status of the power grid. The information about phasors of voltages and currents, time-stamped to UTC time, presents a comprehensive overview of a complete power distribution network from which the behavior of the vast network is easily understood, and necessary corrective actions are initiated. When the power demanded from the grid and the power supplied by it are mismatched, frequency imbalances take place causing interruption of service.

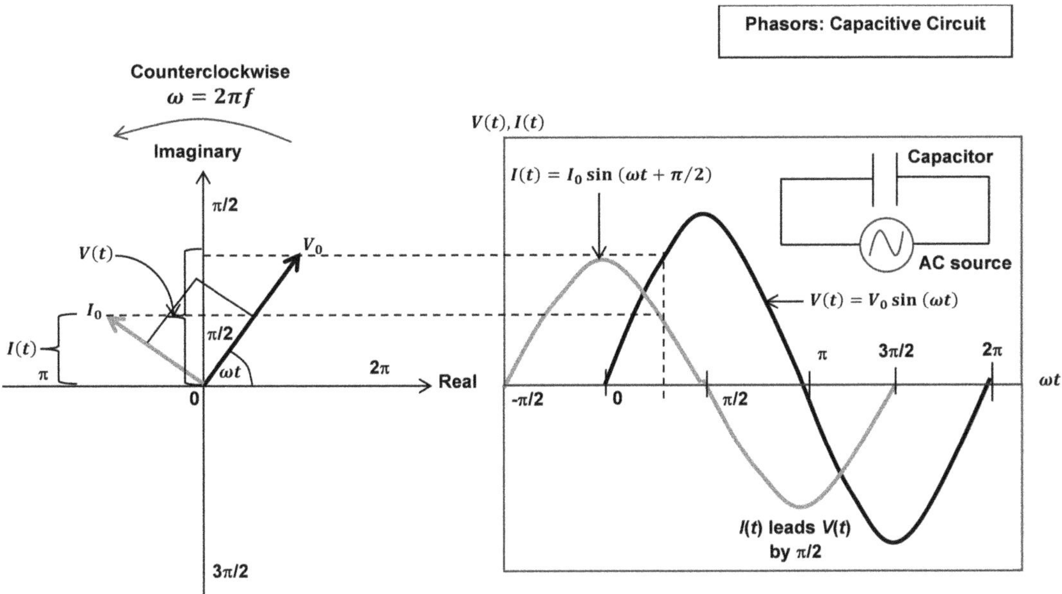

FIGURE 7.12 Phasor diagram of a capacitor circuit with AC source. Right side: A capacitor is connected across an AC source (inset). Instantaneous voltage $V(t)$ and instantaneous current $I(t)$ are plotted along the Y-axis. The angle ωt is plotted along the X-axis. The 0, $\pi/2$, π, $3\pi/2$, 2π, $5\pi/2$ angles are marked. The current leads the voltage in phase and the phase angle $\phi = \pi/2$. The variation of voltage and current with angle are represented by drawing the sine waves

$$V(t) = V_0 \sin(\omega t)$$

$$I(t) = I_0 \sin(\omega t + \pi/2)$$

Left side: The X-axis (real axis), Y-axis (imaginary axis), and the angles 0, $\pi/2$, π, $3\pi/2$, and 2π are marked. The counterclockwise-directed arrow shows the rotational direction of the phasor. The angular frequency

$$\omega = 2\pi f$$

is written. The voltage wave is represented by drawing a vector of length equal to the peak value of voltage (V_0) in a direction subtending an angle ωt with the X-axis. The projection of this voltage vector on the Y-axis is the instantaneous value of the voltage at time t, i.e., $V(t)$. Similarly, the current wave is represented by drawing a vector of length equal to the peak value of current (I_0) in a direction subtending an angle ($\omega t + \pi/2$) with the X-axis because the current leads the voltage in phase and the phase angle $\phi = \pi/2$. The projection of the current vector on the Y-axis gives the instantaneous value of current at time t, i.e., $I(t)$.

7.5.3 MEASUREMENT OF SYNCHROPHASORS

7.5.3.1 Phasor Measurement Units

A PMU can be considered as a high-speed sensor for generating synchronized phasor, frequency, and rate of change of frequency data using voltage or current signals, or both voltage and current signals, along with a signal for time synchronization (Pinte et al. 2014). It is installed at the critical substations of the power network to perform measurements on voltage and current waveforms. Why is synchronization emphasized? The reason is that PMU is the key component of Wide Area

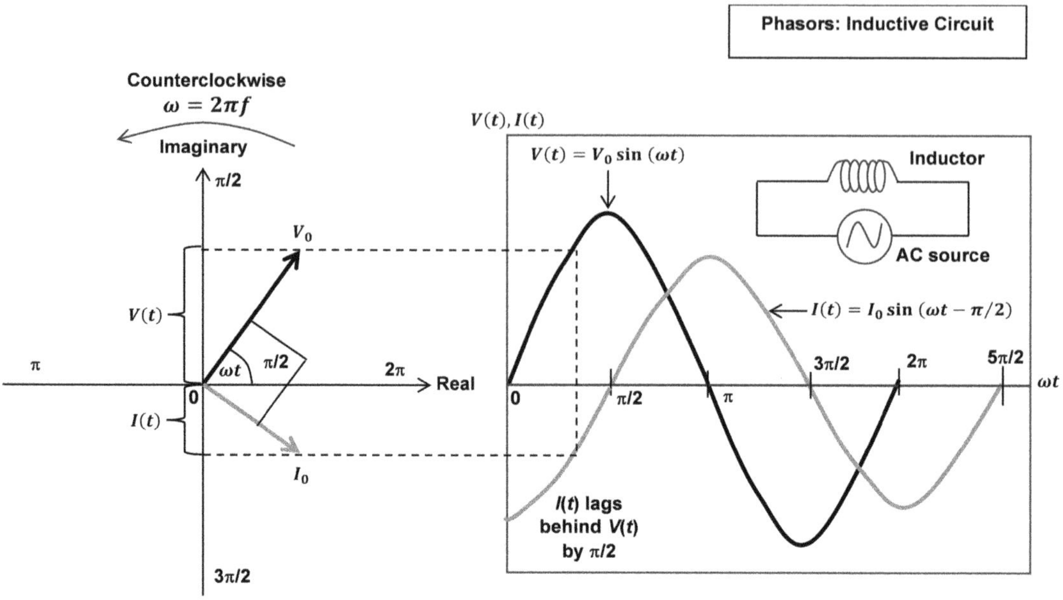

FIGURE 7.13 Phasor diagram of an inductor circuit with AC source. Right side: An inductor (e.g., a coil or choke) is connected across an AC source (inset). Instantaneous voltage $V(t)$ and instantaneous current $I(t)$ are plotted along the Y-axis. The angle ωt is plotted along the X-axis. The 0, $\pi/2$, π, $3\pi/2$, 2π, and $5\pi/2$ angles are marked. The current is delayed behind the voltage in phase, and phase angle $\phi = \pi/2$. The variation of voltage and current with angle are represented by drawing the sine waves

$$V(t) = V_0 \sin(\omega t)$$

$$I(t) = I_0 \sin(\omega t - \pi/2)$$

Left side: The X-axis (real axis), Y-axis (imaginary axis), and the 0, $\pi/2$, π, $3\pi/2$, and 2π angles are marked. The counterclockwise-directed arrow shows the rotational direction of the phasor. The angular frequency

$$\omega = 2\pi f$$

is written. The voltage wave is represented by drawing a vector of length equal to the peak value of voltage (V_0) in a direction subtending an angle ωt with the X-axis. The projection of this voltage vector on the Y-axis is the instantaneous value of the voltage at time t, i.e., $V(t)$. Similarly, the current wave is represented by drawing a vector of length equal to the peak value of current (I_0) in a direction subtending an angle ($\omega t - \pi/2$) with the X-axis because the current is delayed behind the voltage in phase, and the phase angle $\phi = \pi/2$. The projection of the current vector on the Y-axis gives the instantaneous value of current at time t, i.e., $I(t)$.

Monitoring Systems (WAMSs). In a WAMS, several PMUs are installed at the main critical substations of the power network.

From the waveform data, the 50 or 60 Hz phasor components are calculated as complex numbers showing the magnitude and phase of the signals. The phasors obtained from real-time measurements at multiple remote locations of a power grid are synchrophasors because they have been acquired across the power grid with a common timing reference. The voltage and current phasors received from the various substations of the grid can be displayed on the same phasor diagram for comparison and evaluation. In this way, synchrophasors are indicators of power quality in different portions of the grid, giving early warnings about any deteriorations, which may potentially result in major faults (Phadke and Bi 2018).

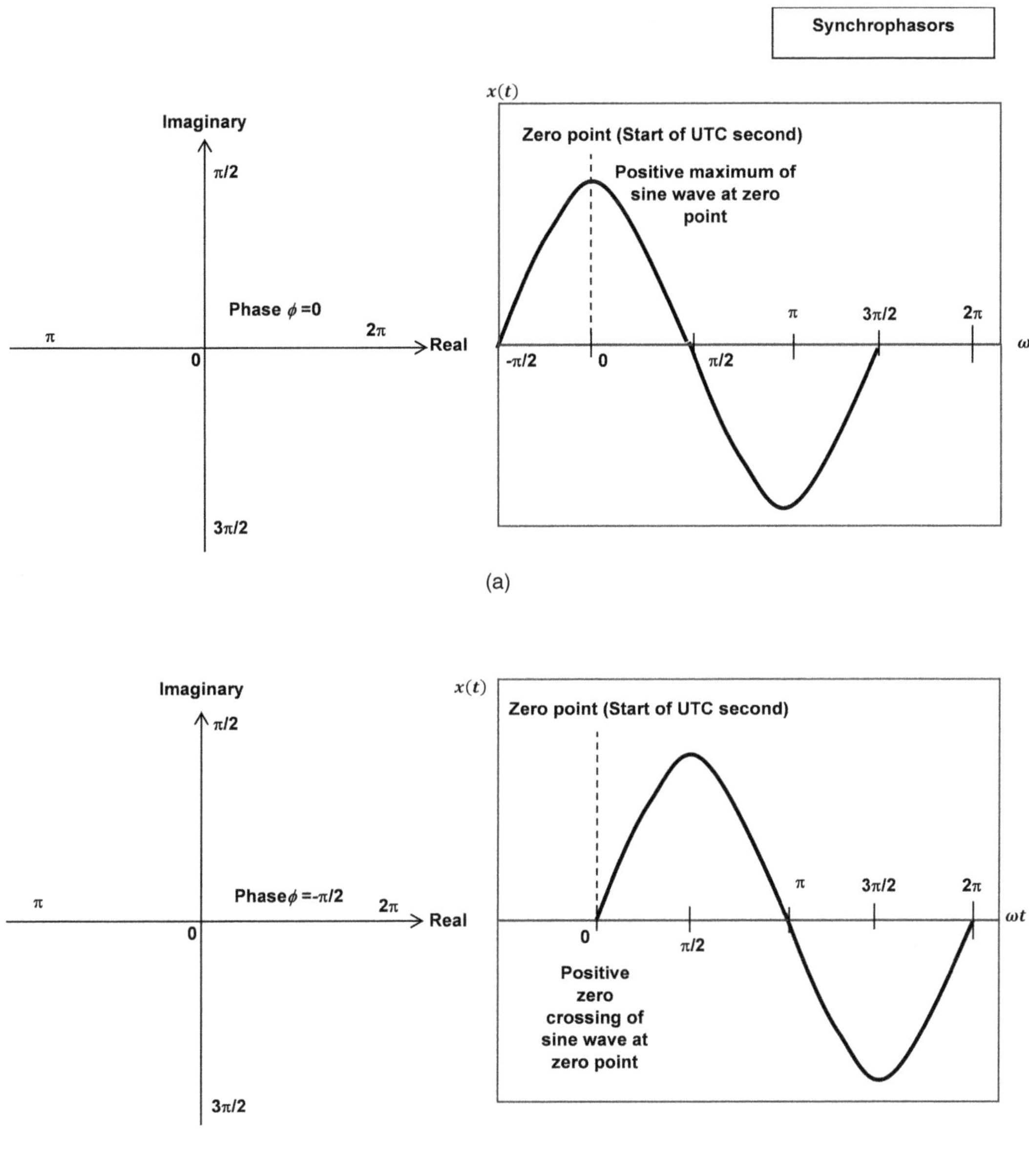

FIGURE 7.14 Synchrophasor diagrams when: (a) the positive maximum of the sine wave is at the zero point, and (b) the positive zero crossing of the sine wave is at the zero point. Part (a): Right side: Instantaneous values $x(t)$ of a sinusoidally varying quantity, voltage, or current, are plotted along the Y-axis. The angle ωt is plotted along the X-axis. The angles $-\pi/2$, 0, $\pi/2$, π, $3\pi/2$, and 2π are marked. The zero point is marked as the start of the UTC second. The positive maximum of the sine wave is located at the zero point. Left side: The X-axis (real axis), Y-axis (imaginary axis), and the 0, $\pi/2$, π, $3\pi/2$, and 2π angles are marked. The phase $\phi=0$ for the positive maximum of the sine wave at the zero point, as shown on the right side. Part (b): Right side: Instantaneous values $x(t)$ of a sinusoidally varying quantity, voltage, or current, are plotted along the Y-axis. The angle ωt is plotted along the X-axis. The angles $-\pi/2$, 0, $\pi/2$, π, $3\pi/2$, and 2π are marked. The zero point is marked as the start of the UTC second. The positive zero crossing of the sine wave is located at the zero point. Left side: The X-axis (real axis), Y-axis (imaginary axis), and the 0, $\pi/2$, π, $3\pi/2$, and 2π angles are marked. The phase $\phi = -\pi/2$ when the positive zero crossing of the sine wave is at the zero point, as shown on the right side.

7.5.3.2 Parts of a Phasor Measurement Unit

The main parts of a PMU (Gharavi and Hu 2017) are illustrated in Figure 7.15:

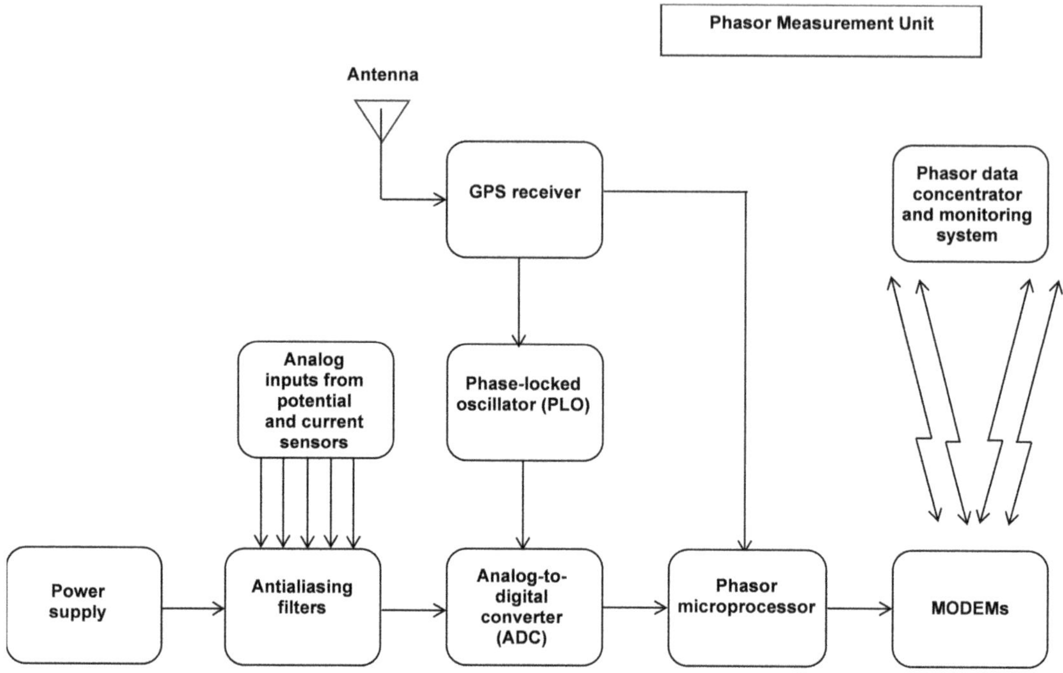

FIGURE 7.15 Block diagram of the phasor measurement unit. Analog inputs from voltage and current sensors are fed to anti-aliasing filters. From the filters, they move to A-to-D converters for digitization. A GPS receiver with an antenna is connected to a phase-locked oscillator and phasor microprocessor to provide the sampling clock for the synchronization of the signals. The phase-locked oscillator produces a stable frequency signal. The phasor microprocessor applies discrete Fourier transform (DFT) methods for calculating the positive-sequence values of voltages and currents. The MODEM modulates the digital signals into analog signals and demodulates the received signals to get the digital data. From the MODEM, the signals are transmitted to the phasor data concentrator and monitoring system.

(i) Voltage and Current Sensors: Voltage and current sense transformers in the substation circuit continuously record voltage and current values.

(ii) Antialiasing Filter: The analog waveforms from the secondary windings of the sense transformers are restricted in bandwidth by passing through an anti-aliasing filter circuit. This is a low-pass filter placed before the signal sampler to remove the spectral content whose frequency is $\geq \frac{1}{2}$ Nyquist rate, for satisfying the Nyquist-Shannon sampling theorem during analog-to-digital conversion. According to this theorem, the unambiguous determination of an analog signal, $s(t)$, containing no frequencies $\geq f$, is achieved by specification of its discrete values produced by sampling at a series of points that are spaced $1/2f$ seconds apart. Nyquist rate is the sampling rate of $2f$ samples per second.

(iii) Analog-to-Digital Converter: This converter takes a snapshot of an analog voltage or current at an instant of time, and yields the digital output code for this analog voltage or current.

(iv) Phase-Locked Oscillator Circuit, and GPS Receiver: An oscillator in which the phase of the output signal is locked to a reference signal is termed a phase-locked oscillator. It works by detecting the phase difference between the oscillator output and a reference signal and controlling the frequency of the oscillator with respect to this phase difference. It thus acts as a fixed-frequency synthesizer, generating a stable output frequency.

The GPS is a satellite-based system providing services for determining position and performing exact time measurements. It works as a reference source with the phase-locked oscillator during synchronized sampling. For synchronization of the signals, the sampling clock provided by the GPS receiver is used to time tag the signals. The GPS time synchronization is obtained via the GPS antenna. See Chapter 10, Sec 10.3 for a detailed description of GPS.

High-speed synchronized sampling is achievable with 1μs accuracy. The sampling clock is phase-locked to one pulse per second (PPS) to provide a common and correct timing referred to as UTC. The purpose of sampling the electrical waveforms using GPS is to clarify doubts about phase and amplitude because these parameters impact the fundamental frequency, which is normally 50 or 60 Hz. The 50/60 Hz AC waveforms can be measured at the typical rate of 48 samples per cycle.

Another method for synchronization of signals is to use the IEEE 1588 synchronization module. Better than 1μs timing accuracy is endowed to devices linked through a network, e.g., Ethernet, which is suggested by IEC 61850-9-2 Ethernet as a favored communication supporting the Precision Time Protocol (PTP).

(v) Phasor Microprocessor: It calculates the positive-sequence values of all the voltages and currents using DFT techniques. DFT methods are widely used for analyzing the fundamental components and harmonics of these signals. The positive sequence components of a balanced power system are sets of three phasors of equal magnitude, 120° apart in phase, and having the same phase sequence as the original phasors. If the magnitude equality and/or 120° phase-shift conditions are not met, the system is unbalanced. The positive sequence components control the production of the rotating magnetic field for the proper functioning of motors and generators. They play a pivotal role in the characterization of unbalancing in three-phase power systems. Unbalanced three-phase motors and other loads show poor performance, and fail prematurely.

(vi) MODEM: The MODEM modulates digital signals into analog signals by encoding the digital data into a carrier signal for transmission over the analog communication channels through radio frequencies or a cable network to the phasor data concentrator (PDC). Upon reception, the signal is demodulated to reconstruct the original digital data.

(vii) Phasor Data Concentrator and Monitoring System: The PDC is a critical link between the PMUs and the applications that utilize the data generated by the PMUs in a network. The data received by the phasor data concentrator (PDC) is time-stamped by the PMU with a time referenced to UTC. It is the absolute time because it is the same irrespective of the place at which we are present in the world. In the PDC, the phasor data from a large number of PMUs is aggregated, and the combined data are transmitted to systems for data archival, visualization, or control. Real-time synchronized comparison of voltage and current data is done to assess system conditions, notably, changes in frequency, MW power, MVARs power, kVolt voltages, etc., for preselected network sites to understand swings in grid stability. The analysis offers a framework for controlling the flow of energy from multiple sources, e.g., coal, wind, nuclear, and others.

7.5.3.3 Applications of PMUs

(i) Automation: PMU is used for automated management of the power grid by controlling voltages, currents, frequencies, and loads.

(ii) Power Outage Detection and Localization: Frequency measurements of high-resolution synchrophasor data are applied for detecting line outages. The location of an outage is pinpointed from the redistribution of active power flow that happens when a line outage occurs (Deng et al. 2019).

(a) Line Outage Detection: The PMU-measured frequency data are analyzed by the detrending method. In statistics, detrending means the manipulation of a data set to show only the differences in values from the trend. This is accomplished by highlighting short-term changes through the removal of the effects of long-term trends. The method consists of finding an approximation to the data set and subtracting it from the original signal.

First, the erratic spikes and high-frequency noise are eliminated from the PMU-measured frequency data by passing the frequency data through a moving median filter. The moving median is a method of calculation of the median value of a subset of data points falling within a specified window or period. The name 'moving median' comes from the fact that the window or period moves from one data point to another.

The resulting filtered frequency data is then passed through a moving mean filter for obtaining the frequency trend data. A moving mean is a calculation of a series of mean values of different subsets of a full data set. For a given series of numbers and a fixed subset size, the first element of the moving mean is the mean value of the original fixed subset of the number series. After finding the first element, a new fixed subset is taken in which the first number of the series is excluded and the next value in the subset is included.

The de-trended frequency data is determined by subtracting the frequency trend data from the filtered frequency data. By statistically analyzing past frequency data, two thresholds are defined. The line outage incident is prompted using the evaluation of the threshold. The time of the incident is found from the GPS timestamp on the frequency data. This time is applied for locating the spot of the incident through active power information.

(b) Outage Location Determination: Immediately after detecting that a line outage has taken place, the active power data are passed through a median filter to remove the noise. Using the timestamp of the event detected, the difference between active power before outage and that after outage is found from the active power data. The active power changes on the transmission lines under consideration are ranked. The outage location is spotted with maximum value.

(iii) Fault Diagnosis: PMU data give a clear picture of the phenomena taking place in a power system. Fault diagnosis is approached through different routes. In one method, a detection algorithm based on voltage threshold value is applied to the positive and zero sequence voltages from PMUs to distinguish between power system disturbances and faults (Jain et al. 2018). After fault detection, the type of fault is categorized by a Support Vector Machine, a supervised learning algorithm used for classification and regression analysis. In another method, a fault detection algorithm based on equivalent power factor angle (EPFA) is formulated (Vishal and Shenil 2021). The algorithm helps in understanding the presence of fault, utilizing which backup protection algorithms can be designed using PMU data. In a third method, the PMU data are transformed into polar plots, and a fault analysis method is developed using the polar PMU data plots with convolutional neural networks (CNNs) (Zhang et al. 2022).

(iv) Congestion Management and Operations Planning: This is done by congestion analysis, and planning operations for the day and hour ahead. Power congestion occurs when the transmission lines are insufficient to transfer the required power to fulfill the customer demands, e.g., due to changes in demand, generator breakdown, and transmission line faults (Pillay et al. 2015). The PMU-gathered information can be applied for accurate calculations of path limits and path flows, thus providing more freedom in managing congested paths (Novosel and Vu 2006).

7.6 DYNAMIC LINE RATING: CONDUCTOR AND AMBIENCE SENSORS

The maximum current value which a conductor can safely carry is termed its ampacity. Static thermal ratings of transmission and distribution conductors are estimated on the expected worst-case local meteorological conditions, giving their theoretical ampacity. However, the thermal capacity of a conductor is higher than the conservative theoretical estimates, as assessed from presumed heating considerations, and varies in certain favorable situations. Under high winds, the conductors cool down more rapidly than in regular conditions. Then the ampacity of conductors increases. Similarly, when the weather is cool and the atmospheric temperature drops to low values, more electricity flow is permitted in the conductors without causing any damage. In both conditions, it is likely that there is unused capability of overhead transmission lines if static ratings are applied (IRENA 2020).

DLR, also called real-time thermal rating (RTTR), is a blanket term for an assortment of technologies and methodologies concerned with varying the thermal capacity of overhead transmission lines dynamically with the help of actual sensor data about the environment in the situation at hand. It allows the integration of increased renewable energy sources such as solar and wind power in the grid. It helps in the optimization of existing grid infrastructure and deferment of expensive upgradation. It reduces congestion on transmission lines and minimizes power shutdown caused by electrical overloading.

The information collected by a team of sensors, e.g., conductor temperature, line angle, ambient temperature, and wind speed sensors, is transmitted via a 920 MHz wireless transmitter to a cloud platform or centralized control unit where the data are fed to advanced software algorithms to determine the current capacity of the line. The DLR software takes into account the type of conductor and the environmental conditions of the terrain through which the line passes for utilization of the line to its full capacity without compromising safety.

7.6.1 DLR Sensors

The sensors used for DLR are (Dynamic line rating 2019) as follows.

Conductor Temperature: Thermocouples or thermistors are fixed on the conductors for direct line measurements. For measurements from the ground, infrared thermometers or infrared cameras are used.

Conductor Tension: The tension in the conductor is measured by a tension monitor which is attached to the conductor to determine the mechanical force between the line and the structure.

Conductor Sagging: Ground-based sensors include cameras loaded with image processing software to find the distance through which the line has sagged. Line-mounted sensors directly measure the inclination and vibration of the conductor.

Conductor Clearance from the Ground: Ground-based sensors are electric field measuring devices to measure the electric field emitted by the line. Line-mounted sensors are SONAR (sound detection and ranging) and LIDAR (light detection and ranging) devices to measure the distance between the line and the ground.

Weather Sensors: Wind speed and direction sensors are used along with temperature and solar radiation measuring devices.

7.6.2 Wind Speed and Direction Sensors

7.6.2.1 Three-Cup Anemometer for Wind Speed Measurement

It consists of a vertical rod at the top of which three horizontal hemispherical or half-cone carbon fiber cups are attached by arms (Figure 7.16). The cups rotate about the vertical rod when the wind blows. The higher the wind speed, the faster the cup rotation. As the cups rotate, the vertical axial rod to which they are connected also rotates. The rod is connected to a multi-tooth rotor and slit-optocoupler assembly. Hence, pulse signals are produced by the optocoupler in accordance with the rotation of the vertical rod. The number of pulses is proportional to the wind speed, which is calculated in meters per second from the pulse signal frequency.

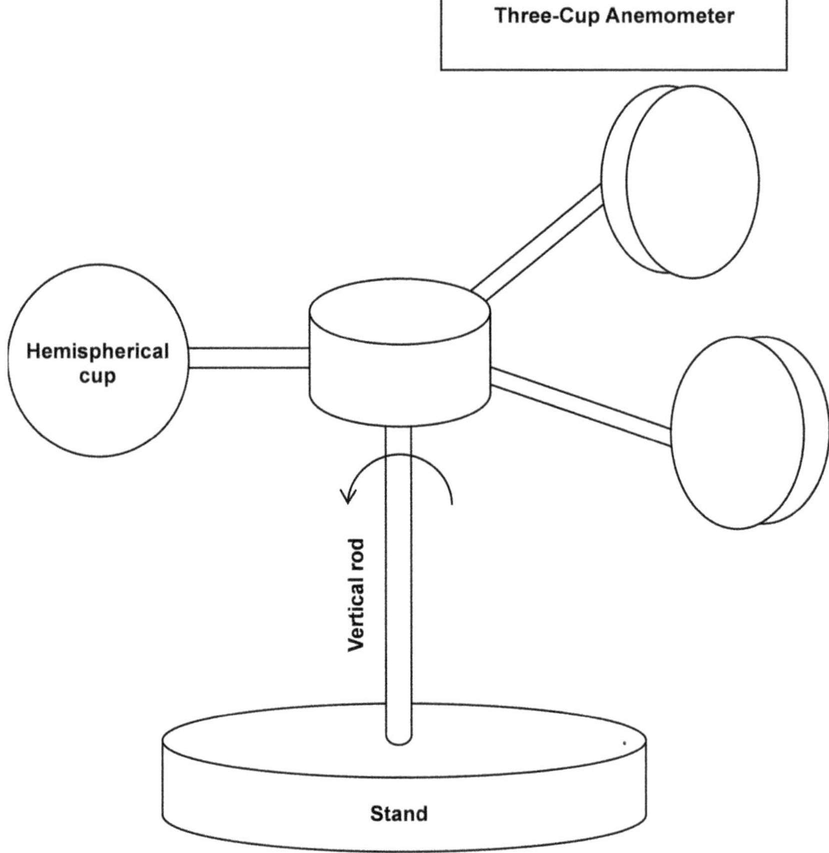

FIGURE 7.16 Three-cup anemometer. A vertical rod mounted on a stand can rotate freely about its axis. Three arms project horizontally from the top end of the rod. Hemispherical cups are attached at the ends of these horizontally protruding arms from the rod. The impact of the blowing wind presses against the cups. Hence, the cups are pushed by the wind to move in its direction. As a result, the rod rotates about the vertical axis at a speed dependent on the wind velocity.

7.6.2.2 Wind Vane Sensor for Finding Wind Direction

It consists of a vertical spindle mounted on bearings for moving freely (Figure 7.17). The spindle is fitted with a horizontal arm. At one end of the horizontal arm, a vertical tailfin is attached. The opposite end of the horizontal arm carries a balance weight in the shape of a pointer to indicate the wind direction. The vane is mounted on an elevated pole. It rotates under the influence of the wind, and the direction in which the pointer of the wind vane is aimed shows the direction of the blowing wind in degrees from true north.

7.6.2.3 Wind Speed Monitoring and Warning System

Using the wind sensor together with weather forecast information, a wind speed monitoring and early warning system is developed (Mardiyono et al. 2021). The system is made of hardware, software, notification, and display units. The hardware unit consists of an anemometer sensor converting the blade revolution into a voltage signal, a microprocessor, and the embedded program. The software unit is loaded with a program for wind speed calculation and for determining the warning thresholds; and the weather harvester program. The notification and display unit produces the alarm signal, displays the wind speed, and issues the warning message.

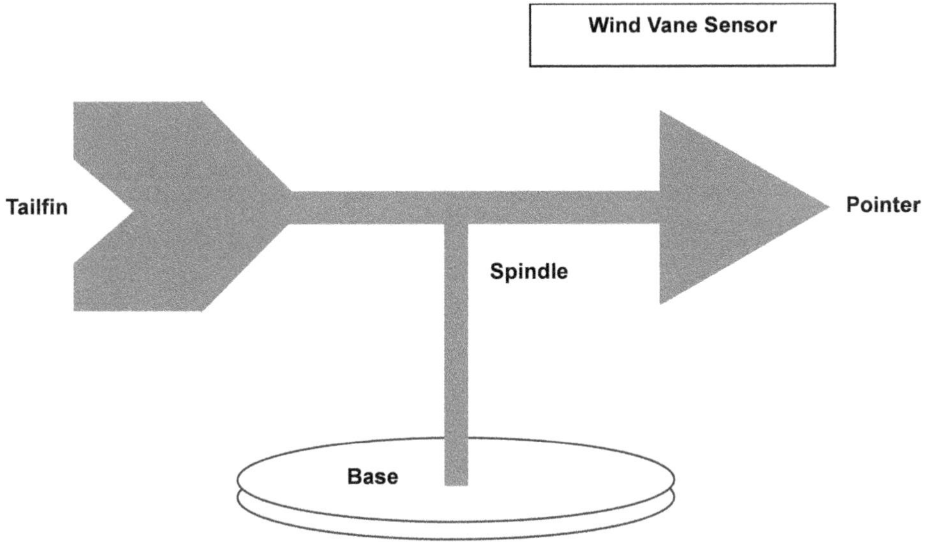

FIGURE 7.17 Wind vane sensor. A vertical spindle with a horizontal arm is mounted on a base and is free to move around its axis. At one end of the horizontal arm, a vertical tailfin is fixed. At the other end of the horizontal arm, a balance weight in the shape of a pointer is attached.

7.6.2.4 Ultrasonic Anemometer

The ultrasonic anemometer (Figure 7.18) consists of four ultrasonic sensors (Kuncara et al. 2020). Each ultrasonic sensor is made of a transmitter and a receiver. The ultrasonic sensors are configured in two mutually perpendicular directions. One pair of ultrasonic sensors is positioned along the east–west direction (X-axis) at a known distance apart. The other pair of ultrasonic sensors is placed along the north–south direction (Y-axis) at the same separation distance. The ultrasonic waves sent at 40 kHz by the transmitter of the sensor placed on the eastern side propagate through the air and fall on the receiver of the sensor on the western side. Similarly, the ultrasonic waves sent by the transmitter of the sensor placed along the west are incident on the receiver of the sensor placed at the eastern point. The same remarks apply to sensors at the northern and southern points (Abichandani et al. 2020).

When no wind is blowing, the time t_{EW} taken by ultrasonic waves sent from the transmitter at the east to reach the receiver at west = the time taken t_{WE} by ultrasonic waves sent from the transmitter at the west to reach the receiver at east= t. But when the wind is blowing in the east-to-west direction, the time $t_{\mathrm{EW}} < t$ because the ultrasonic waves are assisted by the wind. Also, the time $t_{\mathrm{WE}} > t$ because the ultrasonic waves are opposed by the wind. This results in a time-of-flight difference = $t_{\mathrm{WE}} - t_{\mathrm{EW}}$, which depends on the wind speed. In the same way, the ultrasonic wave propagation along the north–south and south–north directions are influenced by wind speed. As the distance between the sensors is prefixed, the wind velocities v_x and v_y along X- and Y- directions, i.e., east–west and north–south directions, are determined from time measurements using the ultrasonic sensor arrangement. Then the resultant wind velocity is given by the equation

$$v = \sqrt{v_x{}^2 + v_y{}^2} \tag{7.15}$$

and the wind direction by the angle

$$\theta = \tan^{-1}\left(\frac{v_y}{v_x}\right) \tag{7.16}$$

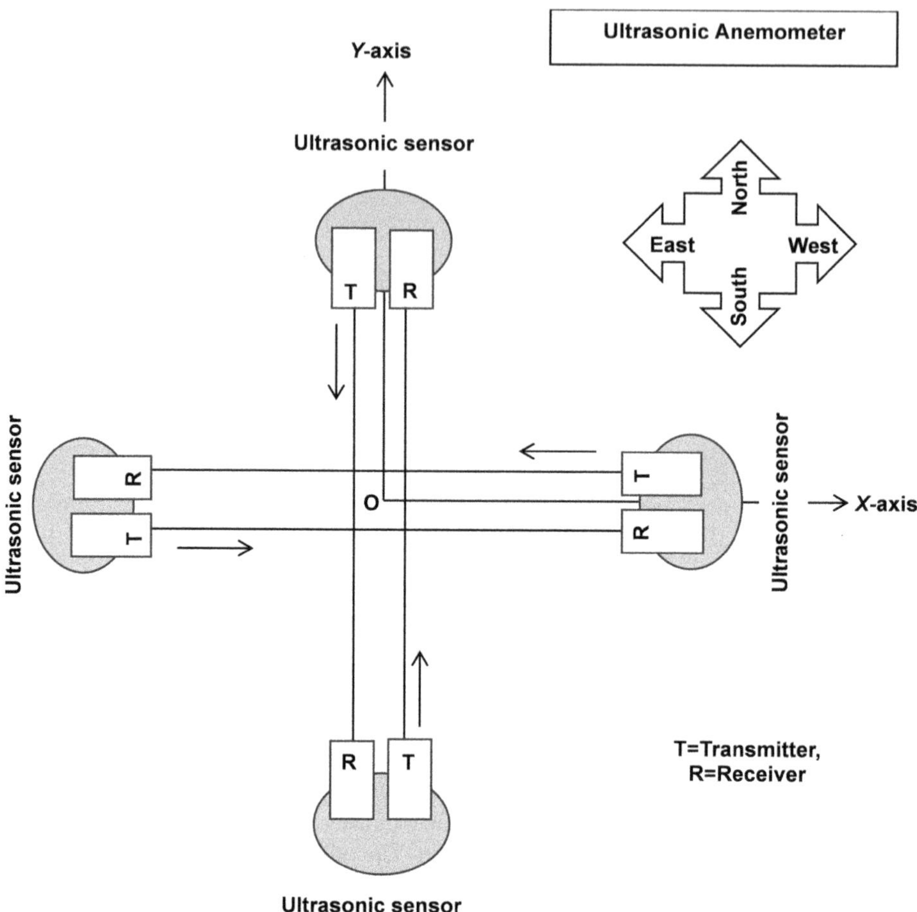

FIGURE 7.18 Ultrasonic anemometer. The origin O, and *X*- and *Y*-axis directions are marked. The north–south–east–west directions are indicated. Four ultrasonic sensors are taken, each having a transmitter and a receiver in one package. They are organized in two pairs, with one pair mounted in the east–west direction, and another pair in the north–south direction. In each pair, the transmitter of one sensor is placed exactly opposite to and facing the receiver of the other sensor.

The sensor is equipped with a data processing unit using a microcontroller. The sensor output is passed through a Kalman filter (linear quadratic estmation algorithm) for noise removal. The results are stored in a secured digital (SD) card. An OLED (organic light-emitting diode) displays the results. They are integrated into the IoT with the Blynk application on a smartphone through the internet. The wind speed is measured with a resolution of ~0.1 m/s and the measurement error is 0.14 m/s. The wind direction is measured at a resolution of 1° and the measurement error is 7° (Kuncara et al. 2020).

7.6.3 Solar Irradiance Sensors

Solar irradiance is the surface power density of electromagnetic radiation received from the sun, i.e., power per unit area expressed in Wm^{-2}. Solar radiation is measured by two types of sensors: pyranometer and pyrheliometer (Widén and Munkhammar 2019).

7.6.3.1 Pyranometer

It measures the global solar radiation, the direct plus diffused solar radiation received from the hemisphere above the horizontal plane of the pyranometer. It is equally responsive to radiation from all

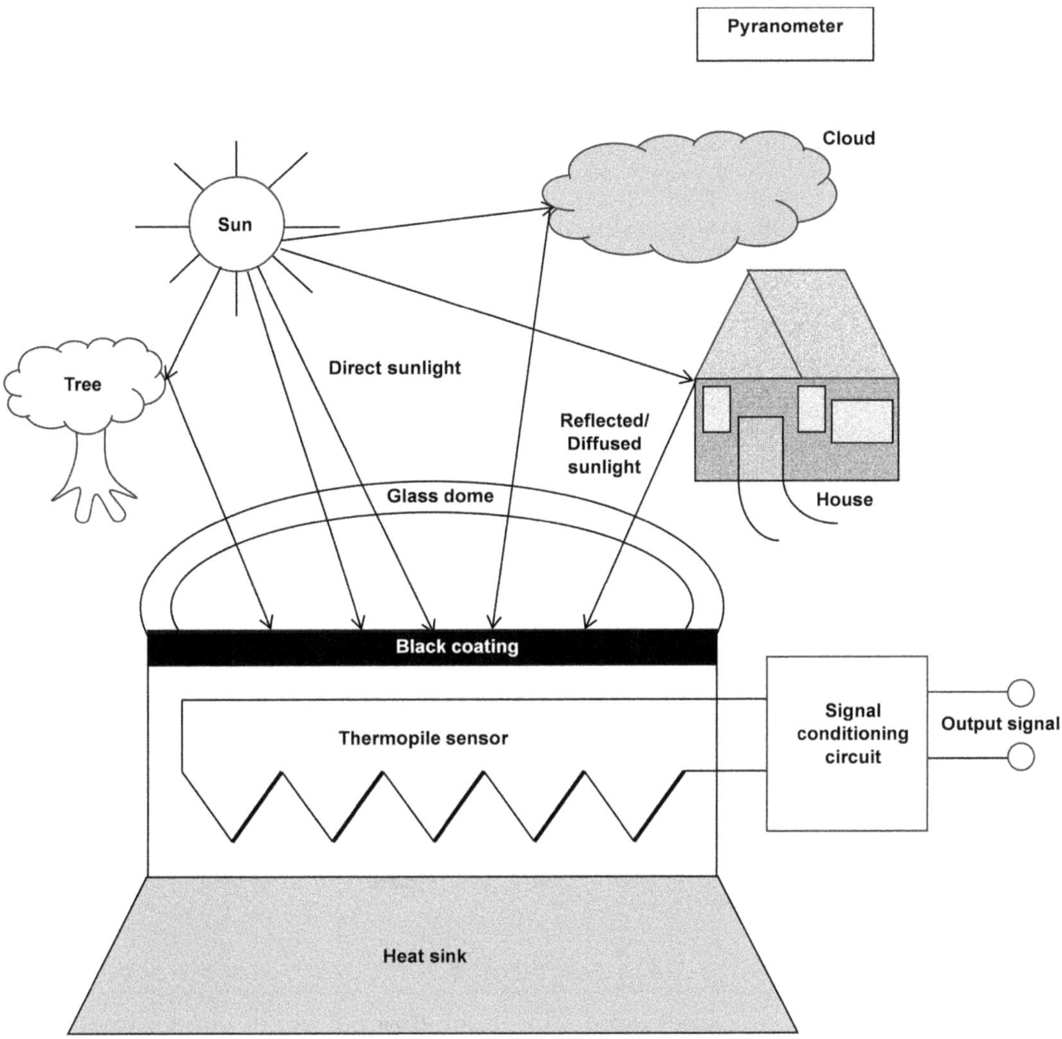

FIGURE 7.19 Pyranometer. A thermopile sensor is shown with the top surface coated with a black film and a heat sink placed below the bottom surface. A glass dome covers the black-coated surface of the sensor. Sunlight falling on the sensor consists of direct sunlight and diffused/reflected sunlight from clouds, houses, and trees. The signal generated by the thermopile is supplied to the signal conditioning circuit, which delivers the output signal in the desired format.

directions within this hemisphere. It consists of a horizontally oriented thermopile sensor with a black coating, mounted on a planar surface, and covered by a glass dome for restricting the wavelength range (300–2,800 nm), preserving the 180° angular view, and also for protecting the thermopile from air convection (Figure 7.19). The output signal ~ a few tens of millivolts is fed to a signal conditioning circuit, yielding an output of 0–1 V.

7.6.3.2 Pyrheliometer

It measures the radiation directly coming from the sun (together with a small portion of the sky around the sun) at normal incidence, giving direct normal irradiance. It is fixed on a pivot for rotation around two axes, the zenith and the azimuth, designed for solar tracking (Figure 7.20). The solar tracking keeps its direction pointing toward the sun throughout the day. The light from the sun enters the pyrheliometer through a quartz window. This window acts as a filter allowing entry to radiation only in the wavelength range 200–4,000 nm {near-infrared + visible + UV(A) + UV(B)}. The window is the

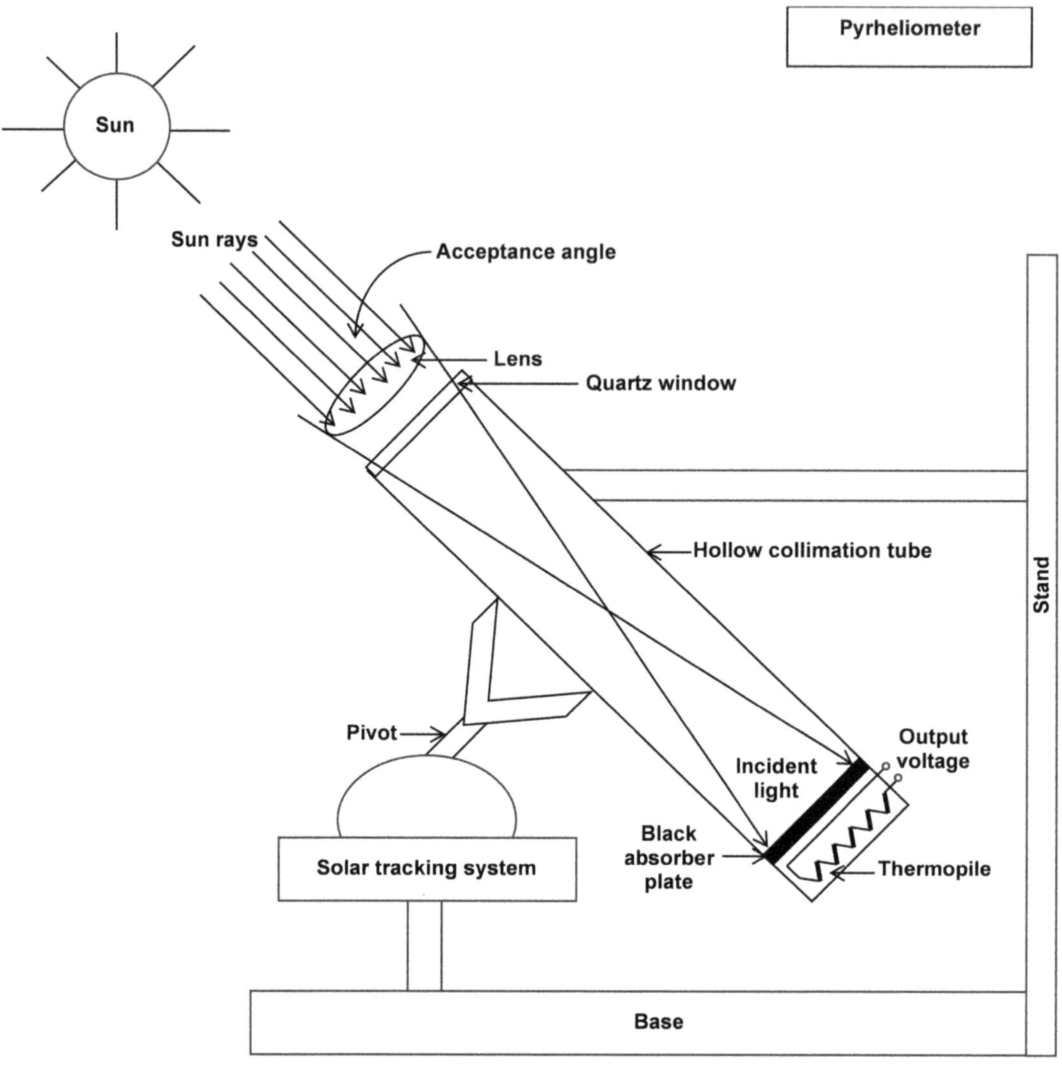

FIGURE 7.20 Pyrheliometer. A supporting stand with a base plate carries a solar tracking system. A hollow collimation tube is mounted on the pivot of this system. The tube is held in its position by the stand. Parallel sun rays received by the collimation tube within an acceptance angle are focused by a lens and enter the collimation tube through a quartz window. These rays fall on the black absorber plate of the thermopile. An output voltage is measured across the thermopile terminals due to illumination with sun rays.

front face of a long collimation tube, which defines the field of view of the instrument. The collimator tube looks like a telescope. The filtered sunlight falls on a black absorber plate, and is converted into heat. A thermopile placed underneath the black absorber plate measures this heat from which the solar energy per unit area per unit time is calculated in Wm^{-2}.

7.7 CONCLUDING REMARKS AND PREPARING FOR THE UPCOMING CHAPTER

We saw that PMUs are high-speed sensors that can capture samples of supply voltage/current waveforms and employ GPS-enabled time synchronization to compare measured data from multiple remote substations. The central supply station is able to know whether the voltage, current, and frequency are within the specified tolerable limits. In case of any deviation from norms, the related mishappening is readily detected, its location is determined and the type of event that occurred is identified. The PMU data provides useful insights into the updated status of the grid. Thus, the grid is made more reliable, and resilient to unexpected power interruptions.

The health of the power grid maintained by controlling its vital functioning parameters, is a high-priority area. But even more crucial is the human welfare and wellbeing to safeguard human life from diseases and injuries. Let us see what sensors can do for human health monitoring in the next chapter.

REFERENCES

Abichandani P., D. Lobo, G. Ford, D. Bucci and M. Kam 2020 Wind measurement and simulation techniques in multi-rotor small unmanned aerial vehicles, *IEEE Access*, 8: pp. 54910–54927.

Analog ICs, Inc. 1999 Energy Metering IC with On-Chip Fault Detection AD7751, Protected by U.S. Patent Nos. 5,745,323; 5,760,617; 5,862,069 and 5,872,469. https://www.analog.com/media/en/technical-documentation/obsolete-data-sheets/1223584ad7751.pdf

Daigle P. 1999 All-electronic power and energy meters, Analog Dialogue 33-2, © 1999 Analog Devices, pp. 1–3.

Deng X., D. Bian, D. Shi, W. Yao, Z. Jiang and Y. Liu 2019 Line outage detection and localization via synchrophasor measurement 2019 *IEEE Innovative Smart Grid Technologies - Asia (ISGT Asia)*, 21–24 May, Chengdu, China, pp. 3373–3378.

Dynamic line Rating, United States Department of Energy, June 2019, pp. 1–27. https://www.energy.gov/oe/articles/dynamic-line-rating-report-congress-june-2019

Gharavi H. and B. Hu 2017 Synchrophasor sensor networks for grid communication and protection, *Proceedings of the IEEE*, 105(7): pp. 1408–1428.

IRENA 2020 Innovation landscape brief: Dynamic line rating, International Renewable Energy Agency, Abu Dhabi, UAE, pp. 1–20.

Jain A., T. C. Archana and M. B. K. Sahoo 2018 A methodology for fault detection and classification using PMU measurements, *20th National Power Systems Conference (NPSC)*, 14–16 December, Tiruchirappalli, India, pp. 1–6.

Kuncara I. A., J. E. Suseno, A. Setyawan and I. Gunadi 2020 Development of ultrasonic anemometer using HC-SR04 with Kalman filter based on microcontroller integrated IoT, *E3S Web of Conferences 202*, 15011, pp. 1–14.

Liu J., Y. Hu, W. Yang, X. Pan, Y. Li, G. Qu, Y. Xie and W. Qiu 2021 The on-line detection technology of oil level in main transformer conservator of an 110kV substation, *Journal of Physics: Conference Series (ICETMS)*, 2076: 012009, pp. 1–7.

Mahanta D. K. and S. Laskar 2015 Power transformer oil-level measurement using multiple fiber optic sensors, *2015 International Conference on Smart Sensors and Application (ICSSA)*, Kuala Lumpur, Malaysia, 26–28 May, IEEE, pp. 102–105.

Mardiyono M., R. E. Sari and O. N. Dini 2021 Wind speed monitoring and alert system using sensor and weather forecast, *International Conference on Innovation in Science and Technology (ICIST 2019)*, IOP Conf. Series: Materials Science and Engineering, 1108, 012029, pp. 1–9.

Markow J. 1999 Microcontroller-based energy metering using the AD7755, *Analog Dialogue*, 33–9: pp. 1–2. https://www.analog.com/media/en/analog-dialogue/volume-33/number-1/articles/microcontroller-based-energy-metering-ad7755.pdf

Mohammad N., A. Barua and M. A. Arafat 2013 A smart prepaid energy metering system to control electricity theft, *International Conference on Power, Energy and Control (ICPEC)*, 6–8 February, Dindigul, India, pp. 562–565.

Novosel D. and K. Vu 2006 Benefits of PMU technology for various applications, International Council on Large Electric Systems-CIGRÉ, Croatian National Committee, *7th Symposium on Power System Management*, Cavtat, 5–8th November, pp. 1–13.

Phadke A. G. and T. Bi 2018 Phasor measurement units, WAMS, and their applications in protection and control of power systems, *Journal of Modern Power Systems and Clean Energy*, 6(4): pp. 619–629.

Pillay A., S. P. Karthikeyan and D.P. Kothari 2015 Congestion management in power systems – A review, *International Journal of Electrical Power & Energy Systems*, 70: pp. 83–90.

Pinte B., M. Quinlan, A. Yoon, K. Reinhard and P. W. Sauer 2014 A one-phase, distribution-level phasor measurement unit for post-event analysis, *Power and Energy Conference*, 28 February–01 March, Illinois (PECI), Champaign, IL, IEEE, pp. 1–7.

Vishal V. P. and P. S. Shenil 2021 PMU based real-time fault detection algorithm for transmission line faults, *Fourth International Conference on Electrical, Computer and Communication Technologies (ICECCT)*, 15–17 September, Erode, India, pp. 1–5.

Widén J. and J. Munkhammar 2019 Chapter 4: Quantifying available solar energy on planar surfaces, In: *Solar Radiation Theory*, Uppsala University, Uppsala, Sweden, p. 28.

Zhang X., Z. Guo, Y. Zheng, J. Liu, P. Yan and L. Zheng 2022 Power grid fault diagnosis using polar PMU data plots, *International Journal of Electrical Power & Energy Systems*, 141: p. 108148.

8 IoT Sensors for Remote Health Monitoring

8.1 INTRODUCTION

"Health is wealth" because earning any amount of money, or acquisition of any level of power and fame becomes meaningless to a sick person. This chapter deals with the applications of sensors in healthcare by prevention, diagnosis, and treatment of diseases, injuries, and mental impairments.

8.2 SMART HEALTHCARE

Smart healthcare is an interconnected medical system of patients, doctors, and paramedics combining sensors such as wearable fitness trackers, health watches, pulse oximeters, biosensors, blood pressure, respiration rate, and ECG devices, with the Internet of Things and specialized software to dynamically transmit information regarding vital physiological parameters of patients to doctors for getting insights into their health status, enabling distant disease diagnosis and proactive treatment by telemedicine, supplementing the routine face-to-face consultation.

Smart healthcare encompasses the broad discipline involving the measurement of physiological parameters using wearable sensors for body temperature (**diode thermometer**), heart rate (**photoplethysmography [PPG] sensor**), respiration rate (**inductance plethysmography, capnography, piezoelectric, or bioimpedance-based sensors**), blood pressure (**oscillometric sensor**), saturation of peripheral oxygen (SpO_2, pulse oximeter), blood glucose and metabolite levels (**glucose sensor**), electrocardiogram (**smartwatch optical ECG sensor**), electromyogram (**surface EMG electrodes**), electrodermal activity (**galvanic skin response GSK sensor**).

Capnography, a routine part of anesthesia and intensive care, entails the CO_2 partial pressure measurements in the exhaled breath of patients, which are graphically illustrated as a function of time by drawing capnograms. The capnograms are pointers toward lung disease conditions.

The electromyogram (EMG) measures the electrical activities of muscles during rest state, and under mild and strong contraction, by inserting needle electrodes to diagnose neuromuscular aberrations, e.g., muscular weakness or disorders of nerves.

8.3 WEARABLE BODY TEMPERATURE SENSOR

Body temperature is a vital parameter monitored for human healthcare. Normal body temperature lies within 97.8°F (36.5°C) to 99°F (37.2°C) for a healthy adult.

A flexible sensor fabricated by printing multi-walled carbon nanotubes (MWCNTs) on a fabric has been developed (Kuzubaşoğlu et al. 2021). For this purpose, a specially formulated aqueous CNT-based conductive ink is deposited in a serpentine pattern by an inkjet printer on a soft taffeta fabric substrate (Figure 8.1). The CNT ink is prepared by dispersal of CNT powder in distilled water with dodecyl sulfate sodium salt (SDS) surfactant, and agitating in an ultrasonic probe sonicator until the mixture shows homogeneity.

The CNT pattern is encapsulated with a translucent polyurethane (PU) welding tape for protection from environmental degradation. The electrical resistance of the printed sensor decreases linearly with temperature. It works as a negative temperature coefficient (NTC) thermistor with a

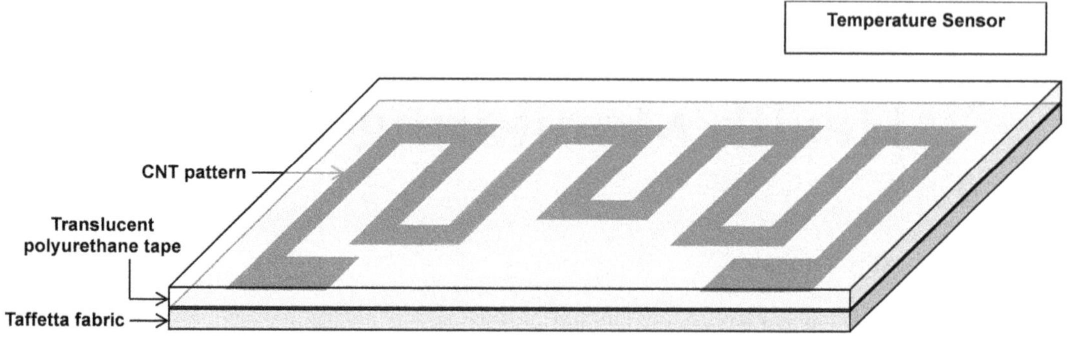

FIGURE 8.1 Body temperature sensor. A CNT pattern formed on a flexible taffeta fabric substrate is covered with a protective polyurethane film.

negative temperature coefficient of resistance (TCR) value ~ −1.04%°C⁻¹. Its TCR is higher than that of a Pt-resistance sensor (0.392%°K⁻¹) The thermal index given by the equation

$$\beta = \frac{\ln\left(R_x/R_0\right)}{1/T_x - 1/T_0}$$ (8.1)

is 1,135 K; R_x is the resistance at temperature T_x and R_0 is the original resistance at temperature T_0. High-temperature sensitivity is observed between room temperature and 50°C (Kuzubaşoğlu et al. 2021).

8.4 HEART RATE PPG (PHOTOPLETHYSMOGRAPHY) SENSOR

The heart rate, also known as the pulse rate, is the number of heartbeats per minute (bpm). It is normally 60–100 bpm for a resting adult; a resting heart rate <60 bpm is referred to as bradycardia while that >100 bpm is called tachycardia.

A PPG sensor consists of a light-emitting diode as the source of light, and a photodiode as the detector of light signal (Figure 8.2). Besides these two main components, the sensor contains the associated circuitry for performing measurements and displaying the PPG waveform.

8.4.1 Transmissive and Reflective Configurations of the PPG Sensor

Looking at the positions of placement of the LED and the photodiode, the PPG sensor is subdivided into two configurations, viz., transmissive and reflective (Park et al. 2022).

8.4.1.1 Transmissive PPG Sensor

As the name suggests, in a transmissive-type PPG sensor, the LED and the photodetector are positioned on the opposite sides of the skin tissues in alignment with each other and along the same straight line (Figure 8.2(a)). Light produced by the LED passes through the skin tissues and is received by the photodiode. Since the light signal is weakened while traveling through the skin tissues, this configuration is suitable for measurement in those parts of the human body where the skin tissues are thin. Such parts are the toes, the fingers, and the earlobes.

8.4.1.2 Reflective PPG Sensor

In a reflective-type PPG sensor, the LED and the photodetector are placed on the same side of skin tissues near each other (Figure 8.2(b)). Light incident on the skin tissues rebounds from the blood vessels and enters the photodiode. Since the measurement is based on scattered light, the optical signal is of low intensity, and also less stable than in the transmission-type PPG sensor. Nonetheless, the configuration is well-suited for application to those parts of the body through which light cannot pass easily, such as the wrist, the forehead, and the esophagus.

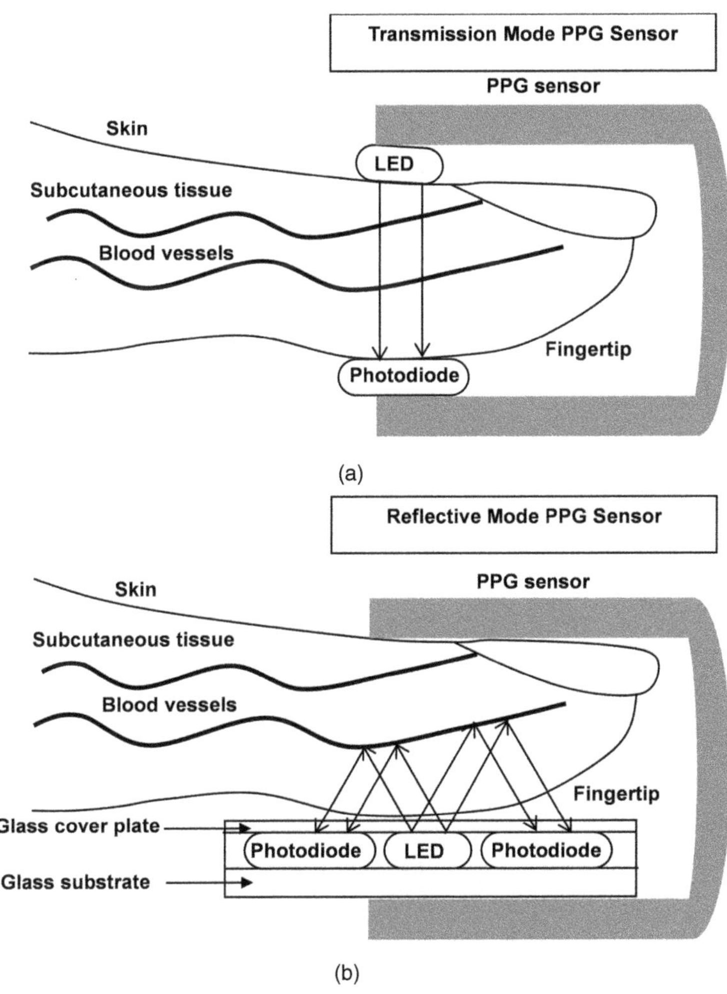

FIGURE 8.2 PPG sensor: (a) transmission mode, (b) reflective mode, (c) PPG signal acquisition circuit, and (d) PPG waveform. Parts (a) and (b) show PPG sensors fixed on the fingertips of a patient. The skin, the subcutaneous tissue, and the blood vessels are shown. In part (a), an LED lies on the top surface of the fingertip and the photodiode on its bottom surface. Light rays from the LED pass through the fingertip, and the emerging rays hit the photodiode. In part (b), there is one LED at the center and two photodiodes, one on each side of the LED. The devices are made on a glass substrate, and fitted with a glass cover plate. Light rays emitted by the LED are reflected by the blood vessels toward the two photodiodes.

(*Continued*)

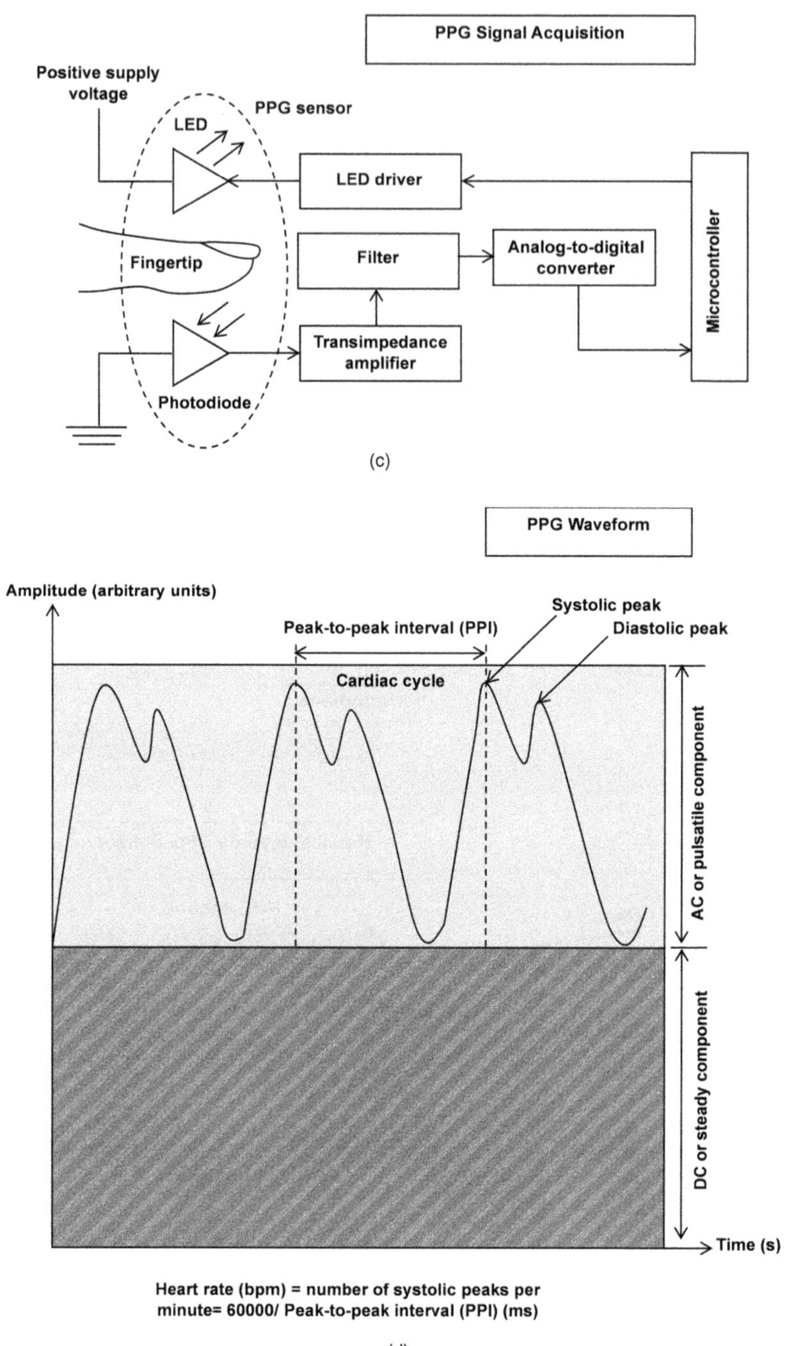

FIGURE 8.2 (CONTINUED)   Part (c) shows that light from the LED falls on the fingertip when the microcontroller sends a signal to the LED driver. The light passing through the fingertip strikes the photodiode. The current produced in the photodiode is fed to a transimpedance amplifier, filtered to remove the DC component/noise, converted into a digital signal by an analog-to-digital converter, and delivered to the microcontroller. In part (d), the amplitude of the PPG signal (in arbitrary units) is plotted against time in seconds. The PPG signal consists of a DC or steady component, and an AC or pulsatile component. The two components are distinguished. Systolic and diastolic peaks are indicated by arrows, and the peak-to-peak interval (PPI) between two consecutive systolic peaks representing a cardiac cycle, is marked. The formula for the heart rate, the number of systolic beats per minute, shows that the heart rate is calculated in beats per minute by dividing 60,000 by the peak-to-peak interval (PPI) in milliseconds.

8.4.2 Relation between the Intensity of Light Incident on the Photodiode of PPG Sensor and the Vascular Blood Volume Changes

Beer-Lambert law for optical absorbance states that the quantity of light absorbed by a substance dissolved in a light-transmitting solvent is directly proportional to:

(i) The concentration of the substance in the solution and
(ii) The length of the path traversed by light in the solution.

Accordingly, the intensity of light falling on the PPG sensor is maximum when the arterial blood volume is minimum. This happens during the diastole period. Conversely, the intensity of the optical signal incident on the PPG sensor is minimal during the period that the arterial blood volume is maximum. This is the systole period. Thus, it is evident that there is an inverse relationship between the intensity of light received by the photodiode of a PPG sensor and the change in the vascular blood volume of the patient.

8.4.3 Components of the PPG Signal

The raw PPG signal is made up of two components:

(i) An AC or pulsatile component and
(ii) A DC or static component.

The two components are ascribed to different biological phenomena. The AC component owes its origin to the changes in the volume of blood flowing through the arteries during the diastolic and systolic phases. This phasic cardiac activity is synchronous with the heartbeat. On the contrary, the DC component arises from the absorption of light by the skin, tissues, and bones of the body. Information on respiration, thermoregulation, and venous blood flow is obtained from this component.

8.4.4 Wavelengths of LEDs Used in the PPG Sensor

Owing to the deep penetrating power of red or infrared LEDs, they are commonly used in transmissive-type PPG sensors. On the other hand, it is found that the maximum pulsatile component of reflected light occurs in the wavelength range of 510–590 nm. For this reason, the LEDs giving green (565 nm) or yellow (50 nm) illumination are more appropriate for reflective-type PPG sensors (Attivissimo et al. 2023).

8.4.5 PPG Sensor Instrumentation

The instrumentation system for PPG signal recording (Figure 8.2(c)) consists of the PPG sensor, a transimpedance amplifier, an LED driver circuit, a filter to remove the DC component and noise, an analog-to-digital converter, and a microcontroller (Lin et al. 2019).

8.4.6 Heart Rate Determination from the PPG Waveform

The peaks in the PPG signal are detected using the algorithm of the moving window integration and adaptive threshold method (Utomo and Nuryani 2021). The heart rate is found from the time interval between the successive peaks, as shown in Figure 8.2(d).

8.5 ELECTROCARDIOGRAMS AND HEART ARRHYTHMIAS SENSING

8.5.1 12-LEAD ECG

The standard 12-lead ECG is a non-invasive test to capture the electrical activity of the heart. In this ECG, six limb leads are attached to the arms and legs of the patient, and six precordial leads are fixed on the chest; 'precordial' relates to precordium, the area of chest wall covering the heart. Using these leads attached to the body, the biopotentials produced by electrical signals controlling the contraction and expansion of the heart chambers are measured with respect to time. The 12-lead ECG provides a comprehensive evaluation of the cardiac condition. Unfortunately, the procedure of performing the 12L-ECG is a cumbersome process.

8.5.2 THE ECG GRAPH

The ECG graph (Figure 8.3(a)) sketches the electrical activity of the heart as a PQRST signal in which the P-wave indicates atrial depolarization, the Q-wave is representative of the initial

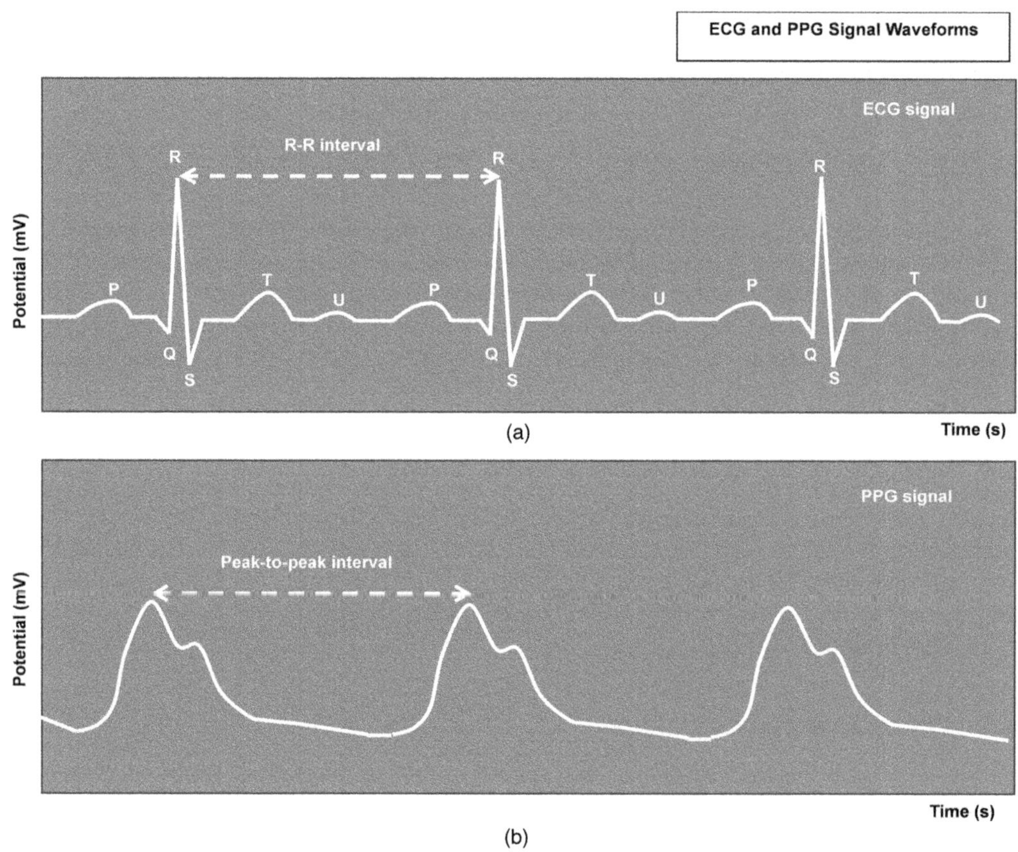

FIGURE 8.3 Waveforms for diagnosing cardiovascular diseases: (a) ECG and (b) PPG. Part (a): Recording of electrical potential generated by the heart from the surface of the body with respect to time. P-wave: Electrical depolarization of left and right atria. Q-wave: Initial depolarization of the interventricular septum. R-wave: Ventricular depolarization. S-wave: Final ventricular depolarization. T-wave: Ventricular repolarization. U-wave: Repolarization of the Purkinje fibers. R-R interval: The time between two consecutive R-waves, representing one complete heart cycle. It measures the variability of heart rate. Part (b): Potential variation with respect to time comprising a pulsatile AC waveform, and a DC baseline corresponding to movement of blood in the vessels of the body. Peak-to-peak interval: The width between peaks of the PPG waveform, representing the heart beat-to-beat interval. It is correlated with the R-R interval of ECG because both represent one complete heart cycle.

depolarization of the interventricular septum, the R-wave signifies early ventricular depolarization, the S-wave points to the final depolarization of the ventricles, and the T-wave stands for ventricular repolarization. The PR interval is the time interval between the time start of the P-wave to the beginning of the QRS complex. The RR interval is the time interval between two R-waves in the QRS signal. Figure 8.3(b) shows the PPG signal for comparison.

8.5.3 Single Lead ECG Sensor

A single lead ECG is done with one lead placed on the chest or wrist (Figure 8.4(a)). It is easily implemented with minimal training. Although limited in accuracy, it provides useful information about heart rate and rhythm facilitating timely diagnosis, particularly for episodes of atrial fibrillation, a heart arrhythmia in which atria, the upper chambers of the heart beat chaotically, and which shows no symptoms in many patients.

An unobtrusive, handheld smartphone-connected, direct-to-consumer, single-lead ECG (1L-ECG) device (AliveCor KardiaMobile) has an in-built AF detection algorithm making it comparatively easier to use and is less time-consuming for the patient (Kardia AliveCor 2017). Smartphone connectivity aids in sharing the ECG between patients and health professionals. The recording is done by holding two metal electrodes of the device loosely with two hands or multiple fingers of each hand (Figure 8.4(b)) (corresponding with lead I for 12L-ECG), preferably after washing with soap or cleaning with alcohol wipes on the fingertips to enhance electrical conduction (Witvliet et al. 2021). Lead I: RA (−) to LA (+), measures the potential difference $V = V_{LA} - V_{RA}$ between the

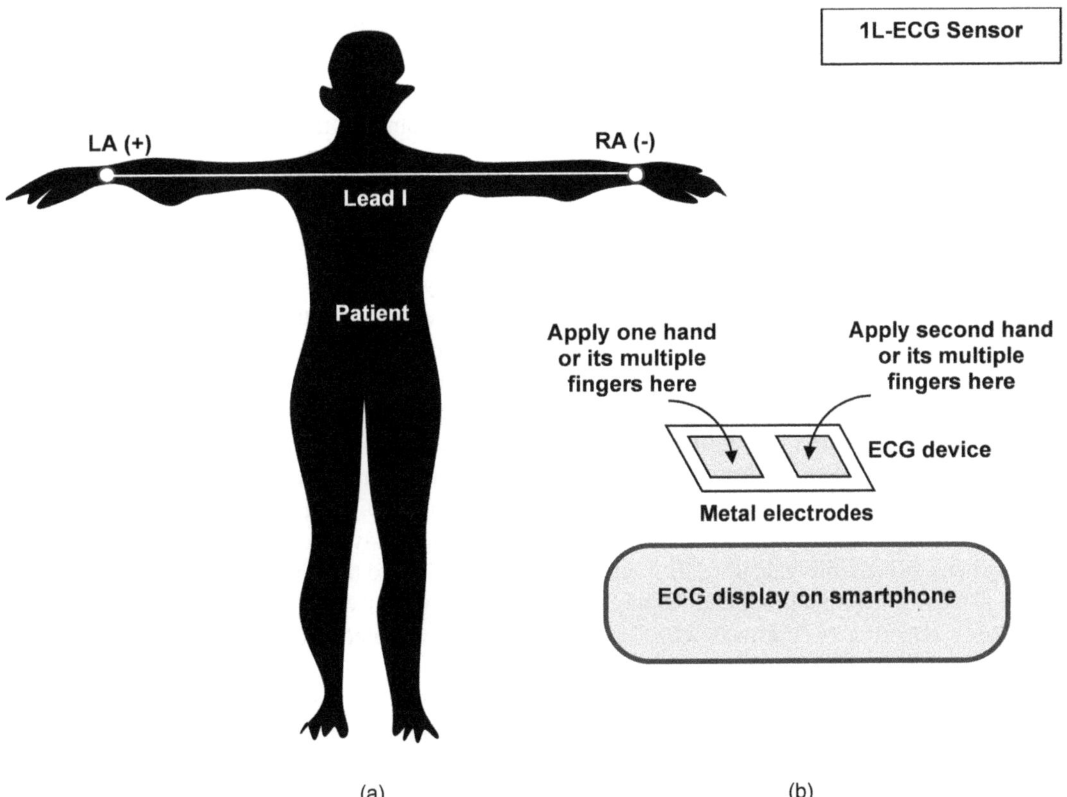

FIGURE 8.4 Single-lead ECG sensor. Left: A person is standing with the positive electrode on the left arm LA (+) connected to the negative electrode on the right arm RA (−) by lead-1. Right: The ECG of a person is displayed on the screen of a smartphone on applying one hand or its multiple fingers to one electrode of the ECG device, and the other hand or its multiple fingers on the second electrode of this device,

negative electrode on the right arm (reference), V_{RA} and the positive electrode on the left arm (exploring electrode), V_{LA}, by observing the heart's electrical activity from an angle of $0°$, which is described in medical practice as looking at the lateral wall of the left ventricle. The patients must support their hands on a surface such as a table because tightly gripping the electrodes or quivering motions introduce noise in the signal. The recorded one-lead 30 s ECG (lead I) is sent to a connected smartphone through a wireless communication protocol and also classified by analysis as a normal sinus rhythm, possible AF, unclassified, or unreadable (Selder et al. 2019). Assessment of the diagnostic capabilities of the 1L-ECG device by cardiologists against simultaneously performed 12L-ECG shows that the 1L-ECG had 100% sensitivity and 100% specificity for AF (atrial fibrillation)/AFL (atrial flutter). It has high sensitivity (90.9%) and specificity (93.5%) for any arrhythmic abnormalities. It provides an accurate diagnosis of AF/AFL and ectopic (skipped or extra) heartbeats but shows less sensitivity to conduction irregularities (Himmelreich et al. 2019).

8.5.4 Cardiac Arrhythmia Detection Smartwatch PPG Sensor with 1-Lead ECG Recording

Cardiac arrhythmias can be detected with a PPG sensor augmented with an algorithm. Smartwatches such as the Apple Watch (Apple Inc.) are widely used for this purpose. Patients can self-diagnose their heart rhythm by placing a finger on their device at any point.

A smartwatch sensor measures the PPG signal to determine the volumetric changes in blood passing through the wrist caused by peripheral pulse, which is the pulse arising from the transport of blood at a high pressure from the heart, palpated as 60–100 pulses per minute. The PPG waveform is generated with the help of a PPG sensor similar to the finger-cuff PPG sensor described above. It consists of an LED as a light source and a photodiode or phototransistor as a photodetector. Light is flashed by the LED on the skin. The photodetector operating either in the reflective or transmissive mode measures the intensity of the light signal transmitted through the tissue or reflected from it. Blood flowing through the arteries of the wrist absorbs light. The blood flow variations according to the cardiac cycle produce proportional light changes reaching the photodetector. Consequently, a pulsatile waveform corresponding to the cardiac cycle is obtained from the wrist cuff. This waveform gives the heart rate as already described for the finger cuff case. The patient must be in a resting position as arm movement gives a noisy signal.

The ECG and PPG are two perspectives of the same phenomenon, the cardiac cycle. Hence the PPG signal can be used to get insights of the heart cycle. The RR interval of the ECG is comparable to the peak-to-peak interval of PPG (Panda et al. 2022). Both these intervals picture a complete heart cycle. The RR interval time series of ECG is used for derivation of heart rate variability (HRV). Hence, the peak-to-peak interval of PPG is incorporated into an algorithm for detecting atrial fibrillation. Thus, the PPG waveform can provide information about the irregularities of heart rhythm, thereby enabling screening for atrial fibrillation (AF). A proprietary algorithm applied to a tachogram, reveals irregularity of pulses and thus AF. The tachogram is a time series graph obtained by plotting beat-to-beat intervals (the RR intervals) on the Y-axis, and time or consecutive beat numbers on the X-axis, showing a graphical display of the flow and velocity of the blood. The sensitivity of arrhythmia detection by smartwatch PPG sensors has been evaluated to be 100%, specificity is 95%, and accuracy 97% (Nazarian et al. 2021).

For recording a 1-lead ECG (iECG), the patient has to complete the circuit between the detector on the backside of the watch and the digital crown hardware input, a rotating button on the lateral side of the watch. For ECG recording, the patient rests his/her arm on the table or lap. Then, with the hand opposite the watch, the patient holds his/her finger on the digital crown during the recording session without pressing it and waits till the recording is completed. The heart rhythm is analyzed within 30 s after recording. It is divided into three classes: sinus rhythm, AF, or inconclusive. The watch recordings are also saved in PDF format (Apple Watch User Guide 2023, Isakadze and Martin 2020).

8.6　BLOOD PRESSURE MEASUREMENT: OSCILLOMETRIC SENSOR

Blood pressure, the pressure exerted by the blood against the wall of the artery, varies between a low diastolic level for heart contraction to a high systolic level for heart relaxation during each heartbeat. The normal diastolic and systolic values are 80 mm and 120 mm Hg, respectively. High blood pressure is often asymptomatic. Also called hypertension, it is a silent killer.

8.6.1　ARM-CUFF OSCILLOMETRIC SENSOR

The time-honored traditional mercury sphygmomanometers using an auscultatory method based on the detection of Korotkoff sounds are being replaced by the simpler easy-to-use oscillometric instruments. There are two reasons for this replacement:

(i) The withdrawal of toxic mercury, and
(ii) The training/practice needed for correct BP measurements by auscultation.

In oscillometric measurement, the pressure oscillations recorded during gradual depressurization of a cuff sphygmomanometer are analyzed. The oscillations commence around the systolic pressure level. They continue below the diastolic pressure level. The point of maximum oscillation is related to the average intra-arterial pressure. An empirical algorithm is formulated for determining the values of systolic and diastolic pressures. The oscillometric blood pressure measurement setup consists of a microcontroller, a pump, a pressure sensor, gas valves, and an inflatable cuff. The measurement procedure, displayed in Figure 8.5, comprises the following steps (Lewis 2019):

(i) Wrapping the Cuff: An occluding cuff is wrapped around the upper arm Figure 8.5 (a) and (b). The pressure within the cuff continuously changes during blood flow. When the heart contracts, blood is pumped into the arteries, causing their expansion. When the heart relaxes, the artery walls return to their normal state. The expansion and contraction of the arteries constitute a pulse. During each arterial pulse wave, the volume of the occluded limb within the cuff slightly increases and decreases. These volume changes produce corresponding increases and decreases in the pressure within the surrounding cuff. A pressure sensor detects these pressure changes.
(ii) Inflating the Cuff: The cuff is inflated with an electronic pump Figure 8.5 (c). As inflation continues, the pressure rises to a level at which the blood flow in the artery is blocked. This is detected by the pressure sensor from the cessation of pressure variations. The pump continues inflation for a few seconds more to ensure a complete stoppage of blood flow.
(iii) Stopping the Inflation and Deflating: The inflation by the pump stops with the simultaneous opening of a mechanical valve to release the pressure built up in the cuff Figure 8.5 (d) and (e). The cuff is slowly deflated, releasing the pressure. So, the blood flow resumes.
(iv) Recording the Deflation Pressure–Time Curve: The pressure variations within the cuff during deflation are carefully watched Figure 8.5 (f). In the initial phase, the pressure falls in a non-pulsating manner. When the pressure in the cuff decreases to a value less than the pressure of the peak of the arterial pulse, a small pressure wave is detected. This pressure wave is a replication of the difference between the pressure in the cuff and pressure in the artery. As the deflation progresses, these pressure differences increase until the cuff begins to loosen from the arm and the volume pulsation becomes weaker. Thus, the instrument records a series of waves, which are initially flat, then pulses with small variations, followed by an increase to a peak value, and then decaying until they become undetectable.
(v) Analyzing the Deflation Curve: The pressure–time variations inside the cuff, as measured with a pressure sensor during deflation, are processed to extract the blood pressure

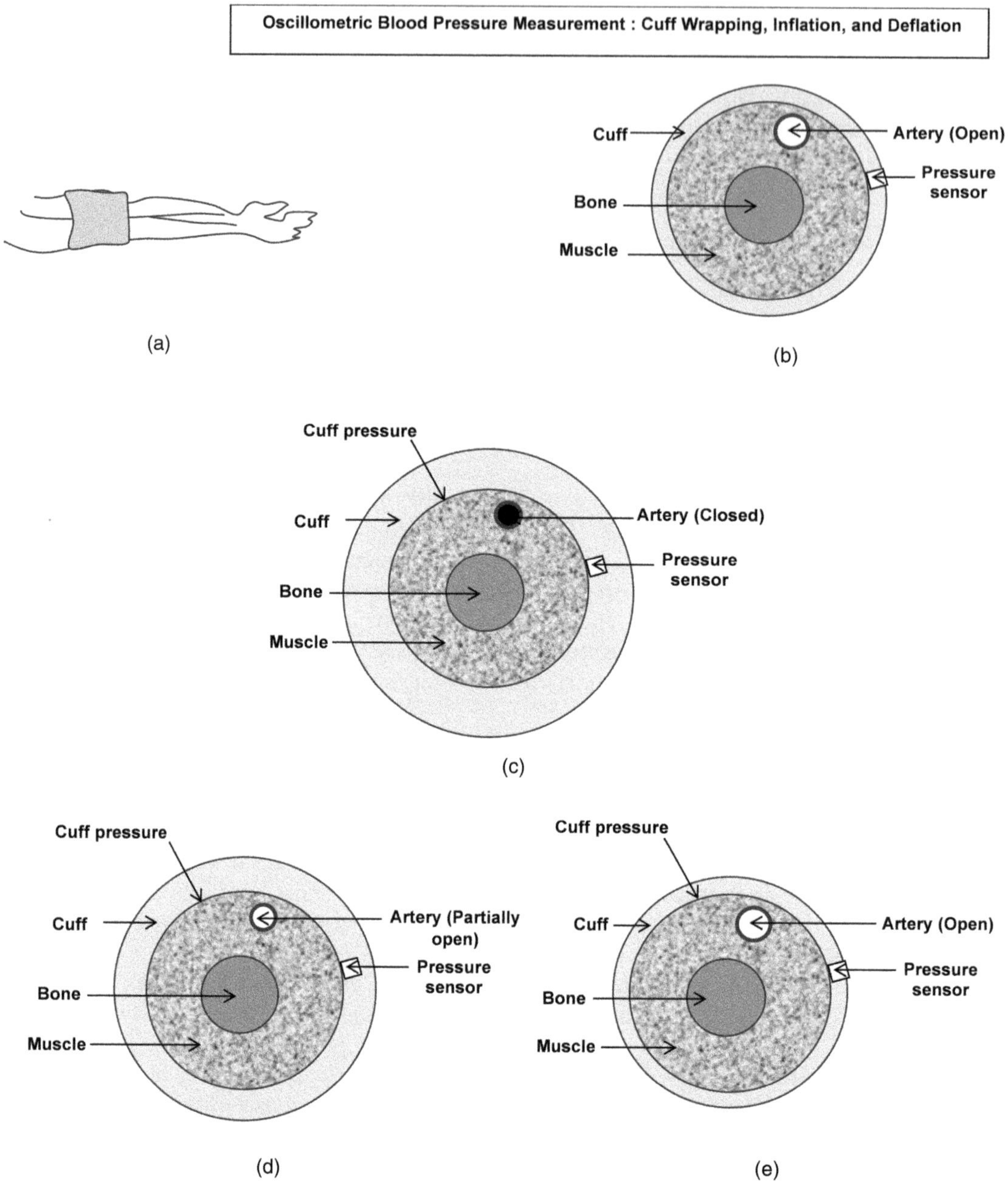

FIGURE 8.5 Oscillometric blood pressure measurement: (a) and (b) cuff wrapping, (c) cuff inflation, (d) and (e) cuff deflation, (f) deflation curve, (g) extracted oscillometric waveform, and (h) constructed oscillometric waveform envelope. Part (a): A cuff is wrapped around the upper arm of a person. Parts (b)-(h): A cuff wrapped around the muscle of the arm of a person. The bone and artery inside the muscle are shown. A pressure sensor is placed between the skin of the person and the cuff. Part (b): The cuff is deflated. No pressure is exerted on the muscle skin. The artery is open. Part (c): The cuff is inflated. A pressure acts on the muscle skin. It is the cuff pressure. The artery is closed. Part (d): Cuff deflation starts. The cuff pressure starts falling and the artery partially opens. Part (e): Cuff deflation is completed. The cuff pressure further decreases, and the artery is fully open.

(Continued)

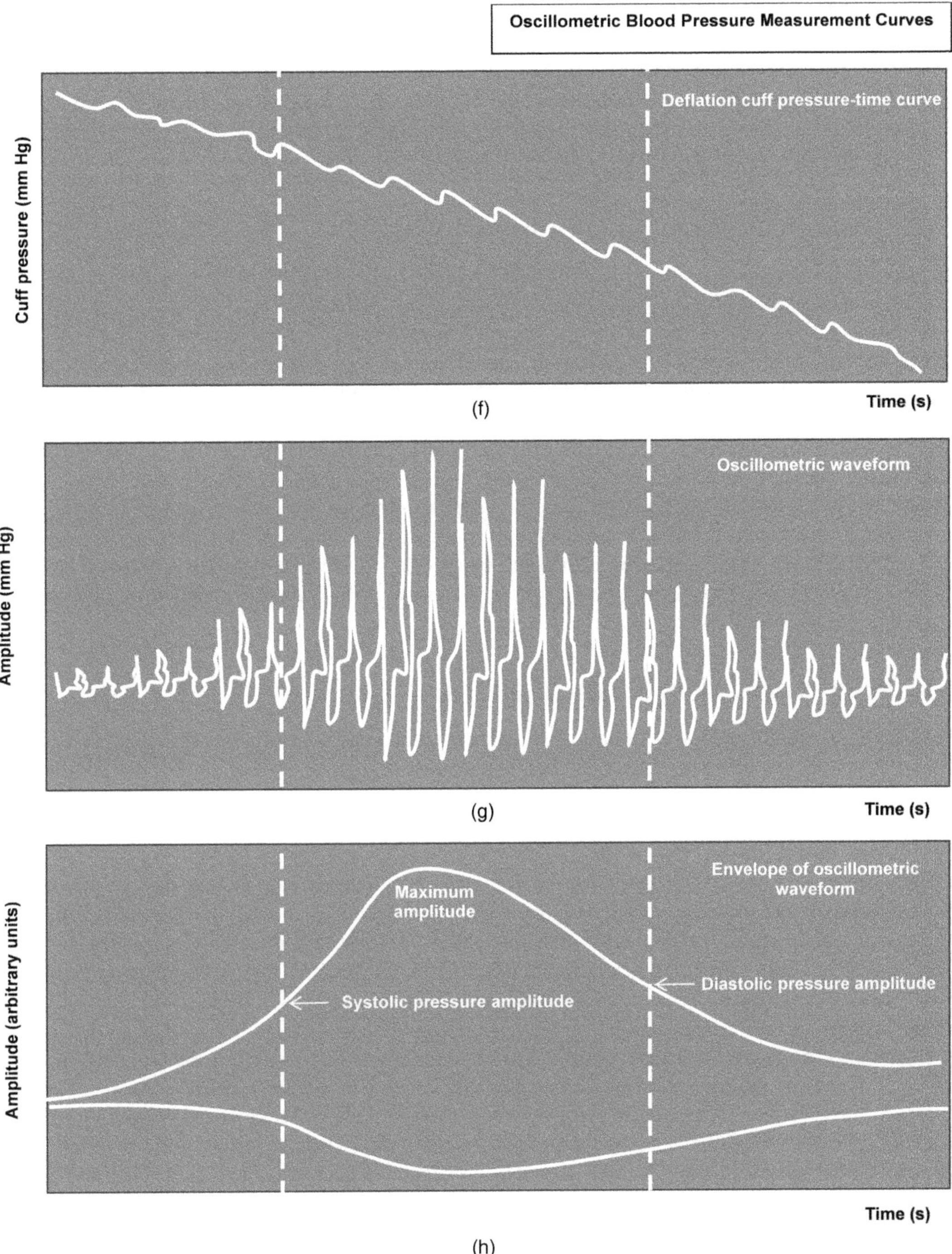

FIGURE 8.5 (CONTINUED) Part (f): Cuff pressure is plotted in mm of mercury against time in seconds. This is the deflation pressure–time curve. Part (g): Amplitude of cuff pressure variations in mm of mercury is plotted with respect to time in seconds. This is the oscillometric waveform. Part (h): The graph presents the envelope of the oscillometric waveform. Envelope of the amplitude of cuff pressure variations in arbitrary units is generated with respect to time. The maximum amplitude region is indicated. The points of amplitudes of systolic and diastolic pressures are shown by the intersections of the dashed lines drawn over figures (f), (g), and (h) with the envelope of the amplitude of cuff pressure.

information Figure 8.5 (g) and (h). Among the several algorithms designed, a popular algorithm to calculate arterial blood pressure from the oscillometric waveform is the maximum amplitude algorithm (MAA). In this algorithm, the average blood pressure is taken as the cuff pressure at which the amplitude of oscillation of pressure is maximum. The systolic and diastolic blood pressures are obtained by combining this average value with the rate of change of pressure wave (Lee et al. 2013, Kumar et al. 2022).

8.6.2 WATCH-TYPE WRIST CUFF OSCILLOMETRIC SENSOR

A watch-type wearable device with a wrist cuff designed for a range of wrist circumferences is used for automatic oscillometric blood pressure measurement (Kuwabara et al. 2019). At the time of measurement, the watch is held at the heart level. It is small in size and has a watch-like display. It can be self-used by the patient at any place and when desired, in a routine work environment or while sleeping, for on-demand BP evaluation, or by programming for defined recording. So, it is an effective tool for the diagnosis of large blood pressure variability experienced by patients in stressful real-life situations, and for detecting daytime masked hypertension, defined as normal BP in the clinic and elevated BP outside.

The device has an inflatable belt. Automatic inflation of the cuff by an electrical pump is followed by its deflation through a valve. It causes less compression of muscles than the arm-cuff devices, thus reducing the inconvenience during pressurization. The pulse wave recorded during the inflation of the cuff is examined to extract the systolic and diastolic blood pressure values by applying an algorithm (Konstantinidis et al. 2022).

8.7 RESPIRATION MONITORING SENSOR

Respiratory or breathing rate is the number of breaths taken by a person per minute. For an adult person at rest, it lies between 12 and 18 breaths per minute.

8.7.1 CAPACITIVE RESPIRATION SENSOR

Capacitive sensors are less sensitive to temperature and humidity effects and also consume low power. A capacitive pressure sensor (Figure 8.6) is fabricated by sandwiching a dielectric layer made of porous Ecoflex (porosity ~36%) between flexible electrodes made of polydimethylsiloxane thin films containing silver nanowires and carbon fibers (Park et al. 2017). It has a sensitivity of 0.161 kPa^{-1} at pressures <10 kPa, an operating pressure range of 0–200 kPa, and can withstand >6,000 cycles.

The sensor is attached to the backside of the buckle of a waist belt worn by the patient. When the patient exhales, the stomach deflates (Figure 8.6(a)). There is no pressure on the sensor, so it is not deformed. The distance between its electrodes remains unaffected. Therefore, its capacitance does not change. But during inhalation by the patient, the stomach inflates, exerting pressure on the sensor which is compressed, resulting in a decrease in distance between the capacitor plates and hence, an increase in the capacitance of the sensor (Figure 8.6(b)).

The capacitance changes are converted into voltage signals using a low-pass filter, an instrumentation amplifier, and an analog-to-digital converter. The respiration pattern is monitored from the voltage-time variations. These curves help in finding the respiration rate, and in detecting abnormalities, if any, caused by diseases such as asthma and sleep apnea (Park et al. 2017).

8.7.2 RFID RESPIRATION SENSOR

RFID sensor works on the backscattering method, a communication method involving a battery-less RFID tag with an antenna, and a powered RFID reader, also with an antenna. Electromagnetic waves

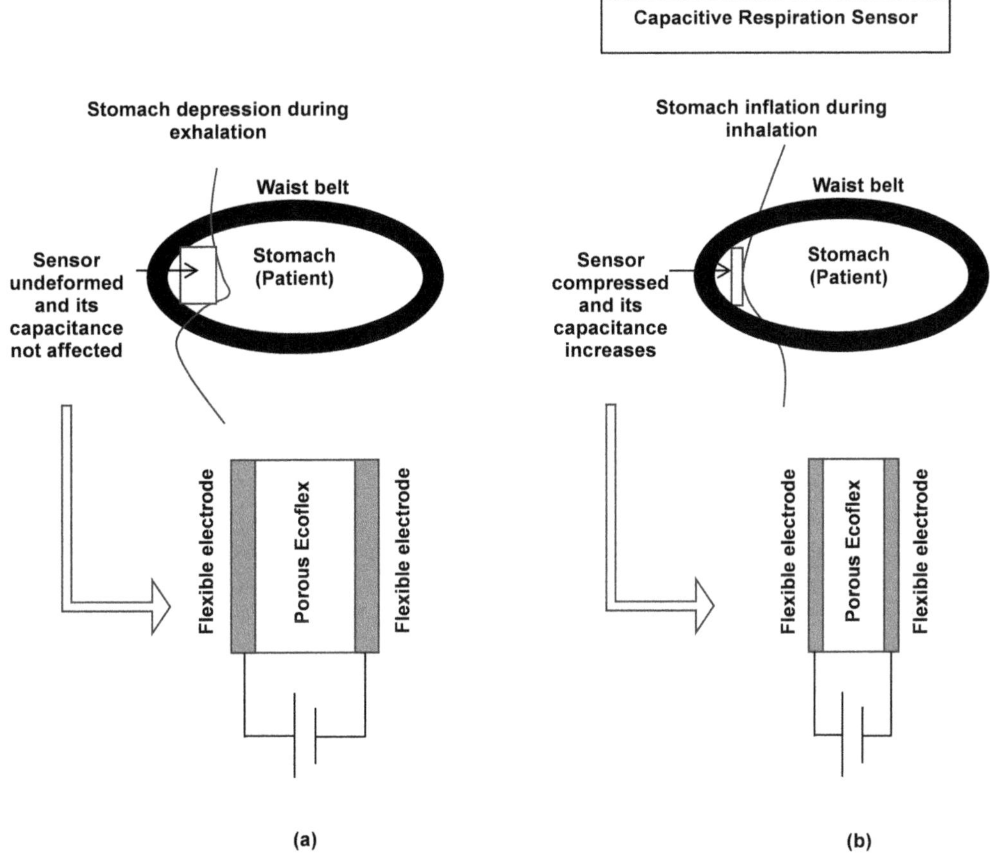

FIGURE 8.6 Capacitive respiration sensor during (a) inhalation and (b) exhalation. Diagrams (a) and (b): Top side: A capacitive sensor is tied to the stomach of a person by a waist belt. Bottom side: The capacitive sensor consists of a porous Ecoflex dielectric sandwiched between flexible electrodes. Part (a): The person is exhaling. The stomach is depressed and away from the belt so that pressure acting on the sensor is low. The porous Ecoflex dielectric is in its naturally enlarged form. The distance between the electrodes of the capacitive sensor is large. Hence, the capacitive sensor has a small capacitance value. Part (b): The person is inhaling. The stomach presses against the sensor, and the sensor is compressed between the stomach and the waist belt. The pressure on the sensor is high. The porous Ecoflex dielectric is squeezed into a thin film. The distance between the electrodes of the capacitive sensor is small. Hence, the capacitive sensor has a large capacitance value.

transmitted from the antenna of the RFID reader are received by the antenna of the RFID tag. The RFID tag is thus energized, and its internal integrated circuit is activated. The remaining energy received by the RFID tag is modulated with the data in the IC to send a reply to the reader as electromagnetic waves are emitted from its antenna. Thus, the RFID tag receives energy from the RFID reader and utilizes the received energy to send back a reply message to the RFID reader.

The RFID respiration sensor (Figure 8.7) works as a strain sensor based on the backscattering principle (Hussain et al. 2023). The respiration sensor is the antenna of the RFID tag placed in contact with the body, e.g., around the belly, using a fabric belt ensuring that a compressive force acts on the sensor. The antenna is specially designed according to the impedance of the integrated circuit inside the RFID tag. It is also made of a flexible material, e.g., a textile for comfortable conformal placement on the body.

During breathing activity, the antenna is subjected to a strain causing its mechanical deformation. Due to this deformation, its resonance frequency is altered, and hence the backscattered signal

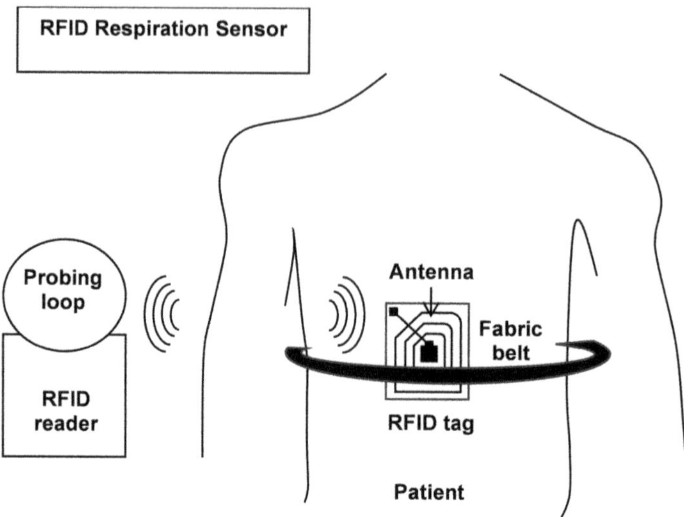

FIGURE 8.7 RFID respiration sensor. An RFID tag with an antenna is tied to the belly of the patient with a fabric belt. An RFID reader with a probing loop is used for interaction with the tag.

changes, which is measured in terms of the received signal strength indicator (RSSI), a metric of the power of the returned signal of the RFID tag. The variation of RSSI is correlated with the deformation of the antenna of the RFID tag fitted on the patient's body, and hence, with the respiratory movements of the patient responsible for antenna deformation. The breathing pattern of the patient is revealed in the RSSI vs. time plots. The RFID tag sensor being a low-cost, lightweight, small form-factor, and maintenance-free passive device, working by harvesting energy from its external reader, serves as a convenient, non-intrusive respiration monitoring device (Hou et al. 2017).

The backscattering measurement approach is demonstrated by fabricating a wearable, compression-sensing antenna knitted with a silver-coated nylon fabric (Tajin et al. 2021). It is called the Bellypatch (Figure 8.8). It is a patch-type antenna due to which it can preserve its radiation efficiency in the vicinity of the human body. Its on-body radiation efficiency is 43.1%. The size of the antenna is 135 mm × 30 mm × 11 mm. Its substrate material is flexible polyethylene foam to enable easy compression and for comfortable wearing. It can be reused.

The Bellypatch antenna acts as an on-body RFID respiration sensor working at ultrahigh frequencies (902–928 MHz). It shows a high sensitivity to respiration-induced compression and relaxation. When the patient inhales, the expanding torso of the patient exerts pressure on the antenna, decreasing its radiation efficiency, and so also the backscattered power level. So, an inhalation event is observed as a valley or a dip in the RSSI–time curve. During the patient's exhalation period, the

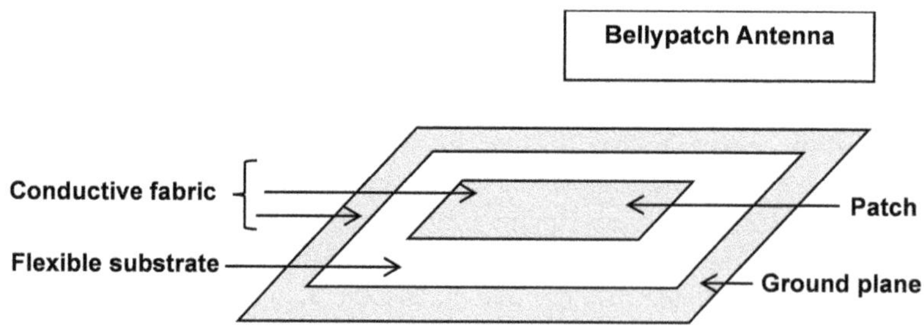

FIGURE 8.8 Bellypatch antenna. A patch-shaped antenna made with a conductive fabric on flexible polyethylene foam is surrounded by the ground plane, also made with the conductive fabric.

contracting torso relaxes the pressure on the antenna. As the antenna recovers its original shape, its radiation efficiency increases, and along with it the backscattered power. The backscattered power (RSSI) variations during respiration range between 6 and 15 dB. The on-body read range of the sensor is 5.8 m (Tajin et al. 2021).

8.8 SATURATION OF PERIPHERAL OXYGEN (SpO$_2$): PULSE OXIMETER

8.8.1 SpO$_2$ LEVEL

The pulse oximeter is a small, lightweight, non-invasive electronic probe used to measure SpO$_2$, the peripheral capillary oxygen saturation, defined as the ratio of oxygen-bound hemoglobin or oxy-hemoglobin, HbO$_2$ in the blood of a person to the total hemoglobin, i.e., [oxy-hemoglobin (HbO$_2$) + deoxy-hemoglobin (Hb)], expressed as a percentage by multiplying with 100 (Mildenhall 2008, Lopez 2012). Deoxy-hemoglobin is also called reduced hemoglobin.

The SpO$_2$ is a vital parameter representing the effectiveness of the cardio-pulmonary system in supplying oxygen-rich blood to the body. An SpO$_2$ level within 95–100% is considered to be normal. A healthcare provider must be immediately called when it falls below 95%.

8.8.2 WORKING PRINCIPLE

The pulse oximeter works on the principle that Hb absorbs more red light than infrared light, permitting passage of more infrared light through it whereas HbO$_2$ absorbs more infrared light than red light, allowing more red light to pass through it. Therefore, a measurement of the transmitted red and infrared lights through the blood establishes a correlation between HbO$_2$ and Hb levels of a patient, giving an estimate of their relative proportions in the blood.

8.8.3 TRANSMISSIVE AND REFLECTIVE-TYPE OXIMETERS

Applying the above difference in absorption and transmission pattern of red and infrared lights by oxygenated and de-oxygenated blood, the pulse oximeter consists of two LEDs:

(i) A red LED emitting light of 660 nm wavelength and
(ii) An infrared LED emitting at 940 nm.

In a transmissive oximeter (Figure 8.9), these LEDs are placed on one side of a translucent part of the human body such as a fingertip, a toe, or an ear lobe. A photodiode is placed on the opposite side of that part facing both LEDs. A clip is used for fixing the probe. In a reflective oximeter (Figure 8.10), the LEDs and the photodiode are placed on the same side of the skin of the body part chosen for measurement.

8.8.4 THE CIRCUIT OF THE OXIMETER

The circuit part of the oximeter probe (Figure 8.11) comprises a dual single-pole/double-throw (SPDT) analog switch actuated by pulse width modulated (PWM) signals from a microcontroller; the signal conditioning circuit for the output from a photodiode; and a digital-to-analog converter for feeding the digital signal from the microcontroller to the LED current control block (Feng 2013–2015). The microcontroller is used for regulating the sequential switching ON/OFF of the LEDs and for calculation of the SpO$_2$ from transmitted red and infrared lights.

PWM is a technique of transforming a digital signal into a series of pulses. It is used for producing an analog signal correlating to a digital value. The PWM pulses generated from the microcontroller are used to control analog devices. In the circuit shown, two PWM signals, PWM1 and

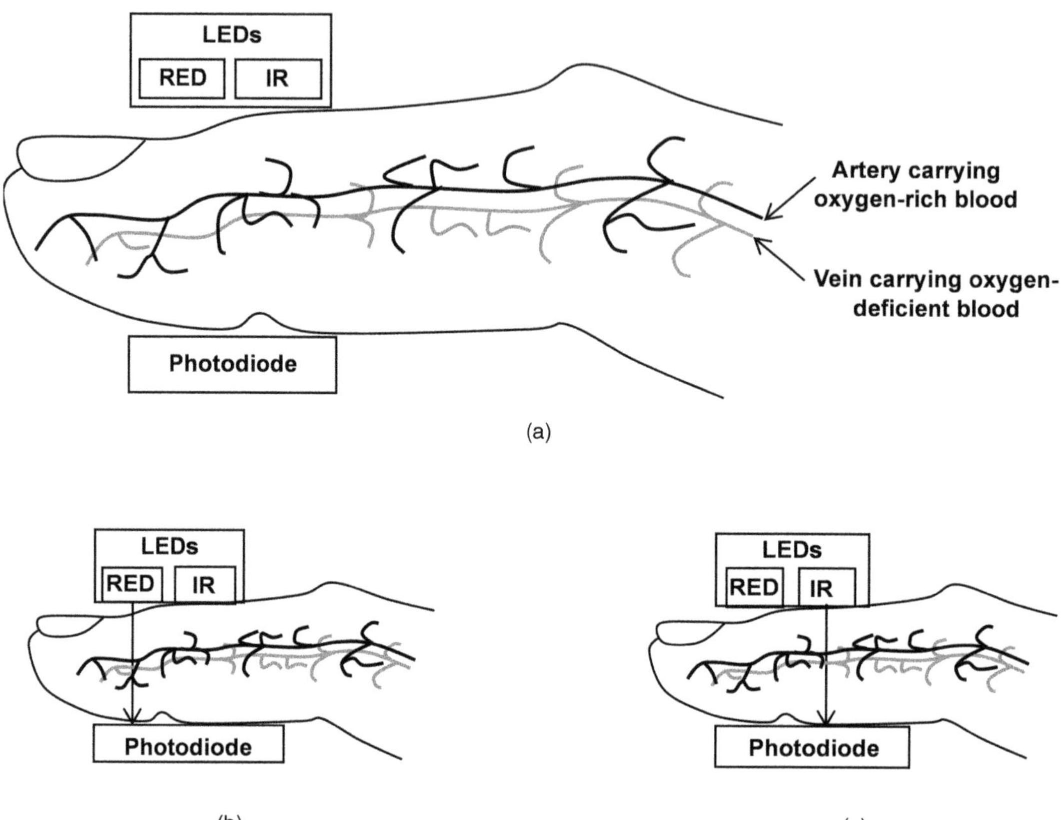

FIGURE 8.9 Transmissive pulse oximeter, and the sequential switching ON of red and infrared LEDs: (a) construction of the transmissive type pulse oximeter, (b) red LED ON, and infrared LED OFF; (c) red LED OFF, and infrared LED ON. Part (a) shows the finger with red and infrared LEDs mounted on the upper skin and the photodiode in touch with the skin on the opposite side. The artery carrying pure blood and the vein carrying impure blood are shown. In part (b), the red light passes through the finger and falls on the photodiode, and there is no infrared light. In part (c), the situation is reversed with the infrared light passing through the finger and falling on the photodiode, and red light is absent.

PWM2, are sent by the microcontroller to drive a dual SPDT switch for alternately turning ON and OFF the red and infrared LEDs according to a timing schedule for acquisition of the desired number of analog-to-digital converter (ADC) samples, and allowing adequate time for data processing prior to the turning on of the next LED. First, one LED is turned ON, red (suppose), then the other, i.e., infrared. This sequence is repeated several times per second.

The red light and infrared light outputs received from the photodiode of the pulse oximeter enter the three-stage signal conditioning circuit. In the signal conditioning circuit, the first stage is a transimpedance amplifier. It is a current-to-voltage converter that changes an input current signal into a proportional output voltage signal. The few microamperes current signal from the photodiode is converted into a few millivolts signal. The second stage is a high-pass filter. This filter is necessary to eliminate the interference effects of the background illumination. The filtered data moves to the

Reflective Pulse Oximeter and Sequential LED Switching

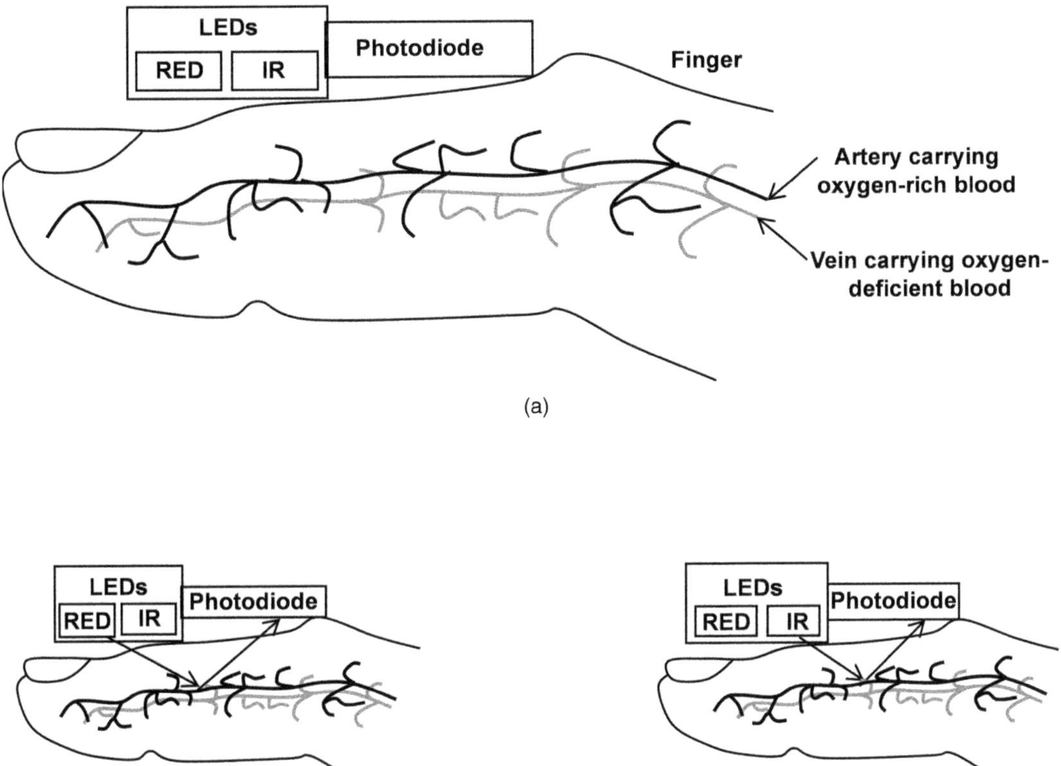

FIGURE 8.10 Reflective-type pulse oximeter, and the sequential switching ON of red and infrared LEDs: (a) construction of the reflective type pulse oximeter, (b) red LED ON, and infrared LED OFF; (c) red LED OFF, and infrared LED ON. Part (a) shows the finger with red and infrared LEDs mounted on the upper skin and the photodiode placed nearby on the same side. The artery carrying oxygenated blood and the vein carrying deoxygenated blood are shown. In part (b), the red light entering the finger is reflected by a blood vessel toward the photodiode, and there is no infrared light. In part (c), the situation is reversed. Infrared light is sent into the finger, and infrared light reflected from a blood vessel is incident on the photodiode. Moreover, red light is absent.

third and final stage, the gain amplifier. It provides sufficient gain to place the signal within the range of the ADC module of the microcontroller. The SpO_2 percentage is calculated. It is displayed and also sent to a wireless network through the communication interface.

8.8.5 SpO$_2$ Calculation and Display

A large fraction of the incident light is absorbed by the skin and bones. This light remains static. The arterial blood circulation provides an alternately increasing and decreasing blood flow underneath

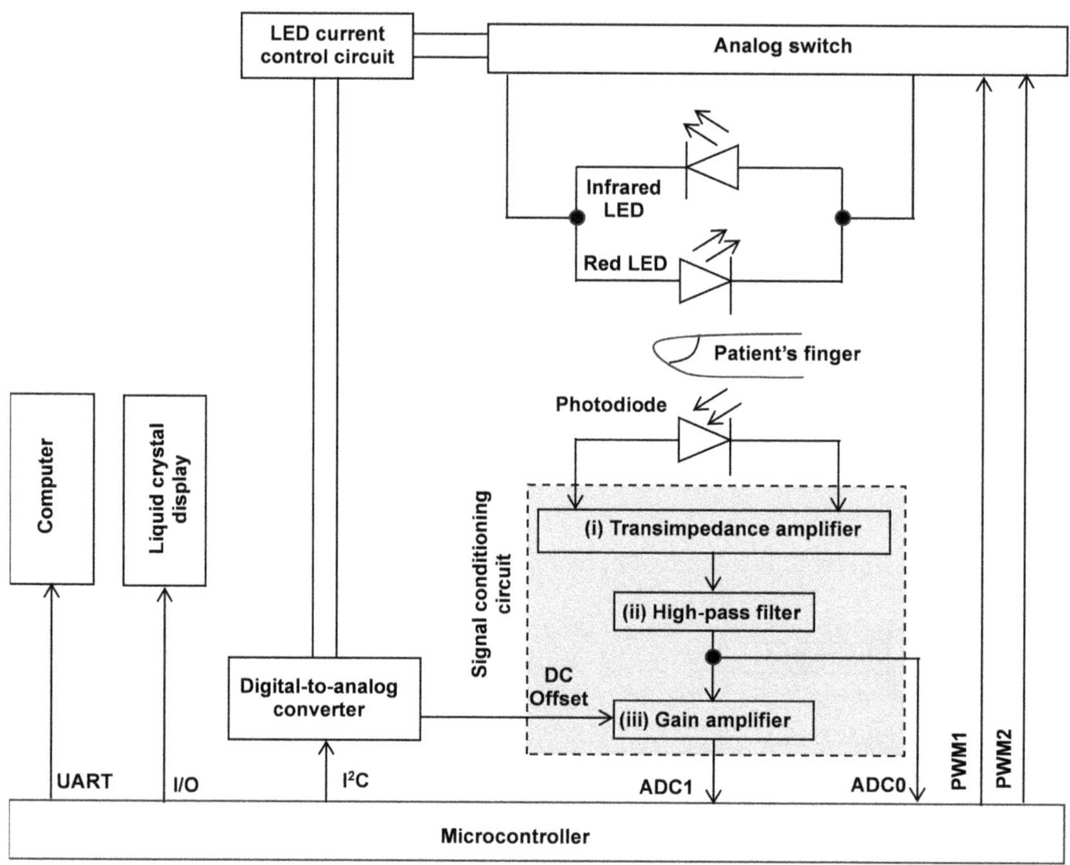

FIGURE 8.11 Block diagram of the read-out circuit for a pulse-type oximeter. A transmissive-type oximeter is illustrated. The pulse oximeter consists of the LEDs and a photodiode. The finger is positioned between the red and infrared LEDs on one side, and a photodiode on the other side. The circuit contains an analog switch with the LED ON/OFF signals entering it. The circuit also contains a microcontroller. The PWM1, PWM2, ADC0, ADC1, I²C, I/O, and UART pins of the microcontroller are marked. The ADC0 and ADC are two in-built analog-to-digital converters in the microcontroller. The I²C or inter-integrated circuit is a communication protocol for interfacing with devices. The I/O is the input/output port for the connection of the microcontroller with the outside world. The UART stands for Universal Asynchronous Receiver Transmitter. The UART is an in-built dedicated integrated circuit in the microcontroller for serial communication with external devices, e.g., Wi-Fi, Bluetooth, or PC. The interfacing is done through the UART port. The signal conditioning circuit, enclosed inside a rectangle made of dashed lines, consists of a transimpedance amplifier, a high-pass filter, and a gain-stage amplifier. The DAC together with the LED current control block, adjusts the currents flowing in the LEDs. The SpO$_2$ level is shown on a display device such as an LCD. The communication interface provides connectivity with Wi-Fi, Bluetooth, or PC.

the oximeter probe following the rhythmic contraction and expansion of the heart. These variations of blood flow cause changes in red and infrared light absorption and transmission.

The ratio of red transmitted light to infrared transmitted light is calculated by the processor, converted into an SpO$_2$ reading via a lookup table prepared according to Beer-Lambert law and stored in the memory. The calculated SpO$_2$ value is displayed by the probe.

8.9 WEARABLE EPIDERMAL GLUCOSE SENSORS

8.9.1 NEED FOR A PAINLESS GLUCOSE TEST

The traditional method of glucose testing involves painful finger pricking. The patient is further annoyed by the agony of multiple tests done for continuous diabetes monitoring. Although glucose information can be obtained from sweat analysis, the availability of sweat from bedridden or aged patients at rest becomes limited. The obvious recourse is to look for a patient-friendly non-invasive approach. Interstitial fluid (ISF) extraction provides the answer.

8.9.2 INTERSTITIAL FLUID

Body fluids are subdivided into intercellular fluid (inside the cells) and extracellular fluid (outside the cells). Extracellular fluid is further divided into plasma (within the blood vessels) and ISF (the fluid present in spaces between blood vessels and the cells but not in capillaries). It is formed via transcapillary exchange during the flow of blood and contains electrolytes, salts, sugars, acids, neurotransmitters, and water. ISF constitutes three-fourth content of the total extracellular fluid. Volumetrically, the ISF is three times the blood in the body, contributing to 15–25% of the body weight. This fluid surrounding the cells of the body is the bridge between blood and the cells, acting as the medium of transportation of nutrients, secretion of cell wastes, and communication through molecular signals. Its diagnostic value originates from the fact that it is rich in biomarkers of diseases.

8.9.3 TRANSDERMAL REVERSE IONTOPHORESIS TECHNIQUE FOR ISF EXTRACTION

It is a needleless method to extract ISF through intact skin using a pair of skin electrodes. Iontophoresis is the process of passing a low-amperage current through the skin for transdermal delivery of a drug. The process used here is not iontophoresis but iontophoresis in the opposite direction, i.e., reverse iontophoresis. A low-level current is passed through the two skin-worn electrodes (Figure 8.12). This current affects the subdermal molecules. Under the applied electric field, the ionic species and neutral molecules are transported across the skin by the following two separate mechanisms.

(i) Electromigration: This mechanism applies to ionic movements. Positive ions move toward the cathode and negative ions toward the anode. The movement of charged ionic species toward electrodes of opposite charge polarity is called electromigration.
(ii) Electroosmotic Flow: This mechanism applies to neutral molecules such as glucose. At the physiological pH, the skin is negatively charged. This makes it perm-selective to positive ions. The charge carriers carrying electricity through the skin are mainly monovalent, positively charged sodium and potassium ions. The flux of Na^+ and K^+ cations toward the ionophoretic cathode induces an electroosmotic solvent flow toward the cathode. This convective solvent flow carries along with it the solute species, including neutral and especially polar molecules such as glucose molecules to the cathode. Electroosmotic flow is the movement of a liquid in response to the Coulomb force induced by the application of an electric field across a membrane, a microcapillary, a microchannel, or a porous structure.

8.9.4 GLUCOWATCH® BIOGRAPHER

It is a small wristwatch device (Figure 8.13) combining reverse iontophoretic ISF sample extraction with two amperometric glucose biosensors, an electronic circuit, and a digital display (Garg et al. 1999; Tierney et al. 1999, 2001; Potts et al. 2002). A galvanostat circuit is used for the reverse iontophoresis. Two independent potentiostat circuits are used for the biosensors.

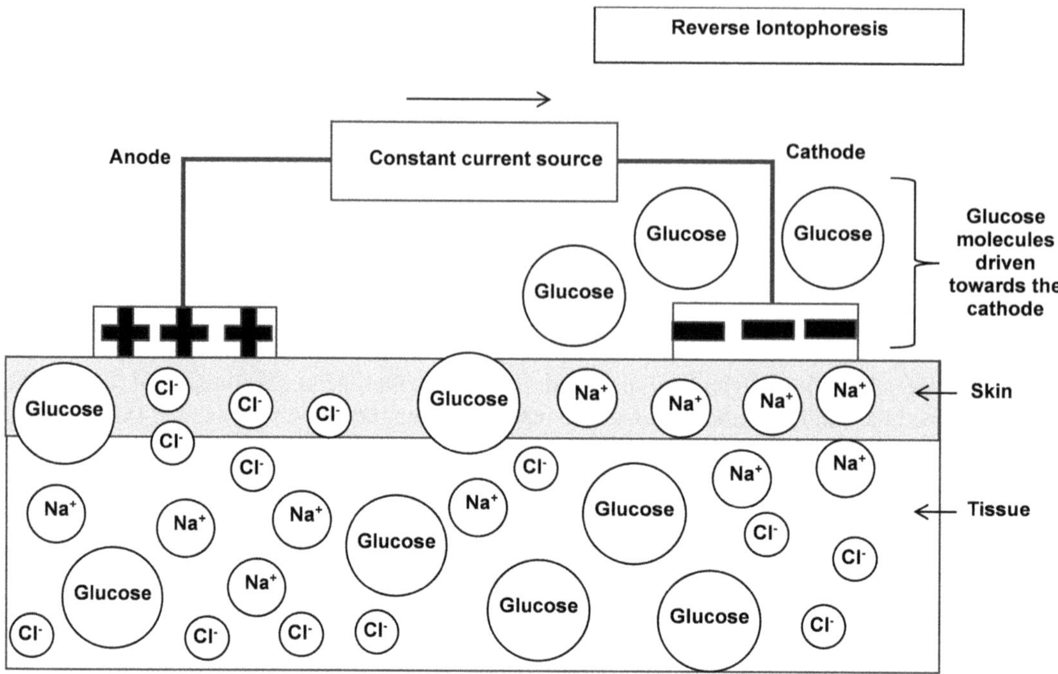

FIGURE 8.12 Reverse iontophoresis. The skin and tissue layers contain Na^+ and Cl^- ions, and glucose molecules. A constant current source is connected between the positive electrode (anode) and the negative electrode (cathode). Current flows from the anode to the cathode, and glucose molecules are driven toward the cathode.

The working electrode, acting as the sensing electrode, is an electrode having a large surface area, made by screen-printing a film of platinum–graphite composite ink. The reference electrode is made from a screen-printed Ag/AgCl layer. The counter electrode, serving as the iontophoresis electrode, is made from a screen-printed Ag layer. The biosensor assembly comprises two biosensors A and B. Hydrogel discs containing dissolved glucose oxidase are placed over each biosensor. These discs being in contact with the skin are the storage containers into which glucose is collected from the skin. They provide the electrolytic solution for analysis.

The oxidation of glucose ($C_6H_{12}O_6$) is catalyzed by the glucose oxidase enzyme producing glucono-δ-lactone ($C_6H_{10}O_6$) and hydrogen peroxide (H_2O_2) in the presence of oxygen from the air. The liberated hydrogen peroxide is measured to determine glucose concentration.

$$\text{Glucose} + \text{Oxygen} \rightarrow \text{Glucono} - \delta - \text{lactone}$$
$$+ \text{Hydrogen peroxide} \left(\text{in the presence of Glucose oxidase} \right) \tag{8.2}$$

The glucose measurement procedure consists of the following steps:

(i) Iontophoresis current (0.3 mA) supply for 3 min, and glucose collection at the cathode.
(ii) Activation of biosensors A and B, and integration of biosensor currents at the cathodes for 7 min.
(iii) Polarity reversal of iontophoresis current, and repetition of the process.
(iv) Addition of integrated biosensor currents at each cathode, and feeding to signal processing algorithm.
(v) A one-point calibration of the system after a 3 h interval was obtained from a single finger-stick blood glucose test.
(vi) Using the calibration factor to give glucose values every 20 min for 12 h.

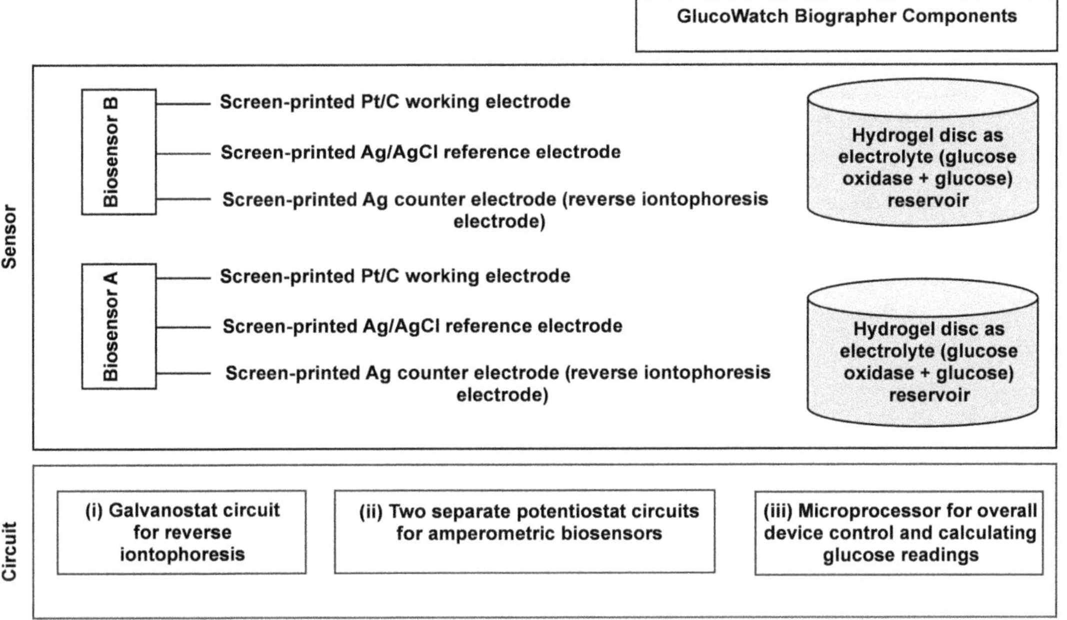

FIGURE 8.13 Main parts of the GlucoWatch Biographer. The biographer is made of two blocks: the sensor block and the circuit block. The sensor block contains two biosensors A and B, each with three screen-printed electrodes, a Pt/C working electrode, an Ag/AgCl reference electrode; and an Ag counter electrode for reverse iontophoresis. Further, each biosensor has a hydrogel disc which serves as the electrolyte (glucose oxidase + glucose) reservoir. The circuit block contains the following circuits: (i) a galvanostat circuit for reverse iontophoresis, (ii) two potentiostat circuits, one for biosensor A and the other for biosensor B, and (iii) a microprocessor which regulates the operation of the entire circuit, calculates glucose concentrations, and displays them digitally.

The device operation is microprocessor-controlled, providing glucose readings from recorded electrical signals. From a single blood glucose calibration value, the blood glucose is closely tracked from 2.2 to 22.2 mmol L^{-1} for 12 h. Evaluation of the performance of the sensor by carrying out large clinical trials showed that the sensor performed equally well irrespective of whether it was worn in a controlled clinical setting or in an uncontrolled home environment (Garg et al. 1999, Tierney et al. 2001).

The daily calibration with the fingerstick (step (v)) is an invasive step. It was shown that this step of calibration with a blood sample is avoidable by refining the method to use sodium ion as an internal standard, rendering the method completely non-invasive (Sieg et al. 2004).

8.9.5 HIGHLY INTEGRATED GLUCOSE MONITORING WATCH

8.9.5.1 The Biosensor Patch

It is a flexible electrochemical biosensor patch containing two glucose sensors (Chang et al. 2022). Figure 8.14 shows this watch. Its flexibility helps in making conformal contact with the skin. One sensor has a Nafion film coating on the top surface. The other does not have this coating. The biosensors are fabricated on a 100-μm-thick polyimide film substrate. Each sensor has three electrodes: a working electrode, a counter electrode, and a reference electrode. Further, each sensor also has two ISF extraction electrodes for reverse iontophoresis-based ISF extraction. The extraction is done transdermally at the user's wrist. For extracting ISF, 50 μA electric current is supplied through one end of each extraction electrode from a constant current source.

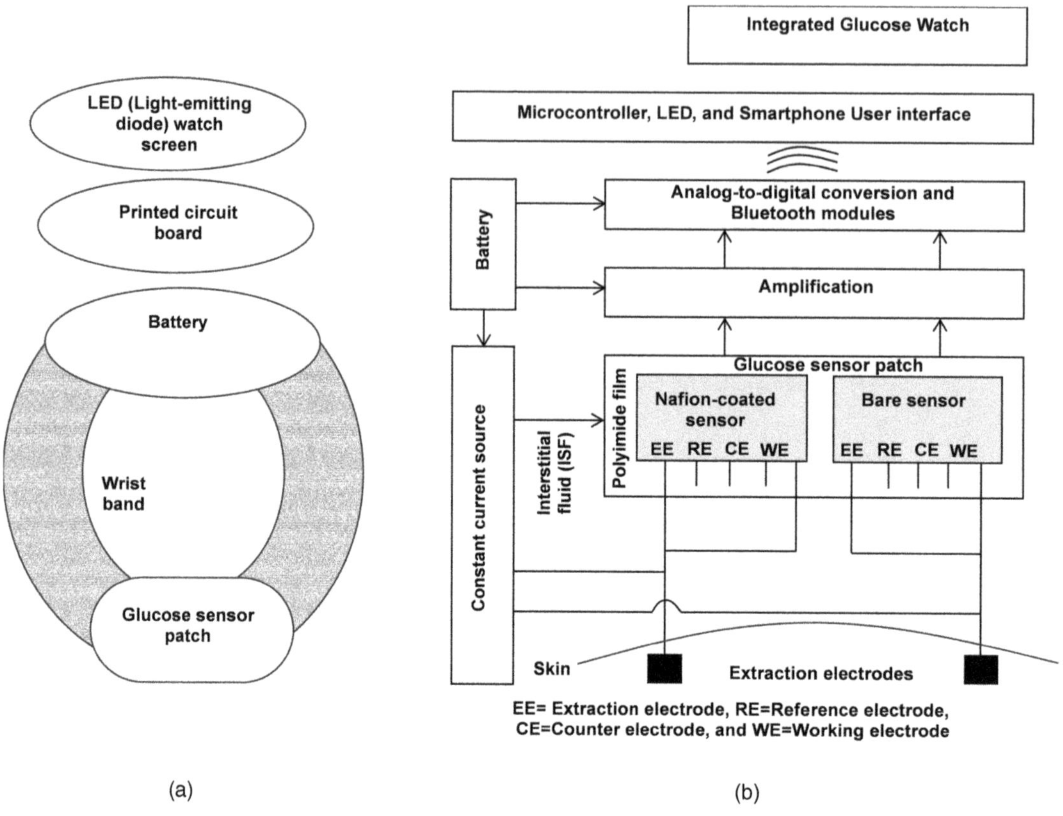

FIGURE 8.14 Integrated glucose watch: (a) constituent parts and (b) circuit. Part (a): Wrist band for tying the watch, glucose sensor patch in contact with the skin for glucose sample extraction as ISF, battery for supplying power to the watch, the printed circuit board on which the various circuit components are fixed, and the LED screen displaying the glucose levels. Part (b): A battery supplying power to the constant current source; A-to-D conversion and Bluetooth modules, and amplifier; the glucose sensor patch comprising a Nafion-coated sensor, and a bare sensor made on a polyimide film substrate, with each sensor having four electrodes, an ISF extraction electrode (EE), a reference electrode (RE), a counter electrode (CE), and a working electrode (WE). The extraction electrodes EE piercing the skin receive current signals from a constant current source to extract ISF. The extracted ISF is delivered by the constant current source to the glucose sensor patch. The glucose testing signal is amplified, converted into a digital format, and conveyed over Bluetooth to the microcontroller, LED, and smartphone interface.

8.9.5.2 Biosensor Electrodes

The electrodes of the biosensors are fabricated by depositing Cr/Au by magnetron sputtering on a polyimide film, and lift-off photolithography. The working electrodes are made by electrodeposition of Prussian blue (PB), $Fe(III)_4[Fe(II)(CN)_6]_3$ film on the Au electrodes, drop-casting a GO_x-containing selective membrane made of chitosan/ single-walled carbon nanotube mixture on PB/Au electrode and drop-casting Nafion solution over the selective membrane. The reference and extraction electrodes are made by screen printing Ag/AgCl on the Au electrodes. The counter electrodes are unmodified Au electrodes.

8.9.5.3 Two-Step Glucose Sensing Mechanism

The glucose-sensing mechanism consists of two steps:

(i) Oxidation of glucose in the presence of oxygen to produce glucono-δ-lactone and H_2O_2 with glucose oxidase acting as a catalytic agent

(ii) Reduction of H_2O_2 in the presence of Prussian blue electrocatalyst. Consumption of an electron by PB during the reduction reaction yields the amperometric response. PB shows a high reactivity and selectivity for H_2O_2 reduction.

8.9.5.4 Glucose Measurement

The biosensor patch is fixed on a circular watchband along together with the PCB circuit containing electronic modules for signal processing and wireless transmission, a rechargeable battery (3.3 V), and an LED screen (Figure 8.14(a)). Detection of the glucose molecules in the ISF by the two sensors produces a current signal (Figure 8.14(b)). From this current signal, a voltage signal is generated. The voltage signal is amplified by an instrumentation amplifier. The amplified voltage signal flows to an analog-to-digital converter. The resulting signal is sent to a microcontroller. The calibration algorithm is executed, and the blood glucose level is calculated. The numerical value of the glucose level is displayed on the LED screen of the watch. It is also conveyed to the smartphone user interface. A Bluetooth module is used for communication. Real-life testing of the watch was done on 23 volunteers. It was found that the results of blood glucose measurements were 84.34% clinically accurate. The Nafion film-coated sensor shows higher sensitivity and a smaller decrease in sensitivity over a two-week stability test than the one without Nafion film (Chang et al. 2022).

8.10 WEARABLE MULTIPLEXED PERSPIRATION ANALYSIS SENSOR ARRAY FOR SIMULTANEOUS DETECTION OF GLUCOSE, LACTATE, SODIUM, AND POTASSIUM IONS, AND SKIN TEMPERATURE

This sensor chip (Figure 8.15) detects a selection of analytes, including glucose (for diabetes) and lactate (a marker of pressure ischemia) metabolites; sodium and potassium electrolytes (for hyponatremia, hypokalemia, muscle cramps, or dehydration testing); and skin temperature (for understanding skin injuries such as pressure ulcers, and calibration of the response of sensors) (Gao et al. 2016). The aim is to assess an individual's physiological condition.

The chip is made on a flexible polyethylene terephthalate (PET) substrate to establish good contact with the skin. The metal lines for the contacts are protected from sweat and skin by a Parylene insulating film. The final electrode area is photolithographically defined, and parylene is removed from the unwanted portions by etching in oxygen plasma. Signal conditioning, processing, and wireless transmission circuitry are integrated into the chip. The glucose and lactate biosensors in the chip show stable sensitivities for more than four weeks. More details about the sensors in the chip are given below.

(a) Glucose and Lactate Amperometric Biosensors: The glucose sensor works with glucose oxidase enzyme while the lactate sensor uses lactate oxidase enzyme. The enzymes are immobilized in the respective sensors inside a permeable film. This film is made of linear

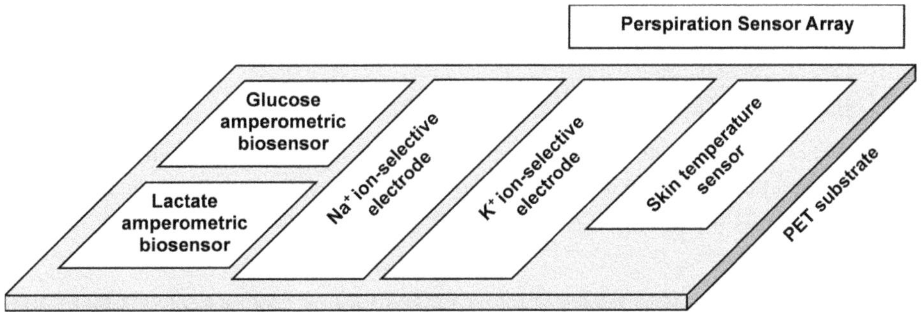

FIGURE 8.15 Perspiration sensor array chip. The chip consists of a substrate on which five sensors are made: glucose and lactate amperometric biosensors, Na^+ and K^+ ion-selective electrodes, and a skin temperature sensor.

polysaccharide chitosan. An Ag/AgCl electrode is shared as reference and counter electrodes between the two sensors. The reduction potentials are reduced to null value with respect to Ag/AgCl by using Prussian blue dye as a mediator, making these biosensors self-powered, hence eliminating the requirement of an external power supply for their functioning. Depending on the glucose/lactate concentration, an output current flows between the working electrode and the Ag/AgCl electrode. The measurements are done chronoamperometrically in the glucose concentration range 0–200 μM and lactate concentration range 0–30 mM. The current varies linearly with glucose concentration, showing a sensitivity of ~2.35 nA μM^{-1}. A linear relationship is also observed between current and lactate concentration with a sensitivity of ~220 nA mM^{-1}.

(b) Na$^+$ and K$^+$ Ion Sensors: These are ion-selective electrodes (ISEs) in which PEDOT: PSS {poly(3,4-ethylenedioxythiophene) polystyrene sulfonate} is used as an ion-to-electron transducer. A stable potential is maintained using a carbon nanotubes-mixed polyvinyl butyral (PVB)-coated reference electrode. The ISEs are potentiometric sensors giving a voltage output with near-Nerstian sensitivity of ~64.2 mV per decade for the Na$^+$ ion sensor and 61.3 mV per decade for the K$^+$ ion sensor.

(c) Temperature Sensor: It is a resistive temperature sensor. The sensitive element is made of Cr/Au metal microwires. Apart from skin temperature measurements, the output of this sensor is applied for temperature compensation of the chemical sensors with the help of a signal processor (Gao et al. 2016).

8.11 FALL DETECTION SENSORS

For elderly persons often living in solitary conditions, accidental fall detection is imperative to provide timely medical help. Two types of fall detection sensors have been developed.

8.11.1 WEARABLE FALL SENSOR

It consists of a triaxial accelerometer which is programmed to detect falling incidents usually characterized by a sudden downward acceleration followed by lack of movement. The sensor is connected to a professional emergency monitoring center. Two-way speech communication is also built into this arrangement. If the person confirms the fall, the emergency team sends the helping personnel to attend to the emergency. The inclusion of sensors like gyroscopes and magnetometers further enhances the capabilities of such systems, reducing the chances of false alarms.

8.11.2 AMBIENT FALL SENSOR

Cameras fitted at various locations in a person's home constantly keep a vigil on a person's movements. In case of a fall, the emergency monitoring center contacts the person. The person responds through a speaker. Immediately, necessary action is taken upon confirmation of the falling incident.

8.12 FITNESS TRACKER SENSORS

Fitness trackers are non-invasive, wearable, autonomous devices for monitoring physical activities that help us track our physiological data and vital health parameters over an extended period of time. Commonly used sensors in fitness trackers are given in further subsections.

8.12.1 GPS SENSOR

A GPS-equipped fitness tracker is very helpful for determining a person's location (Sec. 10.3) and for tracking outdoor workouts, such as running or cycling. It aids in route planning and improves data accuracy. However, for indoor exercise, it is expensive and also unnecessarily drains the battery.

8.12.2 Magnetometer

In conjunction with GPS and compass, the magnetometer helps in accurately finding the coordinates of the user's location.

8.12.3 Triaxial Accelerometer and Pedometer

An accelerometer can sense the inclination, tilt, and orientation of the body. It is used to count the number of steps taken during walking, the distance traveled by a person, and calories burned in bodily activity. A 3-axis MEMS accelerometer interfaced with a microcontroller uses a devised algorithm to calculate the number of steps taken by a person (Sai et al. 2016). The calculation is done by analyzing the acceleration of the person's hand swing during walking, and noting that one full hand swing takes place when two steps are covered. The device displays the total distance traveled by the person. Thus, it functions as a pedometer, an instrument used for measuring the distance walked. The application of a mathematical formula to the data collected gives the total calories burned, which is shown on an LCD display.

8.12.4 Three-Axis Gyroscope

It shows the orientation such as horizontal or vertical, and angular rotation. A MEMS gyroscope is commonly used.

8.12.5 Optical Heart Rate Monitor and Oximetry Sensor

These are based on transmissive or reflective PPG devices. Dual-wavelength PPG is done in pulse oximetry.

8.12.6 ECG Sensor

This sensor records the heart signal through electrodes placed on the skin.

8.12.7 Altimeter

It measures changes in height such as during climbing stairs or a hill or when moving down a slope, by correlating the atmospheric pressure with the altitude of a place. The atmospheric pressure is determined by an aneroid barometer or a MEMS pressure sensor. The aneroid barometer has a flexible evacuated capsule with a corrugated diaphragm around its surface (Figure 8.16). A metal spring and a thin rod are fixed at the top of the capsule. The rod is connected through hinges, a pivot, and a chain to an indicator moving on a scale. When the atmospheric pressure increases, such as on descending from a height, the capsule is compressed and the rod fixed to it is depressed. The motion of the diaphragm rotates the needle showing the pressure reading.

8.12.8 Bioimpedance Sensor for Monitoring Respiratory Rate

The thoracic impedance changes during the expansion/contraction of the chest and abdomen when a person is breathing (Singh et al. 2020). This happens because of variations in the conductivity and volume of underlying tissues by respiratory movements. The bioimpedance sensor used to track the impedance variations consists of electrodes placed on the skin of the patient to create a closed electrical circuit. Two-electrode (bipolar) and four-electrode (tetrapolar) configurations are used (Aqueveque et al. 2020). The configurations are shown in Figure 8.17. An alternating current at 50–100 kHz is injected into the chest. The measurements reveal the respiratory patterns of the person.

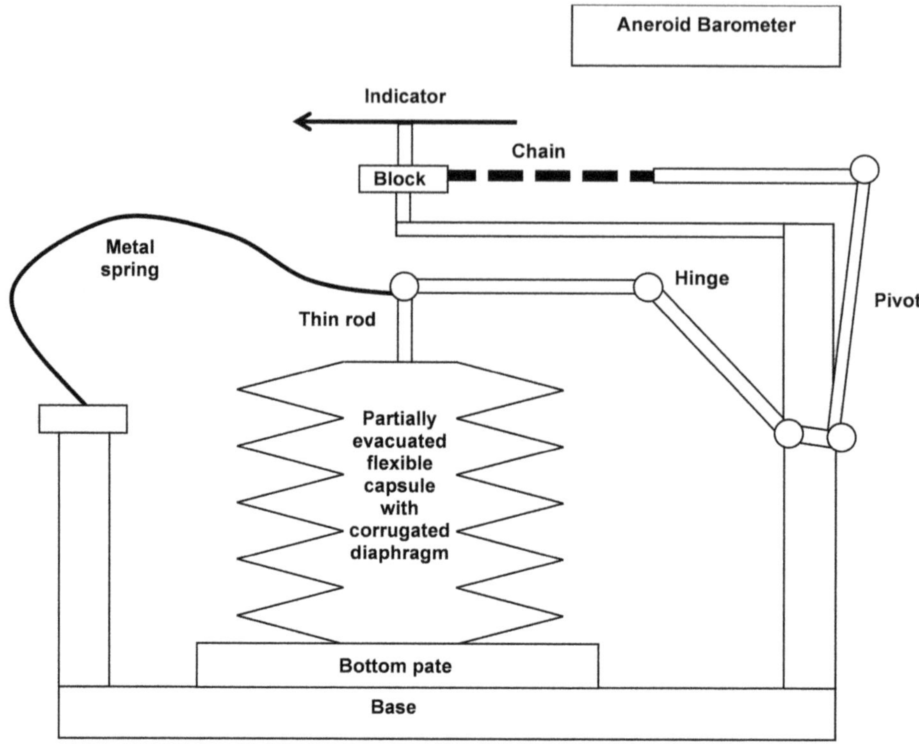

FIGURE 8.16 Aneroid barometer-based altimeter. On the base of the device rests a bottom plate on which a flexible capsule with a corrugated diaphragm is mounted. The capsule has a partial vacuum inside it. At the middle of the top surface of the capsule, a thin rod is fitted. To the upper end of the thin rod, a spring is attached. This end is also connected to hinges, a pivot, and a chain at the opposite extremity of which a block and a rod are fixed for the pointer, as shown by an arrow, for indicating the height after conversion of the pressure reading into the corresponding altitude.

8.12.9 Electrodermal Activity (EDA) Sensor

The calm/distress condition of a person is reflected by his/her electrodermal activity (Zangróniz et al. 2017). The EDA records the changes in conductivity produced in the skin, such as in the palms of the hands or soles of the feet, resulting from enhancements in the activity of sweat glands. Two Ag/AgCl disc electrodes with contact diameters of 10 mm are attached to the medial phalanges in the palm sides of the index and middle fingers, and a small DC current $< 10\ \mu A/cm^2$ is applied to the stratum corneum under the electrodes in order to avoid any damaging effect on the sweat gland ducts.

8.12.10 Skin Temperature Sensor

It is used to measure the skin temperature. Although different from internal body temperature, the skin temperature can indicate potential fever. Normal skin temperature for healthy adults lies between 33°C and 37°C at the wrist.

8.12.11 UV Sensor

This sensor assesses the exposure of the body to UV radiation when we step out in the sun, providing us an estimate of the harmful effect of the UV component of sunlight on our bodies. The UV signal is converted into an electrical signal using a photosensor with ultraviolet transmission material for 250–400 nm wavelengths, in the photovoltaic mode.

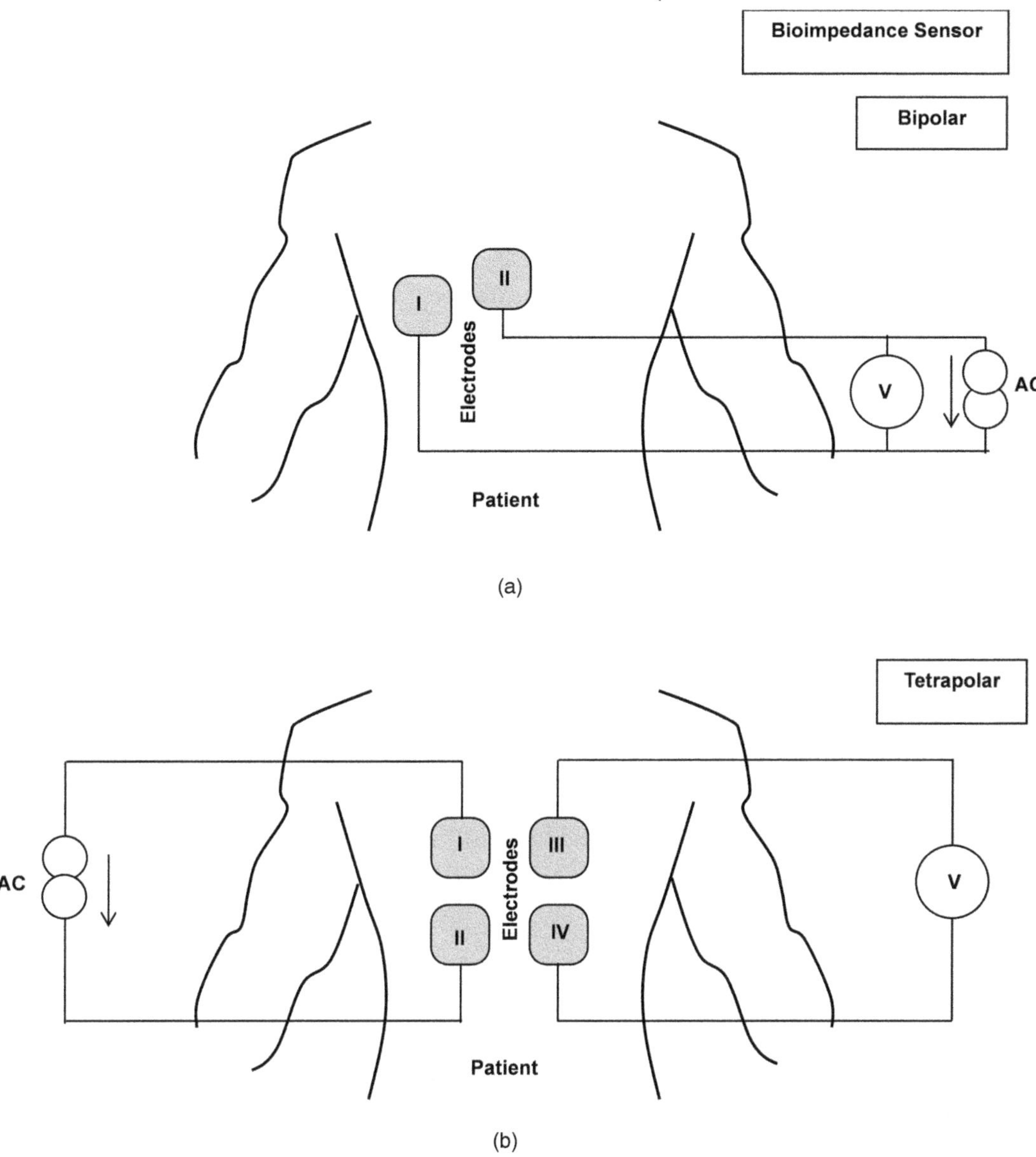

FIGURE 8.17 Arrangements for measurements of bioimpedance for determination of respiratory rate, using: (a) two electrodes, and (b) four electrodes. Both parts (a) and (b) show a person. Electrodes are fixed on the surface of the person's chest at a certain distance apart. Part (a): Bipolar configuration in which the two electrodes are supplied a high-frequency current from an alternating current supply, and the voltage across the same two electrodes is measured with a voltmeter connected between them. Hence, only two electrodes are used, serving the dual function of supplying current and voltage measurement. Part (b): Tetrapolar configuration in which two electrodes placed at a certain distance apart are used for supplying alternating current to the chest surface while the potential difference between two different electrodes placed at a chosen separation distance is measured with the help of a voltmeter. Hence, a total of four electrodes are used in this configuration, two for supplying current and another two for measuring voltage.

8.12.12 Ambient Light Sensor

The ambient light sensor saves the battery life of a device by tweaking the brightness of its display in accordance with the surrounding illumination.

8.12.13 Proximity Sensor

The proximity sensor enables a device to go into sleep mode when no one is nearby, and to awake it when the user comes, thus increasing the battery life. The sensors in Sections 8.12.12 and 8.12.13 are energy-saving sensors incorporated into fitness trackers.

8.12.14 Hand Gesture Recognition Sensor

Infrared and ultrasonic sensors, accelerometers, and cameras are used to capture the spatial position and movement of the hand. The captured data are fed to complex machine learning algorithms, enabling recognition and interpretation of particular hand gestures. Gesture recognition falls under the broad discipline of computer vision.

8.13 CONCLUDING REMARKS AND PREPARING FOR THE UPCOMING CHAPTER

Human healthcare is a subject of paramount importance worldwide, and embellishing health services with sensors and internet connectivity makes these services easily and rapidly accessible to a wider cross section of the population. So, healthcare is receiving increasing attention everywhere.

For a healthy life, we need a healthy diet and this is only possible by taking care of the growth and nourishment of plants, and by providing a conducive water supply and climatic conditions to them, a topic which comes under agricultural science. This topic will be pursued in the upcoming chapter.

REFERENCES

Apple Watch User Guide © 2023 Apple Inc., watchOS 10.2, pp. 1–395. https://help.apple.com/pdf/watch/10/en_US/apple-watch-user-guide-watchos10.pdf

Attivissimo F., L. De Palma, A. Di Nisio, M. Scarpetta, and A. M. L. Lanzolla 2023 Photoplethysmography signal wavelet enhancement and novel features selection for non-invasive cuff-less blood pressure monitoring, *Sensors*, 23:2321, pp. 1–18.

Aqueveque P., B. Gómez, E. Monsalve, E. Germany, P. Ortega-Bastidas, S. Dubo, and E. J. Pino 2020 Simple wireless impedance pneumography system for unobtrusive sensing of respiration, *Sensors (Basel)*, 20(18):5228, pp. 1–16.

Chang T., H. Li, N. Zhang, X. Jiang, X. Yu, Q. Yang, Z. Jin, H. Meng and L. Chang 2022 Highly integrated watch for noninvasive continual glucose monitoring, *Microsystems & Nanoengineering*, 8: pp. 1–9. Article number: 25.

Feng Z. 2013–2015 Pulse oximeter design using microchip's analog devices and dsPIC® digital signal controllers (DSCs), AN1525, © 2013–2015 Microchip Technology Inc., DS00001525B-page 1 to DS00001525B-page 12.

Gao W., S. Emaminejad, H.Y. Y. Nyein, S. Challa, K. Chen, A. Peck, H. M. Fahad, H. Ota, et al. 2016 Fully integrated wearable sensor arrays for multiplexed *in situ* perspiration analysis, *Nature*, 529: pp. 509–514.

Garg S. K., R. O. Potts, N. R. Ackerman, S. J. Fermi, J. A. Tamada, and H. P. Chase 1999 Correlation of fingerstick blood glucose measurements with GlucoWatch biographer glucose results in young subjects with type 1 diabetes, *Diabetes Care*, 22(10): pp. 1708–1714.

Himmelreich J. C. L., E. P. M. Karregat, W. A. M. Lucassen, H. C. P. M. van Weert, J. R. de Groot, M. L. Handoko, R. Nijveldt, and R. E. Harskamp 2019 Diagnostic accuracy of a smartphone-operated, single-lead electrocardiography device for detection of rhythm and conduction abnormalities in primary care, *Annals of Family Medicine*, 17(5): pp. 403–411.

Hou Y., Y. Wang and Y. Zheng 2017 TagBreathe: Monitor breathing with commodity RFID systems *2017 IEEE 37th International Conference on Distributed Computing Systems (ICDCS)*, 5–8 June, Atlanta, GA, pp. 404–413.

Hussain T., S. Ullah, R. Fernandez-Garcia and I. Gil 2023 Wearable sensors for respiration monitoring: A review, *Sensors*, 23(17):7518, pp. 1–28.

Isakadze N. and S. S. Martin 2020 How useful is the smartwatch ECG? *Trends in Cardiovascular Medicine*, 30(7): pp. 442–448.

Kardia AliveCor Inc. 2017 User Manual for Kardia™ Mobile by AliveCor®, 00LB17 Revision 4 © 2011–2017 AliveCor, pp. 1–26. https://alivecor.com/previous-labeling-files/kardiamobile/00LB17.4.pdf

Konstantinidis D., P. Iliakis, F. Tatakis, K. Thomopoulos, K. Dimitriadis, D. Tousoulis and K. Tsioufis 2022 Wearable blood pressure measurement devices and new approaches in hypertension management: The digital era, *Journal of Human Hypertension*, 36: pp. 945–951.

Kumar S., S. Yadav, and A. Kumar 2022 Oscillometric waveform evaluation for blood pressure devices, *Biomedical Engineering Advances*, 4:100046, pp. 1–10.

Kuwabara M., K. Harada, Y. Hishiki and K. Kario 2019 Validation of two watch-type wearable blood pressure monitors according to the ANSI/AAMI/ISO81060-2:2013 guidelines: Omron HEM-6410T-ZM and HEM-6410T-ZL, *Journal of Clinical Hypertension (Greenwich)*, 21(6): pp. 853–858.

Kuzubaşoğlu B.A., E. Sayar, C. Cochrane, V. Koncar, and S. K. Bahadir 2021 Wearable temperature sensor for human body temperature detection, *Journal of Materials Science: Materials in Electronics*, 32: pp. 4784–4797.

Lee S., G. Jeon and G. Lee 2013 On using maximum a posteriori probability based on a Bayesian model for oscillometric blood pressure estimation, *Sensors*, 13(10): pp. 13609–13623.

Lewis P.S. 2019 On behalf of the British and Irish hypertension society's blood pressure measurement working party. Oscillometric measurement of blood pressure: A simplified explanation. A technical note on behalf of the British and Irish, *Hypertension Society, Journal of Human Hypertension*, 33: pp. 349–351.

Lin B., M. Atef and G. Wang 2019 14.85 μW Analog front-end for photoplethysmography acquisition with 142-dBΩ gain and 64.2-pA$_{rms}$ noise, *Sensors*, 19:512, pp. 1–13.

Lopez S. 2012 Pulse Oximeter Fundamentals and Design, Freescale Semiconductor Document Number: AN4327, Application Note, Rev. 2, 11/2012 © 2011 Freescale Semiconductor, Inc., pp. 1–39.

Mildenhall J. L. 2008 The theory and application of pulse oximetry, *Journal of Paramedic Practice*, 1(2): pp. 52–58.

Nazarian S., K. Lam, A. Darzi, H. Ashrafian 2021 Diagnostic accuracy of smartwatches for the detection of cardiac arrhythmia: Systematic review and meta-analysis, *Journal of Medical Internet Research*, 23(8): p. e28974.

Panda A., S. Pinisetty and P. S. Roop 2022 Runtime monitoring and statistical approaches for correlation analysis of ECG and PPG, ArXiv, Vol. abs/2202.00559, https://api.semanticscholar.org/CorpusID:246442402

Park J., H. S. Seok, S.-S. Kim and H. Shin 2022 Photoplethysmogram analysis and applications: An integrative review, *Frontiers in Physiology*, 12:808451, pp. 1–23.

Park S. W., P. S. Das, A. Chhetry and J. Y. Park 2017 A flexible capacitive pressure sensor for wearable respiration monitoring system, *IEEE Sensors Journal*, 17(20): pp. 6558–6564.

Potts R. O., Tamada J. A., and Tierney M. J. 2002 Glucose monitoring by reverse iontophoresis, *Diabetes/Metabolism Research and Reviews*, 18(Suppl 1): pp. S49–53.

Sai R. P., S. Bapanapalle, K. Praveen and M. P. Sunil 2016 Pedometer and calorie calculator for fitness tracking using MEMS digital accelerometer, *2016 International Conference on Inventive Computation Technologies (ICICT)*, Coimbatore, India, 26–27 August, pp. 1–6.

Selder J.L., L. Breukel, S. Blok, A. C. van Rossum, I. I. Tulevski and C. P. Allaart 2019 A mobile one-lead ECG device incorporated in a symptom-driven remote arrhythmia monitoring program. The first 5,982 Hartwacht ECGs, *Netherlands Heart Journal*, 27: pp. 38–45.

Sieg A., R.H. Guy, and M. B. Delgado-Charro 2004 Noninvasive glucose monitoring by reverse iontophoresis in vivo: Application of the internal standard concept, *Clinical Chemistry*, 50(8): pp. 1383–1390.

Singh G., A. Tee, T. Trakoolwilaiwan, A. Taha, and M. Olivo 2020 Method of respiratory rate measurement using a unique wearable platform and an adaptive optical-based approach, *Intensive Care Medicine Experimental*, 8(1):15, pp. 1–10.

Tajin M. A. S., C.E. Amanatides, G. Dion and K.R. Dandekar 2021 Passive UHF RFID-based knitted wearable compression sensor, *IEEE Internet Things Journal*, 8(17): pp. 13763–13773.

Tierney M. J., J. A. Tamada, R. O. Potts, L. Jovanovic, and S. Garg 2001 Clinical evaluation of the GlucoWatch® biographer: a continual, non-invasive glucose monitor for patients with diabetes, *Biosensors and Bioelectronics*, 16(9–12): pp. 621–629.

Tierney M., Y. Jayalakshmi, N. A. Parris, M. P. Reidy, C. Uhegbu and P. Vijayakumar 1999 Design of a biosensor for continual, transdermal glucose monitoring, *Clinical Chemistry*, 45: pp. 1681–1683.

Utomo T. P. and Nuryani N. 2021 Photoplethysmogram peaks detection based on moving window integration and threshold for heart rate calculation on android smartphone, 10th International Conference on Physics and Its Applications (ICOPIA 2020), *Journal of Physics: Conference Series*, 1825:012032, pp. 1–8.

Witvliet M. P., E. P. M. Karregat, J. C. L. Himmelreich, J. S. S. G. de Jong, W. A. M. Lucassen, and R. E. Harskamp 2021 Usefulness, pitfalls and interpretation of handheld single-lead electrocardiograms, *Journal of Electrocardiology*, 66: pp. 33–37.

Zangróniz R., A. Martínez-Rodrigo, J.M. Pastor, M.T. López, and A. Fernández-Caballero 2017 Electrodermal activity sensor for classification of calm/distress condition, *Sensors (Basel)*, 17(10):2324, pp. 1–14.

9 IoT Sensors for Precision Agriculture and Farming

9.1 INTRODUCTION

Agriculture is a broad scientific discipline dealing with the science of soil cultivation for crop production, research and development of crops, and rearing livestock, viz., domesticated animals for food and fiber. Farming is the implementation of agricultural activities. Agriculture and farming constitute the backbone of our society. In this chapter, we seek to delve into the use of sensors in the assessment of the quality of soil, and for keeping a watch on the health of crops and the conditions of livestock.

9.2 PRECISION AGRICULTURE AND FARMING

Smart agriculture refers to an integrated approach to managing farming with the help of modern information and communication technologies such as sensors, the Internet of Things, drones, GPS mapping, robotics, machine learning, and data analytics, developed at the dawn of the Fourth Industrial Revolution for monitoring crops, insects and livestock to increase the quantity as well as quality of crops, all accomplished with the optimum use of human labor and best utilization of scarce resources, and causing minimal environmental impact. It stands on three strong pillars, namely, growth of agricultural productivity and income, reinforcement of resilience capability of livelihoods toward climate change, and reduction/removal of greenhouse gas emissions from the atmosphere.

Precision agriculture and farming are umbrella concepts in which the sensors and IoT-based techniques are applied to control farming in an improved way to achieve the utmost accuracy to boost production, save water, reduce wastages, and keep track of plant growth or animal health anomalies. The crops, the hen, and cattle receive the desired attention, care, and nourishment calculated by machines allowing decision-making per square meter of field or per plant/animal, e.g., artificial intelligence-powered robots give a clear interpretation of field images to aid in the effective use of fertilizers and pesticides. Timely identification and isolation of sick animals from the herd prevents the spreading of communicable or infectious diseases.

9.3 COLLECTING REAL-TIME DATA ON SOIL MOISTURE CONTENT AND COMBINING WITH WEATHER FORECASTS FOR OPTIMUM WATER USAGE IN IRRIGATION: SOIL MOISTURE AND RAIN SENSORS

Soil moisture, the quantity of water within 1–2 m depth below the soil surface is a crucial ecohydrological resource. It is a key variable governing plant growth and health of the ecosystem. It is the medium through which plants acquire nutrients from the soil. It is essential for photosynthesis. It provides turgidity to plants enabling them to stand upright. Information about soil moisture level along with temperature and humidity data, and historical records is fed to machine learning algorithms for predicting plant growth. The analysis helps farmers in scheduling planting and harvesting.

Temperature, humidity, and soil moisture sensors are the main devices used in this analysis. Soil moisture quantification is the focus of attention. Several methods have been used for soil moisture estimation (Zazueta and Xin 1994, Pandey et al. 2020). Main soil moisture measuring techniques are described further.

9.3.1 Neutron Moisture Probe

This probe (Figure 9.1) works on the principle of neutron scattering from hydrogen atoms of the water molecules in soil (Chanasyk and Anne Naeth 1996). It is one of the most accurate methods of soil moisture measurement. A neutron-transparent access tube is installed inside the ground. A source tube containing a source of high-energy neutrons, e.g., californium or americium 241/beryllium pellet (americium in a beryllium matrix), and a detector of low-energy neutrons ~0.025eV (thermal neutrons), is inserted in the access tube and lowered to the required depth. The neutrons ejected from the neutron source collide with the water molecules in the surrounding soil, and are backscattered from them. These backscattered neutrons are still lower energy neutrons because the water molecules

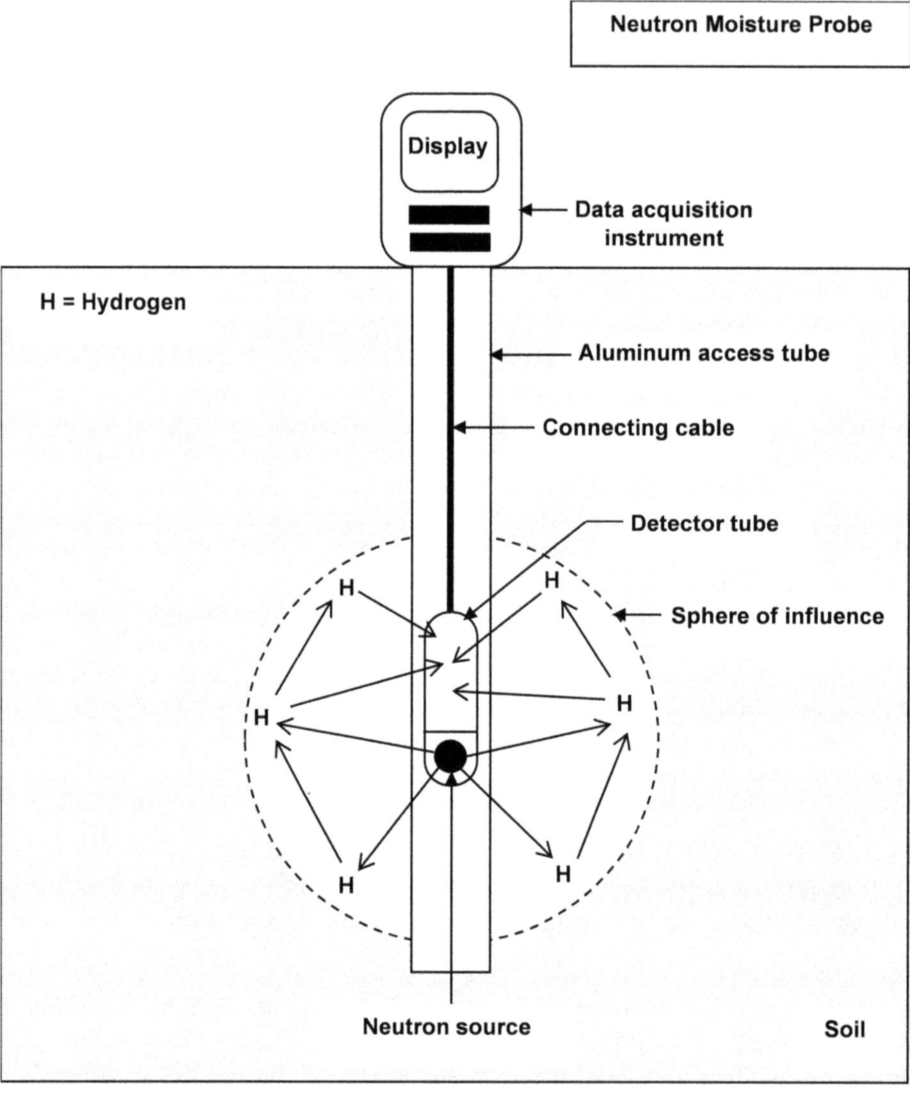

FIGURE 9.1 Constructional details of a neutron moisture probe. A neutron source and a detector tube are suspended at the lower end of a cable inside a long aluminum access tube introduced by penetration into the soil through drilling up to the decided probing depth. The cable is connected to the data acquisition instrument with the display. Paths of neutrons emitted from the source, those scattered by hydrogen atoms in the soil, and those rebounding to the detector tube are shown by arrows. The sphere of influence is drawn.

contain hydrogen, which acts as a neutron moderator, slowing down the neutrons to give reduced kinetic energy neutrons. The backscattered neutrons are counted by the detector of slow neutrons placed inside the source tube. The backscattered flux of low electrons is proportional to the density of hydrogen atoms in the soil around the neutron source. Since these hydrogen atoms are present inside the water molecules, an estimation of soil water content is readily obtained from the measured backscattered neutron flux.

Advantages

(i) The volume of soil sampled is about the size of a volleyball. Hence, the method gives a closer idea about soil moisture content than those sampling smaller volumes.
(ii) Measurements of soil moisture can be performed at different depths in a short time span of a few minutes.

Disadvantages

(i) Measurements at shallow depths below the soil surface are susceptible to errors because some neutrons escape into the air, evading detection.
(ii) The neutron source is a radioactive material and must be handled carefully avoiding any exposure of human operators. The storage and transportation guidelines of the neutron source must be strictly followed. Hence, the method can be used only by trained operators. A license is essential for possessing and using the neutron moisture probe.

9.3.2 Gamma Emission Soil Moisture Sensor

The gamma soil moisture sensor (gSMS) measures the minuscule amount of radiation emitted by naturally occurring radionuclides, which are ubiquitously present in soil samples (van der Veeke et al. 2020). In order for the signal to be a true representation of moisture present in the soil, the influence of gamma radiation falling on the sensor from the radioactive radon gas present in the atmosphere must be excluded from the detected signal. The radon-induced noise is filtered out by performing a spectral analysis.

The sensor has a 50.8 mm × 50.8 mm cesium iodide (CsI) scintillation crystal. This crystal is coupled to a multichannel analyzer. The γ-ray spectra are continuously recorded using an embedded microcontroller. Analysis of the spectra reveals the ^{222}Rn (as ^{238}U), ^{40}K, and ^{232}Th concentrations in real time. Only the concentrations of the more stable ^{40}K and ^{232}Th radionuclides are expected to vary with soil moisture. So, the contribution from ^{222}Rn is separated from the contributions of ^{40}K and ^{232}Th. The spectral data over a given time interval are saved in a thumb drive. Through the internet, they are uploaded to an online platform for user view. A battery charged by a solar panel powers the sensor.

9.3.3 Gamma Attenuation Soil Moisture Sensor

This method is based on the principle that gamma radiation incident on a material is absorbed and scattered in proportion to the density of the material, and a change in the moisture content of the soil affects its saturated density. Therefore, by measuring the absorption/scattering of gamma rays on soil, its density is determined from which its moisture content is calculated.

The setup consists of ^{241}Am and ^{137}Cs γ-ray sources enclosed inside a lead shield (Rasheed et al. 2022). The γ-rays emerge out from a hole in the shield and impinge on the soil specimen after collimation (Figure 9.2). The soil sample is kept in an acrylic chamber on a porous plate from which the water in the sample can ooze out by seepage. The pressure of the chamber is measured with a gauge.

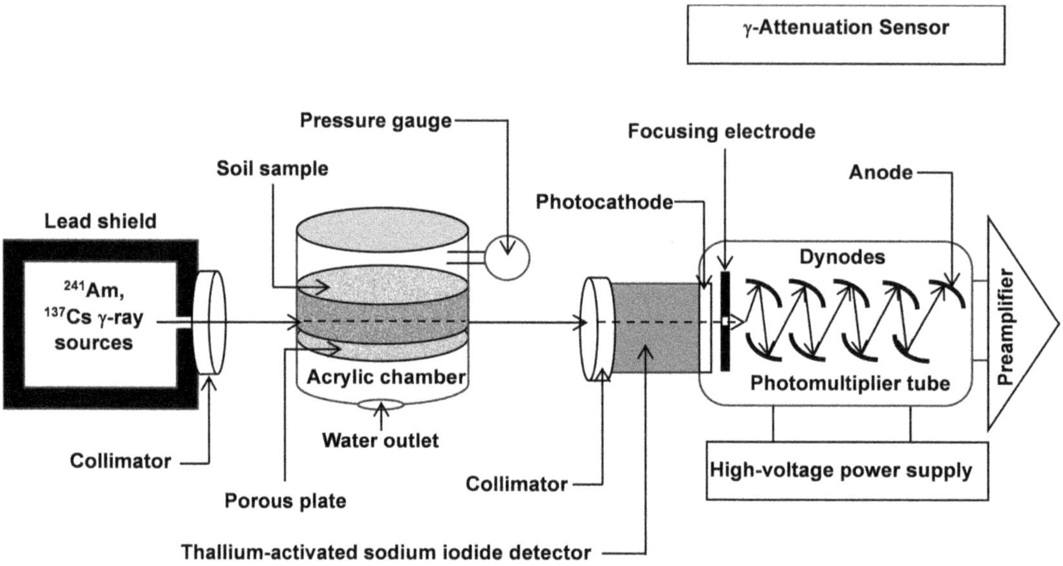

FIGURE 9.2 γ-attenuation sensor. ^{241}Am and ^{137}Cs γ-ray sources are enclosed in a lead shield to prevent radiation leakage. The γ-rays coming out through an opening in the lead shield fall on a collimator. The soil specimen is kept on a porous plate in an acrylic chamber. The base plate of the chamber has an outlet for water discharge. A pressure gauge is fitted in the chamber to measure the pressure in the chamber. After crossing the soil specimen, the γ-rays fall on a collimator and then hit a Tl-activated NaI detector. The scintillations produced in the detector strike the photocathode of a photomultiplier tube operated by a high-voltage power supply. A focusing electrode directs the electron beam to undergo successive reflections from a series of dynodes. Avalanche multiplication of the accelerated electrons produces a large number of secondary electrons. These electrons are collected by an anode, and the output current is fed to a preamplifier.

The γ-rays coming out from the soil are again collimated, and the collimated γ-ray beam strikes a thallium-activated sodium iodide NaI-Tl detector, producing a scintillation. A photomultiplier tube converts this scintillation into an electrical pulse and feeds the pulse to a preamplifier.

Advantages

> The method is non-destructive. It is easy to calibrate and automate. Temporal changes in soil moisture can be mapped by this method.

Disadvantages

> It is restricted to soil samples of thickness less than 2.5 mm. It is expensive and difficult to use because of the hazardous nature of γ-rays.

9.3.4 TDR (Time Domain Reflectometry) Soil Moisture Sensor

The main idea behind this sensor is to utilize the dependence of propagation constants of electromagnetic waves, e.g., their velocity of motion in the soil on the dielectric constant of the soil, which in turn is related to the amount of moisture present in it (Skierucha et al. 2012). So, the moisture content of the soil is inferred from its relative permittivity, and the relative permittivity is determined from the velocity of electromagnetic waves. A smaller value of the wave velocity resulting from a large degree of velocity attenuation by the soil, indicates a higher moisture percentage.

The chief component of the sensor is a waveguide, comprising three or more parallel wires (Figure 9.3). The waveguide is embedded in the soil to be examined. Its other end is connected by a

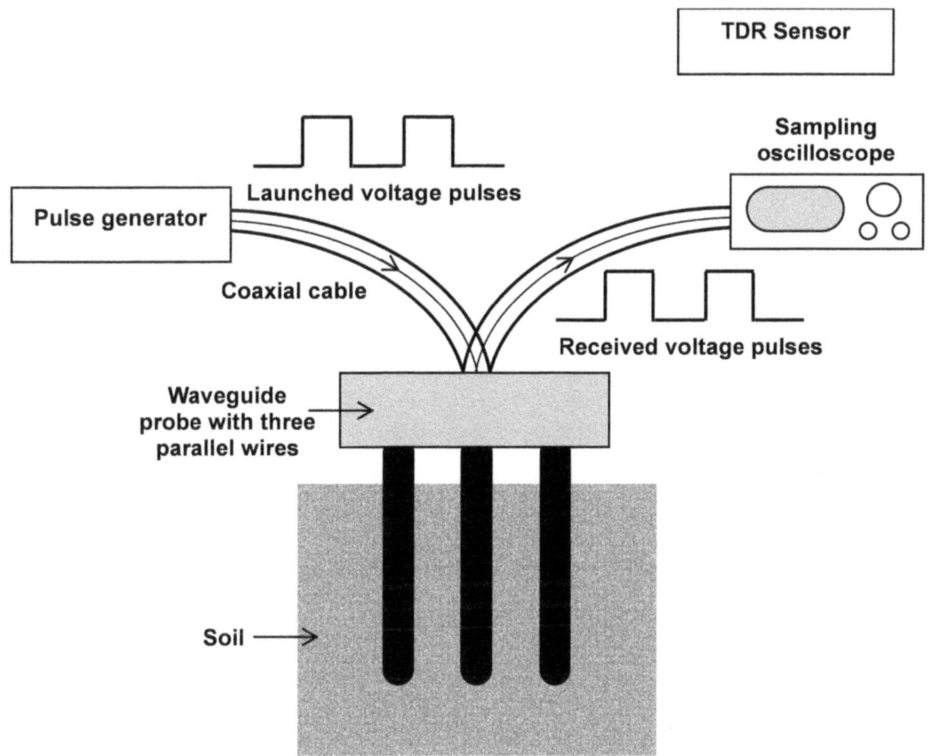

FIGURE 9.3 Time domain reflectometry soil moisture sensor. Voltage pulses are launched from a pulse generator through a coaxial cable toward a waveguide with three probes buried under the soil. The pulses reflected back from the ends of the probes are collected by a sampling oscilloscope for analysis and determination of the permittivity of the soil, indicating its moisture content.

coaxial cable to a pulse generator supplying voltage pulses. Each voltage pulse traverses the waveguide and the velocity of its movement varies with the quantity of moisture in the surrounding soil. When the voltage pulse reaches the end of the waveguide, it is reflected back. A TDR waveform is observed in an oscilloscope connected to the device. The time interval between the launching of the voltage pulse and the receipt of the reflected pulse is measured to find its traveling time and hence velocity. By applying an equivalent circuit model of the sensor, the relative permittivity of the soil sample is evaluated, giving its average volumetric moisture content.

The above method, called the standard waveform analysis, does not give the spatial profile of moisture. A profile analysis is necessary to incorporate spatial information in the obtained value. It is done by modeling the propagation of the voltage pulse in the waveguide. The model is calibrated with reference to measurements carried out in the laboratory. Then the spatial distribution of moisture is deduced by comparison of measurements on actual samples with the model.

Advantages

It is an easy, non-destructive method that gives continuous high-resolution data output, both spatially and temporally.

Disadvantages

The necessary electronic equipment is complex and costly. Moreover, the method has restricted application for saline soils.

9.3.5 FDR (Frequency Domain Reflectometry) Soil Moisture Sensor

Its principle of operation is similar to that of the TDR sensor (Figure 9.4). The distinguishing feature is that the relative permittivity and degree of wetness of soil are determined by the difference in frequency between the incident and returning waves (Shock et al. 2016). After installing a PVC pipe in the soil, the waveguide probe is inserted in that pipe. An oscillator circuit is connected to the waveguide.

The FDR sensor works by sweeping a range of input frequencies and reconstructing the reflected signals by conversion into the time domain using inverse fast Fourier transform (IFFT).

Advantages

It is a non-destructive technique. The FDR sensor is relatively easier to build and inexpensive than the TDR sensor. Unlike the TDR sensor, it can work in high-salinity soils.

Disadvantages

The sensor must be calibrated for each type of soil because the electric field around it is complex. Its use over an extended period of time is uncertain.

9.3.6 Tensiometer

9.3.6.1 Soil Moisture Tension

Soil moisture tension, also called soil water potential is a measure of the firmness or tenacity with which water is grasped by the soil, i.e., the force that the roots of the plants need to apply for extracting water from the soil. When the soil dries, more energy or pressure is necessary for water extraction from it. So, soil moisture tension is high. When the soil is sufficiently wet, the application of much smaller pressure suffices to extract water from the soil. Accordingly, the soil moisture tension

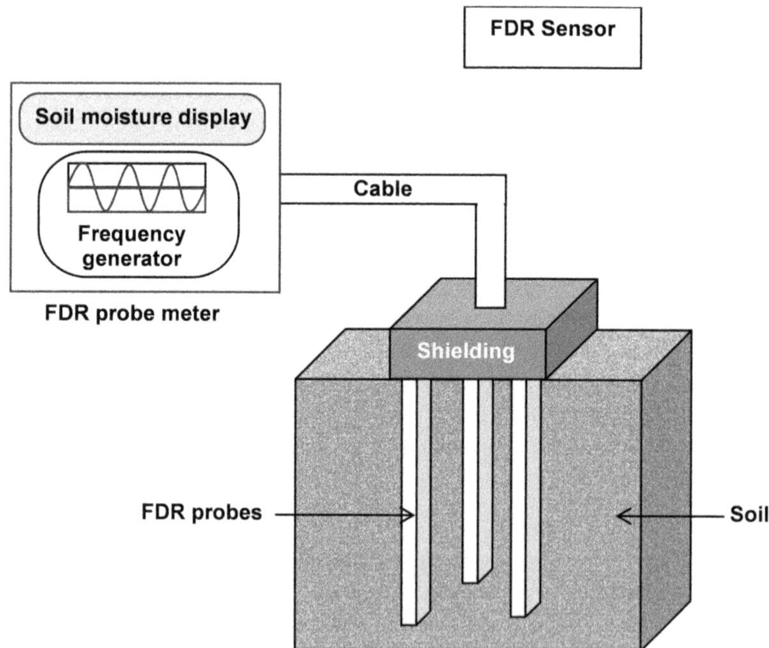

FIGURE 9.4 Frequency domain reflectometry soil moisture sensor. An oscillatory signal is sent through a cable to a waveguide probe with three under-the-soil probes. The frequency difference between incident and reflected waves is measured, soil permittivity is assessed, and soil moisture content is depicted on an LCD screen.

is low. The soil moisture tension around the plants in deserts is ~60 bars while that in a soil saturated with water is 0.1 centibar. Perennial horticultural crops grow in the soil moisture tension range from 8 to 60 centibar.

9.3.6.2 Tensiometer Principle and Use

The soil moisture tension is measured by an instrument known as a tensiometer (Figure 9.5). It is used for scheduling irrigation in intelligent irrigation systems. The tensiometer consists of a sealed,

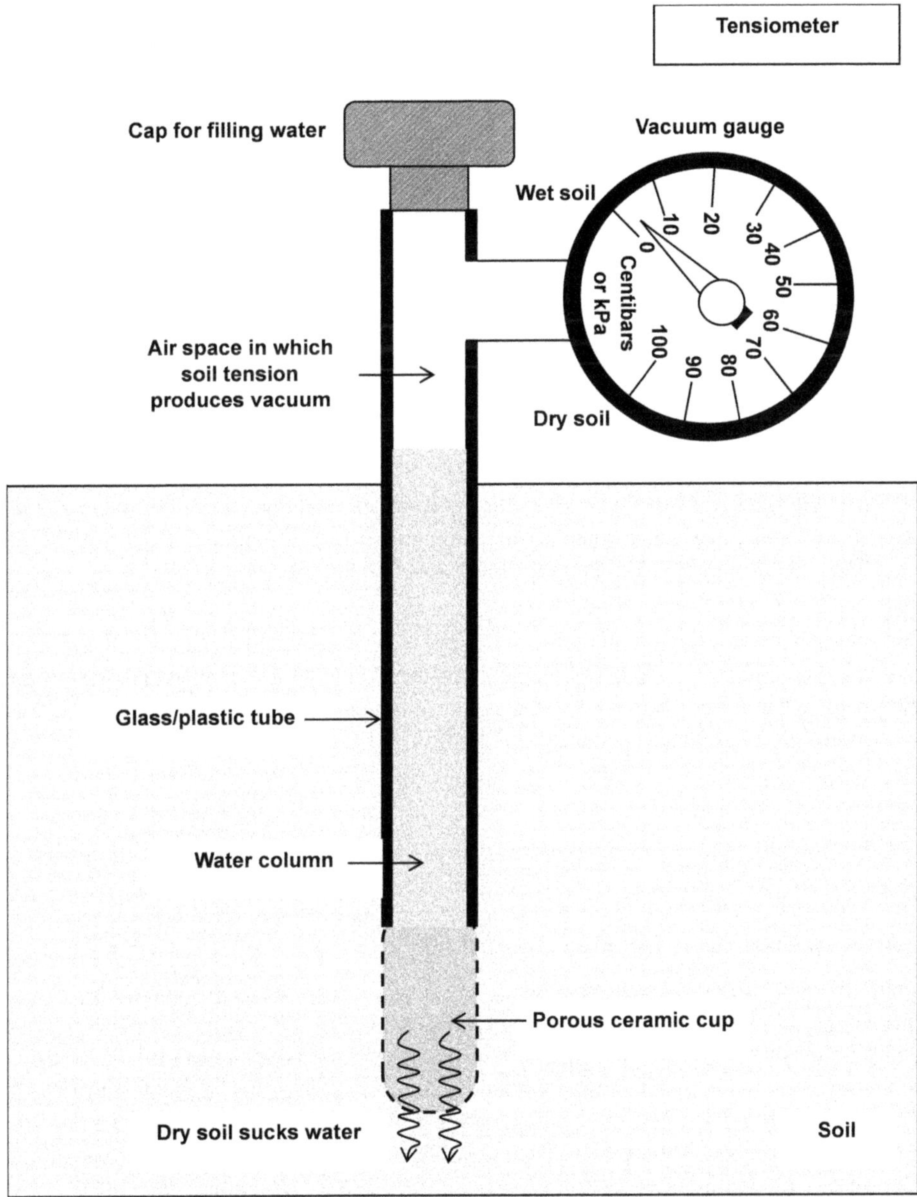

FIGURE 9.5 Sketch of a tensiometer. The drawing shows a glass/plastic tube fixed inside the soil, the water column inside it, a porous ceramic cup at the bottom of the tube, a closure cap at the top of the tube, and a vacuum gauge (0–100cbar) near the upper end of the tube. Sucking of water from the tensiometer by dry soil is seen. It is indicated by the downward directed wavy arrows. A reading near 100 cbar indicates extremely dry soil while a 0 cbar reading points toward very wet soil. The air space whose pressure is measured is marked.

air-tight glass or plastic tube filled with water, and fitted with a porous ceramic cup at the bottom side (Martin 2009). Available in a range of lengths from 15 to 180 cm, it is installed with the ceramic cup at a suitable pre-selected depth below the soil surface. The tube is filled with water up to a level below a vacuum gauge to measure the pressure of the air space in which the soil tension is shown to generate vacuum. After filling with water, the tube is firmly closed with a cap at the top. The vacuum gauge measures pressure from 0 to 100 centibar or kPa. A reading approaching 100 cbar implies very dry soil, which takes away plenty of water from the tensiometer, creating a high vacuum. On the opposite extreme, a reading of 0 cbar means that the soil is wet or saturated with water. So, it does not draw any water from the tensiometer.

Modern models use high-sensitivity electronic transducers in place of the traditional vacuum gauge.

Advantages

(i) It is a low-cost, easy-to-operate, robust, portable, accurate, and reliable instrument that can be interfaced with a datalogger or hand-held readout/display device for fast measurements.
(ii) Its readings are immune to the salinity of the soil.

Disadvantages

(i) It requires periodic maintenance and servicing.
(ii) It is not useful for the examination of dry soils as it tends to lose good contact with the soil.

9.3.7 Resistive Soil Moisture Sensor

Soil resistivity falls as its moisture level rises. The soil resistivity variation is the basis for this sensor. Either resistance between two electrodes fixed in the soil at a certain distance apart is measured, or the resistance of a material in equilibrium with the soil is assessed.

Advantages

It is one of the very low-cost devices among the available soil moisture sensors. It gives a fairly accurate reading when the ionic concentration of soil does not change. The size of the soil sample is adjustable, and so also the depth in the soil at which the experiment is done.

Disadvantages

Its calibration varies mainly because of changes in the ionic concentration of the soil.

9.3.8 Gypsum Block Soil Moisture Sensor

Also known as a soil moisture block, it consists of two electrodes fixed in a small block of a porous material such as gypsum (Figure 9.6). The block is buried under the soil at the chosen depth of measurement (Stenitzer 1993). The wires attached to the electrodes come out of the soil surface and are connected to an alternating current bridge circuit. Water moves in or out of the gypsum block buried under the soil. This water movement continues until the matric potential of the gypsum block = the matric potential of soil. Matric potential is the attractive force acting between a soil particle and pore spaces. It ranges from 0 for water-saturated soil to 1×10^7 hPa for dry soil.

The electrical resistance of the gypsum block obtained from the bridge circuit is related to the quantity of water vapor in the soil.

Advantages

Gypsum blocks are easily installed at low costs.

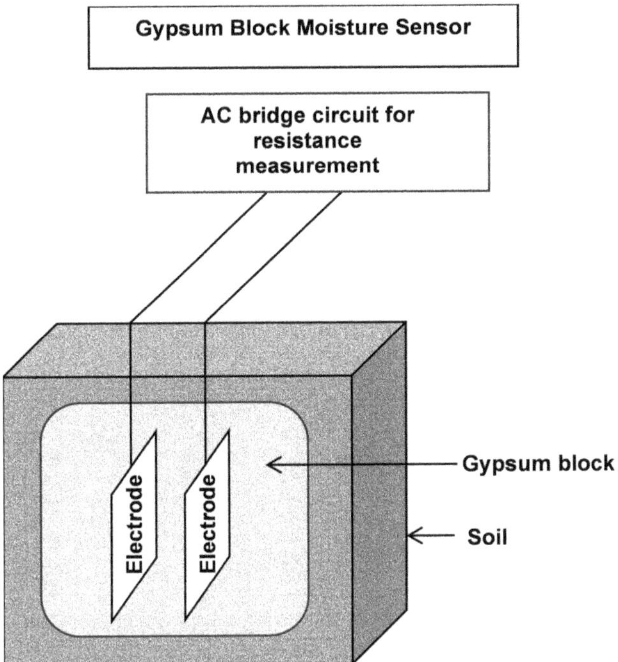

FIGURE 9.6 Gypsum block soil moisture sensor. A gypsum block is buried under the ground at a given depth. There are two electrodes inside the block. These electrodes are connected by wires to an AC bridge circuit placed above the ground to determine the soil resistance, which indicates its degree of wetness.

Disadvantages

Each gypsum block must be individually calibrated. The calibration drifts with time. Further, the gypsum blocks must be periodically replaced as gypsum dissolves and disintegrates.

9.3.9 Capacitive Soil Moisture Sensor

The capacitance of the soil is measured between two electrodes implanted in the soil at a certain distance apart by exciting the electrodes at a given frequency (Figure 9.7). The dielectric constant of the intervening soil medium gives its volumetric water content.

Advantages

The moisture content of soil is measurable at any specified depth and the area of the soil sample can be selected as desired. The method is precise if the ionic concentration of the soil does not change.

Disadvantages

The method is costly and not suitable for long-term measurements.

9.3.10 Rain Sensors

A rain sensor prevents the wastage of water which happens when a sprinkler continues to splash water around during rain. One type of rain sensor consists of a water-absorbing disc which swells on

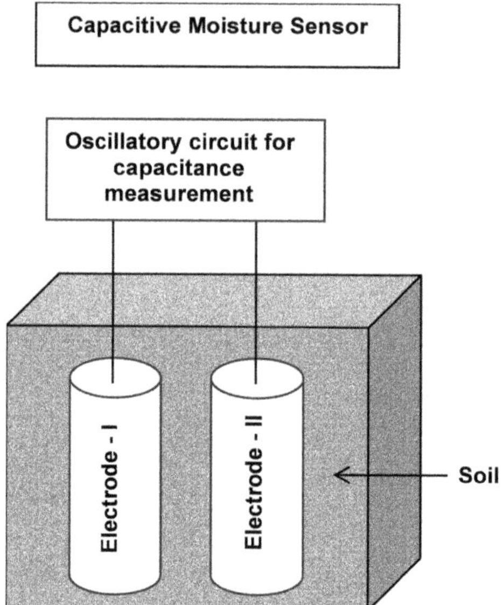

FIGURE 9.7 Capacitive soil moisture sensor. Two electrodes are implanted in the ground at a given depth, and connected by wires to a capacitance meter equipped with an oscillatory circuit. The capacitance meter and the oscillatory circuit are placed above the ground. The capacitance of the soil at a particular frequency is measured using this arrangement. The obtained capacitance value is related to the quantity of moisture inside the soil at that depth below the ground level.

wetting and returns to its normal size on drying. This disc is used to trigger the opening of the sprinkler switch during rain to stop its function. Again, when the rain ceases to fall and the disc dries up, the switch is closed to turn the sprinkler on.

Another type of rain sensor consists of an array of electrodes made on an insulating substrate at a fixed distance apart (Yogesh et al. 2021). When the electrode array is dry, it shows a high resistance. The circuit is designed in such a way as to keep the sprinkler switch ON in this condition. Raindrops falling on the sensor short-circuit the electrodes because the conductive rainwater film built on the interelectrode gaps allows current to flow easily between the electrodes, thereby reducing the resistance of the sensor. The circuit is designed such that the lowering of resistance opens the sprinkler switch, turning it OFF.

9.4 SOIL FERTILITY MANAGEMENT: PLANT MACRONUTRIENT SENSORS

Sensor-aided, on-field soil property monitoring systems help in checking soil fertility status, thereby promoting the efficient use of fertilizers and organic manures (Aarthi et al. 2023).

9.4.1 PLANT MACRONUTRIENTS

Plants get food mostly from the soil. Some nutrients are also obtained through photosynthesis. Plants must be continuously supplied some nutrients in large quantities called macronutrients, and some in smaller amounts known as micronutrients. Macronutrients are subdivided into two classes: primary macronutrients, nitrogen (N), phosphorus (P), and potassium (K), needed in the highest concentrations; and secondary macronutrients, calcium (Ca), magnesium (Mg), and sulfur (S), required in lower concentrations than the macronutrients. As the soil nutrients give an indication of the fertility of the soil, it is essential to monitor their concentrations via *in situ* measurements using sensors. These sensors are of two types: optical and electrochemical.

9.4.2 Optical Plant Nutrient Sensors

9.4.2.1 Visible-Infrared Spectroscopy

The Visible-Infrared (Vis-IR) spectroscopy works by shining light from a visible, and infrared poly-chromatic radiation source on the soil sample (Burton et al. 2020). The visible portion of the electromagnetic spectrum extends from 380 to 780 nm while the IR portion lies between 780 nm and 1 mm consisting of near-infrared (780–2,500 nm), middle infrared (2,500–25,000 nm), and far-infrared (25,000 nm–1 mm) regions. The principal components of the system are a source of light, mechanisms for light isolation, the detector, and sampling devices.

The light energy is absorbed and diffusely reflected by soil nutrients. The reflected light falls on the spectrophotometer, forming a reflectance spectrum representing the distribution of light energy as a function of wavelength. It is a graph of absorbance (or transmittance) of infrared light versus wavelength. Specific molecules are identified by comparing the characteristic peaks or valleys in the recorded spectrum with a reference spectral database.

Vis-NIR spectroscopy is validated as an accurate multi-nutrient availability index evaluator for soil from a study of estimation of the plant-available phosphorous, and identification of responsive sites for phosphorous, potassium, magnesium, calcium, and iron, confirming its value in soil fertility assessment (Recena et al. 2019).

9.4.2.2 Hyperspectral Imaging Sensor

Hyperspectral Imaging (HSI) works by acquisition of images at continuous wavelength bands over a given segment of the spectrum (Figure 9.8). Combining spectroscopy with imaging, it yields both spectral and spatial information. A unique spectral autograph originates from the physical and chemical properties of the material. The HSI setup comprises a light source, a spectrograph, a camera, a mechanical translation stage, and a computer.

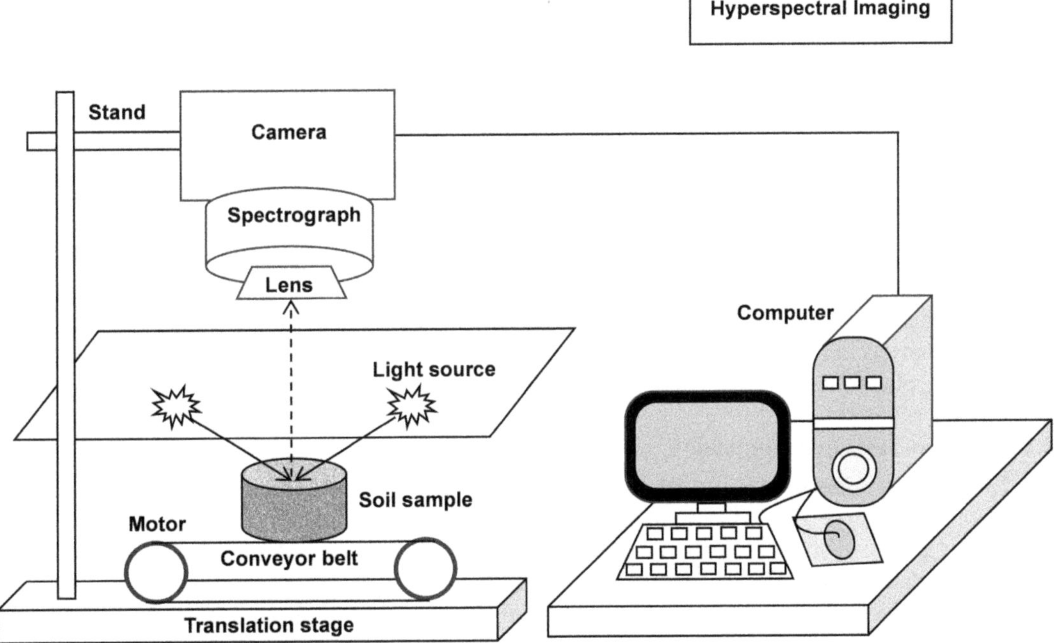

FIGURE 9.8 Hyperspectral imaging sensor. A conveyor belt moves on a translation stage fitted with a stand. The soil sample is placed on the conveyor belt driven by a motor, and illuminated with light sources. A lens is fitted above the soil sample. The light received by the lens falls on a spectrograph. A camera mounted above the spectrograph and supported by the stand records the image signal, which is supplied to the computer.

Acceptability of HSI to excellent predictability of soil properties was observed in a study carried out using hyperspectral remote sensing. Properties such as soil pH, salinity (electrical conductivity), bulk density (BD), together with the presence of elements, e.g., nitrogen(N), zinc (Zn), magnesium (Mg), and boron (B) could be predicted accurately (Mahajan et al. 2020).

9.4.2.3 Attenuated Total Reflectance Spectroscopy

In the attenuated total reflectance (ATR) spectroscopy, the IR energy strikes a diamond or zinc selenide crystal placed in contact with the soil sample. The crystal reflects the IR light producing evanescence. The reflected energy from the crystal falls on the spectrometer generating a reflectance spectrum, which is analyzed by the standard procedure.

ATR-FTIR (Attenuated total reflectance – Fourier-transform infrared) spectroscopy is used for analysis of soil organic matter (SOM) contents and its constituents, as well as for reckoning adsorbed water and minerals (Volkov et al. 2021).

9.4.2.4 Raman Spectroscopy

It is a versatile method of soil testing. Monochromatic light from a laser source at visible/infrared wavelength interacts with the electron cloud and chemical bonds of the soil specimen. A major fraction of light undergoes scattering without any change in energy; this is Rayleigh scattering. However, a small percentage of light is inelastically scattered, either losing energy to molecular vibrations or gaining energy during the process; this is Raman scattering. So, the energy of the light beam increases or decreases. The scattered energy is measured, and the energy shift is correlated with the vibrational modes of the sample. These vibrational energies are specific to the structure and composition of the molecules, serving as their chemical signatures. Therefore, the Raman spectrum of the soil sample gives the necessary chemical and molecular structural information about it.

Effects of specific soil-related factors on Raman spectra are investigated by creating a controlled environment and methodically varying sample compositions to help in the development of new techniques for accurate and reliable soil analysis (Zarei et al. 2023). As soil is a complex granular mixture, the study is performed by analyzing the chemical composition of 48 mixtures consisting of randomized weight proportions of six amino acids: L-histidine, L-glutamic acid, L-aspartic acid, L-(+)-asparagine, DL-alanine, and D-glucosamine-HCl, together with a randomized weight fraction of bentonite clay, and randomized proportions of pulverized quartz and calcite.

9.4.3 Electrochemical Plant Nutrient Sensors

The various nutrients are absorbed by the plant roots in ionic forms, e.g., nitrogen as nitrate ions (NO_3^-), phosphorous as orthophosphate ions, $H_2PO_4^-$ and HPO_4^{2-}, and potassium as potassium ions K^+, and similarly for other nutrients. The relevant ionic species is measured to get an idea of the concerned nutrient concentration. Two electrochemical sensors are widely used for this purpose: the ion-selective electrodes (ISEs) and the ion-sensitive field-effect transistor (ISFET).

9.4.3.1 Ion-Selective Electrode

The ISE is a potentiometric sensor (Adamchuk et al. 2005). It converts the change in effective concentration (activity) of a specific ion in a specimen solution into a potential. The measurement setup consists of an electrochemical cell containing the sample electrolyte solution. Into this electrolyte solution, two electrodes are immersed (Figure 9.9).

One electrode is the ISE consisting of a silver chloride-coated silver wire dipped in a saturated potassium chloride solution, and communicating with the sample solution through a membrane which allows a particular ion to pass through it while restricting the inflow of other ions; hence, it is called the ion-selective membrane.

The other electrode is the Ag/AgCl reference electrode with saturated KCl solution communicating with the sample solution via a porous frit, not the ion-selective membrane as in ISE.

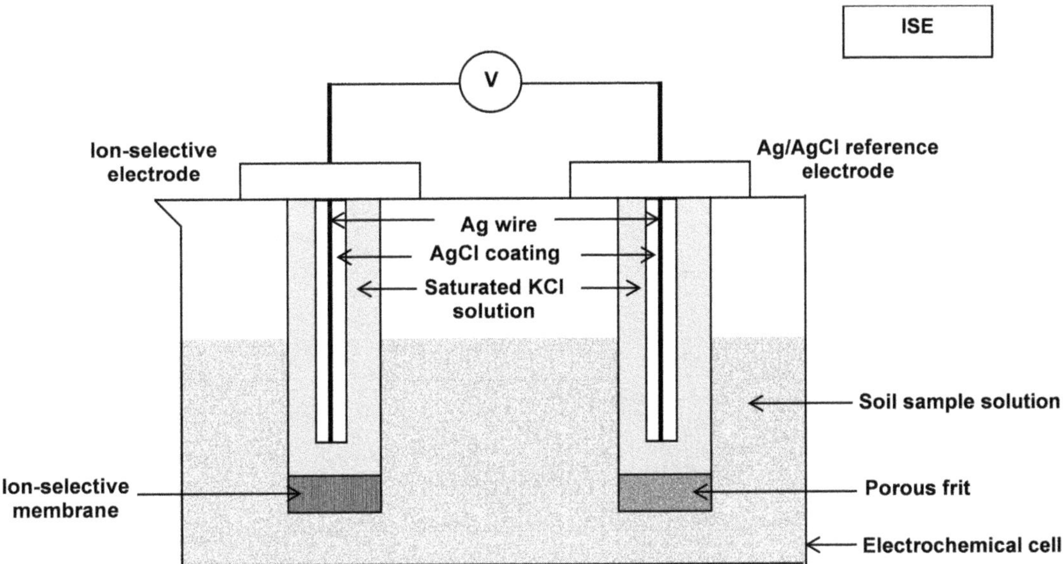

FIGURE 9.9   Ion-selective electrode. The solution of the soil sample is taken in the electrochemical cell. Two electrodes are immersed in the soil sample solution. One electrode is the ion-selective electrode consisting of a silver wire coated with a silver chloride film and dipped in a saturated solution of potassium chloride. The bottom of the electrode is sealed with an ion-selective membrane, which is permeable only to specific ions. The other electrode is the silver/ silver chloride electrode, which has an identical construction to the ion-selective electrode except that its bottom end is closed with porous frit instead of the ion-selective membrane. The potential difference between the two electrodes is measured by connecting a voltmeter across them.

A millivoltmeter connected between the two electrodes shows an electrical potential related to the natural logarithm of the ion concentration in accordance with the Nernst equation. This equation is expressed as

$$E_{\text{Cell}} = E_{\text{cell}}^0 - \frac{RT}{nF} \ln\left(\frac{[X^-]}{[X]}\right)$$

(9.1)

for the reaction

$$X + e^- \leftrightarrow X^-$$

(9.2)

where the symbol E_{cell} represents the electrode potential in Volts, E_{cell}^0 is the standard reduction potential of the couple (X/X$^-$) in Volts, R denotes the universal gas constant $=8.315$ J K^{-1} mol^{-1}, T is the temperature in Kelvin scale, n typifies the number of electrons transferred ($=1$ here), F stands for Faraday's constant $= 96,485$ Coulombs mol^{-1}, $[X^-]$ is the concentration of X$^-$ and $[X]$ that of chemical species X in moles L^{-1}.

Popular ISE membranes for the detection of soil nutrients (N, NO$_3^-$, K$^-$, H$_2$PO$_4^-$, and HPO$_4^{2-}$) are as follows.

   (i) Nitrogen: Nitrogen-doped polypyrrole (N-doped PPy),
  (ii) Nitrate ion: Tetradodecylammonium o-nitrophenyl octyl ether (TDDA-NPOE),
 (iii) Potassium ion: Valinomycin-bis(2-ethylhexyl) sebacate (DOS), and
 (iv) Phosphate ion: Cobalt rod-based (Kim 2006).

9.4.3.2 Ion-Sensitive Field-Effect Transistor

The ISFET is a solid-state potentiometric sensor (Figure 9.10) which transforms the changes in ionic concentrations into corresponding changes in gate potential of an FET device in the same way as the potential changes are created in ISEs (Archbold et al. 2023). The main differences between

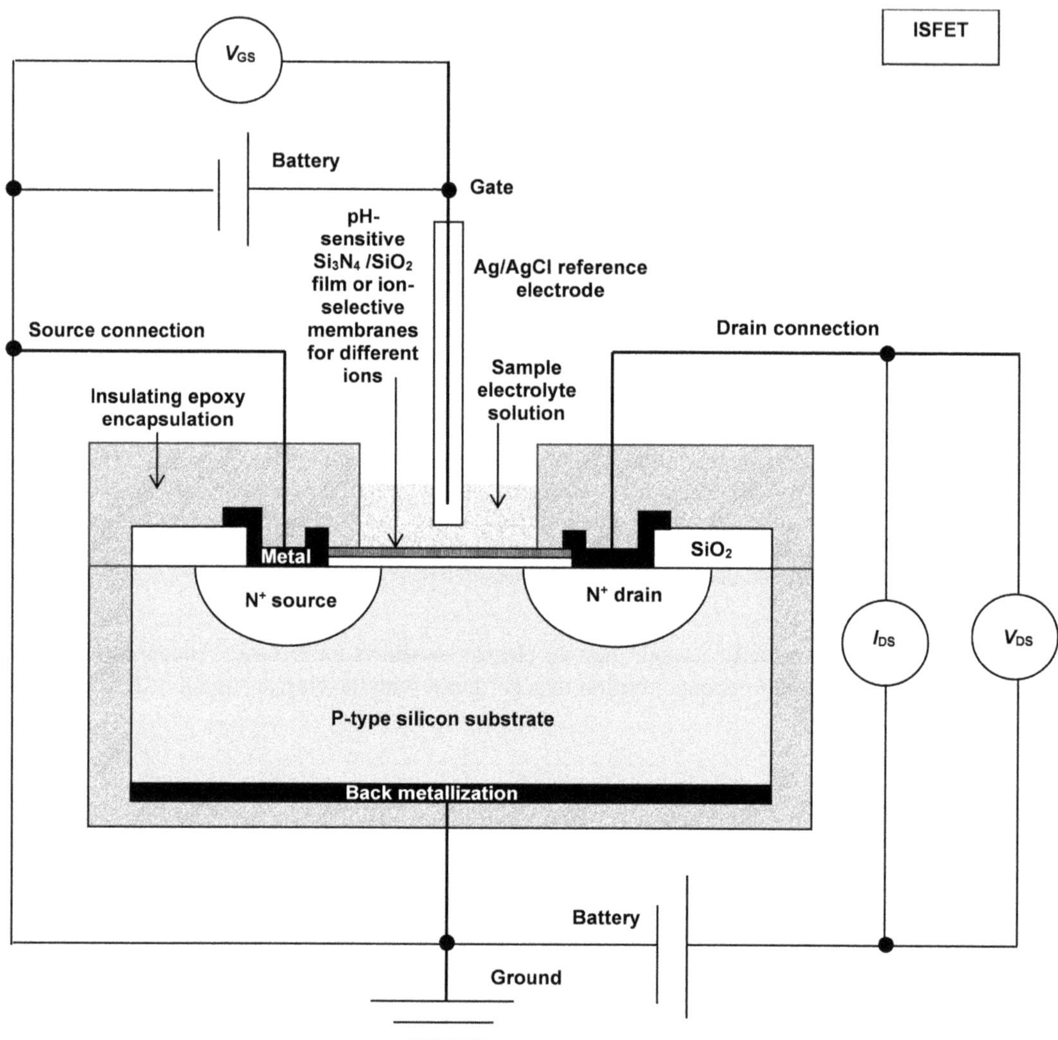

FIGURE 9.10 Ion-sensitive field-effect transistor. A P-type silicon substrate is shown in which N⁺ source and drain regions are diffused. The surface of the Si substrate separating these regions is covered with a thin silicon dioxide gate dielectric layer passivated with a silicon nitride film or the relevant ion-selective membrane chosen for the ion to be detected. Over the N⁺ source and drain regions, a metal film is deposited for contacts while the surrounding regions are covered with thick field oxide. The back surface of the P-substrate is also metalized. Wires are bonded to the metal films for electrical connections and the whole device is coated with an insulating epoxy except for a small portion over the gate dielectric where the sample electrolyte solution is poured for analysis. An Ag/AgCl reference electrode is dipped in the sample solution. It is connected to the positive terminal of a battery for applying the gate–source potential V_{GS}. The source terminal is short-circuited to the back metallization, which itself is grounded. The positive terminal of the battery is connected to the drain while its negative terminal is connected to the ground. The drain–source potential V_{DS} is measured by a voltmeter while the drain–source current I_{DS} is recorded with an ammeter.

the ISFET and the standard metal-oxide-semiconductor field-effect transistor (MOSFET) are as follows:

(i) There is no metal layer on the gate of ISFET.
(ii) The ISFET gate oxide is covered with an insulating layer which is sensitive to the particular ion to be detected, and this layer is exposed to the analyte solution.
(iii) In the ISFET measurements, the gate potential is applied to an Ag/AgCl reference electrode dipped in the analyte solution.
(iv) The source and drain contact regions of the ISFET are protected by covering them with an insulating epoxy through which the connecting wires are taken out.

For measuring the hydrogen ion concentration or pH of a solution Si_3N_4, Al_2O_3, or Ta_2O_5 films are coated on the SiO_2 gate dielectric. For detecting soil nutrient ions, special ion-sensing or blocking membranes are coated over the gate SiO_2 layer as in the case of ISE.

9.4.3.3 Nitrate ISFET

A polyvinyl chloride (PVC) membrane is modified with ethylenediamine solution to get aminated PVC, which is oxidized with sodium tungstate forming a nitrone-coated PVC membrane; nitrone is the N-oxide of an imine. The ISFET fabricated by immobilizing this membrane on the gate shows a nitrate detection limit of 1.20 ppm. Its characteristic is linear between 3 and 20 ppm. The sensitivity is 56 mV decade^{-1} (Chaisriratanakul et al. 2020).

9.4.3.4 Potassium ISFET

The SiO_2/Si_3N_4 gate of ISFET is a polymer made from plasticizer-free poly (n-butyl acrylate) (nBA), containing valinomycin as an ionophore for K^+ ions. A lipophilic additive is included for enhancing the membrane resistance. The K^+ ISFET has a detection limit of 3×10^{-3} mM. Its dynamic linear range is 10^{-2}–100 mM. The sensitivity is 57.8 mV decade^{-1}(Ling et al. 2009). An ionophore is a substance which transports only particular ions across it.

9.4.3.5 Hydrogen Phosphate ISFET

The ISFET with PVC-dibutyl phthalate (DBP) membrane shows a dynamic linear range from 3.0×10^{-5} to 3.0×10^{-2} M with a super Nernstian sensitivity of 69.54 mV-decade^{-1} for monohydrogen phosphate HPO_4^{2-} ions up to a detection limit of 1 μM (Hidouri et al. 2015).

The ISFET with PVC-o-NPOE membrane too shows the same detection limit of 1 μM for HPO_4^{2-} ions with a sensitivity of 58.9 mV-decade^{-1} from 10 μM to 10 mM.

A dihydrogen phosphate ISFET is obtained with a membrane of PVC-BA (benzyl acetate). It has a dynamic range of ~ 0.1 μM to 0.1 M for KH_2PO_4 ions. Its sensitivity is 51.8 mV-decade^{-1}, and the limit of detection is <0.1 μM (Hidouri et al. 2015).

9.4.3.6 Multisensor ISFET Probe for Detecting Soil Nutrients

The probe has four Ta_2O_5 gate ISFETs (Lehmann and Grisel 2014). These ISFETs detect pH, K^+, NO_3^- and $H_2PO_4^-$ ions. They are made by depositing different ion-selective membranes on their gate areas. These membranes are made of a PVC matrix in which ionophores of the selected ions are incorporated. The ISFET sensors are mounted in a flow cell made of LTCC (low-temperature cofired ceramics). A solid-state Ag/AgCl reference electrode is included in the flow cell. The sensitivity for pH detection is 55 mV/decade, that for K^+ ions is 48 mV/decade and for NO_3^- ions is 33 mV/decade. Selective detection of $H_2PO_4^-$ ions is also demonstrated.

Advantages of ISEs/ISFETs

(i) Can be manufactured on a large scale at low costs.
(ii) Provide real-time and accurate measurements.
(iii) Measurements can be performed in unfiltered solutions of soil taking proper precautions to avoid clogging.

Disadvantages of ISEs/ISFETs

(i) Require frequent recalibration, preferably before each experiment.
(ii) The results are prone to errors from interfering ions.
(iii) The values obtained often depend on soil texture and particle size.
(iv) The sensor must be thoroughly rinsed in water after measuring each sample.

9.5 DETECTION OF PLANT DISEASES AND PEST THREATS: VISIBLE IMAGE SENSOR, THERMOGRAPHY, FLUORESCENCE, HYPERSPECTRAL IMAGING AND GAS CHROMATOGRAPHY SENSORS

Plant diseases and pests impact agriculture adversely. Efficient, non-contact, sensors available at affordable costs are useful for monitoring plant diseases and pests over a large terrain, greatly facilitating the protection of plants (Zhang et al. 2019). Disease identification and categorization are done by techniques such as image acquisition, segmentation, feature extraction, and classification (Singh et al. 2020).

9.5.1 Visible Camera Sensor

A visible camera sensor is an image sensor replicating human vision. It creates images by collecting light in the wavelength range of human perception (400–700 nm). This sensor forms colored images of people and objects by capturing light in the red (R), green (G), and blue (B) wavelengths, the three primary colors of light.

Important crops such as peas are damaged by the rust fungi, *Pucciniales*, a diverse group of pathogens. Estimation of the severity of the disease by visual observation is prone to error. So, rust-induced pustules on pea specimens are counted by automated image analysis. RGB images of pea leaves are acquired by a smartphone equipped with Exmor RS Sensor, a back-illuminated CMOS image sensor with high dynamic range (HDR) and excellent low-light performance. Rust phenotyping is done by RGB spectral indices segmentation and pixel value thresholding. Features such as the leaf and pustule lesions are extracted from the images. The fraction of the leaflet area affected by the fungal pathogen and the number of pustules on a leaflet are the indicators of the disease severity (Osuna-Caballero et al. 2023).

9.5.2 Infrared Sensor and Infrared Thermal Imaging

Infrared thermal imaging or thermography, uses an infrared sensor to detect radiation emitted from a body, generally in the spectral range of 0.7–16 μm, and create an image of the temperature distribution in the body called a thermogram.

Apple scab disease is caused by the fungus *Venturia inaequalis*. In this disease, lesions are produced on the surfaces of leaves and fruits. Infrared thermography examination shows that localized decreases in the temperature of the leaves take place before the appearance of symptoms. These decrements cause an increase in the maximum temperature difference (MTD) of the leaves. As the scab grows, the MTD upsurges. The MTD is also related to the size of the leaf area which has been infected. In the final stage of leaf development, the MTD decreases (Oerke et al. 2011).

A serious disease afflicting oil palm plantations is basal stem rot (BSR) caused by the aggressive attack of the white rot fungus *Ganoderma boninense*, reducing the yield by 50–80%. Thermal images captured from healthy and BSR-infected oil palm seedlings are statistically analyzed to recognize their differentiating features. Among the classification models applied, the SVM (fine Gaussian) classification model yielded the highest accuracy at 80.0% showing that thermal imaging can detect the BSR-infected trees at the seedling stage (Johari et al. 2021). SVM stands for support vector machine.

9.5.3 Multispectral and Hyperspectral Imaging Camera Sensors

Multispectral sensors collect light in several wavelength bands, e.g., the R, G, B, and near-infrared wavebands. Hyperspectral sensors look at objects in the wavelength range, e.g., 350–2,500 nm, usually covering a restricted range: visible-near infrared (300–1,000 nm) or shortwave-infrared (1,000–2,500 nm) with a typical spectral resolution of ~ 1–7 nm. RGB sensors work with three spectral bands (R, G, and B), multispectral sensors with < 20 bands, and hyperspectral sensors with hundreds of bands (Cheshkova 2022). Each band in a hyperspectral sensor measures only a few nm of the electromagnetic spectrum, enabling a high spectral resolution. Moreover, hyperspectral sensors simultaneously collect spectral and spatial information about the imaged object by constructing a spectral profile of each singular pixel. Plant disease identification is done from the differences in the spectral profiles of healthy and diseased plants.

Citrus bacterial canker (CBC), a highly contagious disease of citrus instigated by *Xanthomonas citri* subsp. *citri* (*Xcc*), is detected in the asymptomatic, early, and late symptom development stages on tangerine citrus Sugar Belle leaves and immature (green) fruits by an unmanned aerial vehicle (UAV)-borne hyperspectral (400–1,000 nm) image sensor using radial basis function (RBF) and K nearest neighbor (KNN) classification methods (Abdulridha et al. 2019).

9.5.4 Chlorophyll Fluorescence Imaging

Fluorescence is the phenomenon of absorption of light by a substance at a particular wavelength and its re-emission at a longer wavelength. The energy adsorbed by the chlorophyll of plants in excess of that used in its photochemical processes is dissipated by emitting heat. A small percentage of this energy is dissipated through the emission of light at a longer wavelength. This is chlorophyll fluorescence. Chlorophyll *a*, the dominant chlorophyll in green plants and algae, absorbs light in the 430–470 nm (blue region) and 660–670 nm (red region). Its peak fluorescence emission occurs at 685 nm and 740 nm.

When a chlorophyll *a* molecule absorbs a photon, an electron in its molecule is promoted to a higher energy level. This is the excited state of the molecule. Immediately, the electron loses energy falling to the lower energy level. Simultaneously a photon is released as a fluorescence emission. By this mechanism, the chlorophyll *a* molecule returns to its ground state (Sánchez-Moreiras et al. 2020).

Chlorophyll *a* fluorescence excitation is done by illuminating the plant tissue sample with pulsed or continuous light in the visible or ultraviolet wavelength range by either a single LED or an array of LEDs, or by using halogen lamps placed at a defined distance from the sample. The emitted red chlorophyll *a* fluorescence emission is detected by a specially designed fluorometer fitted with a CCD (charge-coupled device) video camera connected to a computer. The camera is fitted with suitable color glass filters to protect it from the light used for fluorescence excitation as well as near-infrared radiation. The captured images are processed by custom-made software to determine the photosynthetic parameters, quantifying the photosynthetic activity of the plant sample. Any changes in photosynthetic activity show that pathogens are present. Analysis of the chlorophyll signature and pattern enables the detection and discrimination of fungal infections before visible symptoms appear. Thus, chlorophyll fluorescence imaging (CFI) serves as a powerful noninvasive tool for rapidly assessing the physiological status and efficiency at all plant levels from individual cells and leaves to entire crop populations.

9.5.5 Gas Chromatography – Mass Spectrometry

It is a non-optical sensor used for detecting plant diseases from the volatile organic compounds (VOCs) released by infected plants. VOC emissions from plants provide useful information about their physiological conditions, and the stresses undergone by them through abiotic (non-living matter) and biotic (living organisms) causes. Biotic stresses are induced in plants by bacteria, viruses, fungi, parasites, harmful insects, and birds. The action of pathogens on plants leads to the liberation of specific volatile organic compounds, characteristic of the type of pathogen. Mechanical damages suffered by plants are also accompanied by VOC emissions. However, these VOCs differ from those due to plant diseases.

Gas chromatography – mass spectrometry (GC-MS) is a technique which can distinguish between the VOCs emitted from diseased and non-diseased plants and can identify the type of infection. GC-MS is a combination of a gas chromatograph and a mass spectrometer. The sample is injected in the gaseous phase or in the vaporized phase (if liquid) into the instrument. It is transported by a helium or nitrogen carrier gas stream (mobile phase) into a separation tube called the column. Under the same driving force, different components of the gas/vapor mixture are retained for different times in the column, depending on their chemical and physical properties, and therefore their interactions with the column lining or filling (stationary phase). Hence, the components of the mixture exit the column in different orders. The effluent components enter the mass spectrometer in which an ion source produces gaseous ions of the sample. An analyzer separates the different molecules according to the differences in their charge-to-mass ratio. The detector reads the ions and determines their relative abundance (Medeiros 2018).

Long-time collection of samples for analysis restricts the on-field use of this technique.

9.6 FARM ANIMAL HEALTH MANAGEMENT, AND ENVIRONMENTAL SENSORS

Wearable sensors for animal healthcare, e.g., thermometer, pH electrode, tri-axis accelerometer, and microphone are used for checking the behavioral patterns of farm animals to recognize vital signs of illness/distress to enable farmers to intervene for medical treatment.

9.6.1 Accelerometers and Pedometers

These sensors are used to gauge animal's mobility, behavioral patterns, and locomotion problems. The efficacy of the pedometer in predicting lameness earlier than the appearance of the clinical signs in a herd of dairy cows was investigated by correlating pedometric activity (PA) with clinical cases of lameness (Mazrier et al. 2006).

9.6.2 Rumination Microphone

A microphone tied around the neck of a cow is used to observe cud chewing in cows. Using a data logger, rumination times and patterns provide insights into cow health status. They are used for the early detection of diseases and for the optimization of reproduction in dairy cows (Paudyal 2021).

9.6.3 VOC Sensors

These sensors are used for breath analysis to identify metabolic and pathologic disease biomarkers. Bovine tuberculosis (bTB) is a chronic disease caused by *Mycobacterium bovis*. For the development of diagnostic strategies in veterinary medicine, a VOC study is done on cattle (Ellis et al. 2014). GC-MS is used for investigating the breath-derived volatile organic compound samples from *M. bovis*-infected and non-infected male Holstein calves, collected onto sorbent cartridges using a mask system. Analysis of the chromatographic data shows that healthy and *M. bovis*-infected cattle can be differentiated by breath VOC analysis, which can serve as a tool for the diagnosis and early treatment interventions for bTB.

9.6.4 Farm Air Quality Monitoring Sensors

Temperature, humidity, carbon dioxide, ammonia, and dust sensors are used for air quality monitoring of enclosed farming facilities, e.g., poultry or swine barns to regulate ventilation systems to provide healthy living conditions to animals.

9.7 CONCLUDING REMARKS AND PREPARING FOR THE UPCOMING CHAPTER

IoT sensors help farmers in making correct and timely decisions regarding crop planning and forecasting.

Almost always all the food products as well as those produced by industrial manufacturing must be supplied to the consumers because they are the end users for whose satisfaction the strenuous endeavors were made in agriculture and industry. In particular, food items need to be supplied fresh because they become rotten after a certain period. Medicines too must be transported taking care of stipulated expiry dates. So, timely delivery of goods to shops for onward transmission to consumers calls for stringent and prompt action maintaining deadlines fastidiously. This is the topic planned for discussion in the ensuing chapter.

REFERENCES

Aarthi R., D. Sivakumar, and V. Mariappan 2023 Smart soil property analysis using IoT: A case study implementation in backyard gardening, *Procedia Computer Science*, 218: pp. 2842–2851.

Abdulridha J., O. Batuman and Y. Ampatzidis 2019 UAV-based remote sensing technique to detect citrus canker disease utilizing hyperspectral imaging and machine learning, *Remote Sensing*, 11:1373, pp. 1–22.

Adamchuk V.I., E.D. Lund, B. Sethuramasamyraja, M.T. Morgan, A. Dobermann and D.B. Marx 2005 Direct measurement of soil chemical properties on-the-go using ion-selective electrodes, *Computers and Electronics in Agriculture*, 48(3):272–294.

Archbold G., C. Parra, H. Carrillo and A. M. Mouazen 2023 Towards the implementation of ISFET sensors for in-situ and real-time chemical analyses in soils: A practical review, *Computers and Electronics in Agriculture*, 209 (C): 107828.

Burton L., K. Jayachandran, and S. Bhansali 2020 Review—The "real-time" revolution for in situ soil nutrient sensing, *Journal of The Electrochemical Society*, 167:037569, 9 pages.

Chaisriratanakul W., W. Bunjongpru, A. Pankiew, A. Srisuwan, W. Jeamsaksiri, E. Chaowicharat, N. Thornyanadacha, et al. 2020 Modification of polyvinyl chloride ion-selective membrane for nitrate ISFET sensors, *Applied Surface Science*, 512: p. 145664.

Chanasyk D. S. and M. Anne Naeth 1996 Field measurement of soil moisture using neutron probes, *Canadian Journal of Soil Science*, 76(3):317–323.

Cheshkova A. F. 2022 A review of hyperspectral image analysis techniques for plant disease detection and identification, *Vavilovskii Zhurnal Genet Selektsii*, 26(2): pp. 202–213.

Ellis C. K., R. S. Stahl, P. Nol, W. R. Waters, M. V. Palmer, J. C. Rhyan, et al. 2014 A pilot study exploring the use of breath analysis to differentiate healthy cattle from cattle experimentally infected with *Mycobacterium bovis*, *PLoS ONE*, 9(2): p. e89280, pp. 1–12.

Hidouri S., Z. Baccar, A. Errachid, and O. Ruiz-Sanchèz 2015 High sensitive and selective hydrogen phosphate ISFET based on polyvinyl chloride membrane, *Portugaliae Electrochimica Acta*, 33(6): pp. 343–351.

Johari, S. N. Á. M., S. K. Bejo, G. A. Lajis, L. D. J. DaimDai, N. B. Keat, Y. Y. Ci and N. Ithnin 2021 Detecting BSR-infected oil palm seedlings using thermal imaging technique, *Basrah Journal of Agricultural Sciences*, 34: pp. 73–80.

Kim H.-J. 2006 Ion-Selective Electrodes for Simultaneous Real-Time Analysis of Soil Macronutrients, A Dissertation presented to the Faculty of the Graduate School, University of Missouri-Columbia, 215 pages.

Lehmann U. and A. Grisel 2014 Miniature multisensor probe for soil nutrient monitoring, *Procedia Engineering*, 87: pp. 1429–1432.

Ling Y. P., I. Syono, and L. Yook Heng 2009 A solid-state potassium ion sensor from acrylic membrane deposited on ISFET device, *Malaysian Journal of Chemistry*, 11: pp. 64–72.

Mahajan G., B. Das, B. Gaikwad, A. Desai, S. Morajkar, K. Patel, R. Kulkarni and D. Murgaonkar 2020 Using hyperspectral remote sensing to monitor the properties of salt-affected soils, *Authorea*, February 19, pp. 1–21.

Martin E. C. 2009 Methods of Measuring for Irrigation Scheduling—WHEN, The University of Arizona Cooperative Extension, AZ1220, Arizona Water Series No.30, pp. 1–7.

Mazrier H., S. Tal, E. Aizinbud and U. Bargai 2006 A field investigation of the use of the pedometer for the early detection of lameness in cattle, *Canadian Veterinary Journal*, 47(9): pp. 883–886.

Medeiros P. M. 2018 Gas chromatography–Mass spectrometry (GC–MS). In: White W.M. (Eds.) *Encyclopedia of Geochemistry*. Encyclopedia of Earth Sciences Series. Springer, Cham, Switzerland, pp. 530–535.

Oerke E.C., P. Fröhling and U. Steiner 2011 Thermographic assessment of scab disease on apple leaves, *Precision Agriculture*, 12(5): pp. 699–715.

Osuna-Caballero S., T. Olivoto, M. A. Jiménez-Vaquero, D. Rubiales and N Rispail 2023 RGB image-based method for phenotyping rust disease progress in pea leaves using R, *Plant Methods*, 19: pp. 86.

Pandey J., V. Chamoli and R. Prakash 2020 A Review: Soil Moisture Estimation Using Different Techniques, In: S. Choudhury et al. (eds.), *Intelligent Communication, Control and Devices, Advances in Intelligent Systems and Computing*, Vol. 989, Springer Nature Singapore Pte Ltd., pp. 105–111.

Paudyal S. 2021 Using rumination time to manage health and reproduction in dairy cattle: A review, *Veterinary Quarterly*, 41(1): pp. 292–300.

Rasheed M. W., J. Tang, A. Sarwar, S. Shah, N. Saddique, M. U. Khan, M. I. Khan, et al. 2022 Soil moisture measuring techniques and factors affecting the moisture dynamics: A comprehensive review, *Sustainability*, 14:11538, pp. 1–23.

Recena R., V. M. Fernández-Cabanás, and A. Delgado 2019 Soil fertility assessment by Vis-NIR spectroscopy: Predicting soil functioning rather than availability indices, *Geoderma*, 337: pp. 368–374.

Sánchez-Moreiras A. M., E. Graña, M. J. Reigosa and F. Araniti 2020 Imaging of chlorophyll *a* fluorescence in natural compound-induced stress detection, *Frontiers in Plant Science*, 11: pp. 1–15. Article 583590.

Shock C., A. Pereira, E. Feibert, C. Shock, A. Akin and L. Unlenen 2016 Field comparison of soil moisture sensing using neutron thermalization, frequency domain, tensiometer, and granular matrix sensor devices: Relevance to precision irrigation, *Journal of Water Resource and Protection*, 8:154–167.

Singh V., N. Sharma and S. Singh 2020 A review of imaging techniques for plant disease detection, *Artificial Intelligence in Agriculture*, 4: pp. 229–242.

Skierucha W., A. Wilczek, A. Szypłowska, C. Sławiński and K. Lamorski 2012 A TDR-based soil moisture monitoring system with simultaneous measurement of soil temperature and electrical conductivity, Sensors, 12(10), 13545–13566.

Stenitzer E. 1993 Monitoring soil moisture regimes of field crops with gypsum blocks. *Theoretical and Applied Climatology*, 48: 159–165.

van der Veeke S., R. Koomans and H. Limburg 2020 Using a gamma-ray spectrometer for soil moisture monitoring: Development of the gamma soil moisture sensor (gSMS), *IEEE International Workshop on Metrology for Agriculture and Forestry (MetroAgriFor)*, 4–6 November, Trento, Italy, pp. 185–190.

Volkov D. S., O. B. Rogova and M. A. Proskurnin 2021 Organic matter and mineral composition of silicate soils: FTIR comparison study by photoacoustic, diffuse reflectance, and attenuated total reflection modalities, *Agronomy*, 11(9): 1879, pp. 1–30.

Yogesh. S, S. Tm and G. T. Bharathy 2021 Rain detection system using Arduino and rain sensor, International Journal of Scientific Research in Science, *Engineering and Technology*, 9(8):64–68.

Zarei M., N. V. Solomatova, H. Aghaei, A. Rothwell, J. Wiens, L. Melo, T. G. Good, S. Shokatian and E. Grant 2023 Machine learning analysis of Raman spectra to quantify the organic constituents in complex organic–mineral mixtures, *Analytical Chemistry*, 95:43, pp. 15908–15916.

Zazueta F. S. and J. Xin 1994 Soil Moisture Sensors, University of Florida, Florida Cooperative Extension Service, Bulletin 292 April 1994, pp. 1–11.

Zhang J., Y. Huang, R. Pu, P. Gonzalez-Moreno, L. Yuan, K. Wu and W. Huang 2019 Monitoring plant diseases and pests through remote sensing technology: A review, *Computers and Electronics in Agriculture*, 165: p. 104943.

10 IoT Sensors for Logistics, Fleet Management, and Retail Stores

10.1 INTRODUCTION

Logistics is a part of the supply chain dealing with the storage and transportation of goods and services from the originating point to the consumer. Fleet management oversees and coordinates commercial vehicles and drivers regarding their operational efficiency and policy compliance. Retailers or shops are establishments where products and services are delivered to customers.

10.2 ROLE OF IoT SENSORS IN SMART LOGISTICS AND FLEET MANAGEMENT

Smart logistics and fleet management is a comprehensive technological approach to goods and cargo supply chain management using sensors, the Internet of Things, and digital methods. It helps companies and organizations in accurate real-time tracking of assets during warehousing and transport by vehicles to apportioned destinations including acquisition of information about vehicle fuel, speed, and driver behavior, whether normal or risky.

Sensors play a crucial role in logistics and fleet management. Notable areas of sensor applications are given below with the relevant sensor(s) enclosed in boldface letters within parentheses:

(i) Robotic tracking of shipped objects (**RFID tags**).
(ii) Warehouse management and real-time pursuit of the location of cargo vehicles and consignments to know about their status updates (**GPS sensor**).
(iii) Planning for optimization of the travel route of the fleet to minimize travel time and to evade crowded roads and traffic jams on highways (**GPS sensor**).
(iv) Following the adherence to compliance of temperature specifications for perishable goods such as food items and medicines during transportation (**temperature sensors**).
(v) Monitoring moisture levels for humidity-sensitive items during the journey (**humidity sensors**).
(vi) Keeping an eye on changes in the position or movement of goods in transit (**motion sensors**).
(vii) Checking package sealing and integrity to find any leakages (**pressure sensors**).
(viii) Detecting unauthorized access and tampering to prevent theft (**light sensors**).
(ix) Looking at driving characteristics like speeding, braking, idling, and vehicle condition for driver safety (**telematics camera**). The telematics camera or dash cam is a video camera equipped with sensors to record driving incidents such as sudden acceleration, hard braking, and dangerous cornering to inform managers about the driver's attitude, to identify driver training deficiencies, and to provide personalized guidance. The video footages of such incidents are uploaded to the internet for examination by the telematics platform experts to issue the necessary advisory.

DOI: 10.1201/9781003374442-10

10.3 GPS SENSOR (GPS RECEIVER)

10.3.1 THREE-BLOCK CONFIGURATION

The global positioning system, popular as GPS, is a satellite broadcast system from which anyone can always acquire information about absolute position, velocity, and time from anywhere in the world (Parkinson and Spilker Jr. 1996, Ko et al. 2005). The basic structure of GPS is a three-block configuration consisting of the following:

(i) A Space Segment (a 24-satellite constellation comprising satellites deployed in 6 orbital planes at an altitude of 20,200 km with 4 satellites in each plane, and encircling the earth in 11 h 58 min),

(ii) A Land-Based Control Segment (ground stations for monitoring and maintaining the orbits of the satellites and adjusting their clocks), and

(iii) The User Segment (GPS receivers).

10.3.2 ALLOCATED FREQUENCY BANDS AND CODES

Two RF bands are used:

(i) L1 (1,575.42 MHz), and
(ii) L2 (1,227.6 MHz).

Provision of two carrier frequencies (L1 and L2) enables the two-frequency user to measure the ionospheric delay. Users of a single frequency (L1 only) determine the ionospheric delay from modeling parameters given in the navigation message.

Communication system: Direct sequence spread spectrum (DSSS)
Modulation technique: Binary phase-shift keying (BPSK)

Pseudorandom noise (PRN) Spreading Codes for L1 Band:

(i) Coarse Acquisition (C/A) Code for civilian use and
(ii) Precision or P-code for military use.

The C/A code is called the Gold code after its inventor. Its PRN sequence is 1,023 chips long; the chip means a bit which does not carry information. The chipping rate of the C/A code is 1.023×10^6 chips s^{-1}. The repetition period of the C/A code is

$$T = \frac{\text{Length of the sequence}}{\text{Chipping rate}} = \frac{1,023\,\text{chips}}{1.023 \times 10^6\,\text{chips s}^{-1}} = 1 \times 10^{-3}\text{s} = 1\text{ms} \tag{10.1}$$

Pseudorandom spreading code for L2 band: P-code only
Each satellite has a unique PRN sequence. The carrier frequencies are modulated by the PRN sequence for each satellite.

10.3.3 NAVIGATION MESSAGE

The navigation information is continuously conveyed to the GPS users by the satellites as a 50bps navigation message consisting of 1,500 bits length and 30-s-duration frames subdivided into

5 subframes of 6s duration with ten 30-bit words in each subframe (Van Dierendonck et al. 1978). The information in different subframes is as follows.

Subframe 1: Date, clock correction parameters of the satellite, status, and health of the satellite.
Subframes 2 and 3: Ephemeris data of the transmitting satellite, viz., its exact orbital information. It is valid for 4 h.
Subframes 4 and 5: The 25-page almanac containing low-resolution information about the orbits of the satellites. It is valid for two weeks.

10.3.4 Code-Division Multiple Access

Although transmission signals from all the satellites have the same frequency, they do not interfere among themselves because of a distinctly different PRN sequence assigned to each satellite. The signals are detected by the code-division-multiple-access (CDMA) technique. For tracking a satellite using CDMA, a GPS receiver produces a replica of the PRN sequence for the concerned satellite, and a replica of the carrier signal including Doppler effects.

10.3.5 Pinpointing the Location of the User by Trilateration Technique

GPS applies the trilateration principle to determine the distance between the satellite and the user (Figure 10.1). In trilateration, the distance of an object is measured from a minimum of three points based on the sphere geometry. The time taken by the signal to reach from the user to nearby satellites is measured. One satellite finds a spherical range of the user's location. Two satellites narrow down the location to the circle formed by their intersection. Three satellites define the user's location as one of the two points of intersection of three circles. The exact location is found by rejecting one point from being practically untenable.

10.3.6 GPS Observables

These are the signals on which measurements are performed to calculate the distance between the GPS satellite and the GPS receiver. Observables are of two types, namely, the codes modulated on the carrier wave and the carrier wave itself without the codes. The codes are used to find the pseudorange, the distance between the satellite transmitter [at the time of emission of the signal from it, the time of transmitter clock (t_T)] and the receiver antenna [at the time of reception of the signal by it, the time of receiver clock (t_R)]. It is given by the equation

$$\rho = c\left(t_R - t_T\right) \tag{10.2}$$

where c is the velocity of light. The pseudorange is determined by the correlation of the code on the modulated carrier wave received from the satellite with a replica of the same code generated in the GPS receiver having the capability for code correlation, and finding the relative shifts between them on the time scale. The pseudorange represents only a rough calculation of the geometrical distance because of errors in the clocks, and those introduced by earth tides, and atmospheric, multipath, and relativistic effects. Nonetheless, it is adequate for low-accuracy applications or where instantaneous positions are needed.

The carrier measurements are done on the phase of the carrier. Here the difference in phase between the oscillators at the transmitter and receiver is the parameter of interest. In many GPS receivers, the pseudorange determined from the carrier code is taken as a first approximation, to be finely adjusted using the carrier phase.

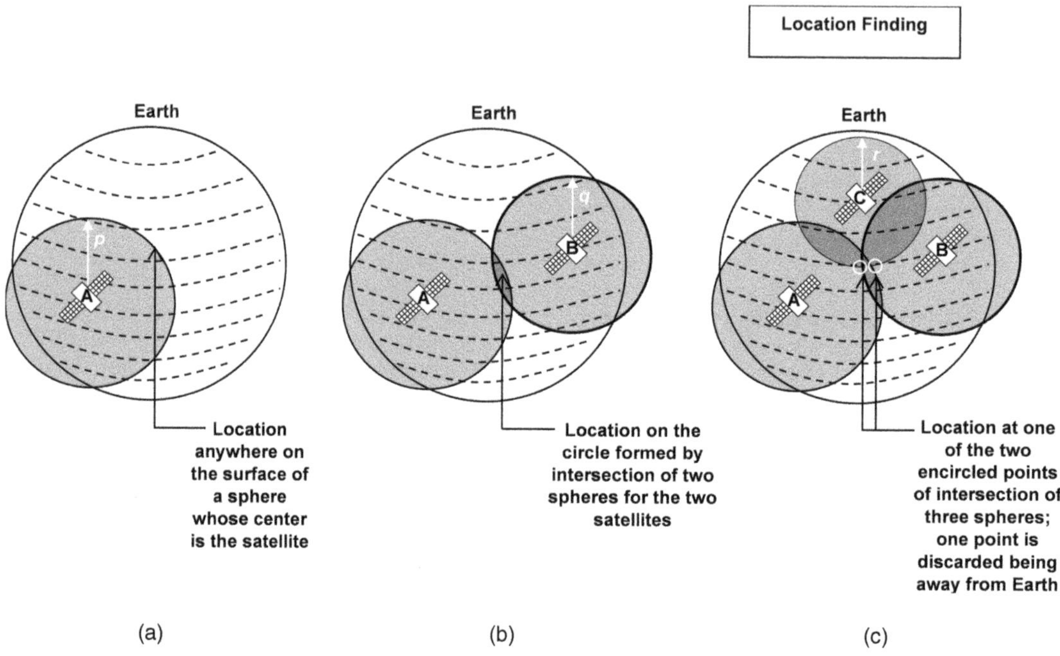

FIGURE 10.1 Determination of the location of a GPS receiver using: (a) one satellite, (b) two satellites, and (c) three satellites. Parts (a), (b), and (c) show the earth and the spheres at the center of which the satellites A, B, and C lie. Part (a): The GPS receiver picks up a signal from a satellite A, and calculates the distance as p km (suppose). The GPS receiver is on the surface of a sphere at whose center lies the satellite and whose radius is p km. Several points on the surface of this sphere are not on earth but in space. Part (b): The GPS receiver picks up a signal from a satellite B, and calculates the distance as q km. So, the GPS receiver is on the surface of a sphere at whose center lies the satellite B and whose radius is y km. This means that the GPS receiver is at the intersection of the spheres for satellites A and B. Since a circle is formed when two spheres intersect, the GPS receiver's location is anywhere on that circle. Part (c): The GPS receiver picks up a signal from a satellite C, and calculates the distance as r km. So, the GPS receiver is on the surface of a sphere at whose center lies the satellite C and whose radius is r km. Three spheres intersect at two points. One of these encircled points represents the location of the GPS receiver. The other point is rejected as it falls in space.

10.3.7 GPS Signal Acquisition

The acquisition and tracking of signal involve its replication in the following two dimensions (Ward 2010).

(i) Code-Phase Dimension: For detection of a satellite signal by tracking the signal in code-phase dimension, a GPS receiver replicates the code of the transmitting satellite and shifts the phase of the replica code until it correlates with the satellite PRN code received at the GPS antenna. Maximum correlation occurs when the phase of the transmitting satellite exactly matches with that of the replica code at the receiver.

In the GPS receiver, cross-correlation is done in two steps. First, the phase of the desired satellite is searched. Then the code state of the satellite is tracked by variation of chipping rate of the replica code generator to recompense for the Doppler effect on the PRN code of the satellite. Code correlation is accomplished by multiplying the phase-shifted replica code with the received satellite code, integrating, and dumping. During integration and dumping, cumulative summation of the discrete-time signal is performed, and the sum is reset to zero according to a planned schedule. The intention is to maintain the prompt code of the replica code generator in maximum correlation with the phase of the desired satellite code.

(ii) Carrier-Phase Dimension: For detection of a satellite signal by tracking the signal in carrier-phase dimension, the GPS receiver replicates the carrier frequency plus Doppler shift, and simultaneously adjusts the replica carrier to match it with the carrier frequency of the desired satellite as received at the receiver.

10.3.8 Block Diagram of the GPS Receiver

RF signals from all the satellites in view are received by a circularly polarized antenna (Figure 10.2(b)). The out-of-band RF interference is removed by passing the signals through a bandpass filter (Braasch. and van Dierendonck 1999). A low-noise preamplifier is used for amplification of the received signals. The amplified signal is down-converted to an intermediate frequency (IF) signal by mixing it with the signal from a local oscillator (LO), which is constructed from a reference oscillator with the help of a frequency synthesizer. From the upper and lower sidebands formed by mixing, the lower sidebands are chosen. A bandpass filter is used to reject the upper side bands. The analog IF signal contains the PRN codes and Doppler signals. This IF signal is subjected to analog-to-digital conversion and automatic gain control functions. The stop-band noise is eliminated using an anti-aliasing filter. The digital IF signals are processed by 1,2,3, …, N digital receiver channels for the given frequency. All the digital receiver channels have a common input. However, each channel extracts its own PRN code signal from the satellite in view of the antenna's sky covered.

In each digital receiver channel (Figure 10.2(b)), the carrier signal is removed from the digital IF signal of the satellite being tracked by the (replica carrier + carrier Doppler) signal. The (replica carrier + carrier

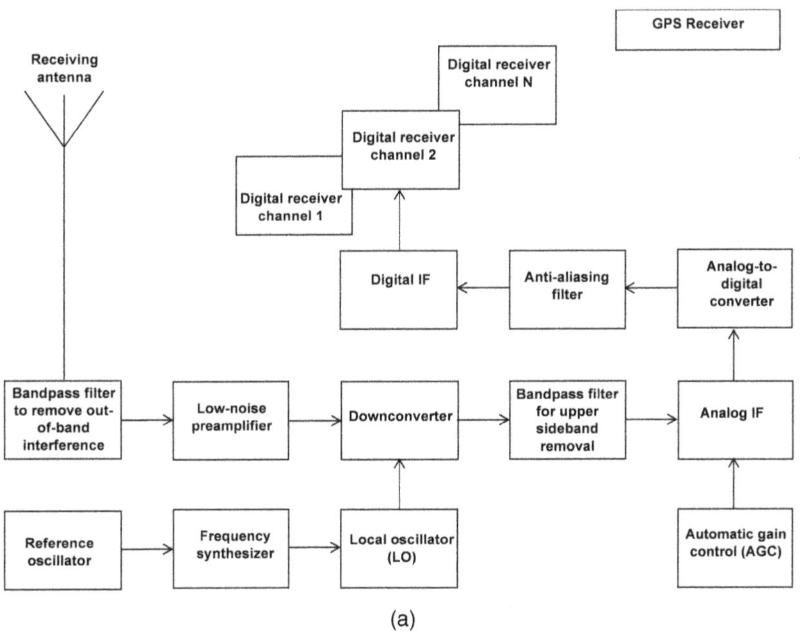

FIGURE 10.2 Block diagram of a digital GPS receiver: (a) stages from the antenna to digital receiver channels and (b) processing of a single digital receiver channel. Part (a) shows that the incoming signal from a satellite is picked up by the receiving antenna, passed through a bandpass filter to get rid of out-of-band interference, amplified by a low-noise preamplifier, and sent to a downconverter receiving the mixing signal from a local oscillator produced by a frequency synthesizer fed by a local oscillator. The downconverted signal is bandpass filtered to discard the upper sideband and get analog IF. The analog IF is put under automatic gain control and fed to an analog-to-digital converter. After the removal of stop-band noise by an anti-aliasing filter, the resulting digital IF signals are processed by 1, 2, 3, …, N digital receiver channels.

(Continued)

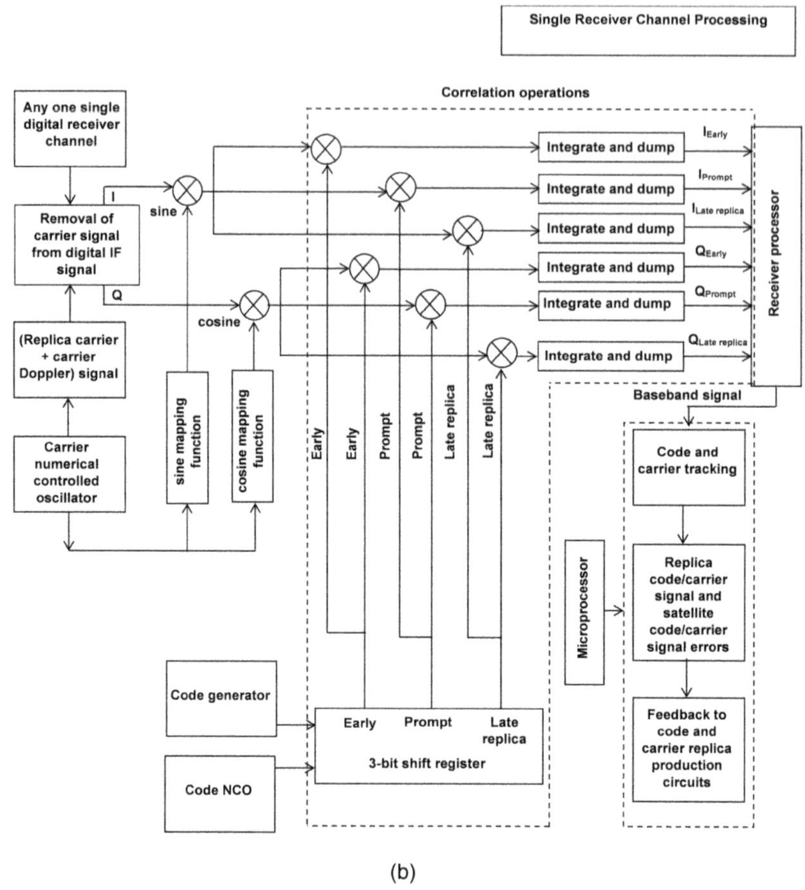

(b)

FIGURE 10.2 (CONTINUED) Part (b) shows the operations performed on the signal from any one digital receiver channel. A carrier numerical controlled oscillator produces the replica carrier plus carrier Doppler signal. By applying this signal to the signal from digital receiver channel, the carrier signal is removed from the digital IF signal, giving in-phase (I) and quadrature-phase (Q) signals. The sine and cosine mapping functions generated by the carrier numerical-controlled oscillator are combined with the I and Q signals of the digital IF signal to get the digital amplitudes of sine and cosine functions. Code correlation is shown for three different replica code phases being correlated at the same time with the same satellite signal. The three replica phases represent the early, prompt, and late replica codes engendered by the code tracking loop in the GPS receiver by the code generator, code NCO, and a 3-bit shift register. The correlated outputs are supplied to the receiver processor after filtration by integrate-and-dump circuits giving three in-phase components I_{Early}, I_{Prompt}, and $I_{Late\ replica}$, and three quadrature-phase components Q_{Early}, Q_{Prompt}, and $Q_{Late\ replica}$. The code and carrier are tracked using a microprocessor. Replica code/carrier signal and satellite code/carrier signal errors are found and fed back to the circuits engaged in code and carrier replica production.

Doppler) signal is synthesized by a carrier numerically controlled oscillator (NCO) producing in-phase (I) and quadrature (Q) signals. It also yields discrete sinusoidal and cosinusoidal mapping functions. These mapping functions combine with I and Q signals to transform the amplitudes of the internal staircase function of NCO into digital amplitudes of sine and cosine functions. The resultant amplitude of the signal is given by the vector sum of the I and Q components. Its phase angle from the I-axis is obtained from $\tan^{-1}(Q/I)$.

Correlation is done between the I and Q signals, and the early, prompt, and late replica codes produced by the code generator, a 3-bit shift register, and code NCO. The output signals of the correlation process undergo filtration by integrate-and-dump circuits, and subsequently go to the receiver processor. Alignment of the phase of prompt replica code with the phase of satellite code is carried out, achieving maximum correlation.

Each pair of I and Q signals fed to the receiver processor is known as a complex input. Each digital receiver channel supplies three complex inputs to the receiver processor. The program inside the receiver processor can process N channels of complex pairs in the form of N code and carrier tracking loops. Two feedback outputs come out from these two tracking loops for each channel.

The I and Q signals supplied by the N digital receiver channels to the receiver signal are called baseband signals. They are so named because in these signals, the carrier signal has been erased and the spread spectrum bandwidth is collapsed. Inside the receiver processor, code and carrier tracking loops are executed on the baseband signals to find replica code/carrier signal, and the incoming satellite code/carrier signal errors. The error signals are sent back to the code and carrier replica production circuits. All the baseband signal processing programs are operated by a microprocessor.

10.4   CARGO SENSORS AND CARGO CONTAINER MONITORING

While the GPS is applied to find the cargo location, sensors are also required for monitoring space utilization in cargo trucks, and the condition of cargo containers.

10.4.1   CARGO SENSORS

Proper space utilization in cargo-carrier trucks is beneficial to save costs and turnaround times. Cargo sensors measure the consumed space behind a truck and give an idea about the unutilized space into which more cargo may be loaded (Cserhat 2023). They also provide information about the loading or unloading of the cargo, which is useful for cargo safety. Ultrasonic and infrared sensors are used for this purpose. An accuracy of up to 1% of the set measuring range is achievable with ultrasonic sensors, meaning 1 mm at 10 cm distance. The sensors are affected by the presence of soft and foamy materials, gases other than air, and vacuum conditions. For the infrared sensor, the infrared cone has a defined field of view, typically ~5° for emission and detection. If an object falls within this field of view, the zone is detected as full, otherwise empty. Apart from these sensors, artificial intelligence-aided high-resolution cameras are viable options.

10.4.2   CARGO CONTAINER MONITORING

Smart monitoring of cargo containers improves efficiency, safety, security, and sustainability. A low-power, cost-effective through-wall ultrasonic communication front-end is non-invasively mounted (Algueta-Miguel et al. 2020). A magnetic case fixes two piezoelectric transducers working at 40kHz to a metallic wall. The low-power technique is essential for the transducers placed inside the container, which are generally powered by batteries or energy harvesters. Among the several carrier modulation schemes evaluated in terms of robustness and bit error rate (BER), a DBPSK (differential binary phase shift keying) modulator using a square digital carrier is selected. A tracking system is designed and experimentally validated.

10.5   ROLE OF IOT SENSORS IN RETAIL STORES

After the cargo is delivered to the retail store, it needs to be continuously followed in the store to know about its status updates. A smart retail store is the enhanced version of a traditional shopping center with modern technologies such as sensors and the Internet of Things to collect and exchange data on various products and customer experience in store rooms and counters, giving the retail owner 360° visibility for greater security; systemizing shelving in warehouses and management of orders; assisting in tracking inventory, assets, and customer behavior/response; making cashier-less payments and checkouts; and automating lighting and temperature control with efficient energy utilization. Sensors perform a multiplicity of tasks in retail stores, as elaborated below.

(i) Smart Shelves and Efficient Inventory Management: Sensors are used for tracking the items in the warehouse and front shelves to ensure that they are replenished in time to prevent running out of stock and checking that the items are not misplaced from their assigned shelves (**CCTV cameras; GPS sensors, RFID tags, readers, and antennas; weight sensors**).

(ii) Supply Chain Tracking Near the Store: Sensors are used for tracking the items during transit from the cargo truck to the retail store (**CCTV cameras; GPS sensors, RFID tags, readers, and antennas**).

(iii) Preventing Food Spoilage in Cold Storage Areas: Sensors are used to ensure that temperature is strictly maintained within prescribed limits in order that the food items remain fresh and food wastage is reduced (**temperature sensors**).

(iv) Studying Customer Interests: Sensors are used for analyzing which section of the store attracts more customer traffic and which products are in greater demand. Premium store areas are marked by the insights acquired from sensors and video analytics. This effort is utilized for improved planning of in-store merchandising, and to provide guided shopping via effective digital display advertisements, calculated allotment of space, and corridor layouts (**proximity/occupancy sensors, CCTV cameras**).

(v) Enhancing Customer's Shopping Experience: Sensors are used for transmitting alerts and notifications about sale discounts, and seasonal offers on smartphones based on a customer's proximity to a section of the shop (**Bluetooth Light Energy (BLE) beacons**). Sensor-based lighting and temperature control are done to create a comfortable environment (**light and temperature sensors**).

(vi) Point-of-Sale Queuing Assistance: Sensors are used for keeping a watch on sales counters. This data is applied to shorten the wait times for optimizing customer service either by diverting the customers to vacant or less crowded counters or deploying additional self-checkouts or more staffing at checkout (**CCD cameras, occupancy sensors**).

(vii) Blending In-Store and Virtual Buying: Real-time availability of stocks is provided to customers. Customers are allowed to load their carts online and enjoy hassle-free in-store shopping in less time avoiding unnecessary crowding (**CCD cameras, RFID tags**).

(viii) Providing Fully Automated, Cashier-Less In-Store Shopping: In checkout automation, the customers enter the stores, select the required items, load the cart, make an online payment, receive a digital receipt on their smartphone, and go home (**CCD cameras, RFID tags**).

(ix) Tracking the Health of Machines in the Shop. The machine's health status is fed to the equipment maintenance section to prevent breakdown (**temperature sensor, vibration sensor**).

(x) Shop Security: Sensors are used for preventing thefts and apprehending shoplifters (**CCD cameras, door and window contact sensors**).

10.6 BLE BEACON, Wi-Fi, AND GPS

Bluetooth Low Energy (BLE) beacon is a wireless radio communication device. It is a one-way communication device that sends information to other devices but cannot request/ receive information from them. All BLE beacon scanners falling within a certain radial distance from the beacon can discover this mini-radio transmission device. The BLE scanners are the smartphones that have been configured with a dedicated BLE app. The BLE beacon, however, cannot perceive these scanners.

Unlike Wi-Fi, it does not require an internet connection. BLE beacon has a high location accuracy of ~±1m. Wi-Fi accuracy is improbable to be better than ±10 m, often near ±15 m, provided by Wi-Fi. A BLE beacon installation is ~20× less expensive than Wi-Fi. It is implemented in less time than Wi-Fi. It needs replacement after 2–4 years while Wi-Fi requires regular calibration and checkups. Both BLE beacon and Wi-Fi are compatible with Android and iPhones ('i' stands for internet, individual, instruct, inform, (and) inspire) but Wi-Fi ranging is not supported by iOS (iPhone operating system). BLE beacon has an operating range of 10–100 m while Wi-Fi has a 50–100 m range. BLE beacon has a data rate of 1 Mbps. The Wi-Fi has a data rate of 2–866 Mbps. BLE beacon works at the 2.4 GHz frequency band. Wi-Fi works at 2.4 GHz and 5 GHz frequency bands. Both BLE and Wi-Fi are equipped with high-level security featuring encryption and authentication.

BLE beacon can be compared to a lighthouse. A lighthouse transmits light waves. Instead, the beacon sends out radio signals. The lighthouse is a navigational aid. The BLE beacon helps in indoor mapping, navigation, asset tracking, and personalized customer experiences in retail stores where GPS is inoperative.

In dissimilarity to GPS tracking based on long-range (hypothetically limitless) radio communication via satellites, BLE beacons transmit short-distance signals <100 m (optimal: 0–25 m). BLE beacons play the same role in indoor positioning as GPS in outdoor positioning. Although GPS has revolutionized outdoor localization, it is unsuccessful in indoor settings owing to deficient signal coverage. The GPS signals from satellites become too feeble in urban canyons, enclosed residential/ commercial/industrial structures, subterranean apartments, spaces adjoining walls, and trees. This happens because they undergo multiple reflections and scattering from walls, floors, and roofs of the building structure and constructional materials. Most of the signals are distorted or absorbed. The attenuated signals become ineffective and unusable for indoor localization.

10.7 BEACON MODULE AND CONCEPT

A BLE beacon module is a portable hardware transmitter consisting of a CPU, a small radio transmitter, and a powering unit. The power source is either lithium batteries or the module is connected through universal serial bus technology for USB power delivery (USB-PD). The BLE beacon is a broadcaster working within a small radius. The usual BLE beacon receivers are smartphones or tablets supporting BLE. The underlying concept is the 'bring-your-own-device' idea.

10.7.1 BLUETOOTH LOW ENERGY TECHNOLOGY

Referring to Section 2.3.2.1, BLE is a low-cost, low-power (0.01–0.5W with peak current consumption <15 mA), easily deployable wireless technology for wireless personal area networks (WPANs) based on the IEEE 802.15.1 standard and optimized to support applications having a relatively low duty cycle. Low power requirement extends battery life. The data rate is 125 kbit/s, 500 kbit/s, 1 Mbit/s, and 2 Mbit/s, wake latency from the non-connected condition is 6 ms, and the minimum total time for sending data is 3 ms.

It operates in the frequency range of 2.402–2.4835 GHz ISM (Industrial, scientific, and medical), an unlicensed radio spectrum but not unregulated. The frequency band is divided into 40 channels. The channel bandwidth is 2 MHz.

It uses frequency-hopping spread-spectrum (FHSS) communication technology. The modulation scheme is Gaussian Frequency Shift Keying (GFSK). FHSS is a robust technique, highly resistant to narrowband interference because interference on any one frequency affects only a small amount of data transmitted at that frequency. Further, in FHSS, the hopping sequence is determined by a pseudo-random sequence generator. So, it counters eavesdropping and hinders jamming because the pseudo-random frequency hopping pattern is known only to the transmitter/receiver, and cannot be easily deciphered. Jamming the transmission at a particular frequency does not upset transmission at other frequencies.

10.7.2 ADVERTISING AND DISCOVERY

BLE beacons operate by a procedure based on broadcasting advertisement packets in a defined format (iBeacon by Apple or Eddystone by Google) on at least one of the three separate frequency channels to minimize interference. The packet is made of a 16-bit header plus 255 octets of payload (maximum). iBeacon uses 30 bytes of payload and Eddystone up to 31 bytes of payload. The broadcast is done at a repetition rate known as the advertising interval. A random delay of up to 10 ms is added to each advertising interval to prevent collisions between packets. The information in a packet includes a universal resource locator (URL), and a universal unique identifier (UUID), besides battery voltage, temperature, and transmit power.

The scanner device listens to the channel. The duration of listening is known as the scan window. The scan window is periodically repeated during every scanning interval.

10.8   INDOOR POSITIONING USING BLE BEACONS

People stay 80–90% of their time indoors, either in homes, offices, shopping malls, airports, etc. Indoor positioning is no less important than outdoor positioning which has been satisfactorily achieved with GPS.

One such application is finding the desired shop in a market. It often takes a long time for a customer to locate a shop of interest when visiting a shopping complex, a market in an urban area with many shops, a shopping mall, or a large multistoried building with shops on different floors. A quarter of the shopping time is wasted in reaching the required shop. The shopping complexes/malls are integral parts of contemporary society, typically featuring a variety of stores, entertainment places, and dining restaurants. They bring together a plethora of retailers close together in a locality or under one roof, making it easier for buyers to find everything they need in a single visit at one conglomeration. These sprawling shopping centers offer a unique retail experience that has evolved over the years, transforming the way people shop and socialize (Samuel et al. 2021).

An indoor positioning system (IPS) is a network of devices for determining the locational coordinates of people or objects in real time in a specific area inside a building by using radio waves, sensors, and mobile devices where GPS becomes imprecise or fails to work. Mobile devices have an advantage over sensors because they already have the requisite technology to process radio waves in the environment around the user.

10.8.1   Receiver Signal Strength Indicator–Based Indoor Positioning

10.8.1.1   Receiver Signal Strength Indicator

Receiver Signal Strength Indicator (RSSI) is the power of the radio signal (in decibel-meter, dBm) measured by the receiver (the smartphone) when it receives packets from the transmitter (the beacon) (Agarwal 2017). RSSI is expressed in −dBm format from 0 to −100 dBm. Let us recall the definitions of dB and dBm.

Decibel (dB) is not a unit. It is a logarithmic comparison of two numbers.

$$1\,\text{dB} = 10\log_{10}\frac{\text{Measured power}}{\text{A chosen reference power}} \tag{10.3}$$

Taking the reference power as 1 mW, a decibel unit of power called dBm ('m' for milliwatt) is defined as

$$1\,\text{dBm} = 10\log_{10}\frac{\text{Measured power}}{1\text{mW}} \tag{10.4}$$

If the power measured at the smartphone is 0 dBm, we can write

$$10\log_{10}\frac{\text{Power received at the smartphone}}{1\text{mW}} = 0 \tag{10.5}$$

or,

$$\log_{10} \frac{\text{Power received at the smartphone}}{1\text{mW}} = \frac{0}{10} = 0 \qquad (10.6)$$

or,

$$\frac{\text{Power received at the smartphone}}{1\text{mW}} = 10^0 = 1 \qquad (10.7)$$

since

$$\log_a b = n \text{ if } a^n = b \qquad (10.8)$$

Eq. (10.7) gives

$$\text{Power received at the smartphone} = 1\text{mW} \qquad (10.9)$$

If the power measured at the smartphone is -100 dBm, we can write

$$10\log_{10} \frac{\text{Power received at the smartphone}}{1\text{mW}} = -100 \qquad (10.10)$$

or,

$$\log_{10} \frac{\text{Power received at the smartphone}}{1\text{mW}} = -\frac{100}{10} = -10 \qquad (10.11)$$

or,

$$\frac{\text{Power received at the smartphone}}{1\text{mW}} = 10^{-10} \qquad (10.12)$$

$$\therefore \text{Power received at the smartphone} = 10^{-10}\text{mW} \qquad (10.13)$$

Similarly, a power of -50 dBm is obtained in mW as

$$\log_{10} \frac{\text{Power received at the smartphone}}{1\text{mW}} = -\frac{50}{10} = -5 \qquad (10.14)$$

from which

$$\text{Power received at the smartphone} = 10^{-5}\text{mW} \qquad (10.15)$$

Noting that power levels 0, -50, and -100 dBm correspond to 1 mW, 10^{-5} mW, and 10^{-10} mW, respectively, it is evident that a higher negative dBm value indicates a weaker signal, i.e., a smaller negative dBm value implies a stronger signal. Thus, RSSI increases as the power approaches 0 dBm from -100 dBm.

10.8.1.2 Multipath Fading Problem

Multipath fading, a major problem in distance estimation from RSSI, occurs when a signal reaches the receiver through multiple paths. These paths are created by reflections of the main signal from the floor and walls of the indoor space. The addition of multiple path signals leads to unexpected signal amplification or attenuation. The problem is mitigated by using the Kalman filter. The Kalman filter models the noise of the signals as a Gaussian distribution. It is conceptualized as made of two distinct phases: prediction and updating. The prediction phase uses the state variables from the previous time step to produce an estimate of the state variables during the present time step (Shue and Conrad 2016).

10.8.2 The Log-Distance Path Loss Model

The log-distance path loss model is a popular model in wireless communications that is used for the conversion of RSSI values into distances. It is a widely acclaimed radio propagation model encompassing shadowing effects. It predicts the path loss incurred by a radio signal inside a building or a densely populated area with respect to the distance traversed. The model is expressed by the equation

$$\text{Path loss}(d) = \text{Path loss}(d_0) + 10n \log_{10}\left(\frac{d}{d_0}\right) + \chi \tag{10.16}$$

In this equation, Path loss (d_0) denotes the path loss in dB at a distance d_0 where d_0 is a reference distance, usually taken as 1m; Path loss(d) is the path loss in dB at an arbitrary distance $d > d_0$ where d is the length of the path to be determined; n is the environment-dependent path loss exponent, e.g., its value inside a building at line of sight is 1.6–1.8; and χ is a zero-mean Gaussian distributed random variable expressed in dB. It has a standard deviation σ. It is only included in conditions in which shadowing effects take place, otherwise it is zero.

Applying this model, the RSSI–distance relationship is given by

$$\text{RSSI}(d) = \text{RSSI}(d_0) + 10n \log_{10}\left(\frac{d}{d_0}\right) + \chi \tag{10.17}$$

where RSSI(d) and RSSI (d_0) are the RSSIs at distances d and d_0, respectively.

Ignoring χ, an approximate formula for distance calculation from RSSI value is derived for $d_0 = 1$ m.

$$\text{RSSI}(d > 1\text{m}) = \text{RSSI}(1\text{m}) + 10n \log_{10}\left(\frac{d}{1}\right) \tag{10.18}$$

or,

$$10n \log_{10} d = \text{RSSI}(d > 1\text{m}) - \text{RSSI}(1\text{m}) \tag{10.19}$$

or,

$$\log_{10} d = \frac{\text{RSSI}(d) - \text{RSSI}(1\text{m})}{10n} \tag{10.20}$$

$$\therefore d = 10^{\frac{\text{RSSI}(d) - \text{RSSI}(1\text{m})}{10n}} \tag{10.21}$$

For $n = 1.64$ (Huang et al. 2019)

$$d = 10^{\frac{RSSI(d)-RSSI(1m)}{10\times1.64}} = 10^{\frac{RSSI(d)-RSSI(1m)}{16.4}} \tag{10.22}$$

Knowing $RSSI(d)$ and RSSI (1m), the distance d can be calculated.

10.8.3 LOCALIZATION TECHNIQUES

We consider a space inside a big building in which beacons are fixed at different known locations and, suppose a person with a smartphone is moving in that building. The position of the person is to be determined.

10.8.3.1 Proximity Detection

This easy-to-implement algorithm indicates the location when the smartphone device comes within a fixed coverage area of a particular BLE beacon. This becomes evident when the RSSI from a single beacon is larger than a preset threshold value. The smartphone is considered to be located with that beacon. This kind of location information is a relative information based on the vicinity of the smartphone to known reference beacons.

Sometimes, the smartphone comes within the coverage area of more than one beacon. In that case, RSSI values from different beacons are compared. The smartphone is collocated with the beacon from which the RSSI is highest of the beacons, i.e., the beacon from which the strongest signal is received.

10.8.3.2 Trilateration

Trilateration is a traditional method to compute the unknown position of a mobile node using three reference nodes of known positions, and the distances between these nodes and the mobile node.

For trilateration, the distances of the smartphone receiver from three pre-known reference beacons are measured with the help of the log-path propagation model for RSSI. A circle is formed around each reference beacon taking the distance of the smartphone receiver from that reference beacon. In this way, three circles are sketched with the distances of the three reference beacons from the smartphone receiver as radii. These circles intersect at a point on a surface in the ideal case. This is the position of the smartphone receiver. The real situation is not idealistic. In practice, errors creep in from various sources. Hence, the intersection of circles is able to define a small region or area where the smartphone is expected to be found.

10.8.3.3 Fingerprinting

The fingerprinting algorithm consists of the following two phases.

(i) Offline Phase: For different coordinates of the smartphone device, the RSSI values of the signals coming from several beacons installed in the building are measured. The measurements are calibrated into specific features, called fingerprints, to describe the indoor environment, Thus, a fingerprint map of the RSSIs of all beacons at different locations of the smartphone is drawn and stored in a database.

(ii) Online Phase: In this phase, the map drawn in the offline phase is utilized for indoor positioning. For a given location of a person with a smartphone inside the building, the smartphone gets signals from various beacons. The RSSI data from the beacons is matched with the fingerprints in the database. The fingerprints in the database that most closely resemble the RSSI data of the person's smartphone are identified, and the person's location is found by reading the location corresponding to those database fingerprints.

Advantages

(i) Higher accuracy and
(ii) Reliable positioning is achieved with fingerprinting-based positioning than trilateration-based positioning if dense fingerprint locations are used.

Disadvantages

(i) Fingerprinting-based positioning needs denser placement of BLE beacons than required for one based on trilateration.
(ii) Fingerprinting-based positioning requires more expensive and time-consuming preparatory work.
(iii) Fingerprint maps are not permanent. The furniture inside a building may be re-organized. The density of people is also variable. The luggage and goods in the building may change with time. All these factors demand that the fingerprint mapping must be done frequently to construct the updated map. Otherwise, the system will lose its accuracy and the locations found will be inexact with the passage of time.

10.8.4 Using BLE Beacons at the Front and Back End of Retail Business

At the front end of a retail store, beacons are used for providing navigational guidance to customers visiting stores, especially useful in the case of vast stores. They are also used for delivery of notifications providing directions to specific items, and for sending personalized promotions, discounts, and in-store offers to customers' smartphones.

At the back end of a retail store, beacons are used as an indoor positioning system for tracking beacon-tagged products or assets in the warehouse, for watching customer behavioral patterns and preferences, and for keeping a vigil on the visitor movements in the store for the security of the premises. Capturing the UUID information transmitted by the beacon, a smartphone can easily display the location of the product/asset to which the beacon tag is affixed.

10.9 CONCLUDING REMARKS AND PREPARING FOR THE UPCOMING CHAPTER

We have seen how innovations in sensors have ushered novelties in methods of logistics and fleet management and helped shop owners in the sale of goods. Unifying the advantages availed by sensor technologies in homes, offices, factories, traffic control, healthcare, and agriculture with fleet management, retail, and customer care, we can virtually imagine what a smart city will look like. We take up this topic in the final chapter of the book because it is a collaborative effort where inputs from all the above sectors are blended together to give a collective meaningful result.

REFERENCES

Agarwal U. 2017 Indoor Positioning with BLE Beacons, © Talentica Software (I) Pvt Ltd., pp. 1–10.

Algueta-Miguel J. M., J. R. García-Oya, A. J. Lopez-Martin, C. A. De la Cruz Blas, F. Muñoz Chavero and E. Hidalgo-Fort 2020 Low-power ultrasonic front end for cargo container monitoring, *IEEE Transactions on Instrumentation and Measurement*, 69(6): 3348–3358.

Huang K., K. He and X. Du 2019 A hybrid method to improve the BLE-based indoor positioning in a dense Bluetooth environment, *Sensors (Basel)*, 19(2):424, pp. 1–15.

Ko J., J. Kim, S. Cho, and K. Lee 2005 A 19-mW 2.6-mm^2 L1/L2 dual-band CMOS, *IEEE Journal of Solid-State Circuits*, 40(7):1414–1425.

Samuel M., N. Nazeem, P. Sreevals, R. Ramachandran and P. Careena 2021 Smart indoor navigation and proximity advertising with android application using BLE technology, *Materials Today: Proceedings*, Vol. 43, Part 6, pp. 3799–3803.

Shue S. and J. M. Conrad 2016 Reducing the effect of signal multipath fading in RSSI-distance estimation using Kalman filters, *CNS '16: Proceedings of the 19th Communications & Networking Symposium*, Article No. 5, SpringSim-CNS, April 3–6, Pasadena, CA © 2016 Society for Modeling & Simulation International (SCS), San Diego, CA, USA, pp. 1–6.

Van Dierendonck A. J., S. S. Russel, E. R. Kopitzke, and M. Birnbaum 1978 The GPS navigation message (space segment), *Navigation: Journal of The Institute of Navigation*, 25(2):147–165.

Ward P. W. 2010 GPS receivers, receiver signals and principles of operation, 2135-1, Second Workshop on Satellite Navigation Science and Technology for Africa, The Abdus Salam International Center for Theoretical Physics, 6-23 April, pp. 1–197.

Braasch M. S. and A. J. van Dierendonck 1999 GPS receiver architectures, and measurements, *Proceedings of the IEEE*, 87(1): 48–64.

Cserhat N. 2023 *Smart Sensors and IoT Applications for Maritime Cargo Tracking and Tracing*, Published in Morpheus. Network, https://news.morpheus.network/smart-sensors-and-iot-applications-for-maritime-cargo-tracking-and-tracing-39beb9bde95e

Parkinson B.W. and J. J. Spilker Jr. (Eds.) 1996 *Global Positional System: Theory and Applications*, Vol. 1, First Edition, American Institute of Aeronautics and Astronautics, Inc., Washinton DC, USA, 632 pages.

11 IoT Sensors for Smart City

11.1 INTRODUCTION

More than 50% of the human population lives in cities (Juraschek et al. 2018) A city is a densely populated geographical region with well-defined boundaries and having its local administrative structure. It usually has several residential and commercial localities, banks, and government buildings along with hospitals and pharmacies for patient care and dispensing of medications. Different places in a city are well-connected by roads. It has a pure drinking water supply system, drainage system, sewer lines, and sanitation facilities. It has well-knit vehicular transport and electronic communication systems, bus and railway stations, and airport(s). Although the city people are primarily engaged in non-agricultural tasks, urban agriculture is becoming popular to maintain greenery and reduce environmental pollution, and for creating sustainable and climate-friendly agro-cities. Despite their adverse effects on the environment, urban factories are needed because industrial value creation is necessary for the economic growth of a society and for making cities sustainable.

Needless to mention, all the contents of the preceding chapters are applicable to smart cities. Notwithstanding, there are some special topics relevant to cities that we shall address in this chapter, among them, the gaseous pollution from the factory and vehicle exhausts, noise pollution, waste disposal issues, drinking water quality testing for the protection of public health, electronic water meters for convenient billing, and many others.

11.2 THE SMART CITY

A smart city is an urban area leveraging sensor and Internet of Things (IoT) technologies to collect data about its inhabitants and infrastructure for an all-inclusive management of its resources, utilities, and civilian services. It provides a smart grid for efficient electricity distribution supported by solar power generation, an assured supply of high-purity drinking water, a city-wide Wi-Fi network, together with air-quality and noise pollution control, efficient garbage collection and cleanliness maintenance, proper sanitation, automated homes and buildings, intelligent street lighting, strict sensor-controlled traffic regulation, excellent public transport facilities, quick parking solutions, digital classrooms in schools, speedy emergency response systems, and electronic city surveillance and governance platforms for transparent functioning. It has a strong commitment to eco-friendly initiatives for the provision of improved quality of life by citizen participation and backed up by citizen-centric sustainable development.

11.3 ROLE OF IoT SENSORS IN SMART CITIES

A few thrust areas demanding sensor support in most cities include the following.

(i) Infrastructure Management: Surveillance of the condition of buildings, roads, and bridges is routinely done to identify maintenance needs in order that the repair and downtime are minimized, and any mis-happenings are avoided (**video cameras, vibration sensors**).

(ii) Public Safety: Tracking criminals quickly, and promptly responding to vehicle or fire accidents is essential (**video cameras**).

(iii) Transportation: Management of traffic to avoid congestion, and facilitating vehicle parking averts accidents, and reducing road rages and quarrels among people (**see Chapter 6 on traffic control**).

DOI: 10.1201/9781003374442-11

(iv) Preserving Energy Efficiency: This is ensured by controlling power grid utilization (**see Chapter 7 on energy-efficient power grid**).

(v) Environmental Pollution Control: Keeping a vigil on the gaseous emissions from automobiles and factories is very helpful (**gas sensors, temperature and humidity sensors, anemometers**)

(vi) Noise Pollution Control: Listening to noise levels to restrict them within safe limits calms down people's anger and irritations (**noise sensor**).

(vii) Waste Management: Reducing garbage collection time prevents the bins form overflowing with waste matter, which may be littered on the streets (**waste bin sensor**)

(viii) Water Distribution and Quality Control: By maintaining water supply at the correct flow rate and pressure in different parts of the city, and by continuously monitoring the purity of supplied water, people remain comfortable and healthy (**water flow sensor, chlorine sensor, turbidity sensor, conductivity sensor, etc.**).

(ix) Smart Streetlighting: Switching On/Off and lamp brightness adjustment according to requirement saves power wastage (**light sensor, occupancy/motion sensor**)

(x) Health and Wellness: Watching the health trends of people and issuing proper advisories prevents illnesses, particularly, the spread of contagious diseases (**see Chapter 8 on human health monitoring**).

(xi) Citizen Participation: Reporting any problems quickly and giving proper feedback on services delivered aids in readily improving services and systems for the general public.

11.4 AIR QUALITY INDEX

11.4.1 MAJOR AIR POLLUTANTS

The particulate matter (PM) pollution is a serious public health threat, especially the pollution owing to $PM_{2.5}$ particles (diameter $\leq$ 2.5 μm) and PM_{10} particles (diameter $\leq$ 10 μm). The tolerable safety limit of $PM_{2.5}$ particles has been decreased by the World Health Organization (WHO) from 10 μg/m^3 to 5 μg/m^3 and that for PM_{10} particles from 20 μg/m^3 to 15 μg/m^3 (Lavanyaa et al. 2023).

The air quality index (AQI) is the yardstick for measuring air pollution. For PM_{10} particles, the air quality conditions are categorized into the following: AQI value 0–50 (good); 51–100 (fair, moderate, and acceptable); 101–150 (poor and adverse for sensitive groups of people); 151–200 (very poor and unhealthy for general public); 201–300 (extremely poor state calling for a very unhealthy alert); and >301–500 (hazardous, and warning for a health emergency). Besides particle pollution, AQI also includes major air pollutants such as ground-level ozone, carbon monoxide, nitrogen dioxide, sulfur dioxide, ammonia, and airborne lead.

11.4.2 AIR PARTICULATE MATTER (PM) SENSOR

An air pump pulls the particles into the particle counter sensor (Figure 11.1) at a controlled flow rate which is maintained at a constant value by a feedback loop and mass flow controller. The laser beam from a laser source crosses the view volume of the particle counter sensor. The view volume is the region of the intersection of the laser beam with the particle flow path. The light scattered by the particles in the view volume is collected by an optical lens assembly. The collected light is directed toward a photodetector for converting the scattered light into an electrical signal in millivolts. The large-size particles scatter more light than the smaller particles. The bigger the particles, the higher the resulting millivolt signal, as seen from the height of the voltage pulse. The captured millivolt signals are fed to threshold comparators to subdivide the particle counts into different size ranges. The comparators are pulse height analyzers that examine the voltage pulses received and distribute them into bins according to the pulse heights. The pulse heights correspond to particle sizes. Therefore, the number of pulses received in each bin gives the number of particles of the particular size to which that bin is allocated. These results are displayed on the particle counter as particle

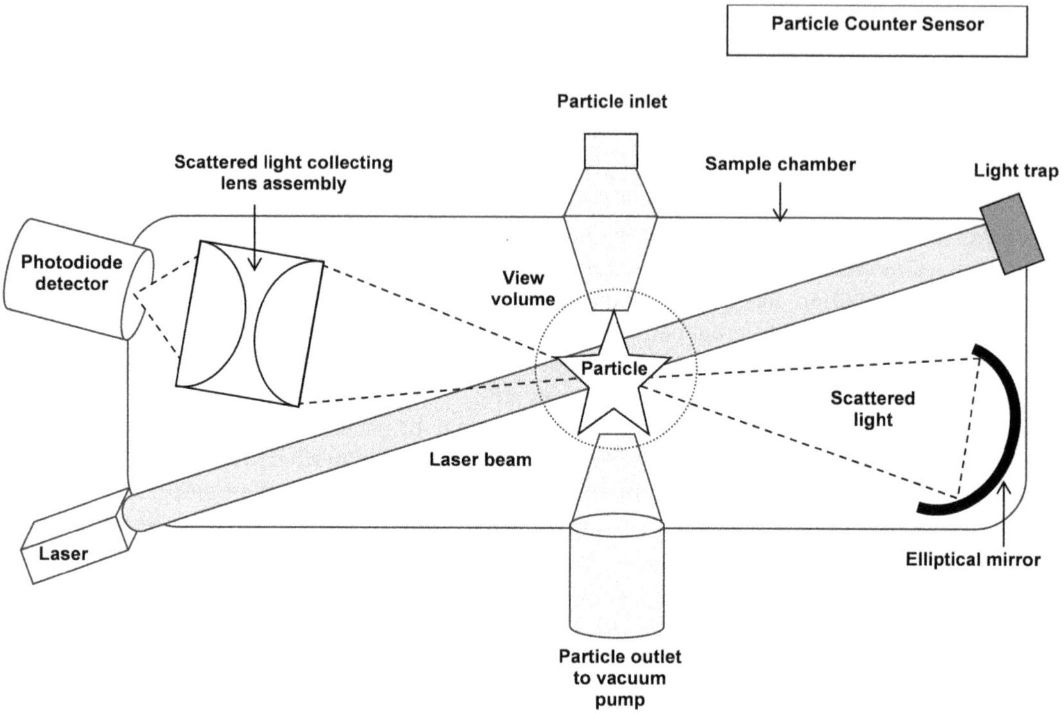

Particle Counter Sensor

Particle inlet

Scattered light collecting
lens assembly

Sample chamber

Light trap

Photodiode
detector

View
volume

Particle

Scattered
light

Laser beam

Laser

Elliptical mirror

Particle outlet
to vacuum
pump

FIGURE 11.1 The airborne particles counter sensor consists of a sample chamber. This chamber has a particle inlet on the top side and a particle outlet on the exactly opposite bottom side. A laser diode directs a beam of light across the path of the incoming stream of particles. Interaction of the incoming particles with the laser beam causes scattering of light. The scattered light falls on an elliptical mirror. From this mirror, it rebounds toward a light-collecting lens system. The system focuses the light on a photodetector.

count data of various sizes. Thus, particle sizes are quantified and the number of particles of specific sizes in a given volume is counted.

11.5 SEMICONDUCTOR METAL OXIDE (SMO$_x$) GAS SENSORS

The resistance of these sensors changes with the concentration of the target gas. Hence, they are called chemiresistors or conductometric sensors. A salient feature of these sensors is the integration of a heater element in their structure to heat the metal oxide film. Heating the oxide film is necessary because the specific chemical reaction between the target gas and the metal oxide film utilized for the detection of the target gas takes place at an enhanced rate at a particular temperature, which is a unique characteristic of the gas.

11.5.1 Types of Metal Oxide Gas Sensors

Three forms of these sensors, namely, tubular ceramic, planar ceramic, and MEMS technology-based, are illustrated in Figure 11.2.

11.5.1.1 Tubular Ceramic Metal Oxide Gas Sensor

It consists of the following parts (Figure 11.2(a)).

   (i) A hollow ceramic tube made of aluminum oxide (Al_2O_3).
  (ii) A nichrome (NiCr) heater coil provides the required high temperature at which the device has a high sensitivity to the target gas. The coil is placed inside the hollow ceramic tube.

(iii) A gas-sensing film is coated over the outer surface of the ceramic tube. The film is made of tin oxide or some other metal oxide, selected according to the gas to be detected.

(iv) Electrode and Electrode Wire: Gold electrodes are formed over the sensing film. Platinum wires are bonded to the gold electrodes for connection.

Working: The sensor is placed in the environment of the target gas. The NiCr heater is turned on, and adjusted to reach the required temperature at which the sensing film exhibits a high sensitivity to the gas. The molecules of the target gas interact with the sensing film, causing a change in its resistance. The resistance change takes place in proportion to the concentration of the target gas, and so the gas concentration can be inferred from the magnitude of the change in resistance.

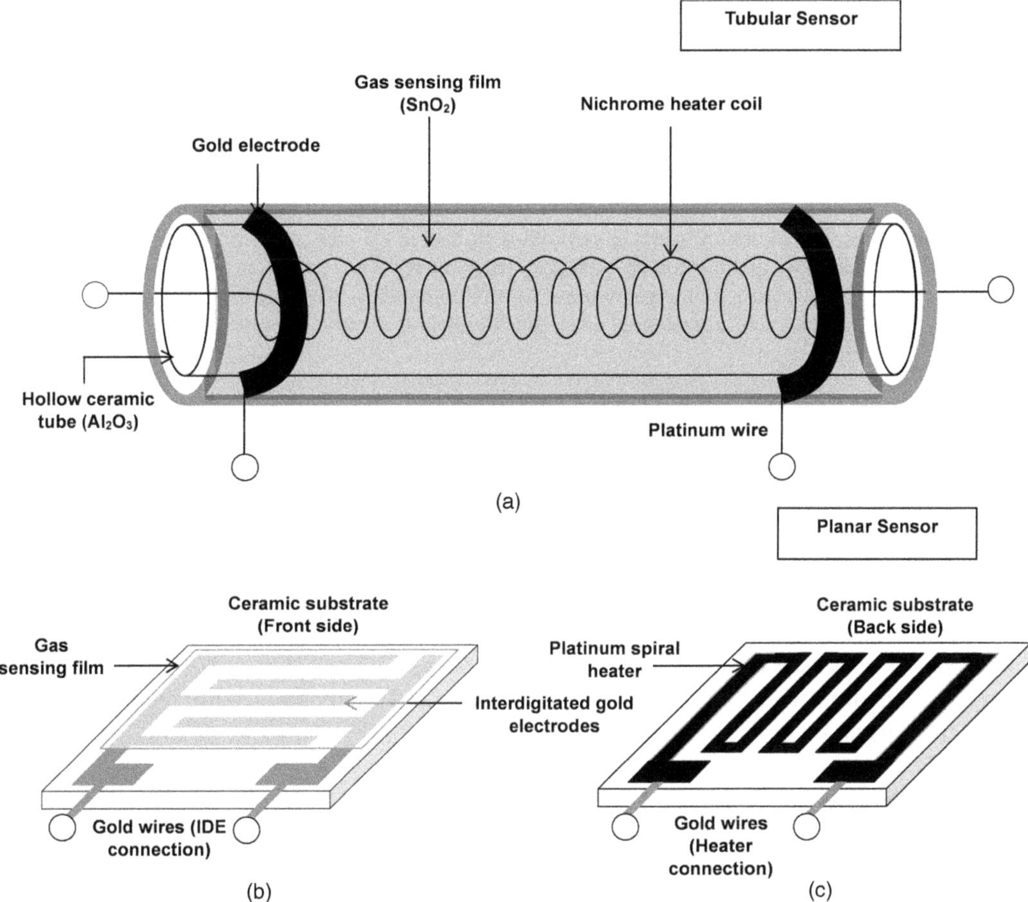

FIGURE 11.2 Metal oxide gas sensors: (a) Tubular ceramic gas sensor, (b) and (c) planar ceramic, and (d) MEMS technology-based. Part (a): Inside a hollow ceramic tube, a nichrome heating coil is fixed. On the surface of the ceramic tube, the gas-sensing SnO_2 film is deposited. Near the left and right ends of the ceramic tube, gold electrodes are made to contact the gas-sensing film. Platinum wires are joined to the gold electrodes. Part (b): Left side: The front side of a ceramic substrate is shown. Interdigitated gold electrodes are patterned on the substrate. The electrodes are terminated with contact pads to which gold wires are bonded for electrical connections. The electrodes are coated with the gas-sensing film by protecting the contact pads so that they are left uncovered. Part (c): Right side: The back side of the same ceramic substrate is shown. A heating coil is made with a platinum film in the shape of a spiral. The coil has contact pads at its ends to which gold wires are bonded for connections.

(Continued)

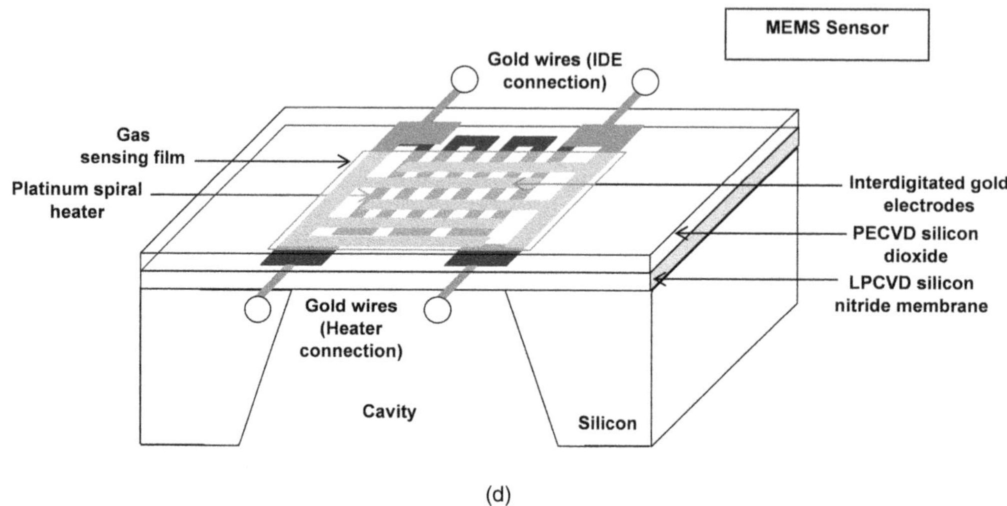

(d)

FIGURE 11.2 (CONTINUED)   Part (d): An LPCVD silicon nitride membrane is suspended over a cavity in silicon. A spiral-shaped pattern of platinum heater is formed over the silicon nitride membrane. It has contact pads at its two ends. The platinum heater is covered with a PECVD silicon dioxide film. Over the silicon dioxide film, gold electrodes are made in an interdigitated geometry. They have contact pads at their two ends. The gold electrodes are coated with the gas-sensing film. Wire bonding of gold wires is done to the platinum heater spiral as well as the interdigitated gold electrodes for making electrical connections.

11.5.1.2   Planar Ceramic Metal Oxide Gas Sensor

A ceramic substrate is taken (Figure 11.2(b) and (c)). On the upper surface of the substrate, a gold film is deposited by thermal evaporation or sputtering. Photolithography is performed to define a pattern of interdigitated Au electrodes. On the opposite surface of the ceramic substrate, platinum is thermally evaporated in vacuum, followed by photolithography to define a spiral-shaped heater. Then, Au wires are attached by wire bonding to the bonding pads of the interdigitated electrodes on the upper surface of the substrate as well as the bonding pads of the platinum heater on its opposite surface. Finally, the sensing film is deposited over the interdigitated electrode pair.

The planar gas sensor works in the same way as the tubular one. The heater is connected to a power supply, and the interdigitated electrodes are connected to an ohmmeter to measure the resistance of the sensing film exposed to the gaseous environment.

11.5.1.3   MEMS Metal Oxide Gas Sensor

This sensor (Figure 11.2(d)) is made on a silicon substrate of <100> crystallographic orientation. It consists of the following:

(i)   A silicon nitride membrane is formed by low-pressure chemical vapor deposition (LPCVD) on the silicon substrate.
(ii)   A meander-shaped platinum heater is formed over the membrane, and embedded in a PECVD silicon dioxide film.
(iii)   A pair of interdigitated gold electrodes is defined over the silicon nitride film covering the platinum heater.
(iv)   A gas-sensing film is deposited over the interdigitated electrodes.

A cavity is formed under the silicon nitride membrane to reduce the power consumption to reach the high temperature because the main chunk of silicon is removed, leaving only a small mass of membrane material together with the heater and the sensing film to be heated.

The operation of the MEMS sensor is similar to that of the other two sensors. It offers the advantages of small size and low power consumption.

11.5.2 CO Sensor

Figure 11.3 shows the CO sensor. The TiO_2 nanowires are grown by vapor-phase growth (Lee et al. 2016). Tri-layer Ti(1 μm)/Pt(200 nm)/Ti(50 nm) metal is deposited on the SiO_2-grown Si substrate. The bottom Ti and Pt layers are formed by sputtering whereas the topmost Ti layer is deposited by electron-beam evaporation. The bottom Ti layer improves adhesion with the substrate. The topmost Ti layer serves as a catalyst for the growth of TiO_2 nanowires.

Lithography is done to make a pattern of interdigitated electrodes. To selectively grow networked nanowires, the patterned Si wafer is loaded in a horizontal tube furnace. The Ti powder is placed 4 cm away from the wafer. The furnace is evacuated to 1×10^{-3} torr. Then, it is maintained at 1,050°C for 1 h. The O_2 and Ar gases are introduced in the furnace at controlled flow rates. Consequently, 90-nm-diameter TiO_2 nanowires grow and join between neighboring electrodes by the accumulation of Ti vapor over the Ti catalytic layer and its oxidation to TiO_2, resulting in a network of chemiresistive electrical pathways.

Variation of resistance of the nanowires is studied with respect to temperature from 300°C to 500°C by exposing to 1, 5, and 50 ppm CO. The resistance follows the start and closure of the CO supply. From an investigation of the sensor response at 1 ppm CO, the response at 400°C is found to be maximum (Lee et al. 2016).

11.5.3 NO₂ Sensor

The NO_2 sensor is shown in Figure 11.4. A pair of comb-type electrodes is screen-printed on an alumina substrate using an Au paste and calcined at 850°C (Kida et al. 2009). Sodium tungstate (Na_2WO_4) is acidified with strongly acidic solutions of pH = −0.5, −0.8 using acids such as H_2SO_4, HCl, and HNO_3. Lamellar-structured WO_3 particles (thickness = 20–50 nm, lateral size = 100–350 nm) are formed in the form of crystalline plates. A paste is made by mixing the WO_3 particles with water. The paste is applied over the comb-type Au electrodes. Calcination is done in air at 300°C for 2 h. The WO_3 sensing layer is obtained as the precursor $WO_3·2H_2O$ is removed during calcination. The sensor response

$$S = \frac{\text{Resistance in air containing } NO_2 \text{ gas}\left(R_g\right)}{\text{Resistance in dry air}\left(R_a\right)} = 150 - 280 \text{ in } 50 - 1,000 \text{ ppb } NO_2 \text{ in air at } 200°C$$

$$(11.1)$$

11.5.4 Gas Sensing Mechanism of SMO$_x$ Sensors

The metal oxide film is a porous network of tiny grains. Some metal oxides are N-type in character while others are P-type. Therefore, the sensing mechanism is discussed relative to both types of oxides.

11.5.4.1 N-type Metal Oxide

Consider an N-type metal oxide, e.g., SnO_2 and TiO_2 (Ponzoni et al. 2017). Figure 11.5(a) shows the metal oxide grains in the as-prepared state. When the metal oxide film is exposed to the atmosphere, the phenomenon is visualized in terms of the energy-band diagram model of the metal oxide grains, in which the conduction band edge bends upward close to the surfaces of the grains (Figure 11.5(b)). Oxygen molecules from the air are chemisorbed on the grains (Figure 11.5(c)). This happens because the

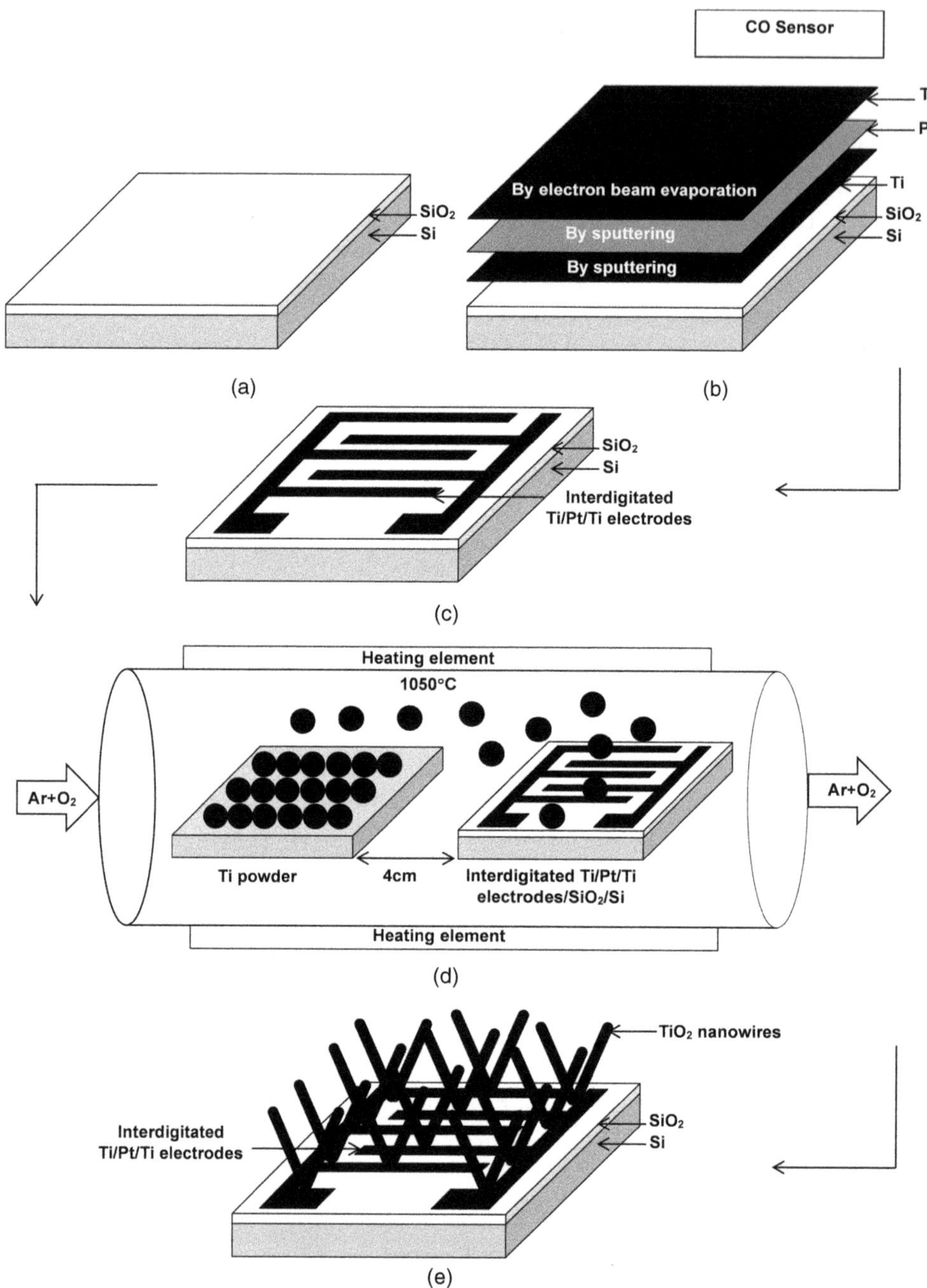

FIGURE 11.3 Carbon monoxide sensor: (a) SiO$_2$/Si substrate, (b) Ti/Pt/Ti trilayer deposition, (c) interdigitated Ti/Pt/Ti electrode pair patterning, (d) vapor phase growth of TiO$_2$ nanowires in a furnace, and (e) the completed sensor chip. Part (a): A silicon substrate with thermally grown silicon dioxide is shown. Part (b): A bottom Ti film is sputter-deposited over the substrate, then a Pt layer is deposited on the Ti film by sputtering, and finally a Ti layer is deposited over the platinum film by electron-beam evaporation. Part (c): A pattern of interdigitated Ti/Pt/Ti electrodes is etched in the multilayer metal film deposited on the substrate. It has contact pads at its two ends. Part (d): In an evacuated furnace, titanium powder is loaded 4 cm ahead of the Ti/Pt/Ti/SiO$_2$/Si structure. Then, Ar+O$_2$ is flowed at a temperature of 1,050°C. Titanium vapors are carried by the gas stream toward the structure. As (Ar + oxygen) is passed in the furnace, the titanium vapors react with oxygen to form TiO$_2$, and TiO$_2$ nanowires grow to sufficient lengths to join the adjacent electrodes. Part (e): The diagram shows the CO sensor consisting of an SiO$_2$/Si substrate with a pattern of Ti/Pt/Ti interdigitated electrodes, and a network of TiO$_2$ nanowires joining the neighboring electrodes to serve as a gas-sensitive film.

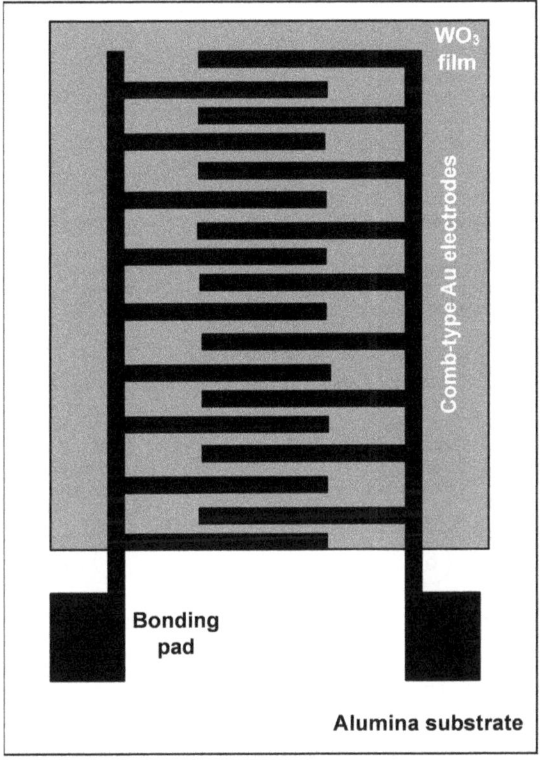

FIGURE 11.4   Nitrogen dioxide sensor. A comb-type pattern of gold electrodes with contact pads is created over an alumina substrate. A film of tungsten oxide lamellar particles is formed over the comb-type electrodes, leaving the contact pads uncovered.

chemisorbed oxygen captures and traps electrons from regions near the surfaces of the grains. Depending on ambient temperature, the adsorbed oxygen exists in one of the three forms: O_2^-, O^-, or O^{2-} with the help of trapped electrons. Below 150°C, the oxygen is in molecular form O_2^- while above this temperature, it is atomic states O^{2-} or O^- (Saruhan et al. 2021):

$$80°C - 150°C : O_{2\,Gas} + e^- \left(\text{from SMO}_x \right) \rightarrow O_{2\,Adsorbed}^- \tag{11.2}$$

$$150°C - 300°C : O_{2\,Adsorbed}^- + e^- \left(\text{from SMO}_x \right) \rightarrow 2O_{Adsorbed}^- \tag{11.3}$$

and

$$300°C - 500°C : O_{Adsorbed}^- + e^- \left(\text{from SMO}_x \right) \rightarrow O_{Adsorbed}^{2-} \tag{11.4}$$

As a consequence of the migration of electrons from the interior of the metal oxide toward surface-bound oxygen, depletion regions are formed around the grains up to small depths below the grains. The depths of the depletion regions formed in the grains are determined by the dielectric constant of the metal oxide, the amount of band bending, and the charge carrier concentration in the oxide.

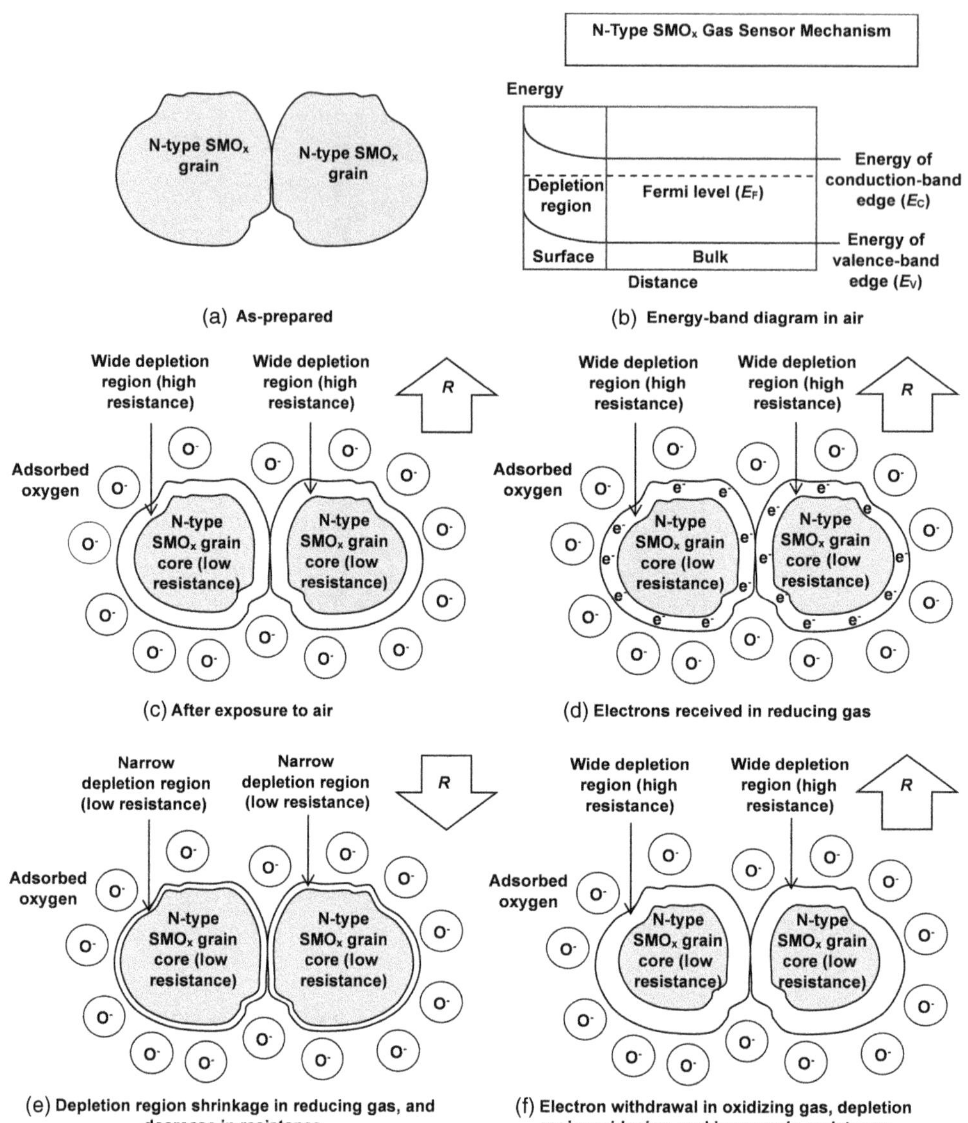

FIGURE 11.5 Gas sensing mechanism of N-type metal oxide: (a) initial as-prepared condition, (b) energy-band diagram after exposure to the atmosphere, (c) condition after exposure to the atmosphere, (d) after exposure to a reducing gas, and electron receipt from the gas, (e) effect of reducing gas on depletion region thickness, and (f) effect of an oxidizing gas ambiance on depletion layer width. Part (a): Two SMO_x grains are shown touching each other. Part (b): Energy-band diagram of an N-type metal oxide in the air showing the formation of a depletion layer by extraction of electrons by the high-electronegativity (electron-attracting tendency) oxygen species adsorbed upon it. Part (c): The diagram shows adsorbed oxygen anions surrounding the SMO_x grains, and the resulting depletion layer produced around the grains due to the grabbing of electrons from the grains by oxygen atoms to form negatively charged ions. The cores of the grains are low-resistance regions but the wide depletion regions built near the grain surfaces have a high resistance due to loss of electrons. Resistance of the grain film increases. Part (d): A reducing gas itself loses electrons and the lost electrons are transferred to the SMO_x grains. They move in the depletion regions of the grains, and are represented by the symbol e^-. The cores of the grains are not affected. Part (e): As the incoming electrons from the reducing gas spread in the depletion region, they thin down the depletion region, and lower its resistance. Hence, the resistance of the SMO_x film decreases. However, the cores of the SMO_x film are still low-resistance regions. Part (f): An oxidizing gas itself gains electrons. These electrons migrating to the oxidizing gas are taken from the SMO_x grain outer surface regions. Consequently, the depletion region surrounding the SMO_x grain broadens. The resistance of the SMO_x film thereby increases. The cores of the grains remain unaffected and behave as low-resistance zones.

Due to the depletion regions surrounding the grains, electrons experience difficulty in flowing from one grain to another. This hindrance in electron transport across the grains is observed as an increase in the resistance of the metal oxide. Thus, the metal oxide exhibits an increase in resistance arising from the adsorption of oxygen atoms, as shown by an downward directed arrow with resistance written inside it.

The target gas, the one to be detected, can be a reducing or an oxidizing agent. Suppose the target gas is a reducing agent, e.g., H_2, NH_3, and CO. The target gas is adsorbed on the metal oxide surface. Then, the reaction takes place between the target gas and the surface-bound oxygen on the metal oxide. In this reaction, the target gas is oxidized (loses electrons) while the metal oxide surface is reduced (gains electrons) (Figure 11.5(d)). These electrons move to the depletion regions in the metal oxide grains, thereby decreasing the depths (or thicknesses) of the depletion regions, and hence lowering the resistance of the metal oxide (Figure 11.5(e)), as shown by an downward directed arrow with resistance written inside it.

Chemical reactions between a reducing gas CO and the oxygen on the metal oxide surface are (Afzal 2019) as follows:

$$2CO(Gas) + O_{2\,Adsorbed}^- \rightarrow 2CO_2(Gas) + e^- (\text{Given to } SMO_x) \tag{11.5}$$

$$CO(Gas) + O_{Adsorbed}^- \rightarrow CO_2(Gas) + e^- (\text{Given to } SMO_x) \tag{11.6}$$

$$CO(Gas) + 2O_{Adsorbed}^- \rightarrow CO_3^{2-} \rightarrow CO_2(Gas) + \frac{1}{2}O_2(Gas) + 2e^- (\text{Given to } SMO_x) \tag{11.7}$$

In all these cases, electrons are given to SMO_x, which is responsible for the fall in its resistance.

How does the metal oxide respond to an oxidizing target gas? Exactly the reverse happens in the presence of an oxidizing gas, e.g., NO_x. The target gas itself is reduced (gains electrons) while the metal oxide surface is oxidized (loses electrons). The loss of electrons by the metal oxide results in an enlargement of the depletion region width, and an increase in the resistance of metal oxide.

Chemical reactions between an oxidizing gas NO_2 and the metal oxide surface lead to electron loss from the metal oxide:

$$NO_2(Gas) + e^- (\text{Surface}) \rightarrow NO_2^- (\text{Adsorbed}) \tag{11.8}$$

$$NO_2(Gas) + O_2^- (\text{Adsorbed}) + 2e^- (\text{Surface}) \rightarrow NO_2^- (\text{Adsorbed}) + 2O^- (\text{Adsorbed}) \tag{11.9}$$

The effect of an oxidizing gas on N-type metal oxide is depicted in Figure 11.5(f).

11.5.4.2 P-type Metal Oxide

In the case of such oxides (Figure 11.6(a)), e.g., CuO and NiO, the absorbed atmospheric oxygen takes away the electrons from the conduction band of metal oxide, and a hole accumulation layer builds up surrounding the grains (Figure 11.6(b)). Therefore, hole transport between the grains becomes easier, and the resistance of metal oxide is lowered (Figure 11.6(c)).

Exposure of a P-type metal oxide to a reducing gas environment leads to a reaction between the reducing gas and oxygen-bound metal oxide. The reducing gas is oxidized (loses electrons) while oxygen-bound metal oxide is reduced (gains electrons). These electrons are released to the hole accumulation layer (Figure 11.6(d)). As these electrons recombine with the holes, the thickness of the hole accumulation layer decreases. As a result, the resistance of the metal oxide increases (Figure 11.6(e)).

Bringing an oxidizing gas near the metal oxide initiates the reverse chain of phenomena, viz., the target gas is reduced by gaining electrons from the oxygen-bound metal oxide surface, and the

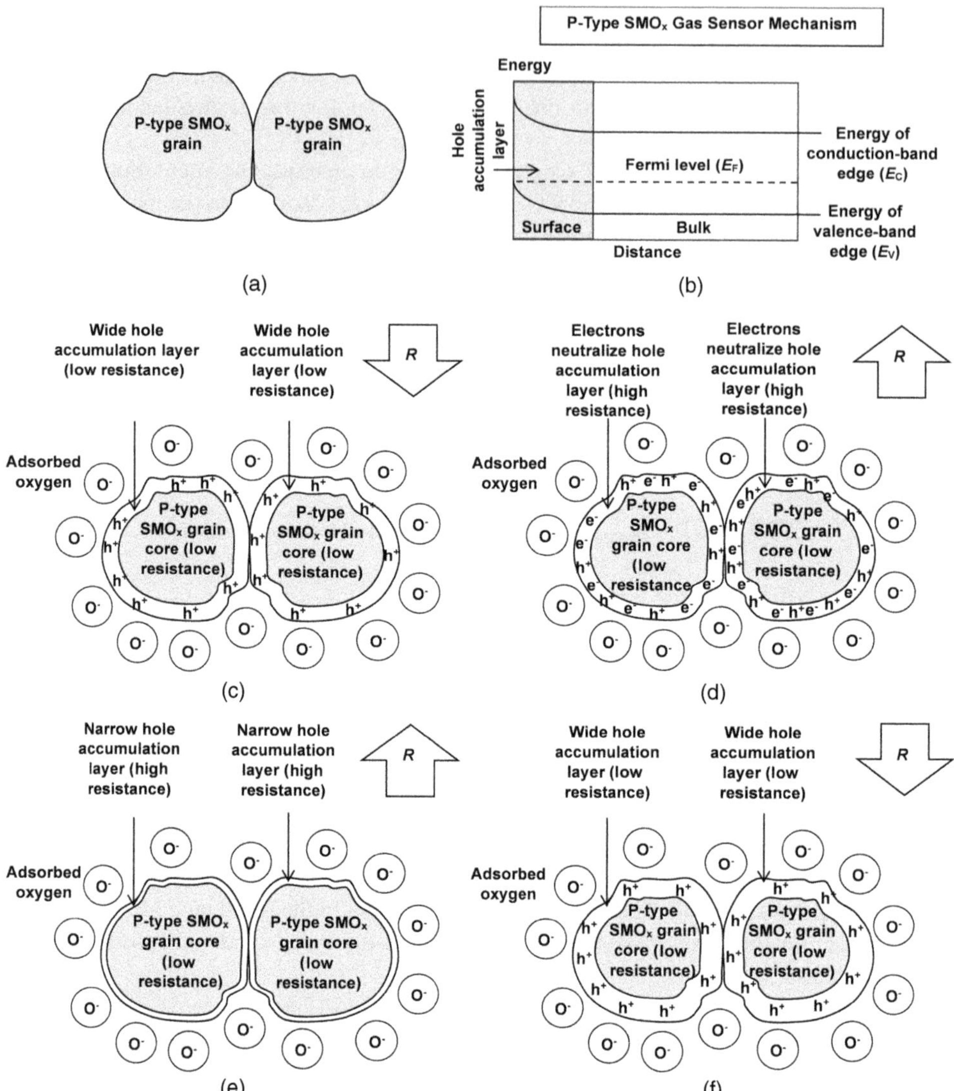

FIGURE 11.6 Gas sensing mechanism of P-type metal oxide: (a) starting as-prepared condition, (b) energy band diagram after exposure to the atmosphere, (c) condition of metal oxide grains after exposure to air, (d) after exposing to a reducing gas, and electron receipt from the gas, (e) influence of a reducing gas on accumulation layer thickness, and (f) effect of an oxidizing gas ambiance on accumulation layer width. Part (a): Two SMO_x grains in contact with each other are shown. Part (b): Energy-band diagram of a P-type metal oxide in air showing the formation of a hole accumulation layer by extraction of electrons by the high-electronegativity oxygen species adsorbed upon it. Part (c): The diagram shows adsorbed oxygen anions surrounding the SMO_x grains, and the resulting hole accumulation layers produced around the grains due to the loss of electrons from the grains to oxygen atoms to form negatively charged ions. The holes are denoted by the symbol h^+. The cores of the grains are low-resistance regions. The wide accumulation layers built near the grain surfaces too have low resistances due to an increase in hole populations. Therefore, the resistance of the grain film decreases. Part (d): A reducing gas itself loses electrons, and the lost electrons are transferred to the SMO_x grains. They move in the accumulation layers of the grains. The cores of the grains are not affected. Part (e): As the incoming electrons from the reducing gas spread in the accumulation layers of the grains, they neutralize the holes. As a result, the free-carrier hole density in the accumulation layers decreases, elevating the resistance of the accumulation layers. Hence, the resistance of the SMO_x film increases. Nevertheless, the cores of the SMO_x film are still low-resistance regions. Part (f): An oxidizing gas itself gains electrons. These electrons migrating to the oxidizing gas are taken from the outer surface regions of the SMO_x grains. Consequently, the accumulation layers surrounding the grains broaden. The resistance of the SMO_x film thereby decreases. Notwithstanding, the cores of the grains remain unaffected, and behave as low-resistance zones.

oxygen-bound metal oxide surface is oxidized by losing electrons, accompanied by the creation of more holes produced from lost electrons, and an increase in thickness of the hole accumulation layer. The wider hole accumulation layer facilitates hole transport between successive P-type grains. Therefore, the resistance of metal oxide diminishes (Figure 11.6(f)).

11.6 ELECTROCHEMICAL GAS SENSORS

This sensor is essentially an electrochemical cell (Figure 11.7). An electrochemical reaction, either oxidation or reduction, takes place inside this cell when the target gas enters it. This reaction leads to the production of an electric current. The current produced is proportional to the concentration of the gas, allowing it to be measured.

The principal components of the sensor are as follows.

(i) An anti-condensation membrane to prevent any dust particles from contaminating the cell.
(ii) A capillary-shaped opening to limit the diffusion of gas into the cell.
(iii) A filter, such as activated charcoal, to restrict the inflow of any unwanted gas, and to provide selectivity to the target gas.

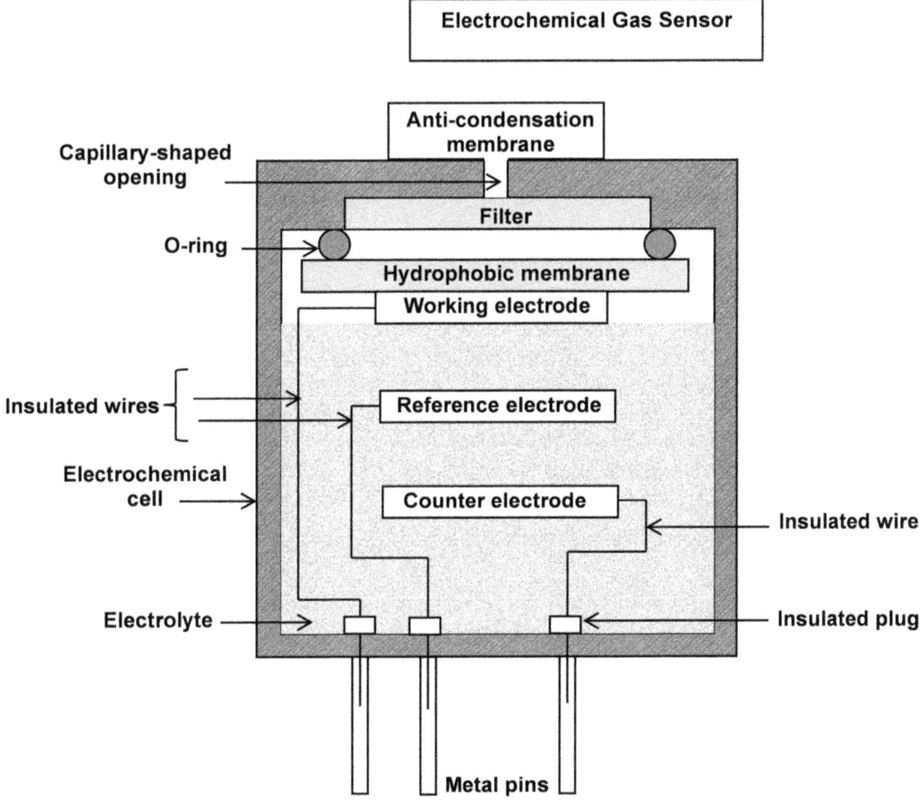

FIGURE 11.7 Generic electrochemical gas sensor. The diagram shows the main parts of the sensor. It consists of an electrochemical cell filled with the electrolyte solution. The roof of the cell has a capillary-shaped opening covered with an anti-condensation membrane. On the bottom side of the capillary opening, a filter is fixed. A hydrophobic membrane is tightened below the filter with an O-ring seal. To the lower surface of the hydrophobic membrane, the working electrode is attached. This electrode is dipping inside the electrolyte. Two more electrodes are immersed in the electrolyte, viz., the reference and counter electrodes. Insulated wires from all three electrodes pass through insulated plugs, and are connected to metal pins outside for connections.

(iv) A hydrophobic membrane, to ensure that the gas of correct molecular weight reaches the working electrode.

(v) A working electrode, which is made of a catalytic material, to promote the electrochemical reaction for the gas being measured. It shows a reactivity with the target gas and inhibits undesirable responses to interfering gases. Precious metals, e.g., platinum or gold are used for this electrode. The working electrode is also called the sensing electrode.

(vi) A reference electrode to maintain the potential of the working electrode at a stable and constant value.

(vii) A counter electrode: The electric current flows through the electrolyte between the working and counter electrodes.

(viii) An electrolyte in which the reaction is carried out, and through which the ions are transferred to the electrodes.

Working: All the components of the cell are enclosed in a housing with the electrodes immersed in the electrolyte. The target gas can enter the cell but the electrolyte cannot flow out. As soon as the target gas reaches the working electrode, it is oxidized or reduced at the active catalyst sites. In this reaction, electrons are released/consumed, resulting in the flow of a current between the working and counter electrodes, and hence across the cell output terminals. If an oxidation reaction has taken place, electrons move from the working electrode to the counter electrode. If the reaction is a reduction reaction, the electrons move in the reverse direction. By amplification and processing of the current, the gas concentration is displayed in parts per million (ppm). Since the sensor produces an output current, its operation is similar to a battery.

11.6.1 SO_2 SENSOR

Atmospheric sulfur dioxide is detected by an electrochemical sensor made by Hodgson et al. 1999. In this sensor, the electrochemical cell is made of flexiglass (Figure 11.8). The reference electrode is mercury/mercurous sulfate. The counter electrode is a gold wire. The working electrode is porous Au/Nafion SPE (solid polymer electrolyte). The electrolyte solution is 1M H_2SO_4 or 1M NaOH.

Cells are made using porous Au-Nafion SPE with a strong acid as the electrolyte or porous Au-Nafion SPE with a strong base as the electrolyte. In a strong acid electrolyte solution, the SO_2 is oxidized at the working electrode with SO_4^{2-} (Aqueous) as the dominant electroactive species while in a strong base electrolyte solution, this species is SO_3^{2-} (Aqueous). Low SO_2 detection limits are achieved in both cases with values in the low-ppb range.

11.6.2 H_2S SENSOR

Near-ppm levels of hydrogen sulfide are detected using a suitable material for the counter electrode (Kroll et al. 1994). The working electrode is made of a platinum wire net coated with Pt by electrolysis. This is the electrode on which H_2S is oxidized. The electrolyte is 35–39% dilute sulfuric acid. Due to its low freezing point (−60°C), the operating range of the sensor is −50°C to +60°C. The counter electrode must remain stable in the electrolyte over the functional life of the sensor, extending over years. It is made by mixing macrocyclic or macrobicyclic polyaza cobalt (III) complexes as catalysts in charcoal in 1:10 weight ratio. The H_2S oxidation reaction taking place in the cell is

$$H_2S + 2O_2 = H_2SO_4 \tag{11.10}$$

The sensor shows a high sensitivity of ~0.1 ppm H_2S.

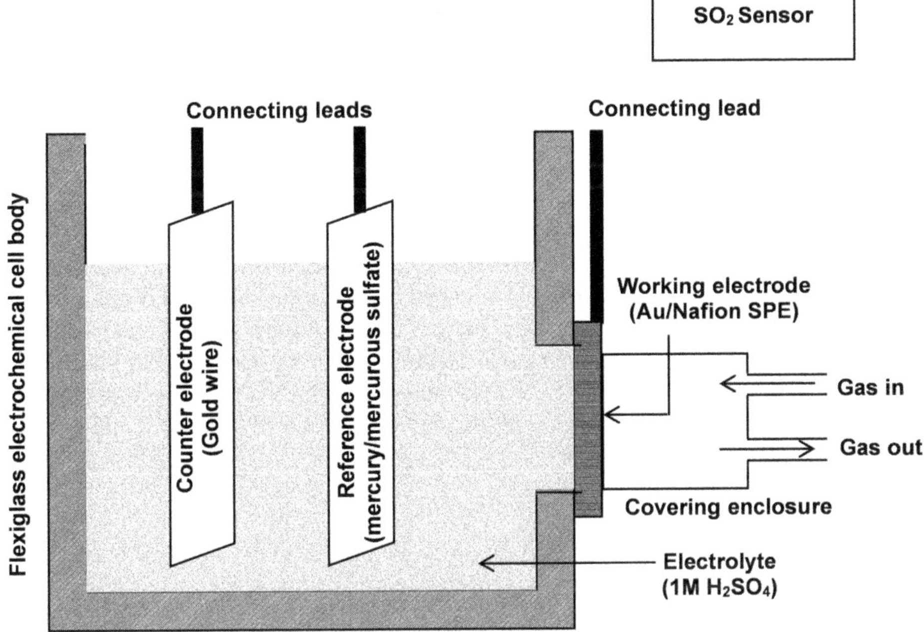

FIGURE 11.8 Sulfur dioxide electrochemical gas sensor. The body of the cell is made of flexiglass. The electrolyte filled in the cell is 1M H_2SO_4. The counter electrode is a gold wire. Mercury/mercurous sulfate reference electrode is dipped in the electrolyte. The working electrode is made of Au/Nafion solid polymer electrodes. It is fitted on an opening in the wall of the cell, touching the electrolyte on the left-hand side, and interacting with the gas through the inlet/outlet tubes in a covering enclosure. Connecting leads are joined to all the electrodes.

Hydrogen is also detectable via the oxidation reaction

$$2H_2 + O_2 = 2H_2O \tag{11.11}$$

11.6.3 AMMONIA SENSOR

This sensor works in the amperometric mode (Kitzelmann 2000). It is a three-electrode electro-chemical cell. It uses a catalytic carbon working electrode made by mixing gold powder and fine graphite powder with a tetrafluoroethylene dispersion, and applying the mixture on a porous polytet-rafluoroethylene tape, followed by drying, pressing, and sintering. The counter electrode is also made of carbon. The reference electrode uses ruthenium as a catalyst, coated on a perforated tape substrate. The electrolyte is a hydrous solution of manganese (II) salt, and manganese (II) nitrate or manganese (II) sulfate in propylene carbonate and/or butyrolactone solvent.

When ammonia reaches the catalyst layer of the working electrode, the OH^- concentration increases by the reaction

$$NH_3 + H_2O \rightarrow NH_4^+ + OH^- \tag{11.12}$$

Manganese (II) in the electrolyte is electrochemically oxidized to manganese (IV) by the action of the catalyst layer according to the equation

$$Mn^{2+} + 2H_2O \rightarrow MnO_2 + 4H^+ + 2e^- \tag{11.13}$$

The H⁺ ions react with the OH⁻ ions, forming water. At the counter electrode, the reaction taking place is

$$O_2 + 2H_2O + 4e^- \rightarrow 4OH^-$$ (11.14)

The output current varies linearly with the concentration of gaseous ammonia in the range 0–100 ppm. The reaction time is 30–60 s, and the signal magnitude of commercial devices is 80–160 nA-ppm⁻¹.

Many ammonia sensors operate by measuring the pH shift caused by ammonia by the reaction given in eq. (11.12). These sensors consist of a pH glass electrode and a reference electrode. The potential difference between these electrodes changes as a consequence of the pH shift occurring due to ammonia. These sensors suffer from the disadvantage that the output signal varies logarithmically with ammonia concentration, the time to attain equilibrium is long, and the pH variations induced by the presence of other interfering gases such as SO_2 and CO_2 vitiate the readings.

11.6.4 Airborne Lead Particles Sensor

Figure 11.9 shows the functional block diagram of the Pb particles sensor. The air sample is passed through a filter (ESTCP2001, n.d.). From the collected particles, the lead contents are extracted and dispersed in an aqueous solution consisting of dilute HCl and other extractants. Lead concentration is determined by voltammetric analysis. The technique used is anodic stripping voltammetry in which the lead is first plated on a working electrode. Subsequently, a reversal of electrode polarity causes the stripping of lead form the electrode. The current consumption in the stripping step provides an estimate of the lead concentration.

The volume of the air sample is found by an air flowmeter. The voltammetric and flowmeter data are combined together to calculate the lead concentration in micrograms per cubic meter. The instrument then automatically resets itself to get ready for analyzing the next air sample.

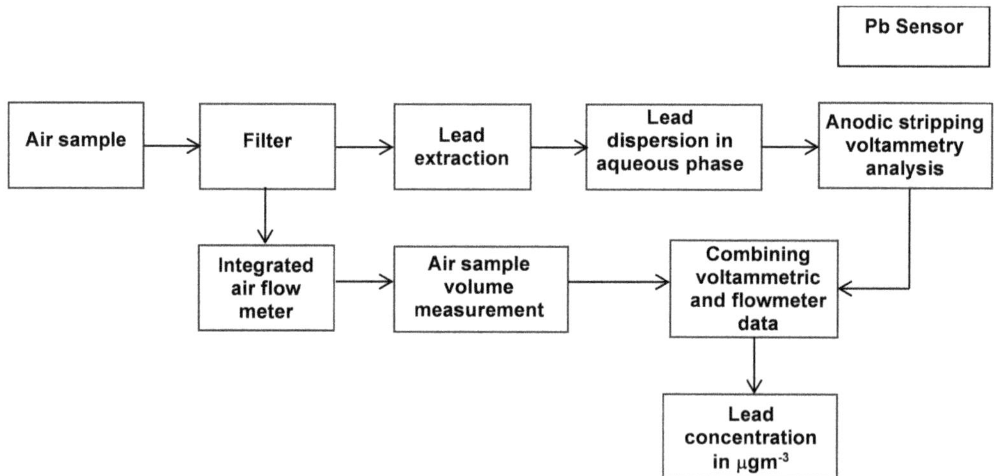

FIGURE 11.9 Block diagram of the instrument for airborne lead measurements consisting of a filter with an integrated air flow meter through which the air sample is passed for lead analysis. The filter extracts the lead particles from the sample. An aqueous phase dispersion of the extracted lead particles is prepared for the anodic stripping voltammetry analysis. At the same time, the flow meter measures the volume of the air sample examined. The outputs from the ASV system and the air volume set up are merged together to provide the final result of lead concentration in micrograms per cubic meter.

11.7   NON-DISPERSIVE (NDIR) INFRARED GAS SENSORS: CO, CO$_2$, NH$_3$

A specific wavelength or wavelength range of infrared light is absorbed by different gases, making it possible to make gas sensors utilizing this phenomenon. The sensor is called a nondispersive IR sensor because no prism or diffraction grating is used to split the light into narrow regions. Instead, a broadband light source is used together with an optical filter to choose the narrow portion of the spectrum overlapping with the particular absorption band of the light for the target gas.

When absorption takes place, the incident IR is attenuated. The concentration of gas is estimated from the decrease in intensity of IR received on the IR detector, as revealed by the transmittance, the ratio of transmitted energy through the gas to incident energy on the gas. There are several absorption wavelengths for various gases. The absorption band for a gas is selected in such a way that it is non-interfering with other gases. For the detection of CO, the absorption band at 4.6 μm is used (Dang et al. 2021), for CO$_2$ strong isolated absorption peak at 4.26 μm is taken (Ottonello-Briano et al. 2020), while ammonia has an absorbance peak at 1.53 μm (Henderson 2021).

IR radiation sources are tungsten halogen lamps and IR light-emitting diodes. Detectors include thermocouples, thermopiles and bolometers. There are two schemes of NDIR sensors: single beam and double beam (Figure 11.10).

11.7.1   SINGLE BEAM IR GAS SENSOR

This sensor uses the following:

(i) A single source of IR radiation,
(ii) A single gas chamber called a cell through which the target gas flows,
(iii) A single thermopile detector, and
(iv) An optical filter placed in front of the detector to eliminate all the light except that at the IR wavelength that is absorbed by the target gas (Figure 11.10(a)).

Working: When the sample gas contains target gas, some wavelengths of the target gas are absorbed and converted into heat so that the gas exiting from the chamber is depleted of those wavelengths. However, the sample gas may be a mixture of gases, comprising some gases apart from the target gas, which may also absorb IR. To prevent misperception about the identity of the gas, an optical filter is placed before the IR detector to cut off all light except that at the wavelength relevant to the target gas detection. Then, the weakening of received light is ascribed only to the target gas. Thus, light is decreased in intensity and we know that the gas responsible for the reduced intensity is the target gas. This light falling on the thermopile is converted into heat producing a voltage. The voltage generated is related to the concentration of the target gas.

Drawbacks of single-beam sensors include the following.

(i) Zero Error: As the IR source weakens in intensity with aging, the intensity of the light source on the day a measurement is performed will differ from that on the day the instrument was calibrated. This will introduce a zero shift in the measured values. The zero shift can be corrected by recalibration of the instrument. Therefore, periodic recalibration of the instrument is essential to avoid errors.
(ii) Effect of Ambient Temperature on the IR Detector: The IR detector being a thermopile is sensitive to variations in temperature of the surroundings besides the IR from the IR lamp. These temperature changes will vitiate the readings.

Drawbacks of single-beam sensors are avoided by adopting a double-beam sensor.

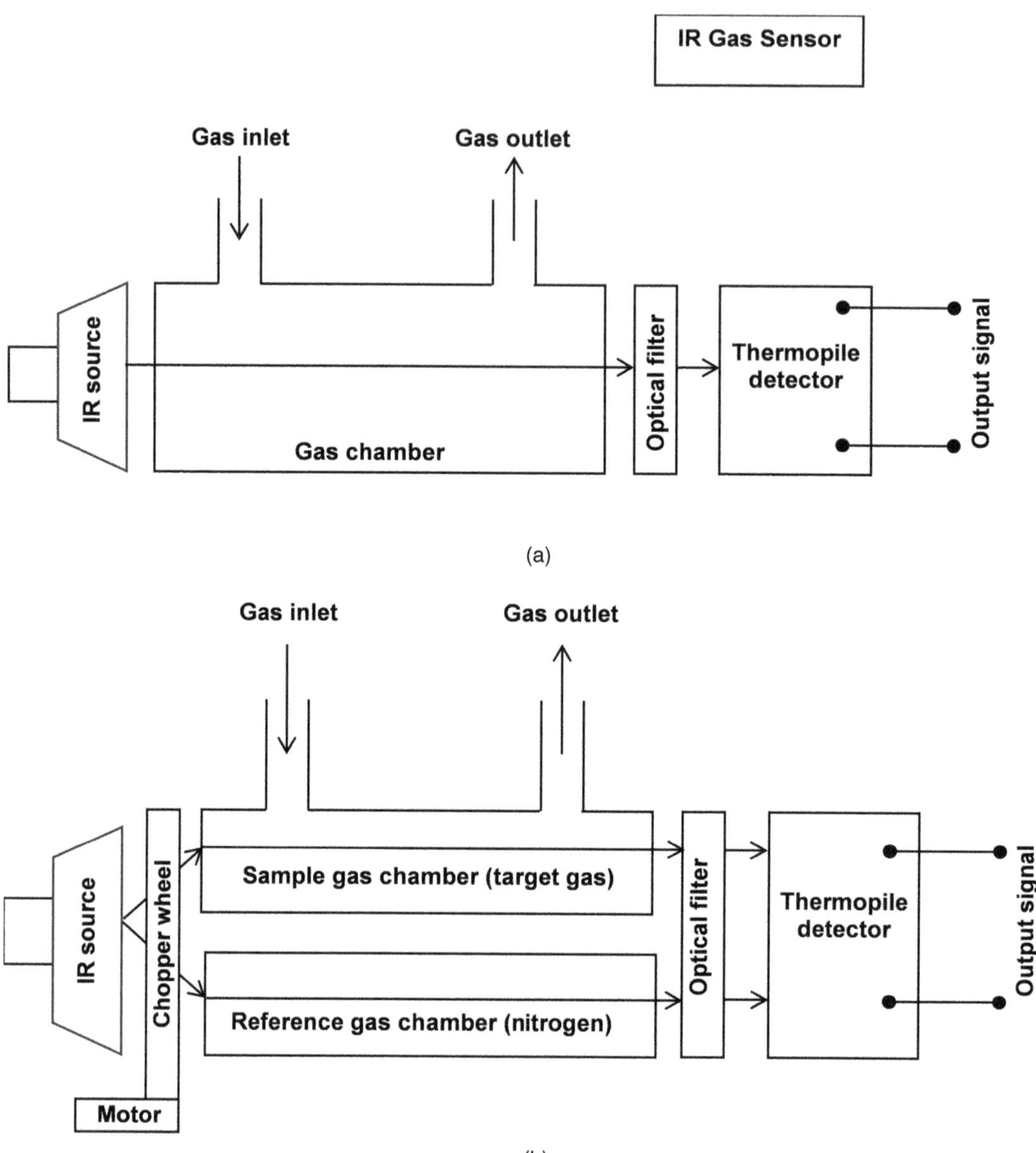

FIGURE 11.10 NDIR infrared gas sensors: (a) single-beam and (b) double beam. Part (a): An infrared beam from an infrared lamp enters the gas chamber with inlet and outlet ports, and the emergent beam falls on a thermopile infrared detector after passage through an optical filter, producing an output signal across the detector leads. Part (b): An infrared beam from an infrared lamp is broken down into two beams of equal intensity, which pass through a rotating chopper wheel energized by a motor. One beam enters the target gas chamber with inlet and outlet ports. The emergent beam falls on the thermopile infrared detector after traversing an optical filter to produce an output signal across the detector leads. The other beam is incident on a gas chamber filled with a reference gas such as nitrogen. Moving further, it travels through a filter to strike the thermopile detector leading to the generation of the reference signal across the terminals of the detector.

11.7.2 DOUBLE BEAM IR GAS SENSOR

This sensor is made of several components (Figure 11.10(b)) such as the following:

(i) A single infrared source but the IR illumination is split up into two beams of equal intensity feeding two parts of the setup.

(ii) Two gas chambers or cells, one containing the target gas to be analyzed, hence known as the sample gas cell; and the other containing some gas like nitrogen which does not absorb IR, hence called the reference gas cell.

(iii) Two IR detectors, which are well matched to each other and placed in series connection with opposite polarities facing each other.

(iv) An optical filter placed before the detectors to pass the relevant IR wavelength absorbed by the target gas and exclude all other wavelengths.

Suppose the sample gas cell is filled with the same gas as in the reference gas cell. Then, the voltages of the two IR detectors are equal and opposite, and cancel each other. The total voltage is zero; hence the detectors constitute a matched pair of IR detectors.

When the target gas flows in the sample gas cell, the intensity of IR received by the detector for this cell is less than the intensity of light falling on the IR detector for the reference gas cell. This difference will be an indicator of the target gas concentration. Electrically, it will be observed by an inequality in the voltages of the two IR detectors, causing a flow of current whose magnitude is proportional to the concentration of the target gas.

Overcoming the Zero-Error Problem: The IR lamp will undoubtedly become dimmer with aging. However, the light from this lamp is decomposed into two halves which are used for a differential measurement. Both halves of the setup, namely, the sample and reference sides, will be equally affected. Resultant changes in both halves will be the same, and therefore the difference signal will be error-free.

Solving the Problem Caused by Ambient Temperature Variations: Both the thermopile detectors are equally influenced by temperature. Therefore, any thermally induced drifts will equally occur in both detectors. As a difference between the outputs of two detectors is being recorded, there will be no error in the output signal. However, if one thermopile is at a different temperature, e.g., from convective heat transferred from the hot target gas, then error is likely to creep in. This kind of error is avoidable by inserting a rotating chopper wheel in the path of the incoming light beam. This chopper switches the light beams at a low frequency between the sample gas cell and reference gas cell, thus removing the error caused by the temperature difference of IR detectors.

11.8 UV ABSORBANCE OZONE SENSOR

The germicidal properties of ozone are utilized for sterilization of water (ozonization). However, ozone has harmful effects too. Long-term exposure to ozone causes respiratory problems and metabolic disorders. Therefore, ground-level ozone is a natural hazard, and its accurate detection is necessary (Aoyagi et al. 2012). Chemical reactions between gaseous effluents from vehicles and factories are the main sources of atmospheric ozone.

An air sample is sucked from the atmosphere into the absorption cell of the ozone sensor through a 3-way solenoid valve by a vacuum pump (Figure 11.11). In one cycle, the ozone is scrubbed from the sucked air, and ozone-free air enters the cell. In the next cycle, the ozone scrubber is by-passed and the ozone-containing air gets entry into the cell. A 254-nm-wavelength UV beam from an Hg LED or low-pressure mercury vapor lamp illuminates the sucked air stream. The brightness of the UV beam exiting the cell is measured by a photodetector. The sensor works on the Beer–Lambert law according to which the absorption of UV radiation is proportional to ozone concentration $[O_3]$

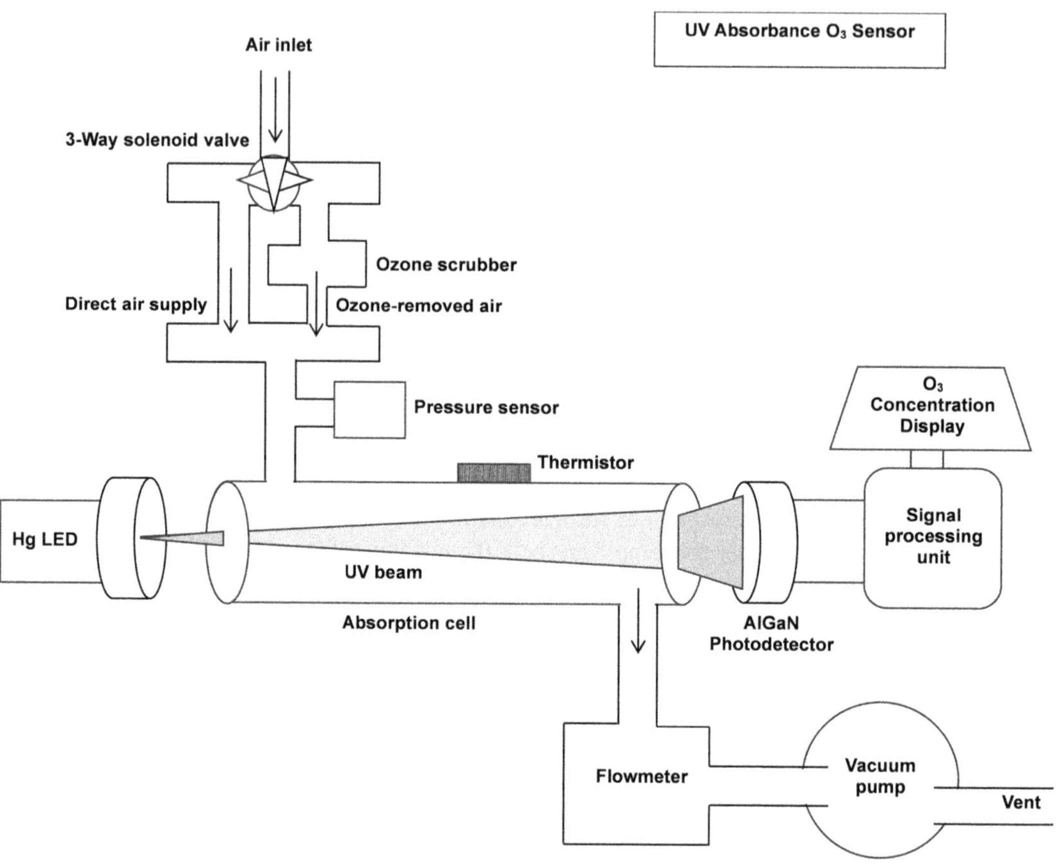

FIGURE 11.11 The UV-absorbance-based ozone monitor consists of an absorption cell on the left side of which an LED lamp is placed and on the right side is mounted an AlGaN photodetector coupled to a signal processing circuit with a display unit for ozone concentration. A thermistor is used to measure the temperature of the cell. Air enters the cell through a 3-way solenoid valve in which one way is used for air inlet, the second way for passing input air directly into the absorption cell, and the third way for passage of air into the cell after ozone removal by a scrubber. A pressure sensor is used to record the pressure of airflow. On the outlet side, the gas passes through a flowmeter giving its flow rate, then through the vacuum pump, and escapes via the vent valve.

in the cell. The ozone concentration is determined by measuring the intensities of UV radiation without ozone and with ozone in the cell with the help of absorption cross section for ozone (1.5×10^{-17} cm² molecule⁻¹), and the path length of the detection cell taken (Turnipseed et al. 2017):

$$\left[O_3\right] = \frac{1}{\text{Absorption cross section for ozone} \times \text{Path length of the cell}} \ln\left(\frac{\text{Intensity}_{\text{Ozone absent}}}{\text{Intensity}_{\text{Ozone present}}}\right) \quad (11.15)$$

The sensor has a pressure sensor and a thermistor to take into consideration the influence of air pressure and temperature on the measurements.

11.9 RADON DETECTORS

Radon is a noble radioactive gas originating from radium (^{226}Ra), a decay product of Uranium (^{238}U). It is a naturally occurring colorless, odorless, tasteless gas, which accumulates from rocks and soils in underground mines and infiltrates into houses through cracks and holes in foundations, basement floors, and walls, acting as an indoor environmental hazard. It infiltrates into workplaces, especially in occupations such as mining as well as in residential buildings (Miles 2004). It is implicated as a causative agent of lung cancer in the general non-smoking public, who breath in a radon-contaminated space. It has the maximum contribution toward the exposure of people to radiation, approximately 40% of the radiation to which the human population is exposed. Radon detectors are of two types (WHO 2009):

(i) Passive detectors, which do not require any power source for functioning, and
(ii) Active detectors, which are connected to a power supply, and often use a pump for sample collection.

The latter also allows us to create charts of radon concentrations, and record any variations in radon concentration over the measurement period. On the other hand, passive detectors work silently in homes collecting radon data.

11.9.1 Passive Radon Detectors

11.9.1.1 Alpha-Track Detector

An alpha-track detector (ATD) consists of a small piece of a polyallyl diglycol carbonate, cellulose nitrate or polycarbonate plastic substrate enclosed within a filter-covered diffusion chamber to allow entry to radon gas but prevent any radon decay products from entering inside. Microscopic regions of damage called latent alpha tracks are generated in the substrate by the impact of alpha particles emitted by radon decay. The alpha tracks are enlarged in size by electrochemically etching the substrate. They are inspected microscopically and counted manually or by an automated counter. The number of tracks per unit surface area of the substrate is calculated, and the background radiation counts are subtracted. A conversion factor determined at a calibration facility is applied to obtain the integrated radon concentration in Becquerel-hours per cubic meter (Bqh/m^3) over a period ranging from 1 month to 1 year. To avoid cross-sensitivity to thoron, a diffusion chamber with a high diffusion resistance to thoron gas is used. Temperature, humidity, and background β and γ-rays do not affect the measurements.

11.9.1.2 Activated Charcoal Detector

Activated charcoal detector (ACD) works by adsorption of radon on the activated sites of carbon. The detector consists of a bed of granulated charcoal. It is clasped by a metal mesh in a metal canister. The canister has a removable lid. Before use, it is heated to remove the adsorbed gas and moisture. After sampling for 2–7 days, the detector is sealed by tightening the lid to allow for equilibration of radon decay products with radon. A waiting time of 3 h is allowed to elapse. Gamma spectrometry of the emissions from Pb-214 and Bi-214 is done to measure the adsorbed radon. Alternatively, preparations are made for liquid scintillation counting. For this purpose, the exposed charcoal is mixed with a liquid scintillation cocktail in which radon is dissolved. Then, a liquid scintillation counter is used for counting. Being sensitive to humidity variations, the ACDs require calibration at different humidity levels.

11.9.1.3 Electret Ion Chamber

Electrets are dielectric materials exhibiting a quasi-permanent electrical polarization which makes them capable of charge storage for relatively long periods of ~100 years. A statically charged Teflon disc is an example of an electret. The electret ion chamber (EIC) allows entry to radon gas, forbidding any radon decay products by passive diffusion of filtered gas through an inlet. The air inside the chamber is ionized by the entering radon gas as well as by any decay products formed within the chamber. The positive electret at the bottom of the chamber collects the resulting negative ions and hence gets discharged. The degree of electret discharge is a measure of the ionization produced during the sampling period, which is typically 5 days to 1 year. The extent of discharging is measured in volts by a non-contact battery-operated electret reader. The radon concentration is determined using a calibration factor and the duration of sampling.

11.9.2 ACTIVE RADON DETECTORS

11.9.2.1 Electronic Integrating Device

An electronic integrating device (EID) is operated with a typical sampling period of ~2 days to a year(s). It comprises a silicon detector enclosed within a diffusion chamber. The EID counts the alpha particles emitted by the radon decay products. Its sensitivity is enhanced by the application of a high voltage during the collection of the charged radon decay products. A drawback of EID is that the measurements are influenced by high humidity levels.

11.9.2.2 Continuous Radon Monitor

The continuous radon monitor (CRM) with a typical sampling period of ~1 h to year(s) has a scintillation cell or ionization chamber. For sample collection, either a small pump is used, or air is allowed to diffuse into the cell or chamber.

In a scintillation cell, a scintillator, e.g., ZnS: Ag covers the internal surfaces of the cell. Three alpha particles are emitted inside the cell during every radon decay event. These alpha particles interact with the scintillator to produce optical pulses, which are recorded by a photomultiplier tube. Integrated radon concentration is calculated by in-built circuitry for the specified period.

In the ionization chamber, an electric field is set up between two electrodes, and the ionization current produced by the radon decay is measured.

11.10 WIRELESS NOISE-MONITORING SENSOR UNIT

Continuous exposure to high noise levels is a primary factor contributing to poor mental health besides blood pressure and heart-related problems. Law enforcement agencies keep a strict vigil on the noise sources and ticket the offenders for curbing urban noise pollution. A noise-monitoring network is made of a collection of noise sensors through which the data of real-time decibel level noise distributions in the 30–120 dB range in a locality are continuously streaming with reference to the noise creator culprits, for rapid tracking and restraining action.

A wireless, self-powered, internet-connected sensor unit for environmental noise level measurement has been developed (Anachkova et al. 2020, 2021). The principle of the sensor unit is displayed in Figure 11.12. The sensor unit is intended to be used to monitor and analyze noise through such units installed at multiple locations in a city. It comprises the following.

(i) The Grove sound sensor, an electret microphone with an amplifier, sensitivity of 52–48 dB at 1 kHz, and a measurement range of 42–90 dB.

(ii) An Arduino Uno microcontroller, the ATmega328P microcontroller (MCU) microchip-based open-source microcontroller board.

(iii) The Lo-Ra click module is a compact add-on board from Microchip Technology containing a low-power, RF technology-based short-range device (SRD) transceiver operating at the frequency of 433/868 MHz. A compliant antenna is used for the connection.

(iv) The cloud storage to the global TTN (The Things Network) platform: It is a global collaborative IoT ecosystem that creates networks using LoRaWAN, operating in sub-GHz industrial, scientific, and medical (ISM) 433, 470–510, 779–787, 863–870, and 902–928 MHz radio bands worldwide, and is used for long range, low power, and low data rate, 250 bps to 22 kbps.

The noise level at a particular location is measured by the microphone. The noise information is processed by the microcontroller, and the processed noise information is transmitted to the TTN cloud server for storage and analysis. The ThingSpeak platform on the server is used for numerical and graphical visualization of the received noise data. This platform is an IoT analytics platform service, allowing aggregation and visualization of live data streams. It can create instant

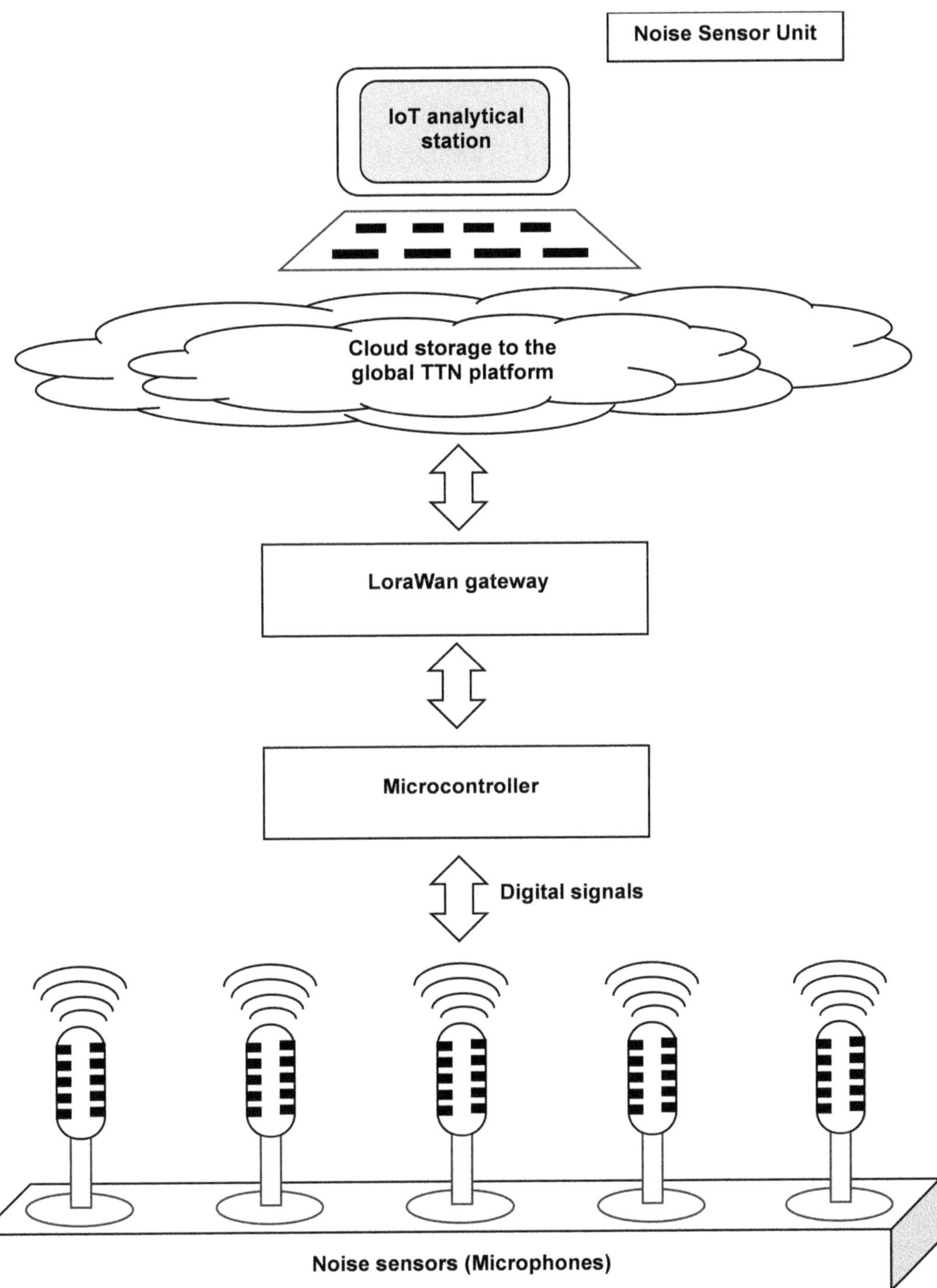

FIGURE 11.12 Sensor unit for noise pollution monitoring. An array of noise-sensing microphones converts the incident acoustic noise into electrical signals, which are digitized and processed by a microcontroller. They are transmitted through the LoraWan gateway to the cloud storage, finally reaching the IoT analytics station. At the IoT analytics station, the noise signals are examined, and necessary correction orders are delivered to parties breaching the prescribed noise levels to maintain silence.

visualization of live data, and send alerts about noise changes. The overall perspective of noise results helps in understanding the noise level at the concerned location for appropriately monitoring and controlling it (Anachkova et al. 2020, 2021).

11.11   SMART WASTE BIN WITH ULTRASONIC WASTE BIN SENSOR

Smart bins (Figure 11.13) are the key components of an intelligent waste management system. The special feature of these bins is the use of ultrasonic fill-level sensors to measure the amount of bin filling. The bin fill data is continuously updated and supplied through an IoT network to a cloud-based platform. The platform analyzes the data and accordingly alerts the garbage collection vehicle about the filling status of the bin. Based on this input information, the garbage collection services can coordinate their vehicles, routes, and frequency of sending vehicles to a given locality for waste pickup.

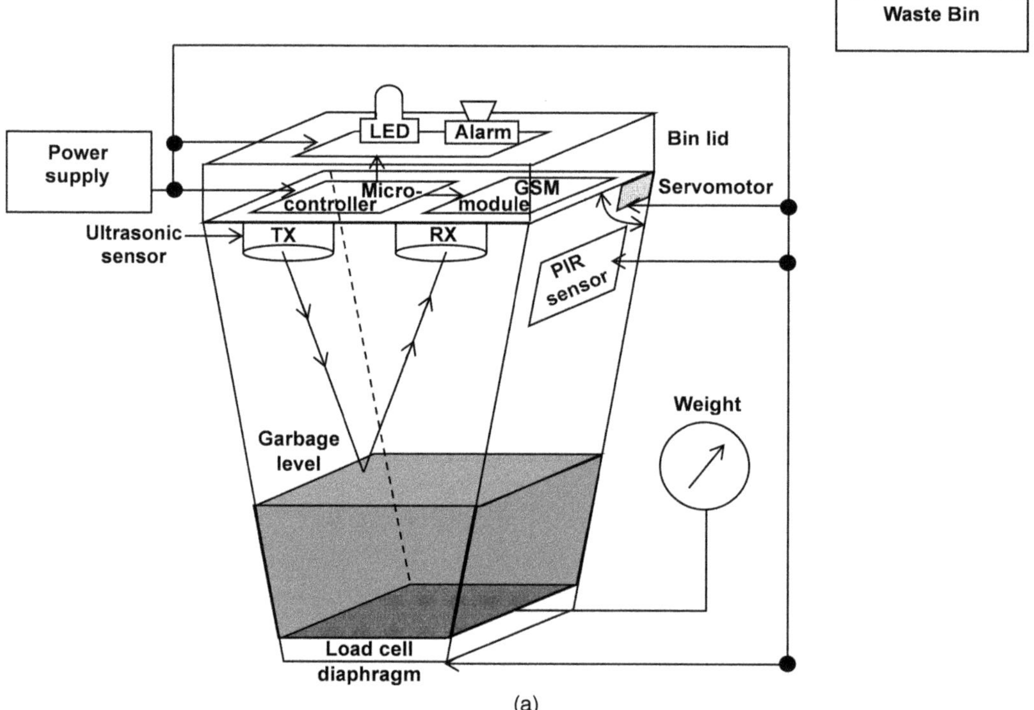

FIGURE 11.13   Smart waste bin: (a) closed lid in the absence of a person and (b) opened lid when a person approaches the bin. Both parts (a) and (b) display a smart waste bin mounted on a load cell platform for continuously monitoring its weight. An ultrasonic transducer is mounted on the under surface of its roof cover. It consists of a transmitter TX and a Receiver RX in the same package. The ultrasound beam emitted by the transmitter rebounds from the waste material in the dust bin, and the time taken for the to-and-fro journey by ultrasound, i.e., to the waste from TX and back from waste to RX, is measured. The measured data is conveyed to a microcontroller which calculates the height up to which the waste bin is filled at a particular instant of time. The data about the waste bin filling status is transmitted over the GSM (Global system for mobile communication), a digital cellular network, to the central garbage management agency, which sends garbage collection trucks for waste collection from bins. The agency collects data from various waste bins in a locality to decide about sending waste collection trucks to a specific locality depending on the updated needs. On the top cover of the waste bin, there is an LED indicator, which becomes red when the bin is full. Also, there is an alarm that alerts users about the fullness/emptiness of the bin. The power supply and wiring connections are shown. Part (a): There is no human visitor near the bin. A passive infrared (PIR) sensor detects human absence near the bin. The lid of the bin is kept closed.

(Continued)

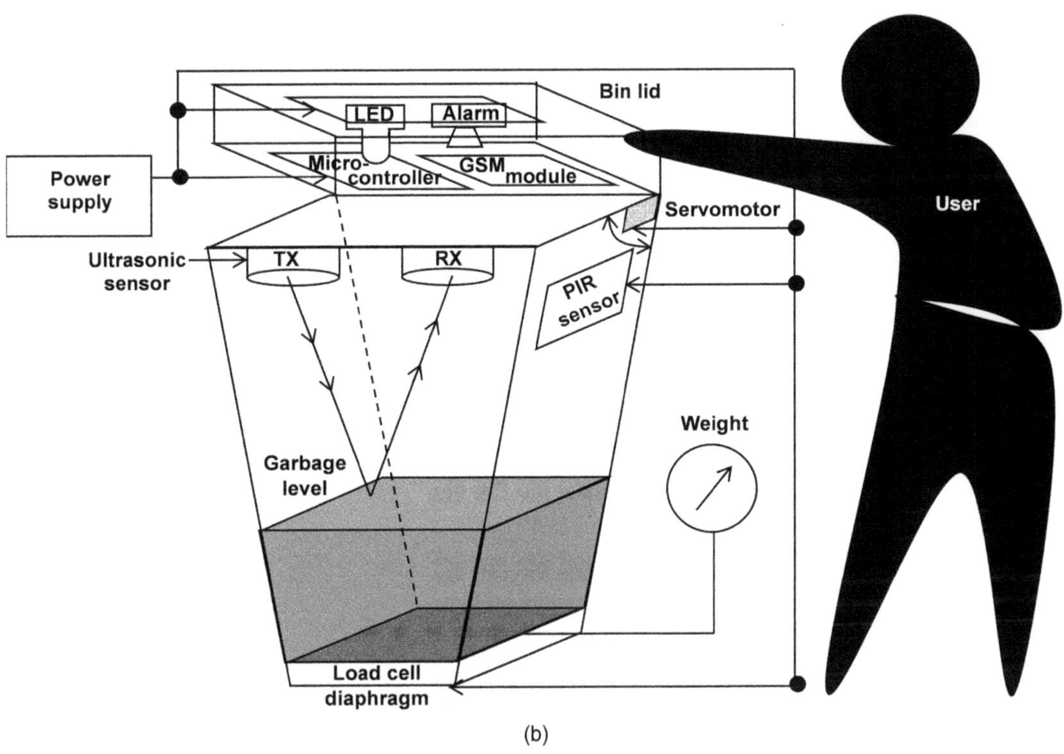

FIGURE 11.13 (CONTINUED)   Part (b): When any person approaches the bin, the PIR sensor of the bin immediately detects human presence, and knows that a visitor has come to dump garbage. A servomotor comes into action, opening the lid of the bin. The person need not pull the lid of the bin to open. Avoidance of human contact prevents the risk of any dirty material from contaminating the hands of the person, and so averts the risk of any infection or harm to the person. The person drops the waste in the bin and goes away. As soon as the person moves away from the bin, the PIR sensor detects that the person is absent, and the lid of the waste bin falls down, closing the waste bin, as in part (a) to stop any foul smell emitted by the garbage from spreading to the neighborhood.

11.12   ELECTRONIC WATER METERING

This topic was briefly discussed in Section 3.11.1. Here, we give more details about it.

11.12.1   Ultrasonic Flow Sensors for Water Metering

Conventional water meters are mechanical in nature having moving parts. Ultrasonic flow sensors containing no moving parts offer long-life, maintenance-free, reliable operation with low power consumption (Shepard 2022). They consist of two piezoelectric transducers mounted in holes in the pipe, protected from water with only the transmitting/receiving faces of the transducers exposed to water for launching or receiving ultrasonic waves (Figure 11.14). Besides the transducers, there are two acoustic reflectors. A reflector is placed below each transducer at 45° angle to reflect the ultrasonic pulses from one transducer to the other reflector and then to the second transducer. The ultrasonic pulses emitted by transducer X undergo reflections at reflectors A and B and reach transducer Y traveling along the path XABY. Similarly, the ultrasonic pulses emitted by transducer Y are directed by reflectors B and A to transducer X along the path YBAX. Thus, in one case, the ultrasonic pulses propagate in the direction of the flow of water, and their velocity is enhanced, and, in the other case, they move against the direction of the flow of water with a consequent reduction in their velocity. The difference between the effective velocities of ultrasonic pulses in the two cases is applied to calculate the flow rate of water.

To begin the measurement, an excitation signal is applied to transducer X at its resonance frequency so that it is activated and sends a burst of pulses. At this moment, a START signal is produced. As soon as the signal is received by the transducer Y, a STOP signal is generated to mark the reception of pulses from transducer X. The difference between START and STOP signals is the time of flight from X to Y ($TOF_{X \to Y}$). Now, the transducer Y switches over to the transmission mode and sends a burst of pulses to transducer X. By the same procedure as for $TOF_{X \to Y}$, the time of flight from Y to X ($TOF_{Y \to X}$) is measured. When there is no water flowing through the pipe,

$$TOF_{Y \to X} = TOF_{X \to Y} \tag{11.16}$$

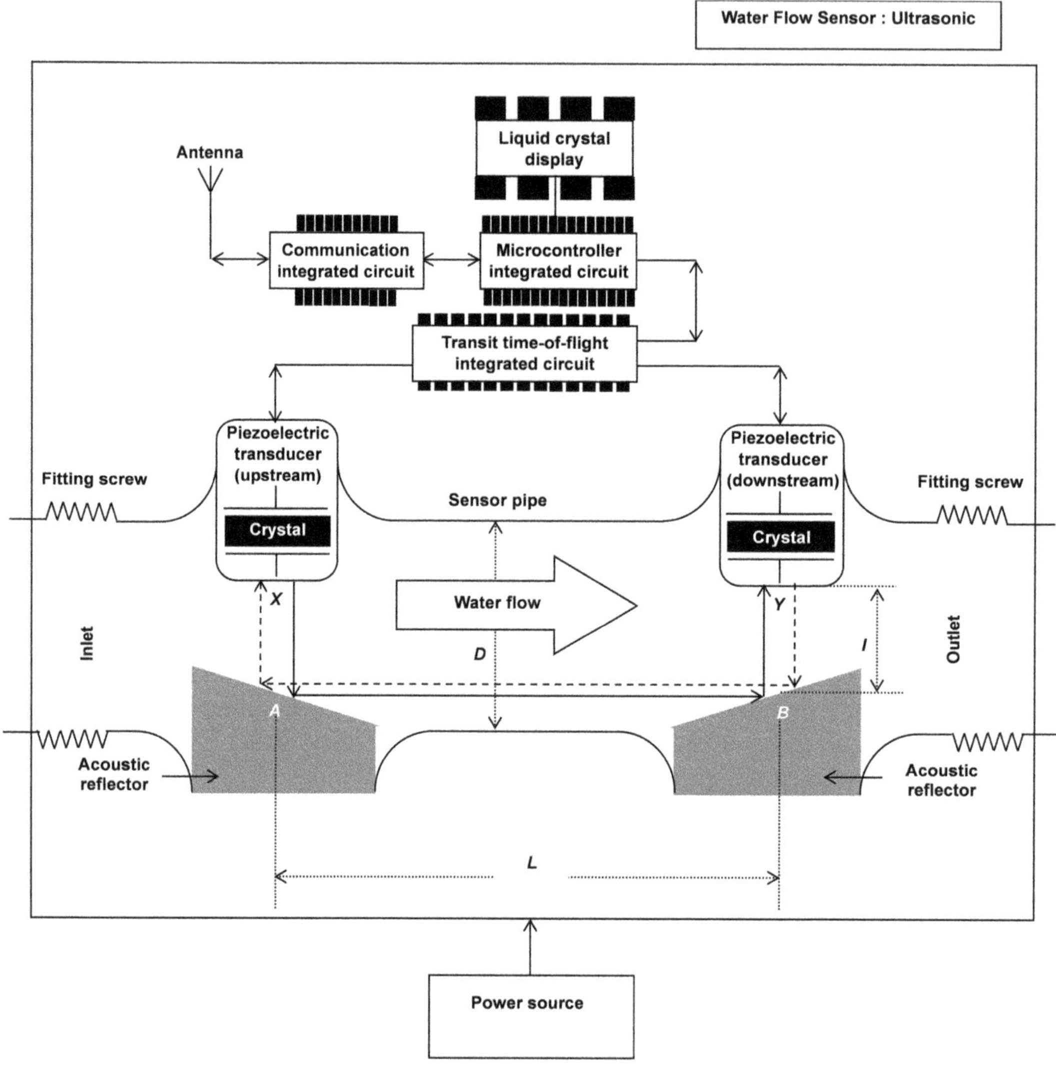

FIGURE 11.14 Ultrasonic water flow sensor. The diagram shows the upstream piezoelectric transducer X, the downstream piezoelectric transducer Y; two acoustic reflectors A and B, one below each transducer; the inlet and outlet ports, the water flowing through the sensor pipe of diameter D, the transit time-of-flight IC, the microcontroller, the communication IC, the antenna liquid crystal display, and the power source. The length of the path traveled by ultrasound between the two transducers is L while the path length for ultrasound travel from a transducer to the acoustic reflector is $l = D/2$.

However, during water flow,

$$TOF_{Y \to X} - TOF_{X \to Y} = \Delta TOF \text{ is proportional to the velocity of water flow} (v) \tag{11.17}$$

We can write the time of flight from the left transducer to the right transducer in the direction of water flow as (Mirshab 2015)

$$\begin{aligned}
TOF_{X \to Y} &= \text{Time taken by ultrasound to travel from } X \text{ to } A \text{ (left transducer to left reflector)} \\
&\quad + \text{Time taken by ultrasound to travel from } A \text{ to } B \text{ (left reflector to right reflector)} \\
&\quad + \text{Time taken by ultrasound to travel from } B \text{ to } Y \text{ (right reflector to right transducer)} \\
&= \frac{\text{Distance from } X \text{ to } A}{\text{Velocity of ultrasound in water}} \\
&\quad + \frac{\text{Distance from } A \text{ to } B}{\text{Effective velocity of ultrasound in the direction of water flow}} \\
&\quad + \frac{\text{Distance from } B \text{ to } Y}{\text{Velocity of ultrasound in water}} \\
&= \frac{l}{c} + \frac{L}{c+v} + \frac{l}{c} = \frac{2l}{c} + \frac{L}{c+v}
\end{aligned} \tag{11.18}$$

Similarly, the time of flight from the right transducer to the left transducer against the direction of water flow is

$$\begin{aligned}
TOF_{y \to x} &= \text{Time taken by ultrasound to travel from } Y \text{ to } B \text{ (right transducer to right reflector)} \\
&\quad + \text{Time taken by ultrasound to travel from } B \text{ to } A \text{ (right reflector to left reflector)} \\
&\quad + \text{Time taken by ultrasound to travel from } A \text{ to } X \text{ (left reflector to left transducer)} \\
&= \frac{\text{Distance from } Y \text{ to } B}{\text{Velocity of ultrasound in water}} \\
&\quad + \frac{\text{Distance from } B \text{ to } A}{\text{Effective velocity of ultrasound aginst the direction of water flow}} \\
&\quad + \frac{\text{Distance from } A \text{ to } X}{\text{Velocity of ultrasound in water}} \\
&= \frac{l}{c} + \frac{L}{c-v} + \frac{l}{c} = \frac{2l}{c} + \frac{L}{c-v}
\end{aligned} \tag{11.19}$$

In these equations, c is the velocity of ultrasound in water and v is the velocity of water flow. Subtracting eq. (11.18) from eq. (11.19), we get

$$\Delta TOF = \frac{2l}{c} + \frac{L}{c-v} - \left(\frac{2l}{c} + \frac{L}{c+v} \right) = \frac{L}{c-v} - \frac{L}{c+v} = \frac{L(c+v) - L(c-v)}{(c-v)(c+v)}$$

$$= \frac{Lc + Lv - Lc + Lv}{c^2 - v^2} = \frac{2Lv}{c^2 - v^2} \simeq \frac{2Lv}{c^2} \tag{11.20}$$

since

$$c \gg v \tag{11.21}$$

and

$$\therefore c^2 - v^2 \simeq c^2 \tag{11.22}$$

Rewriting eq. (11.20),

$$\Delta TOF = \frac{2Lv}{c^2} \tag{11.23}$$

from which

$$v = \frac{c^2}{2L} \Delta TOF \tag{11.24}$$

The volumetric flow rate of water is expressed as

$$V = KvA = KA \frac{c^2}{2L} \Delta TOF \tag{11.25}$$

where K is the calibration factor of the pipe for the sensor used, and A is the cross-sectional area of the sensor pipe.

11.12.2 Electromagnetic Water Flow Meter

The electromagnetic flow meter, also called a magmeter, works on Faradays' law of electromagnetic induction. According to Faraday's law, the movement of a conducting fluid in a magnetic field causes a change in magnetic flux linked with the conducting fluid. The change in magnetic flux results in the induction of an electromotive force across the fluid (Hofmann 2003). Further, the strength of induced EMF is proportional to the rate of change of flux. Hence, the flow of water through a pipe placed in a magnetic field is accompanied by the generation of a voltage between two electrodes mounted on the opposite sides of the pipe. The magnitude of the voltage produced by flowing water is proportional to the velocity of the flow of water. Therefore, the measurement of this voltage provides an estimate of the velocity of water flow. The voltage produced due to the flowing water is in the millivolts range, which is converted into 5–20 mA current.

The flowmeter consists of a non-magnetic pipe with an insulating lining, a pair of magnetic coils with an excitation source, and a pair of electrodes penetrating the pipe and its insulating lining to contact the water (Figure 11.15). When the magnetic coils are energized with an AC source, the flow signal obtained is a sinusoidal waveform together with the noise voltages in the electrode loop. Although the out-of-phase noise is easily removed by filtration, the in-phase noise removal requires a stoppage of water flow. Therefore, this noise is difficult to remove making frequent re-zeroing necessary for accurate measurements. The DC excitation is done with an ~20 Hz pulsed signal. During the pulse period, the output signal is a combination of the actual signal plus the noise signal. Between any two pulses, the output contains only the noise signal. Therefore, the noise signal can be removed in each cycle. Zero drift and zero stability problems are thereby avoided. Additionally, DC-excited sensors are less bulky, can be easily installed, and consume less power than the AC excited devices.

11.13 CHLORINE-IN-WATER SENSORS

11.13.1 Three-Electrode Integrated Electrochemical Sensor using BDD Film for Online Chlorine Monitoring

A silicon-based sensor for free chlorine is fabricated using a liquid-conjugated Ag/AgCl reference electrode, a Pt counter electrode, and a boron-doped diamond working electrode (Yin et al. 2022). The sensor (Figure 11.16) is produced by MEMS technology, which is suitable for bulk manufacturing.

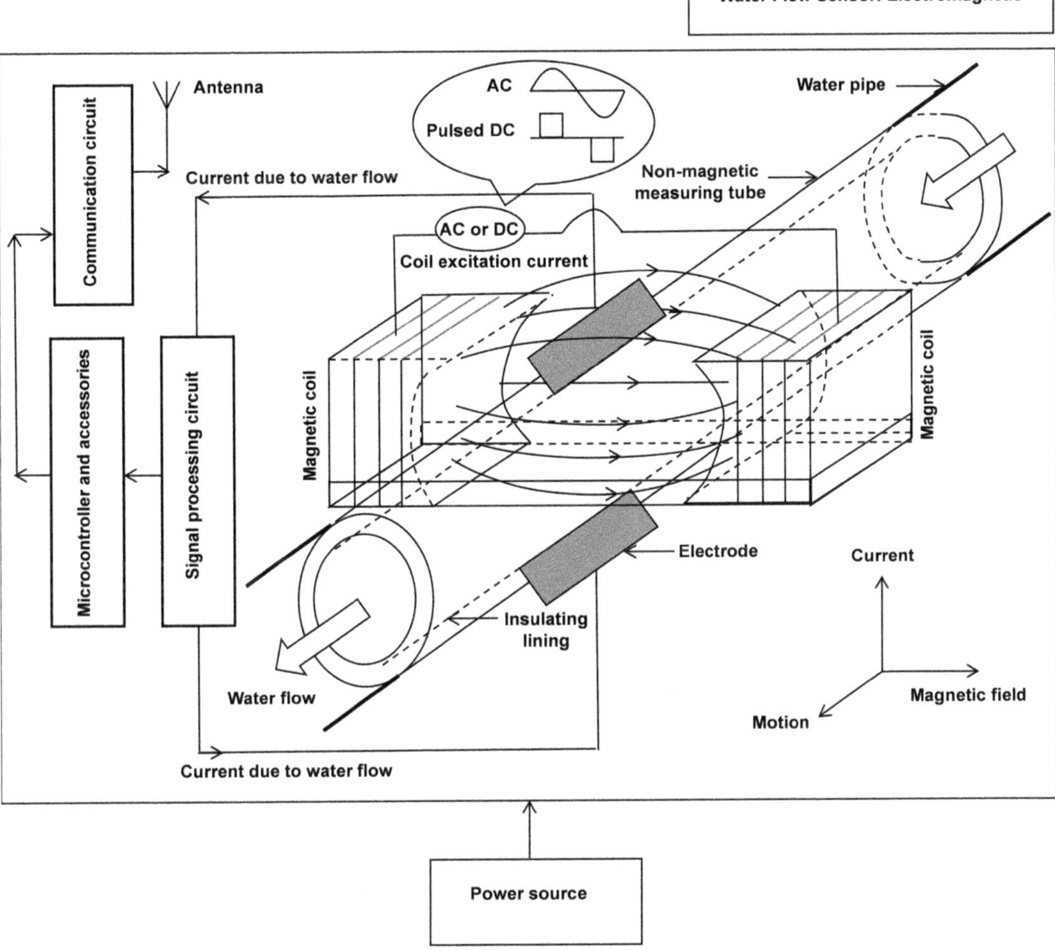

FIGURE 11.15 Electromagnetic water flow sensor. The diagram shows the water pipe in which the non-magnetic tube portion, housing the sensor, is inserted. The sensor consists of a pair of magnetic coils with AC or DC excitation; the AC and pulsed DC waveforms are shown. Also shown are the electrodes (gray-shaded) for carrying current to the main circuit consisting of a signal processing circuit, a microcontroller, a communication circuit, an antenna, and a power source.

11.13.1.1 Making the Boron-Doped Diamond Film Electrode

The process consists of the following two steps.

(i) Formation of Seed Crystals on an Oxidized Silicon Substrate: This step is essential for improving the nucleation density of the diamond in the next step. The oxidized silicon wafer is ultrasonically cleaned with alcohol, ground with diamond grinding paste, cleaned with DI water, and ultrasonicated for ½ h after placing in 60–100 nm diamond nanopowder suspended in acetone.

(ii) Hot-Filament Chemical Vapor Deposition (HFCVD) of the BDD Film: The boron-doped diamond film is formed at 2,200–2,400°C temperature and 1–3.5 kPa pressure in a 2–4% carbon/hydrogen mixture for 4.5 h using (B_2O_3 + C_2H_5OH + CH_4) mixture as a source of boron/carbon. The BDD film thickness is 3–4 µm.

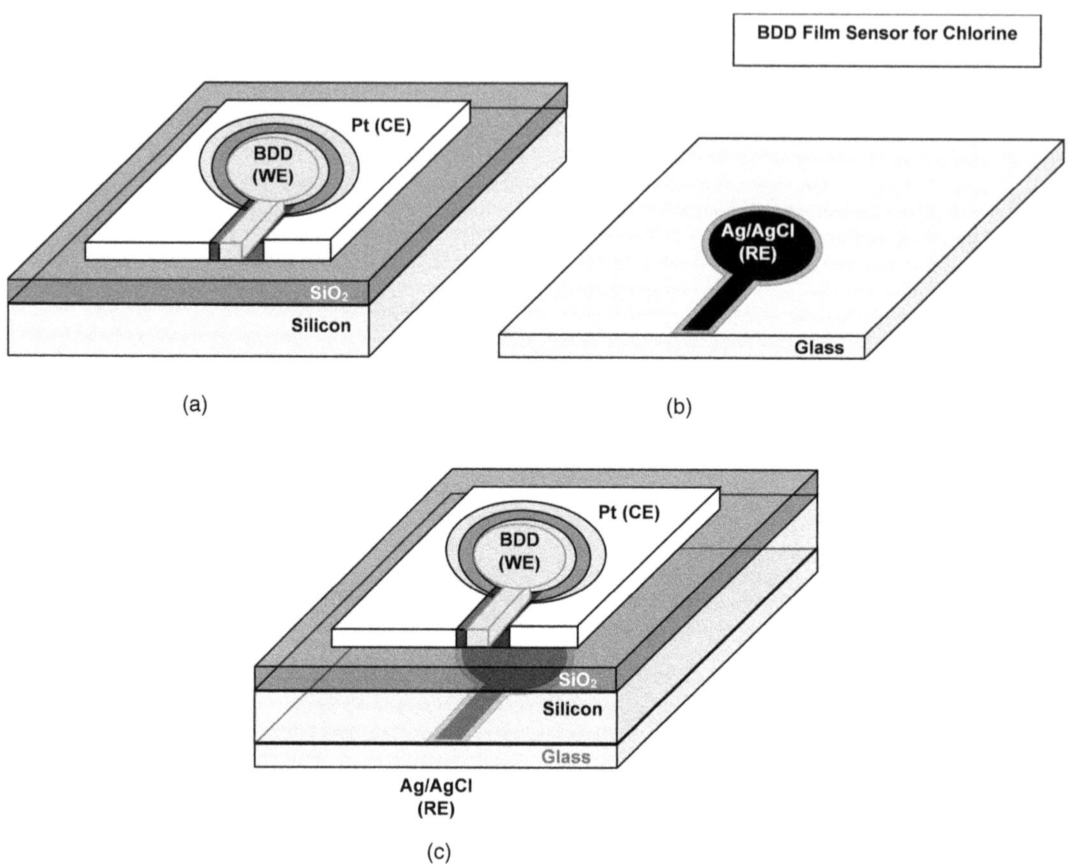

FIGURE 11.16   Boron-doped diamond film sensor for online measurement of free chlorine in tap water: (a) the microfabricated sensor chip, (b) the Ag/AgCl reference electrode, and (c) the complete device assembly with sensor chip of part (a) bonded with the Ag/AgCl reference electrode of part (b). Part (a) shows an SiO_2/Si substrate. At the center of the substrate is the circular-shaped boron-doped diamond (BDD) film with a rectangular projection. This is the working electrode (WE). The boron-doped diamond film is sensitive to chlorine. Surrounding this central BDD film, there is a platinum electrode acting as the counter electrode (CE). The platinum electrode is concentric with the boron-doped diamond working electrode. Part (b) shows a glass substrate on which Ag/AgCl is deposited to form the reference electrode (RE). Part (c) shows the two parts (a) and (b) attached to each other. Some portion of the Ag/AgCl electrode is visible below the silicon.

11.13.1.2   Sensor Chip Fabrication

A photoresist is spin-coated on the BDD film. Lithography is performed, and a pattern is formed over which aluminum is sputtered. The photoresist is lifted off to obtain a patterned aluminum film to be used as a mask for oxygen reactive ion etching (O-RIE) of BDD. After O-RIE, a circular-shaped BDD film is defined. The aluminum mask is removed. A sputtered Pt electrode is formed by a lift-off process. The backside SiO_2 is removed in a buffered oxide etch. A cavity and a groove are formed on the backside for 3M KCl solution, and the lead wire by etching silicon in KOH (Figure 11.16(a)).

11.13.1.3   Making the Ag/AgCl Reference Electrode

On a pyrex glass substrate, a 0.4-μm-thick silver electrode is formed by sputtering, and defined by lift-off lithography. Then, AgCl is electrochemically deposited in 0.25M HCl solution at 4 V (Figure 11.16(b)).

11.13.1.4 Assembly of Complete Electrochemical Sensor

The silicon wafer over which the BDD and Pt electrodes are made, and the glass substrate over which the Ag/AgCl reference electrode is made, are brought together, aligned with each other, and bonded (Figure 11.16(c)).

11.13.1.5 Chlorine Sensing Behavior

The sensor is operated in the cyclic voltammetry (CV) mode to optimize the applied bias potential. Maximum response is obtained by applying a bias voltage of −0.35 V on the BDD working electrode by reduction of ClO⁻. For chlorine measurements, the sensor is operated in the amperometric mode by applying a potential of −0.35 V vs. the Ag/AgCl reference electrode, and the current is measured at different chlorine concentrations using sodium hypochlorite (NaClO) solutions after the current value stabilizes at each concentration. In the ClO⁻ concentration range 5–200 mg-L⁻¹, the peak value of electric current I is linearly related to the ClO⁻ concentration $c(CLO^-)$, representing the concentration of free chlorine, by the equation

$$I(\mu A) = -\left\{1.144c\left(ClO^-\right) + 71.193\right\}, \text{ correlation coefficient } R^2 = 0.995 \qquad (11.26)$$

The chlorine sensitivity of the device is 9.108 µA/mg-L⁻¹/cm². The limit of chlorine detection is 0.056 mg-L⁻¹ (Yin et al. 2022).

11.13.2 LOW-COST GRAPHITE-BASED AMPEROMETRIC SENSOR FOR CHLORINE IN TAP WATER

An inexpensive pencil lead-based graphite sensor is developed for testing free chlorine in tap water (Pan et al. 2015).

11.13.2.1 Graphite Modification for Making the Working Electrode

The electrolyte solution is prepared by adding 0.1M ammonium carbamate solution to 0.1M sodium phosphate buffer of pH = 7.0. The addition of ammonium carbamate solution is continued until the pH of the buffer becomes 8.9. This electrolyte solution is filled in the electrochemical cell (Figure 11.17). Pencil lead is cleaned with lab tissue, rinsed with D.I. water, and loaded in the electrochemical cell as the working electrode. The potential for modifying the graphite surface is adjusted at 1.0 V vs. the Ag/AgCl electrode, and graphite is modified at this potential to make it suitable for testing chlorine.

11.13.2.2 Free Chlorine Measurements

These are performed by chronoamperometry, done in the same electrochemical cell shown in Figure 11.17. For chronoamperometry, the potential is set at 0.1 V vs. the Ag/AgCl reference electrode. Chlorine concentration is increased by adding NaClO to sodium phosphate buffer in the beaker. The current–time profile is recorded after successively adding 1.076 ppm of free chlorine in steps. The cathodic current increases by equal amounts after each chlorine addition step. The change in current varies linearly with the quantity of chlorine added. The sensitivity is 0.303 µA ppm⁻¹ cm⁻². Response time is <3 s. The maximum hysteresis in the 0–6 ppm range is 0.04 ppm (Pan et al. 2015).

11.14 WATER TURBIDITY SENSORS

Turbidity is the degree of cloudiness or haziness of water, i.e., the extent of loss of transparency of water caused by the solid particles suspended in it. It indicates the presence of particulate matter in water, e.g., clay, silt, algae, plankton, pathogens, bacteria, and contaminants such as mercury and lead in water, which are harmful to aquatic life and human health (Wang et al. 2018). While heavy mud particles settle down in calm water, smaller colloidal particles keep hanging and produce

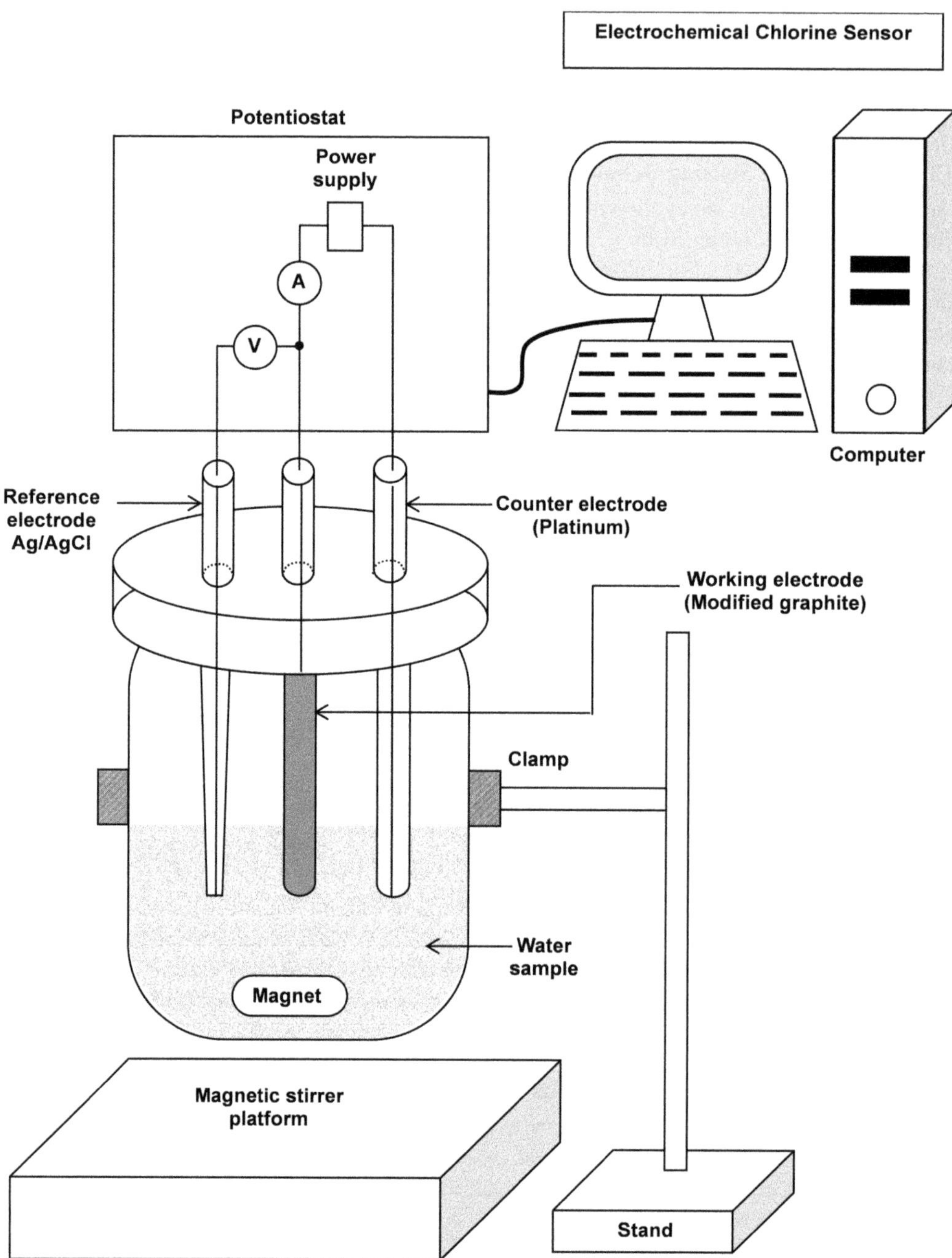

FIGURE 11.17 Three-electrode electrochemical cell for chlorine measurement. The same cell is used for graphite modification to make chlorine-sensing electrodes. A beaker is suspended from a clamp fixed to a stand at a small height over a magnetic stirrer. The beaker contains the test sample in which chlorine concentration is to be determined. At the bottom of the beaker is a magnet which continuously rotates when the magnetic stirrer is switched on. The top surface of the beaker is covered with a plate through which three electrodes are passed into the beaker. These electrodes are immersed in the solution contained in the beaker. The electrodes are the Ag/AgCl reference electrode, the modified graphite working electrode, and the platinum counter electrode. The measurements are done using a potentiostat in which a voltmeter, an ammeter, and a power supply are shown. The voltmeter is connected between the reference and working electrodes. The power supply and the ammeter are connected between the working and counter electrodes. The potentiostat is connected to a computer.

opacity and murkiness. Therefore, turbidity is an important parameter showing water clarity and quality calling for carrying out proper settling and decantation processes (Hussain et al. 2016, Fay and Nattestad 2021). Turbidity is measured using optical sensors, which are classified into three major groups, named as nephelometric, absorption, and total suspended solid sensors (Figure 11.18).

11.14.1 NEPHELOMETRIC TURBIDITY SENSOR

In a nephelometric turbidity sensor (Figure 11.18(a)), the light diverging from the LED is collimated by a lens. The parallel beam of light strikes an obstacle with an aperture. A narrow beam passes through the aperture and falls on the water sample taken in a glass cell. The light incident on water is scattered and transmitted. The scattered light falls on a photodetector placed at 90° angle with respect to the incident beam. The intensity of light measured by the photodetector indicates the turbidity in nephelometric units.

A low-cost nephelometric turbidity sensor is made using an LED as the light source, and a light-dependent resistor as the receiver (Azman et al. 2016). The PIC 16F777 microcontroller is used. RS232 (RS = recommended standard) to TTL (transistor-transistor-logic) serial interface module provides signal conversion for serial communication between the RS 232 and TTL ports.

A nephelometric turbidity sensor is used for measuring water turbidity in low-turbidity specimens, such as in drinking water.

11.14.2 ABSORPTION TURBIDITY SENSOR

The absorption turbidity sensor (Figure 11.18(b)), is similar in construction to that of Figure 11.18(a), except that the photodetector is removed from its position at 90° angle to incident light, and placed behind the water sample opposite to the LED lamp. The light beam passing through the water sample undergoes attenuation in accordance with the number and size of particles in water. Turbidity is measured from the weakening of light intensity on passage through water.

An absorption turbidity sensor is suitable for turbidity measurements in water samples experiencing regular and frequent fluctuations in turbidity.

11.14.3 TOTAL SUSPENDED SOLIDS TURBIDITY SENSOR

The total suspended solids turbidity sensor (Figure 11.18(c)) contains four photodetectors: a 90° light scattering detector, a transmitted light detector, a forward light scattering detector, and a backward light scattering detector. These photodetectors give the intensities of light scattered in forward, backward, and 90° directions together with that of light passing through the water sample. The overall turbidity is calculated from the light intensity signals received from all these detectors by using an algorithm.

This kind of turbidity sensor is employed for turbidity measurements of high turbidity samples.

11.14.4 UNITS OF TURBIDITY

Turbidity data recorded by various probes are available in several different units, which are not comparable to each other. These differences arise principally from the wavelength of light used for measurement, the angle of detection of scattered light, the number of detectors used, and often the specific signal processing applied for the purpose.

(i) Consider the case in which a single broadband or white light source with peak spectral output in the wavelength range of 400–680 nm is used for measuring turbidity. Further, suppose the measurement is done by placing the detector at 90° angle to the incident beam. In this situation, the turbidity is expressed in nephelometric turbidity units (NTU).

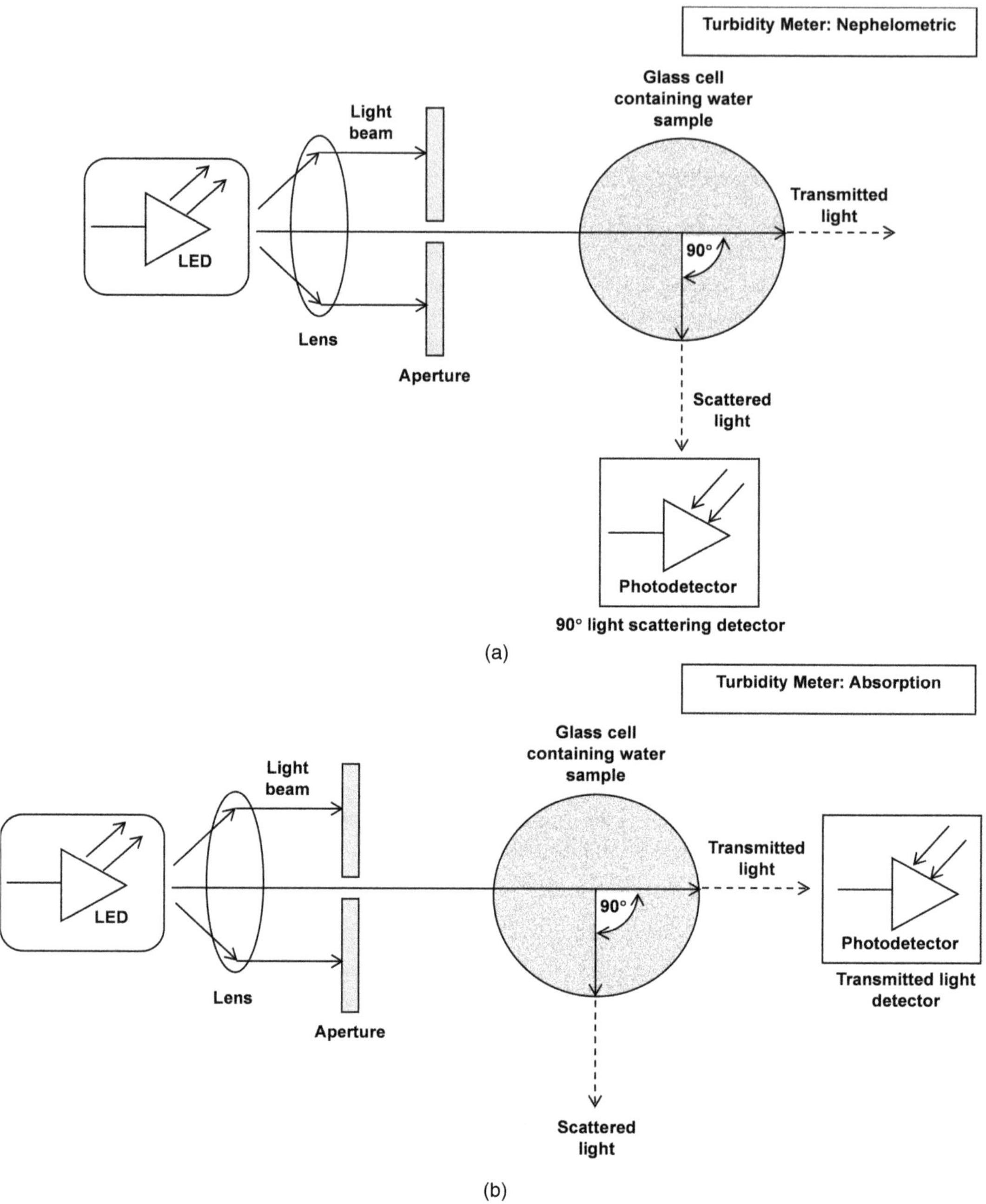

FIGURE 11.18 Three types of turbidity meters: (a) nephelometric, (b) absorption, and (c) total suspended solid type. Part (a) shows an LED, a lens, an aperture, the glass cell containing the sample of water whose turbidity is to be assessed, and a photodetector placed at 90° angle relative to incident light direction. The directions of incident, scattered, and transmitted light beams are shown. Part (b) shows an LED, a lens, an aperture, the glass cell containing the water sample of unknown turbidity, and a photodetector placed to receive light coming out from the water sample. The directions of incident, scattered, and transmitted light beams are indicated.

(Continued)

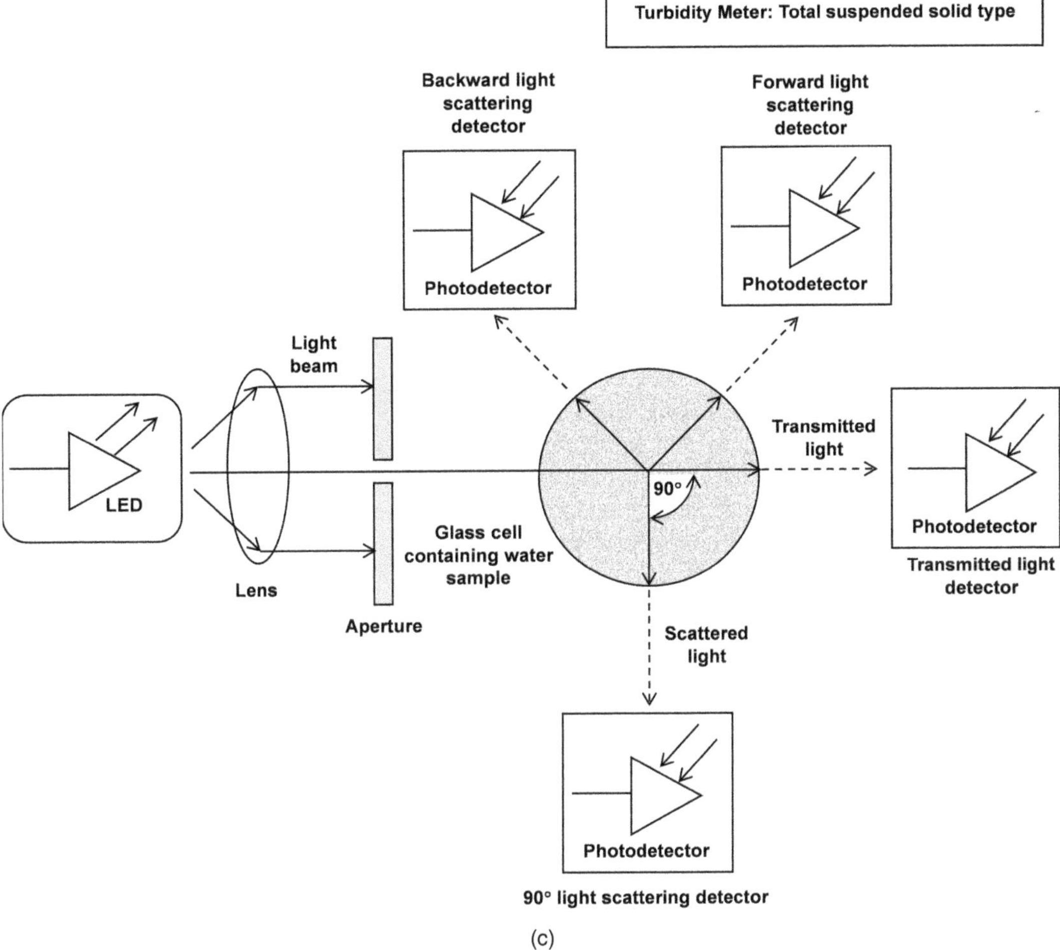

FIGURE 11.18 (CONTINUED)   Part (c) shows an LED, a lens, an aperture, the glass cell filled with water, a 90°light scattering detector, a transmitted light detector, and the forward and backward light scattering detectors. The incident, 90°-scattered, forward-scattered, backward-scattered, and transmitted light directions are shown.

(ii) When the same white light source is used by placing the detector at 180° angle to the incident beam, the turbidity is given in attenuation units (AU).

(iii) When the same white light source is used with multiple detectors by placing the detectors at 90° and other angles to the incident beam, the turbidity is given in the nephelometric turbidity ratio units (NTRU).

(iv) When a single infrared monochromatic light source with output in the wavelength range 780–900 nm is used for measuring turbidity by placing the detector at 90° angle to the incident beam, the turbidity is expressed in formazin turbidity units (FTU).

(v) When the same infrared source is used by placing the detector at 180° angle to the incident beam, the turbidity is given in formazin attenuation units (FAU).

(vi) When the same infrared source is used with multiple detectors by placing the detectors at 90° and other angles to the incident beam, the turbidity is given in formazin turbidity ratio units (FTRU).

The WHO prescribes the upper limit of turbidity of drinking water to not exceed 5 NTU. It is preferred that the turbidity value should be less than 1 NTU.

11.15 MEASURING TOTAL DISSOLVED SOLIDS IN WATER WITH CONDUCTIVITY CELLS

Total dissolved solids (TDS) in water, expressed in mg-L^{-1}, are estimated by measuring its electrical conductivity σ. The dissolved solids are mainly ionic species. At low concentrations, a linear relationship between the TDS concentration and electrical conductivity can be assumed as (Rusydi 2018)

$$\text{TDS}\left(\mu\text{gL}^{-1}\right) = \text{Constant} \times \sigma\left(\mu\text{Scm}^{-1}\right) \tag{11.27}$$

For measuring the conductivity σ of water, a conductivity cell or meter is used (Figure 11.19). The conductivity cell consists of two electrodes insulated from each other. Each electrode has a conducting wire covered with an insulating sheath. A flat rectangular strip made of platinum is fixed at the end of the wire. The open end of the conducting wire coming out from the sheath is used for electrical connection. The water sample whose conductivity is to be measured is taken in a glass cell, either in the shape of a glass bulb Figure 11.19(a) or a cylindrical tube Figure 11.19(b). The cell has an alternating current source, an ammeter connected in series with this source for current measurement, and a voltmeter connected between the electrodes for reading the potential difference between them. The use of alternating current prevents electrolysis.

An alternating voltage is applied between two flat platinum electrodes of cross-sectional area A immersed in water at a distance l apart. The resistance R of the water column between the two electrodes is obtained by measuring the current I for a given voltage V applied across the electrodes as

$$R = \frac{V}{I} \tag{11.28}$$

The resistivity of water is given by

$$\rho = \frac{RA}{L} = \frac{V}{I}\frac{A}{L} \tag{11.29}$$

and its conductivity σ is expressed as the reciprocal of resistivity

$$\sigma = \frac{1}{\rho} = \frac{1}{\dfrac{VA}{IL}} = \frac{L}{A}\frac{I}{V} = K\frac{I}{V} \tag{11.30}$$

where

$$K = \frac{L}{A} = \text{Cell constant} \tag{11.31}$$

11.16 SMART STREETLIGHTING

Streetlight operation consumes enormous power as it is done around the year during the night and on cloudy days. These lights account for heavy electricity bills in municipalities. To cut down on these bills, it is necessary to switch off or at least dim the lights when there is no pedestrian or traffic at a place. Keeping the lights on at full intensity always involves wasteful expenditure of money. Two types of motion sensors, the passive infrared sensor and the RADAR sensor are employed to

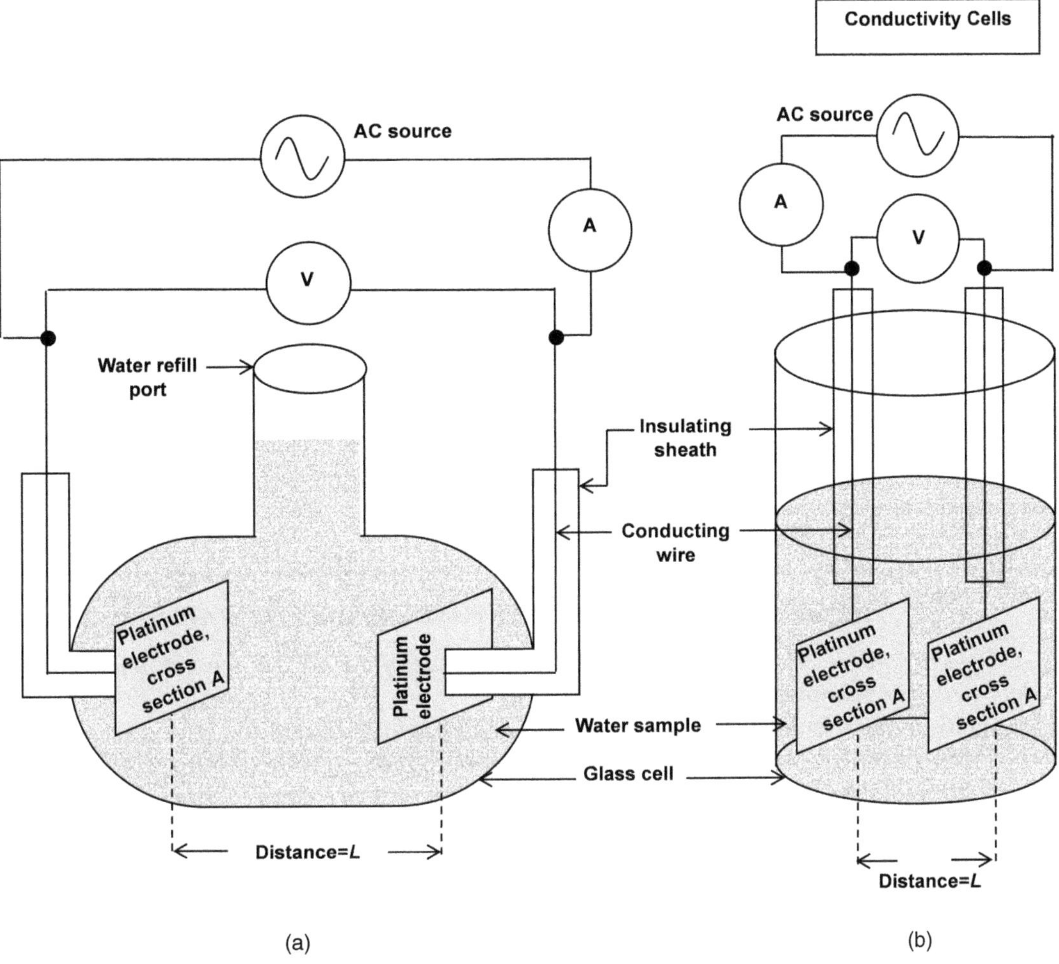

FIGURE 11.19 Two types of conductivity cells: (a) bulb-shaped and (b) cylindrical. Part (a) shows a bulb-shaped glass cell filled with water, the water port, two platinum electrodes (cross-sectional area A) dipped in water, the protective insulation-covered conducting wires for external connections of platinum electrodes, the voltmeter, the ammeter, and the AC source. The distance between the platinum electrodes is L. Part (b) shows a cylindrical tube-shaped glass cell. The remaining components are the same as in part (a).

efficiently control the streetlights in accordance with pedestrian movement and vehicular traffic (see also Section 3.15.2). The sensors are mounted either on a street pole or on the luminaire itself. Upon the appearance of a person or vehicle within the detection range of the sensor, the sensor triggers a command to the luminaire to increase the brightness level to a pre-decided value. PIR sensors are very effective in detecting human motion, and can also be corrected to avoid false alarms created due to pests or animals passing nearby. These sensors also decrease the brightness level when the pedestrian walks away, and therefore not much brightness is needed for illumination. Standalone sensors are self-supported with circuits to make necessary light intensity changes while sensors which form part of a network are connected to a central management system. They can be configured in such a way that the sensor can send a signal to the next lamp post about the approaching pedestrian. Thus, the light intensity can be increased from 20% on a vacant street to 80–100% in the presence of the pedestrian. By adopting such light intensity controlling strategies, the life of the luminaires is prolonged, and less frequent maintenance is required.

11.17 CONCLUDING REMARKS AND FUTURE PERSPECTIVES

Smart city dreams are becoming true. Purity of the atmosphere and plentiful supply of good quality drinking water are basic needs of citizens, two conditions which presently some cities of the world are able to comply with satisfactorily. The trends of smartness are visible. Electric cars and buses are replacing conventional fuel vehicles. Houses are thermally protected. People are able to adjust the temperature and humidity levels to their needs. Houses are equipped with sensors for smoke and fire alarms, door and window opening alarms, human presence detection devices, and much more. These cities have an efficient power distribution network and intelligent transport management system. Early disease diagnosis and real-time health monitoring with personalized patient care are commonplace. Health provision facilities are more streamlined, leading to a reduction in wait times. Sensor technologies are being interwoven with the internet and data analytics to ensure that citizens lead a happy and healthy life. Currently, only a few cities fulfill the essential criteria of smartness. The list of smart city sensors is endless, and human requirements and aspirations are too high. A lot needs to be done and that too at a large scale so that most cities across the world are made smart. Hopefully, the future is bright with the widespread accessibility of the internet in both urban and rural locations, and the availability of low-cost, reliable sensors.

THE SMART CITY WHERE EVERYONE DREAMS TO STAY!

A city where homes and offices are furnished with state-of-the-art IoT sensors, and where human comfort parameters are strictly regulated.

A city in which factories and manufacturing plants are fully automated.

A city of minimal accidents with controlled vehicular traffic within prescribed speed limits.

A city with smart energy meters and a power distribution grid supported with phasor measurement units.

A city where human health is of supreme importance, and supported by patient–doctor interaction, quick responses, and answers.

A city where agriculture is done by inspecting soil conditions, and plant macronutrient status, and IoT networks look after plant diseases and farm animal well-being via the internet and sensors.

A city where air and water quality are of premium level, noisy sounds are never heard, and where cleanliness is maintained every day.

A city where crime, fire, and emergency alerts are attended with great dexterity.

A city where people's safety and security receive the topmost priority
is the city about whom the people are delighted to say
'The smart city where everyone dreams to stay!'

REFERENCES

Afzal A. 2019 β-Ga$_2$O$_3$ nanowires and thin films for metal oxide semiconductor gas sensors: Sensing mechanisms and performance enhancement strategies, *Journal of Materiomics*, 5: pp. 542–557.

Anachkova M., S. Domazetovska, Z. Petreski and V. Gavriloski 2020 Noise exposure level detection using the internet of things (IoT) concept, Forum Acusticum, 7–11 December, Lyon, France. pp. 691–696.

Anachkova M., S. Domazetovska, Z. Petreski, and V. Gavriloski 2021 Design of low-cost wireless noise monitoring sensor unit based on IoT concept, *Journal of Vibroengineering*, 23(4): pp. 1056–1064.

Aoyagi Y., M. Takeuchi, K. Yoshida, M. Kurouchi, T. Araki, Y. Nanishi, H. Sugano, Y. Ahiko and H. Nakamura 2012 High-sensitivity ozone sensing using 280 nm deep ultraviolet light-emitting diode for detection of natural hazard ozone, *Journal of Environmental Protection*, 3(8): pp. 695–699.

Azman A. A., M. H. F. Rahiman, M. N. Taib, N. H. Sidek, I. A. Abu Bakar and M. F. Ali 2016 A low cost nephelometric turbidity sensor for continual domestic water quality monitoring system, *IEEE International Conference on Automatic Control and Intelligent Systems (I2CACIS), Selangor*, Malaysia, 22 October, 2016, pp. 202–207.

Dang J., N. Wang and H. K. Atiyeh 2021 Review of dissolved CO and H_2 measurement methods for syngas fermentation, *Sensors*, 21: 2165, pp. 1–32.

ESTCP2001 n.d. Final Report Field Validation of Real Time Airborne Lead Analyzer Naval Facilities Engineering Service Center Environmental and Life Support Technology, CP-9905-FR-01, pp. 1–3.

Fay C.D., and A. Nattestad 2021 Advances in optical based turbidity sensing using LED photometry (PEDD), *Sensors (Basel)*, 22(1):254, pp. 1–13.

Hodgson A. W. E., P. Jacquinot, and P. C. Hauser 1999 Electrochemical sensor for the detection of SO_2 in the low-ppb range, *Analytical Chemistry*, 71: pp. 2831–2837.

Henderson R. E. 2021 Understanding ammonia sensors, Gas Alarm Systems. https://www.gasalarmsystems.co.uk/latest-news/understanding-ammonia-sensors/#:~:text=Infrared%20(IR)%20Sensors&text=The%20bonds%20in%20a%20particular,after%20the%20transfer%20of%20energy

Hofmann F. 2003 *Fundamental principles of Electromagnetic Flow Measurement*, 3rd Edition. KROHNE Messtechnik GmbH & Co. KG, Duisburg, Germany, pp. 1–72.

Hussain I., K. Ahamadb and P. Nath 2016 Water turbidity sensing using a smartphone, *RSC Advances*, 6: pp. 22374–22382.

Juraschek M., M. Bucherer, F. Schnabel, H. Hoffschröer, B. Vossen, F. Kreuz, S. Thiede, and C. Herrmann 2018 Urban factories and their potential contribution to the sustainable development of cities, *Procedia CIRP*, 69: pp. 72–77.

Kida T., A. Nishiyama, M. Yuasa, K. Shimanoe, and N. Yamazoe 2009 Highly sensitive NO_2 sensors using lamellar-structured WO_3 particles prepared by an acidification method, *Sensors and Actuators B: Chemical*, 135: pp. 568–574.

Kitzelmann D. 2000 Electrochemical sensor for ammonia and volatile amines, EP1 183 528B1, Filing: 12.05.2000, Publication and mention of the grant: 15.12.2010, pp. 1–13.

Kroll A.V., V. I. Smorchkov and A. Y. Nazarenko 1994 Electrochemical sensors for hydrogen and hydrogen sulfide determination, *Sensors & Actuators B: Chemical*, 21: pp. 97–100.

Lavanyaa V. P., K. M. Harshitha, G. Beig, and R. Srikanth 2023 Background and baseline levels of PM2.5 and PM10 pollution in major cities of peninsular India, *Urban Climate*, 48: pp. 101407.

Lee J.-S., A. Katoch, J.-H. Kim and S. S. Kim 2016 Growth of networked TiO_2 nanowires for gas-sensing applications, *Journal of Nanoscience and Nanotechnology*, 16: pp. 1580–11585.

Miles J. 2004 Methods of radon measurement and devices. Proceedings of the 4th European conference on protection against radon at home and at work Conference programme and session presentations, (p. 377). *Czech Republic, International Nuclear Information System (INIS)*, Vol. 36(1), Issue 3(1), Reference No. 36010924.

Mirshab B. 2015 Ultrasonic sensing for water flow meters and heat meters, Texas Instruments, Application Report, SNIA020–April 2015, Copyright © 2015, Texas Instruments Incorporated, pp. 1–14.

Ottonello-Briano F., C. Errando-Herranz, H. Rödjegård, H. Martin, H. Sohlström and K. B. Gylfason 2020 Carbon dioxide absorption spectroscopy with a mid-infrared silicon photonic waveguide, *Optics Letters*, 45: pp. 109–112.

Pan S., M. J. Deen, and R. Ghosh 2015 Low-cost graphite-based free chlorine sensor, *Analytical Chemistry*, 87(21): pp. 10734–10737.

Ponzoni A., C. Baratto, N. Cattabiani, M. Falasconi, V. Galstyan, E. Nunez-Carmona, et al. 2017 Metal oxide gas sensors, a survey of selectivity issues addressed at the SENSOR Lab, Brescia (Italy), *Sensors*, 17, 714, pp. 1–27.

Rusydi A.F. 2018 Correlation between conductivity and total dissolved solid in various type of water: A review, Global Colloquium on GeoSciences and Engineering 2017, *IOP Conf. Series: Earth and Environmental Science*, 118:012019, pp. 1–6.

Saruhan B., R. L. Fomekong and S. Nahirniak 2021 Review: Influences of semiconductor metal oxide properties on gas sensing characteristics, Frontiers in Sensors, 2: pp. 1–24. Article 657931.

Shepard J. 2022 How to use ultrasonic sensing in smart water meters, Digi-Key's North American Editors, https://www.digikey.com/en/articles/how-to-use-ultrasonic-sensing-in-smart-water-meters

Turnipseed A. A., P. C. Andersen, C. J. Williford, C. A. Ennis, and J. W. Birks 2017 Use of a heated graphite scrubber as a means of reducing interferences in UV-absorbance measurements of atmospheric ozone, *Atmospheric Measurement Techniques*, 10: pp. 2253–2269.

Wang Y., S. M. S. M. Rajib, C. Collins and B. Grieve 2018 Low-cost turbidity sensor for low-power wireless monitoring of fresh-water courses, *IEEE Sensors Journal*, 18(11): pp. 4689–4696.

WHO 2009 *WHO Handbook on indoor radon: A public health perspective.* © World Health Organization, Geneva, Switzerland, pp. 40–41.

Yin J., W. Gao, W. Yu, Y. Guan, Z. Wang and Q. Jin 2022 A batch microfabrication of a self-cleaning, ultradurable electrochemical sensor employing a BDD film for the online monitoring of free chlorine in tap water, *Microsystems & Nanoengineering*, 8:39, pp. 1–12.

Index

Pages in *italics* refer to figures.

For Product Safety Concerns and Information please contact our EU
representative GPSR@taylorandfrancis.com
Taylor & Francis Verlag GmbH, Kaufingerstraße 24, 80331 München, Germany